Springer Handbook of Auditory Research

Series Editors: Richard R. Fay and Arthur N. Popper

Springer Handbook of Auditory Research

Volume 1: The Mammalian Auditory Pathway: Neuroanatomy
Edited by Douglas B. Webster, Arthur N. Popper, and Richard R. Fay

Volume 2: The Mammalian Auditory Pathway: Neurophysiology
Edited by Arthur N. Popper and Richard R. Fay

Volume 3: Human Psychophysics
Edited by William Yost, Arthur N. Popper, and Richard R. Fay

Volume 4: Comparative Hearing: Mammals
Edited by Richard R. Fay and Arthur N. Popper

Volume 5: Hearing by Bats
Edited by Arthur N. Popper and Richard R. Fay

Volume 6: Auditory Computation
Edited by Harold L. Hawkins, Teresa A. McMullen, Arthur N. Popper, and Richard R. Fay

Volume 7: Clinical Aspects of Hearing
Edited by Thomas R. Van De Water, Arthur N. Popper, and Richard R. Fay

Volume 8: The Cochlea
Edited by Peter Dallos, Arthur N. Popper, and Richard R. Fay

Volume 9: Development of the Auditory System
Edited by Edwin W Rubel, Arthur N. Popper, and Richard R. Fay

Volume 10: Comparative Hearing: Insects
Edited by Ronald Hoy, Arthur N. Popper, and Richard R. Fay

Volume 11: Comparative Hearing: Fish and Amphibians
Edited by Richard R. Fay and Arthur N. Popper

Volume 12: Hearing by Whales and Dolphins
Edited by Whitlow W.L. Au, Arthur N. Popper, and Richard R. Fay

Volume 13: Comparative Hearing: Birds and Reptiles
Edited by Robert Dooling, Arthur N. Popper, and Richard R. Fay

Volume 14: Genetics and Auditory Disorders
Edited by Bronya J.B. Keats, Arthur N. Popper, and Richard R. Fay

Volume 15: Integrative Functions in the Mammalian Auditory Pathway
Edited by Donata Oertel, Richard R. Fay, and Arthur N. Popper

Volume 16: Acoustic Communication
Edited by Andrea Simmons, Arthur N. Popper, and Richard R. Fay

Volume 17: Compression: From Cochlea to Cochlear Implants
Edited by Sid P. Bacon, Richard R. Fay, and Arthur N. Popper

Volume 18: Speech Processing in the Auditory System
Edited by Steven Greenberg, William Ainsworth, Arthur N. Popper, and Richard R. Fay

Volume 19: The Vestibular System
Edited by Stephen M. Highstein, Richard R. Fay, and Arthur N. Popper

Volume 20: Cochlear Implants: Auditory Prostheses and Electric Hearing
Edited by Fan-Gang Zeng, Arthur N. Popper, and Richard R. Fay

Volume 21: Electroreception
Edited by Theodore H. Bullock, Carl D. Hopkins, Arthur N. Popper, and Richard R. Fay

Continued after index

Geoffrey A. Manley
Richard R. Fay
Arthur N. Popper
Editors

Active Processes and Otoacoustic Emissions in Hearing

Springer

Geoffrey A. Manley
Lehrstuhl für Zoologie
Technische Universität München
Lichtenbergstrasse 4
85747 Garching
Germany
geoffrey.manley@wzw.tum.de

Arthur N. Popper
Department of Biology
University of Maryland
College Park, MD 20742
USA
apopper@umd.edu

Richard R. Fay
Parmly Hearing Institute and
Department of Psychology
Loyola University Chicago
Chicago, IL 60626
USA
rfay@luc.edu

Series Editors:
Richard R. Fay
Parmly Hearing Institute and
Department of Psychology
Loyola University Chicago
Chicago, IL 60626
USA

Arthur N. Popper
Department of Biology
University of Maryland
College Park, MD 20742
USA

ISBN: 978-0-387-71467-7 e-ISBN: 978-0-387-71469-1

Library of Congress Control Number: 2007922744

Printed on acid-free paper.

9 8 7 6 5 4 3 2 1

springer.com

Dedication

This book is dedicated to two colleagues of Geoff Manley: Graeme Yates and Des Kirk, who worked in the Auditory Laboratory of the Department of Physiology of the University of Western Australia. Their untimely deaths within two years of one another robbed this research field of a dedicated and clever research team. Their publications over many years represent very significant contributions to the fields covered by this book, and their intellect, humanity, and humor are sorely missed by those who loved and respected them.

Contents

Contributors

WILLIAM E. BROWNELL

Bobby R. Alford Department of Otolaryngology, Department of Neuroscience, Structural & Computational Biology & Molecular Biophysics, Baylor College of Medicine, Houston, TX, 77030, USA, E-mail: brownell@bcm.tmc.edu

NIGEL COOPER

MacKay Institute of Communication and Neuroscience, Keele University, Keele, Staffs. ST5 5BG, UK, E-mail: n.p.cooper@keele.ac.uk

PIM VAN DIJK

Department of Otorhinolaryngology, University Medical Centre Groningen, 9700 RB Groningen, The Netherlands and School for Behavioral and Cognitive Neurosciences, University of Groningen, 9700 RB Groningen, The Netherlands, E-mail: p.van.dijk@med.umcg.nl

THOMAS A.J. DUKE

Cavendish Laboratory, Cambridge University, Cambridge CB3 0HE, UK, E-mail: td18@cam.ac.uk

MARTIN C. GÖPFERT

Volkswagen-Foundation Research Group, Institute of Zoology, University of Cologne, D-50923 Cologne, Germany, E-mail: m.gopfert@uni-koeln.de

MICHAEL P. GORGA

Boys Town National Research Hospital, Omaha, NE 68131, USA, E-mail: gorga@boystown.org

JOHN J. GUINAN

Massachusetts Eye and Ear Infirmary, Eaton-Peabody Laboratory, Boston, MA 02114, USA, E-mail: jjg@epl.meei.harvard.edu

RICHARD HALLWORTH

Department of Biomedical Sciences, Creighton University, Omaha, NE 68178, USA, E-mail: hallw@creighton.edu

THOMAS JANSSEN
Hals-Nasen-Ohrenklinik, Technische Universität München, D-81675 Munich, Germany, E-mail: T.Janssen@lrz.tum.de

HEATHER JENSEN-SMITH
Department of Biomedical Sciences, Creighton University, Omaha, NE 68178, USA, E-mail: Heather7@creighton.edu

TIFFANY A. JOHNSON
University of Kansas, Speech-Language-Hearing: Sciences and Disorders, Dole Human Development Center, Room 3001, 1000 Sunnyside Ave., Lawrence, KS 66045-7555, USA, E-mail: tiffany-johnson@ku.edu

FRANK JÜLICHER
Max Planck Institute for the Physics of Complex Systems, 01187 Dresden, Germany, E-mail: Julicher@pks.mpg.de

DAVID T. KEMP
Center for Auditory Research, UCL Ear Institute, University College London, London WC1X 8EE, UK, E-mail: d.kemp@ucl.ac.uk

DUCK O. KIM
Department of Neuroscience, University of Connecticut Health Center, Farmington, CT 06030-3410, USA, E-mail: kim@neuron.uchc.edu

BRENDA L. LONSBURY-MARTIN
Research Service (151), Jerry Pettis Veterans Memorial Medical Center, Loma Linda, CA 92357, USA, E-mail: Brenda.Lonsbury-Martin@va.gov

ANDREI N. LUKASHKIN
School of Life Sciences, University of Sussex, Falmer, Brighton BN1 9QG, UK, E-mail: A.N.Lukashkin@sussex.ac.uk

GEOFFREY A. MANLEY
Lehrstuhl für Zoologie, Technische Universität München, 85747 Garching, Germany, E-mail: geoffrey.manley@wzw.tum.de

GLEN K. MARTIN
Research Service (151), Jerry Pettis Veterans Memorial Medical Center, Loma Linda, CA 92357, USA, E-mail: Glen.martin2@va.gov

PASCAL MARTIN
Laboratoire Physico-Chimie Curie, Unité Mixte de Recherche 168 du CNRS, Institut Curie F-75248 Paris Cedex 05, France, E-mail: pascal.martin@curie.fr

JÖRG MÜLLER
Hals-Nasen-Ohrenklinik, Technische Universität München, D-81675 Munich. Germany, E-mail: J.Mueller@lrz.tum.de

STEPHEN T. NEELY
Boys Town National Research Hospital, Omaha, NE 68131, USA, E-mail: neely@boystown.org

ANDREW J. OXENHAM
Department of Psychology, University of Minnesota, 75 East River Parkway, Minneapolis, MN 55455, USA, E-mail: oxenham@umn.edu

JAMES O. PICKLES
Vision, Touch and Hearing Research Centre, Department of Physiology and Pharmacology, University of Queensland, 4072, Qld. Australia, E-mail: j.pickles@uq.edu.au

DANIEL ROBERT
School of Biological Sciences & IRC in Nanotechnology, University of Bristol, Bristol BS8 1UG, UK, E-mail: D.Robert@bristol.ac.uk

IAN RUSSELL
School of Life Sciences, University of Sussex, Falmer, Brighton, BN1 9QG, UK, E-mail: I.J.Russell@sussex.ac.uk

CHRISTOPHER A. SHERA
Eaton-Peabody Laboratory, Massachusetts Eye & Ear Infirmary, Boston, MA 02114, USA, E-mail: shera@epl.meei.harvard.edu

Series Preface

Springer Handbook of Auditory Research

The Springer Handbook of Auditory Research presents a series of comprehensive and synthetic reviews of the fundamental topics in modern auditory research. The volumes are aimed at all individuals with interests in hearing research, including advanced graduate students, postdoctoral researchers, and clinical investigators. The volumes are intended to introduce new investigators to important aspects of hearing science and to help established investigators to better understand the fundamental theories and data in fields of hearing that they may not normally follow closely.

Each volume presents a particular topic comprehensively, and each serves as a synthetic overview and guide to the literature. As such, the chapters present neither exhaustive data reviews nor original research that has not yet appeared in peer-reviewed journals. The volumes focus on topics that have developed a solid data and conceptual foundation rather than on those for which a literature is only beginning to develop. New research areas will be covered on a timely basis in the series as they begin to mature.

Each volume in the series consists of a few substantial chapters on a particular topic. In some cases, the topics will be ones of traditional interest for which there is a substantial body of data and theory, such as auditory neuroanatomy (Vol. 1) and neurophysiology (Vol. 2). Other volumes in the series deal with topics that have begun to mature more recently, such as development, plasticity, and computational models of neural processing. In many cases, the series editors are joined by a coeditor having special expertise in the topic of the volume.

RICHARD R. FAY, Chicago, IL
ARTHUR N. POPPER, College Park, MD

Volume Preface

Most sciences undergo occasional smaller or greater revolutions, such as the Darwinian revolution in biology. In most cases, these revolutions are conceptual in nature, providing a new framework for understanding a host of data, and this new concept plays a dominant role in later discourse in that field. In the early 1980s, hearing science went through a revolution that was at the same time both conceptual and technical and that dramatically changed the face of auditory neuroscience. It is the aim of this book to summarize what is currently known about the remarkable phenomenon of active processes and their manifestations and provide a comprehensive resource for readers to understand and implement this phenomenon in their work.

There is no doubt that in the early 1980s, auditory science was ripe for a revolution. There was a veritable mountain of anatomical and physiological findings that were not clearly reconcilable and with no adequate models to describe them. Physiological studies showed that the movement of the organ of Corti and the behavior of auditory nerve fibers were clearly nonlinear and physiologically sensitive. Anatomically, inner and outer hair cells were very different. What did all this mean, and how could these findings be reconciled? Within the first five years of the 1980s, technical advances permitted studies of outer hair cells and of the newly discovered otoacoustic emissions that together clearly demonstrated that hair cells were not just passive receivers of sound stimuli but that they are able to produce mechanical energy themselves. These technical breakthroughs permitted the development of a new conceptual framework of the function of the hearing organ as an active, highly nonlinear system that modifies the responses to sound such that very sensitive, highly frequency-selective hearing becomes possible. Later studies showed that the active ear originated early in evolution and thus has a long history. We now know that it is based on not just one but several interlocking mechanisms.

For research purposes, one great advantage of the active processes in the inner ear is that some of the energy produced by hair cells is lost out of the hearing organ, passes the middle ear, and produces measurable sounds in the ear canal. These otoacoustic emissions can be spontaneous or induced by different sounds or combinations of sounds. Their study has opened up extremely useful new methods not only for "remote sensing" of how the ear works, but also for more direct comparisons of hearing in animals and humans and, importantly, very useful new clinical procedures in audiology and otolaryngology. Also importantly, the techniques are both noninvasive and objective.

The finding that vertebrate hearing depends on active processes in the inner ear has increasingly dominated and revolutionized hearing research over the last 25 years. The time is now ripe to bring together all of these ideas. This book begins with a discussion of the history of the discovery of otoacoustic emissions by David Kemp (Chapter 1), and a companion chapter by Cooper et al. (Chapter 2) traces the history of research findings and ideas that provided the historical framework that led to the discovery of the active cochlea. These chapters make it easier to understand the origin of this research field and especially the context in which important discoveries were made.

These introductory and historical chapters are followed by three chapters that deal with the concepts and the probable cellular mechanisms underlying active processes of hair cells. Duke and Jülicher (Chapter 3) discuss the theories underlying the concept of active processes, how these may be integrated into a sensory system, and how they are controlled—without knowing necessarily what the actual cellular mechanisms are. Martin (Chapter 4) and Hallworth and Jensen-Smith (Chapter 5) then discuss the main types of motor systems that drive active processes at the hair-cell level. Martin discusses the evidence for an involvement of the hair-cell bundle in both nonmammalian and mammalian ears and reveals mechanisms that are both ancient (clearly also present in vestibular hair cells) and complex. Their complexity rests on the fact that there is good evidence for the involvement both of myosins that control sensory adaptation and of the transduction channels themselves. Hallworth and Jensen-Smith present the evidence for a unique additional molecular motor in the lateral walls of mammalian outer hair cells, based on the protein prestin, and discuss how this system might work.

There follow three chapters that describe the evidence for active processes in three groups of animals. Göpfert and Robert (Chapter 6) describe the active movements of the auditory organ (the antennae) of flies and mosquitoes and reveal a system of sensitivity enhancement with similar functional principles as those seen in vertebrates and that are possibly even evolutionarily related at the level of the sensory cells. Manley and van Dijk (Chapter 7) give a broad overview of nonmammalian land vertebrates, such as amphibians, lizards, and birds, that lack the prestin motor but from which there is a plethora of data that clearly show that their hearing is also driven partly by active processes. Interestingly, amphibians and especially lizards are particularly reliable producers of spontaneous otoacoustic emissions, and this makes them very useful animals for studying these phenomena. Clearly, the performance of many of these ears rivals that of mammals and makes one wonder what advantage mammals gain from also having prestin-driven motility. Lonsbury-Martin and Martin (Chapter 8) then give an overview of otoacoustic emission measurements from laboratory mammals. Although these species tend not to show spontaneous emissions, their sound-induced emissions provide the most directly comparable data to measurements on humans and are therefore of clear interest for the interpretation of human data and of clinical studies.

Shera and Guinan (Chapter 9) discuss the problem as to why otoacoustic emissions in mammals, which are based on the activity of all hair cells, manifest themselves as distinct spectral patterns with energy peaks only at certain frequencies and offer a conceptual model to expain these patterns. Russell and Lukashkin (Chapter 10) then discuss the centrifugal control exerted on the ear by the efferent system and its effects as seen in otoacoustic emissions. Clearly, active systems of hair cells will also be under the influence of efferent fibers, and it is of especial interest for clinical studies to understand the extent to which this occurs and what characteristic activity patterns are produced.

Neely and Kim (Chapter 11) describe the mathematical modeling of active processes and otoacoustic emissions in the historical context. The predictive value of modeling has benefited enormously from the concept of active processes, and the authors show how some features that were incorporated into early models as necessary to explain the phenomena can now be given names and even identified at the cellular level.

Two chapters then discuss the relevance and importance of active processes and otoacoustic emissions for the study of human hearing. Johnson et al. (Chapter 12) bring together the results of previous techniques (such as psychoacoustical methods) used to study the human ear with those recently gained using otoacoustic emissions. Janssen and Müller (Chapter 13) then provide an overview of the use of the measurement of otoacoustic emissions in a clinical context: Which techniques have been applied, what have we learned from them, and what are appropriate protocols for examining different kinds of hearing problems?

Finally, Manley and Brownell (Chapter 14) summarize the main questions that have arisen out of all of these studies and still remain to be answered. Hearing researchers now have an unprecedented array of methodologies at their fingertips, backed up by the fastest and most sensitive equipment ever available. Future work will no doubt solve most of these problems and reveal new and fascinating details about the functions of this quite remarkable sensory system.

We trust that this book will be of interest to a very wide audience of auditory researchers but also clinicians, sensory physiologists, and teachers of neurobiology.

As is frequently the case, there are other chapters and volumes in the Springer Handbook of Auditory Research that provide additional insight into the issues raised in this volume. A discussion of the structure, function, and physiology of sensory hair cells is found in volume 27, *Vertebrate Hair Cells* (edited by Eatock, Fay, and Popper). The mechanics of the mammalian cochlea, particularly as related to active processes, can be found in a chapter by Patuzzi in volume 8, *The Cochlea* (edited by Dallos, Popper, and Fay), while other chapters in that volume provide further insight into the structure and function of the mammalian cochlea. An earlier discussion of otoacoustic emissions was presented by Whitehead et al. in volume 7, *Clinical Aspects of Hearing* (edited by Van De Water, Popper, and Fay)

September, 2007
Geoffrey A. Manley, Garching, Germany
Richard R. Fay, Chicago, Illinois
Arthur N. Popper, College Park, Maryland

1
Otoacoustic Emissions: Concepts and Origins

David T. Kemp

1. Introduction

Otoacoustic emissions (OAEs), that is, "sounds generated by the ear," have been scientifically researched since 1977, but clues to their existence in human ears can be traced back in the literature for decades before that. The fact that this complex phenomenon is still not fully understood, however, has not prevented a number of applications of OAEs from being developed. OAEs are useful for the detection and characterization of peripheral auditory function and dysfunction. In this role, OAEs have helped facilitate universal newborn hearing screening by providing an objective means of confirming healthy cochlear and middle ear function in less than a minute. They have also become an essential component of the audiological diagnostic test battery, helping to differentiate sensory from neural disorders and to monitor objectively changes in cochlear status. In the laboratory, too, OAEs have become an important source of information about how the cochlea interacts with sound stimulation and about how that interaction is changed as a result of various challenges to the auditory system. Numerous experiments have demonstrated that OAEs are intimately associated with a key feature of the cochlear mechanism that has become known as the "active process" or more specifically as the "cochlear amplifier." This chapter deals only with OAEs in mammals and especially in humans, since it was mainly observations in human ears that led to their discovery. However, it needs to be noted that the active process behind OAEs and the cochlear amplifier appears to be universal in vertebrate hearing (Manley 2001), but this is dealt with in other chapters.

OAEs can be used to perform important experiments on the intact ear simply by observing changes in OAE amplitude under specific stimulation conditions. With OAEs it is possible to demonstrate cochlear mechanical frequency selectivity and nonlinearity and to observe the depression of cochlear activity caused by noise, drugs, and medial olivocochlear stimulation. Thus OAEs form a bridge between the clinic and the laboratory and between research on human and nonhuman ears. This exciting aspect of OAEs is often overlooked because much of the OAE literature is concerned with recording technology or with the complexities of OAE phenomenon that have little to do with the primary hearing process.

So, while on the one hand, OAEs promise us a direct insight into hair cell activity from the vantage point of the ear canal, the literature can intimidate and distract the researcher with a wealth of complex information that we cannot yet adequately interpret.

What are OAEs? How were they discovered? How can they help us understand and study the cochlea? This chapter tries to answer these questions by introducing otoacoustic emissions from basic principles and primary observations. It looks at the key issues and concepts that led to the experimental identification of OAEs in 1977 and traces how those ideas evolved during the early exploration of OAEs (Kemp 2003). The various techniques in use today—transient evoked (TEOAEs), stimulus frequency (SFOAEs), distortion product (DPOAEs), and spontaneous (SOAEs; see appendix to this chapter)—were all developed early in the "discovery phase" of OAEs. Early research drew its data mainly from human subjects, but OAE techniques were quickly translated into laboratory experiments, including those on nonmammalian species. The initial laboratory exploration of OAEs coincided with, and contributed to, the rapid development of a new understanding of cochlear function (see Cooper et al., Chapter 2). New concepts introduced at that time to try to explain the OAE phenomenon helped tip the balance of opinion and to force a change in thinking about the cochlea as a sensory mechanism. The change was from "passive and linear" to "active and nonlinear" (see Neely and Kim, Chapter 11).

Some of those original OAE concepts have now been discarded, many others remain highly relevant and have been given a firm theoretical basis (see Shera and Guinnan Chapter 9, Duke and Jülicher, Chapter 3), while yet others continue to be hotly debated and still pose challenging questions about the peripheral auditory system (such as the mechanism for reverse energy transmission in the cochlea).

1.1 The Very Basics

OAEs are just air pressure fluctuations in the ear canal caused by vibration of the ear drum driven by the cochlea. It is wrong to think of OAE as sounds "emitted" by the cochlea itself, because sound pressure is not produced in any measurable quantity until the eardrum, driven by the cochlea, vibrates against the adjacent air in the ear canal. The physical construction of the middle ear positively helps us to observe OAEs. The efficient coupling that the middle ear provides between the low impedance of the thin light eardrum and the high impedance of the closed fluid-filled capsule that is the inner ear operates equally well in reverse. It matches cochlea to eardrum just as well as eardrum to cochlea. A horn can be used either to enhance hearing or in reverse as a trumpet to strengthen the voice. In the same way, the middle ear works in reverse as a kind of stethoscope, helping us to record vibrational activity deep inside the cochlea.

But with an open ear canal the air displaced by eardrum motion would just slide in and out of the ear canal. Very little actual sound pressure would result except at the highest frequencies (> 8 kHz for human ears), where the ear canal begins to act like a horn. Sealing the ear canal as illustrated in Figure 1.1 provides a small confined volume of air that eardrum vibrations can work against to produce sound pressure right down to low frequencies. Sealing the ear canal is an essential part of the OAE recording technique. It is achieved in humans by inserting the OAE measurement probe into the ear canal using a soft plastic tip to ensure a good seal. The probe either contains or is coupled to the recording microphone and acoustic stimulator. An additional practical benefit of sealing

FIGURE 1.1. In otoacoustic emission recording, a probe P containing a microphone M records the sound pressure in the closed ear canal C. Ear canal sound pressure can arise from sounds created by the probe loudspeaker L (the stimulus) or from vibration of the eardrum D driven by the cochlea I through the ossicular chain, which links the stapes S to the ear drum D. In stimulated otoacoustic emission recording, sound from the probe causes oscillatory motion of the stapes, a corresponding displacement of cochlear fluid, and complementary motion of the round window R. The basilar membrane of the cochlear partition receives energy from this cochlear fluid motion, causing a wave to travel along the basilar membrane BM in the forward direction F. The traveling wave shape T is shown schematically for a single pure-tone stimulus, and this wave delivers excitation to the sensory cells within the spiral organ of Corti (shown excised from an electron micrograph image as O). In the single-tone-stimulation case, there is a distinct place where the maximum vibration of the BM occurs just basal to a region where the energy is absorbed and the wave is halted. Typically, from somewhere around the peak some wave energy is retransmitted back to the base of the cochlea, probably as a reverse traveling wave in direction R. The stimulation process then works in reverse. The stapes and round window are moved in a complementary manner by cochlear fluid vibration causing a vibration of the ossicles and finally of the eardrum, leading to the creation of new sound pressure in the ear canal at the same frequency as the stimulus. For two tones or any other complex sound there is an additional way for energy to escape back to the ear canal if intermodulation distortion is created. (Thanks to Andy Forge, Ade Pye, and Martin Robinette for permission to use these images.)

the ear canal is that it attenuates room noise. This is particularly valuable in clinical applications when testing if done outside of a sound booth.

The scale of middle ear motions involved in delivering otoacoustic emissions to the ear canal is incredibly small. Taking the sealed human ear canal volume as 1 cm^3 and air pressure as 100,000 pascals (1 bar), the ear drum needs to compress or expand this trapped air volume by only one part in 1,000,000,000 to generate a peak sound pressure of 100 μPa or 14 dB SPL, a level typical of a good adult OAE. To achieve this amount of compression, the eardrum would have to move only 1/1,000,000,000 of a centimeter, or 0.01 nanometers. This is a subatomic motion: one-tenth the diameter of a hydrogen atom. Many OAE instruments can record sounds 30 times smaller than this (less than −10 dB SPL, or 0.3 picometers of drum vibration). The miniature microphones used in OAE probes are typically the same as those used in hearing aids. It is remarkable that this small scale of vibration is observable with such microphones, a scale that is beyond the current limits of the laser interferometers used for invasive observation of motion inside the cochlea. At present, therefore, OAE techniques offer the most sensitive method of recording cochlear mechanical activity.

How does the cochlea vibrate the eardrum to create sound, i.e., an OAE? Moving the eardrum from within the virtually sealed fluid-filled cochlea requires an imbalance of vibrational fluid pressure on the oval and round windows. This requires (1) the diversion of vibrational energy away from the primary hearing process and (2) the transmission of that energy back to the base of the cochlea in a way that can physically create this pressure difference and cause middle ear motion. The initial *source* of vibrational energy needed to generate OAEs in mammals is generally considered to be the electromotility of the three outer rows of the cochlea's sensory hair cells, the outer hair cells (OHCs). Motility of OHCs is believed to provide the engine of the "cochlear amplifier." The means of transmission of this energy to form OAEs is most commonly believed to be by a reverse traveling wave along the basilar membrane. This would provide a means of conveying vibrational energy back to the middle ear in a reversal of the process that normally delivers auditory stimulation to the hair cells. This explanation is far too simplistic. It does not address the issue of how the hair cell invokes a reverse traveling wave, and it overlooks the fact that the involvement of a reverse traveling wave in OAEs generation has still not been directly demonstrated experimentally. However, it is a good starting point from which to discuss OAEs.

1.2 Otoacoustic Emissions

OAEs are sounds found in the ear canal as a result of physiological activity in the cochlea. In a substantial minority of healthy ears, the otoacoustic emission process feeds back on itself and maintains a small continuous oscillation within the cochlea at one or more precisely defined frequencies. The sound generated by these oscillations is recordable in the ear canal without any evoking stimulus, and such sounds are known as spontaneous otoacoustic emissions (SOAEs). But most mammalian ears create OAEs only in response to sound stimulation and then only after a delay of a few thousandths of a second. OAEs are not a simple reemission, or "echo," of the sound stimulus. Although the emitted sounds are closely related to the acoustic characteristics of the applied stimulus, they are quite a complex transformation of it. This is illustrated with original data in Figure 1.2 for the specific cases of click and tone-burst stimuli. The various frequency components present in the applied stimulus are reemitted but dispersed in time due to their differing trajectories in and out of the cochlea. When a single tone stimulus is applied, the emission is of precisely the same frequency as the stimulus, but this makes it difficult to separate the response from the stimulus. When a short tone-burst stimulus is applied, the time delay before the emission appears makes the separation of the input and output easier. The range of frequencies emitted in response to a short tone burst is centered on the frequency of the stimulus but can be slightly broader than that actually present in the stimulus due to nonlinear intermodulation distortion between the frequency components of the tone burst (Yates and Withnel 1999). The effects of intermodulation are most clearly seen when two close tones (f_1 and f_2) are applied. In this case, emissions occur at several new frequencies not present in the stimulus but nevertheless precisely defined by various arithmetic combinations of the stimulus frequencies (e.g., $2f_1 - f_2$). These are known as distortion product emissions (Figure 1.3). All these different phenomena can be observed using an otoacoustic emission probe in the ear canal (as shown in Fig. 1.1) and by applying the appropriate stimulation and analysis. The type of analysis needed to extract the emission from the probe microphone signal, which of course also registers the stimulus signal, is determined by the type of stimulation applied.

The combination of "stimulus type" and "analysis method" constitutes a mode of OAEs observation. The key technical aspects of the most common observation modes are summarized in Appendix 1. These modes are often confusingly referred to as the "types of OAEs." Although there is more than one "type" of emission process (see below and Manley and van Dijk, Chapter 7, Lonsbury-Martin and Martin, Chapter 8, Shera and Guinnan Chapter 9), in reality, the type of emission is not determined so much by the method of stimulation and analysis as by the properties of the cochlea. For the purposes of this chapter and of cochlear research, it is more useful to have a description of OAEs not based on observation method but based on origin and process, which we can then use to understand and interpret OAEs data.

1.3 What Causes OAE and What Function Does It Perform?

So far as we know, otoacoustic emission has no function in the hearing process. It is a by-product of peripheral auditory function, an escape of stimulus-related energy that is of little if any consequence for hearing itself. But despite being only a by-product, OAEs do provide us with a unique "window" on cochlear

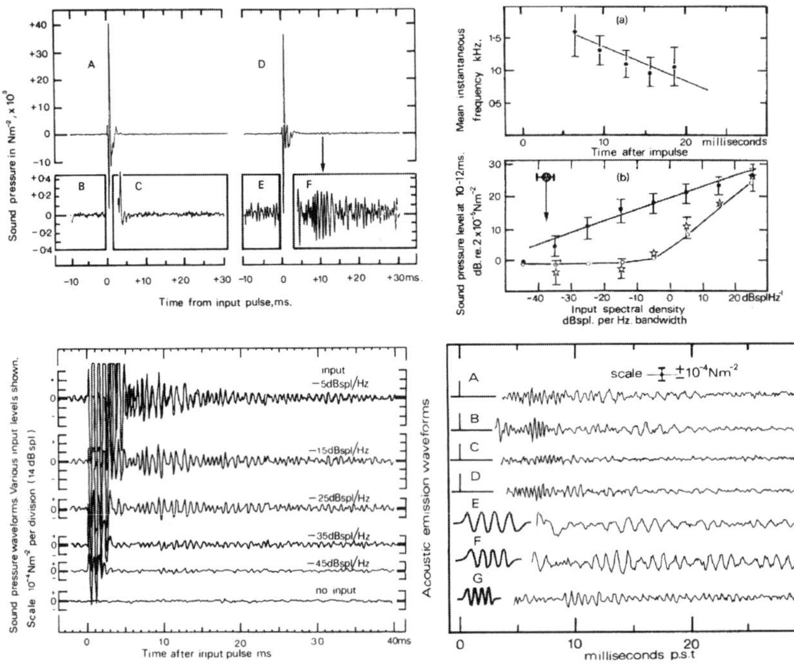

FIGURE 1.2. The first published examples of otoacoustic emissions and demonstrations of their properties from Kemp (1978). *Top left*: On the left side is the acoustic response of a closed ear-canal-sized cavity to an acoustic transient A (a click). The response lasts about 3 ms. Magnification of time sections B before and C after this transient shows only noise. In contrast, when the same measuring system is fitted to a healthy human ear (*right side of top left panel*), a delayed response (a transient-evoked OAEs or TEOAEs) is seen to be maximum around 10 ms after the stimulus (*arrow*). Magnification F clearly shows oscillations lasting more than 25 ms that are considerably larger than the "noise" in the preceding period E. In fact, the signal in E was also due to OAE remaining from the previous stimulus period.

Top right, upper half: Dominant frequency of the click-evoked emission decreasing with time after the click. This is the original evidence of OAE frequency dispersion that linked OAEs to the cochlea, because frequency dispersion is also a fundamental property of the cochlear traveling wave. Top right, lower half, shows the nonlinear relation of emission intensity to stimulus intensity (*filled points*). The response increases by approximately 20 dB for a 60-dB increase in stimulus level, which is a cube-root relation of sound pressures. Data from cavity measurements are shown by unfilled circles. These closely match data from human ears with cochlear deafness (*stars*). The rise in cavity response at higher stimulus level is the result of stimulus invading the measurement and is linear.

Bottom left: Effect of stimulus intensity on TEOAE waveform for one ear. For a linear system, each 10-dB increase in input would cause an approximately threefold increase in response amplitude, but OAEs increase by only about 50%. *Bottom right*: A–D show how different individual click-evoked OAEs tend to be. A–D show click-evoked OAE from four different ears. E–G show OAEs from the same ears evoked not by a click but by a tone burst (shown). The response echoes the stimulus periodicity.

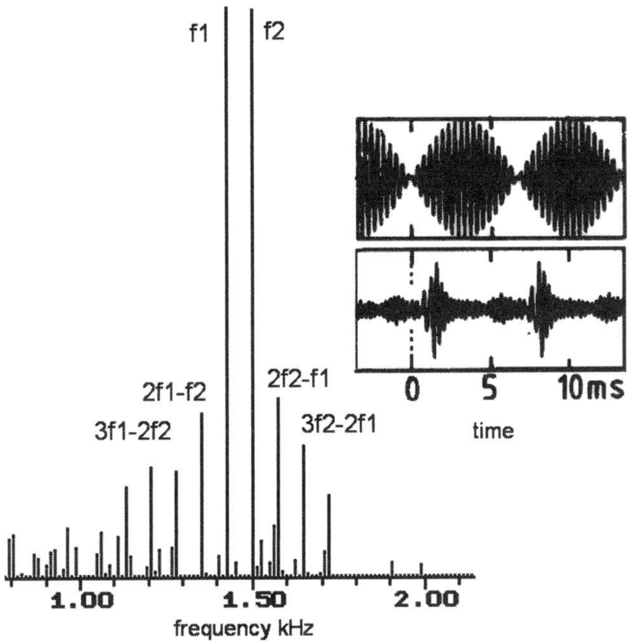

FIGURE 1.3. Schematic representation of distortion product OAEs adapted from Kemp et al. 1986. *Left*: Spectrum of human ear canal sound during stimulation by two tones, f_2 and f_1, with a close frequency ratio of 1.05. Multiple distortion products are seen both below f_1, e.g., $2f_1 - f_2$, $3f_1 - 2f_2$ (lower side band) and above f_2, e.g., $2f_2 - f_1$, $3f_2 - 2f_1$ (upper side band). Insert, right, shows the data in the time domain from a similar experiment in gerbil. The upper trace shows the full ear canal sound waveform. Periodic beats between the two stimuli dominate the waveform. In the lower trace, the stimuli have been removed and the waveform amplified to reveal the distortion-product OAE waveform. This comprises periodic bursts for energy synchronized to the beats. With wider stimulus separation, the $2f_1 - f_2$ component grows to dominate the spectrum.

activity. The potential of this "window" needs to be understood in relation to the ear's proper hearing function.

In the broadest terms, the role of the ear is to collect acoustic energy, to transform it into fluid vibration, and then to direct that fluid motion across the stereocilia of the sensory hair cells. The transformation from sound to fluid flow happens at the base of the mammalian cochlea with the complementary motion of oval and round windows. Through the dynamic interaction of this fluid motion with the graduated stiffness and width of the basilar membrane together with the inertia of the surrounding fluid, and finally with the micromechanics of the organ of Corti, the vibrational energy is selectively distributed over the array of approximately 3000 sensory inner hair cells according to its frequency of vibration. In what can be thought of as a separate (and evolutionarily older and less specialized) stage, the inner hair cells detect, rectify, demodulate, and smooth the narrow band of stimulus frequencies that each one receives. This process preserves only the lower-frequency (< 2 kHz) information, which includes the intensity of the vibration reaching that place and its temporal fluctuations but not the actual stimulus-frequency oscillations unless they are of a low enough frequency. The combination of frequency-band filtering and the inner hair cell properties leads to a reduction in information density at the auditory nerve synapse to that which is within the limited capacity of auditory nerves to transmit to the brain for processing and recognition. Little is lost.

Both stages in cochlear function—the frequency-selective transmission and the mechanical-to-neural transduction stages—are *active* sensory processes that exhibit specific but different vulnerabilities. It follows then that there are two possible components of cochlear sensory hearing loss: "sensory transmissive" loss due to unselective and inefficient transportation of the stimulus energy to the inner hair cells, and "sensory transductive" loss caused by a lack of effectiveness of the inner hair cells to transduce and activate auditory nerve fibers. On its own, sensory transmissive loss can never result in severe or profound deafness, as intense enough sounds will always be able to directly excite inner hair cells, whereas the less-prevalent sensory transductive loss can potentially cause total deafness.

Otoacoustic emissions arise from, and only inform us about, the first stage of cochlear function, namely the active and frequency-selective transmission mechanism. OAEs are a leakage of energy from that cochlear mechanism. Since OAEs test only half the cochlea's systems, it might appear that this rules out their use as a test for deafness. However, it has been empirically established that in the vast majority of sensory hearing losses there is a substantial sensory transmissive component, as inferred from the absence of OAEs. This is the justification for the widespread use of OAEs in newborn hearing screening programs, where their high sensitivity has been fully confirmed. The absence of OAEs in most sensory hearing losses indicates that it is the active transmission stage of cochlear function that is the most vulnerable aspect of the cochlea. OAEs give no direct information about the status of the hearing transduction mechanism of the inner hair cells. Strictly speaking, therefore OAEs should not be referred to as providing a "test of hearing." Pathologies affecting the inner hair cells or neural system but not the outer hair cell can leave OAEs intact with hearing being absent.

In so far as a good hearing threshold depends on efficient transmission of vibratory energy to the inner hair cells, it is to be expected that there will be a substantial correlation between OAE and hearing threshold. This is the case, but even with optimized measurement procedures designed to determine the threshold for DPOAE production, this correlation is limited to about 60 % (see Boege and Janssen 2002; Janssen and Müller Chapter 13). In most cases, this is insufficient to estimate the subjective threshold of individuals. Attempts to "calibrate" OAEs further to define precisely hearing threshold in the individual are doomed to failure simply because the mammalian OAE generation mechanism does not involve the inner hair cells that determine the local excitation threshold for activation of auditory nerve fibers. Auditory threshold is determined by the inner hair cells' transduction threshold multiplied by the cochlear transmission gain, and it follows then that the *difference* between an individual's derived OAE "threshold" and subjective hearing threshold may be a function of the health of the transduction mechanism. This parameter could have important clinical applications.

1.4 Questions Raised by OAE

This necessarily brief and inadequate introduction to OAEs leads on to a number of interesting questions to be addressed by the remainder of this chapter and elaborated on in others:

- Why was sound emission by the ear and its significance overlooked until 1978?
- Why does the cochlea allow valuable stimulus energy to be ejected and lost?
- How and when does this lost energy escape from the cochlea to form OAEs?
- How can we explain the complex characteristics of OAEs?
- How can we best record OAEs?
- What more can OAEs tell us about the auditory system?

2. Historical Review: The Discovery Phase of OAE Research

2.1 Early Pointers to OAEs: Gold's Contribution in 1948

In so far as tinnitus is a common auditory affliction, it can be assumed that the idea of "ears making sound" has been with us for millennia. Not long after coming into the field of audiology as a physicist in 1971, the author of this chapter came across a doctor (R. Hinchcliffe) and a physicist (J.J. Knight) who were trying to record tinnitus with a 1-inch condenser microphone placed close to a tinnitus patient's ear. Sometimes, when the tinnitus was of a "clicking" nature, presumed to be created in the Eustachian tube, the recording was successful. But with most patients, even those reporting loud tonal sounds or roaring sensations in their ears, no sound at all seemed to be emitted from their ears. Their tinnitus was deemed to be "subjective" and assumed not to be a real sound at all. That attempt to record "objective tinnitus" had been in part inspired by a report of multiple high-frequency (> 6 kHz) tones, albeit inaudible to their owners, being emitted by several members of one family with otherwise normal hearing (Glanville et al. 1971). It was concluded at the time that the "emissions" were caused by vascular oscillations, but it now seems likely that Glanville et al. (1971) observed true spontaneous otoacoustic emissions from abnormally overactive ears. The discovery of OAEs led to the recognition that the cochlea could generate such tonal emissions, but this did not herald the breakthrough in tinnitus research that was initially hoped for (see Wilson 1980a; Wilson and Sutton 1981). Even with today's OAE technology, it remains the case that pathological tinnitus is only very rarely associated with any actual sound made by the cochlea. From this we can deduce that the great majority of pathological tinnitus does not originate in the active mechanical cochlear process. Since tinnitus is commonly associated with hearing loss, it is probably the case that tinnitus most often occurs in patients with a mechanically *inactive* cochlea. Nevertheless, the healthy cochlea can generate sound.

The case for there being a *functional* "sound-generating capability" in the healthy cochlea causing tonal tinnitus as a by-product of the hearing process had been made 23 years before Glanville's observations by Thomas Gold (1948). He published his idea of "active" enhancement of the natural resonance of structures in the ear. In 1948, Gold knew that von Békésy had demonstrated the ability of the basilar membrane to crudely separate frequencies by a traveling wave and had discussed it with him (Gold 1989; Gold and Kemp 2003). He predicted that von Békésy-type passive tuning would be very limited (i.e., broadly tuned) by viscous damping, as von Békésy's data appeared to confirm. Neither Gold nor von Békésy actually developed a theoretical model of the traveling wave, but it was generally agreed that poor tuning in a frequency-selective system was an inevitable consequence of natural damping. It can be deduced from Gold (1948) that he conceived of the damping in the cochlea as acting on simple resonant structures and not on a traveling wave. We now know that in mammals and other animals the cochlear traveling wave provides frequency selectivity in a more complex and subtly different way than that of a simple resonance, but many of Gold's arguments still apply.

With his "discrete cochlear resonator" model in mind, Gold conducted psychoacoustic experiments with Pumphrey (Gold and Pumphrey 1948) related to the physical mechanism of frequency discrimination. Gold and Pumphrey erroneously, but fortuitously, convinced themselves that in order to explain the observed precision of frequency discrimination, the cochlea must contain very sharply tuned resonances with the ratio of frequency to half power bandwidth (Q_{3dB}) of the order of hundreds. Such sharp tuning clearly required the almost loss-free oscillation of cochlear structures. It had already been discovered some 15 years earlier that sound induces synchronous modulation of the electrical endocochlear potential, which was easily recordable as the "cochlear microphonic" response. Borrowing the technique of positive feedback from early radio engineering, Gold proposed that some piezoelectric tissue in the organ of Corti responded mechanically and in exact synchrony with the cochlear microphonic potential. He supposed that mechanical linkages through the organ of Corti delivered a synchronous force to the basilar membrane that could counteract viscous forces, potentially eliminating the effects of damping and leading to an extremely sharply tuned mechanical resonance. Gold knew from his own experience in building regenerative radio receivers as a boy (Gold and Kemp 2003) that the sharper you tried to make a tuned circuit resonate by applying positive feedback, the more critical became the amount of feedback required. The ultimate penalty was instability and self oscillation; if fractionally too much feedback was applied, the radio set began oscillating and radiating interference to every radio in the neighborhood. Automatic regulation of feedback is essential for operational stability in such a system, but this had been quite beyond the electronics engineering of the day. Well before Gold adopted the positive feedback idea for his hearing theory, radio engineering had totally abandoned the technique in favor of the multistage filter "superheterodyne" method. In contrast, Gold felt confident that nature would have solved the stability problem, except that, he reasoned, sometimes regulation might fail, resulting in instability and mechanical oscillation inside the cochlea. This, he felt, provided an explanation for the common experience of "transient tonal tinnitus", which would, he deduced, "radiate" tonal sound from the ear canal.

It is interesting to list Gold's far-reaching predictions about the cochlea in his 1948 paper and compare them with our present understanding:

- That there needs to be a mechanism in the cochlea that converts electrical signals into mechanical oscillation, in order to reduce damping and achieve high sensitivity and high frequency selectivity. We now recognize this mechanism as electromotility, and we attribute it, in mammals, to the outer hair cells.
- That the cochlea must contain a control system to precisely regulate the amount of positive feedback applied so as to maintain stable performance. Today we accept that there must be the capacity for self-regulation within the organ of Corti supplemented by central control via the medial cochlear efferent system's innervation of the outer hair cells.
- That there is, within the cochlea, the potential for mechanical instability and self-sustained internal mechanical oscillation and that this will be perceivable as tonal tinnitus. Today we would call this physiological tinnitus.
- That there should be an external manifestation of internal cochlear oscillation in the form of an emission of a tone from the ear canal, which we now know (from spontaneous OAEs) occurs in a substantial minority of healthy human ears, occasionally in other mammals and also frequently in other animals.

Gold faced fierce criticism at the time, particularly for his suggestion that the cochlea contained physically "ringing" resonant elements with little or no damping. Those who understood a little felt sure this would result in the continual emission of ringing sounds from healthy ears during normal stimulation, which today we call TEOAEs (see Fig. 1.2). But in 1948, common sense and experience told Gold's critics that no such sound is emitted. Gold countered this criticism by explaining the theoretical point that an ear with a perfectly smoothly graded continuum of such oscillating elements would emit sounds only if the continuum became unbalanced (see Gold 1989). In other words, Gold argued against the idea of stimulated acoustic emissions and instead conceded the possibility of quasipathological spontaneous emission of tonal sound due to excess feedback around one element. To try to prove this, he sealed a carbon telephone microphone to his subject's ear with wax and searched for pure tone signals during episodes of spontaneous transient tonal tinnitus. Rather than employ a narrowband audio frequency analyzer, which we would do today, Gold adopted the "listening-in" technique via headphones. He had previously used this method to monitor the cochlear microphonic signal from cats by connecting a cochlear electrode to headphones via an amplifier (Gold and Kemp 2003). He tried to precipitate recordable tinnitus in his subjects with drugs such as aspirin, which is known to induce subjective tinnitus. Unfortunately, he failed to detect any tonal or other sound emissions during these tinnitus-observation experiments. Maybe this was because the carbon microphone he used was too noisy or not sensitive enough. Maybe it was because he failed to apply narrow-band signal analysis to the microphone signal. Maybe it was because the spontaneous tinnitus events that Gold chose to study were purely subjective tinnitus and not actually related to spontaneous emissions at all. But if he had been fortunate enough to find a true spontaneous tonal emission, OAE research would have commenced in 1948. As it was, Gold answered his critics to his own satisfaction, but he failed to produce experimental evidence in support of his "active cochlea" hypothesis.

Gold's original candidate for electromotility in the cochlea was the tectorial membrane. This is in physical contact with the stereocilia of the outer hair cells, which at the time were erroneously thought to be the ear's most sensitive "hearing cells." The tectorial membrane is also directly exposed to the local cochlear microphonic potential. Interestingly, Gold updated this suggestion after reading a review by Pye (1980) of the nonlinear cochlear observations presented at the 1979 symposium on "Active and Nonlinear Mechanical Processes in the Cochlea" (Kemp and Anderson 1980; see also later). In a letter to Pye commenting on reports that the motile kinocilium of hair turtle hair cells participated actively in reptilian hair cell tuning (Fettiplace and Crawford 1980), Gold wrote that the kinocilia were "in just the location (he) expected" to support his proposed feedback mechanism, and significantly, he added that "where no kinocilium has been found could it not be that one of the rows of hair cells is nevertheless a mechanically active one?" (Gold 1980). Outer hair cell electromotility was actually demonstrated just three years later (Brownell 1983; Brownell et al. 1985).

It is appropriate to summarize what, in retrospect, Gold missed or got wrong in 1948.

- Gold considered only linear oscillating systems, even though it was well known that biological and even electronic systems are often nonlinear and that nonlinear systems not only produce distortion but also can have very interesting self-stabilizing properties.
- He failed to address the psychoacoustic phenomenon of subjective distortion or combinations tones, well known at the time, as possibly being a by-product of his active process and a possible source of emissions.
- He considered active input only in the context of positive feedback to an independent resonator and did not develop his theory in relation to a traveling wave system, even though he met and discussed his ideas with von Békésy. In a traveling wave system, wave amplification can be achieved by the synchronized release of energy from many active elements spread over an extended

area that become automatically synchronized to the passing excitatory wave, as for example in a laser. Such distributed, synchronized energy release naturally travels *with* the excitatory wave, i.e., traveling-wave amplification. It completely avoids the local instability that Gold recognized would be a problem in his positive-feedback model, and it also does not force high resonant Q values in order to achieve high gain (Kemp 2001, 2007).

- He predicted otoacoustic emissions only as a result of system maladjustment. He did not foresee and even argued against what we call *stimulated* acoustic emissions, i.e., the emission of sound as a by-product of *normal* cochlear function. On this topic, he later expressed surprise that stimulated OAEs existed at all (Gold 1989). In Gold's ideal cochlea, there would be a continuous array of critically stable and sharply tuned oscillators with the bandwidth of each overlapping that of its neighbor (compare with Duke and Jülicher, Chapter 3). Although each elemental resonator in isolation would transmit some of its activity back to the middle ear, as an ensemble there would be perfect mutual cancellation of this effect. He was quite correct about this, and he correctly predicted (Gold 1989) that *linear* reemission of the stimulus energy required "inaccuracies" in the transmission system (see Zweig and Shera 1995). However, as will be discussed later, it is now recognized that irregularities always exist and also that in a nonlinear system, emission can also occur without physical irregularities (Shera and Guinnan, Chapter 9).
- Finally, of course, Gold grossly overestimated the sharpness of tuning needed to explain human hearing. However, had he *not* adopted an extremely narrow resonant bandwidth (i.e., high Q) as the target for his model, Gold may never have needed to address the possibilities of an active cochlea, instability control, self-oscillation, or spontaneous acoustic emissions.

The failure of Gold's attempts to demonstrate spontaneous acoustic emissions in 1948 and so to validate his active cochlea model opened the way to widespread acceptance of von Békésy's description of a poorly tuned, passive linear cochlea (reviewed in von Békésy 1960). This in turn led to a 30-year period of disinterest in active cochlear models and in sound emission by the cochlea. Perhaps as a consequence, the first evidence of low-level nonlinearity in the cochlea (Rhode 1971) was also largely discounted (see Cooper et al., Chapter 2), and other indications (e.g., Kim et al. 1977) received very limited attention. Gold's idea of looking for cochleogenic sounds in the ear canal was also not exploited, with one notable exception. Dallos (1966) used an acoustic ear-canal method to record high level subharmonic distortion, which he attributed to nonlinearity in the cochlea.

2.2 Early Pointers to OAEs: Psychoacoustic Clues from 1931 to 1977

What prompted the first successful attempts to record and validate otoacoustic emissions (Kemp 1978) was not Gold's positive-feedback hypothesis, but a model developed quite independently to explain a number of strange psychoacoustic observations. Since the start of modern psychoacoustics, the literature has contained tenuous evidence that some ears host real—albeit normally inaudible—internal tonal oscillations. Wegel (1931) reported audible "beats" or "double sounds" when listening to a single pure tone at moderate-to-low stimulus levels at certain frequencies. The boundary of the sound levels that gave rise to the beat effect had a V-shaped contour as a function of frequency, and Wegel reported that one of the "V" center frequencies coincided exactly with "the pitch of a tinnitus." Flottorp (1953) reported that he experienced a triggerable tonal tinnitus sensation, which he called an "idiotone" and which occurred when an applied tonal stimulus, falling within a V-shaped level-frequency boundary, was turned off. The "idiotone" frequency was exactly at the center of the "V" (see Fig. 1.4a). "Beats" result from interference between two oscillatory signals, and if the phenomenon were physical, it would imply beats between the applied stimulus and an internal "phantom" vibration. The V-shaped boundary seen by Wegel and Flottorp could be compared to a tuning curve, but in fact, its shape is more like that of a cancellation "notch." The possibility of the "phantom tone" being an internal oscillation of the cochlear filtering system was well understood by Ward (1955), who observed the same V-shaped boundary of the "beat" phenomenon as did Wegel and Flottorp. He analyzed in detail the multiple tonal sensations (monaural diplacusis) that accompanied a single pure-tone stimulus at certain levels and frequencies. He traced these to combination tones (subjective distortion products) between the applied tone and an internal tone at the center of the V-shaped boundary. They fitted the formula $2f_1 - f_2$, where the external tone was f_1 and the V center frequency was f_2 or vice versa.

In retrospect, Ward's conclusion seems as significant for auditory science as Gold's hypothesis. In his 1955 paper, Ward wrote that "monaural diplacusis can be explained in terms of a basic instability of some part of the auditory system at the 'idiotone' frequency... [but] we can go no further until we know a good deal more about the normal production of combination tones." He did not go so far as to cite *mechanical* instability, and he did not reference Gold's work. Had Ward realized the connection between his and Gold's work, it is highly likely that otoacoustic emissions would have been discovered soon after 1955. Suitable measurement instrumentation was commonplace by that time. As it was, these psychoacoustic clues were treated by most workers as just curious hearing aberrations. It was not recognized that these three experienced psychoacousticians had documented phenomena very closely linked to the cochlear amplifier and were pointing the way to otoacoustic emissions.

Another series of psychoacoustic observations suggested a subtle periodicity in normal threshold, loudness, and even pitch as a function of frequency. Elliot (1958) reported "ripples" in the normal audiogram between 600 and 3000 Hz of up to 6 dB and occurring with a spacing of roughly 100 Hz. van den Brink (1970) reported pitch differences between left and right ears when presented with the same frequency stimulus. The pitch discrepancy (binaural diplacusis) also varied periodically with frequency on the same scale as Elliot's threshold ripple, and van

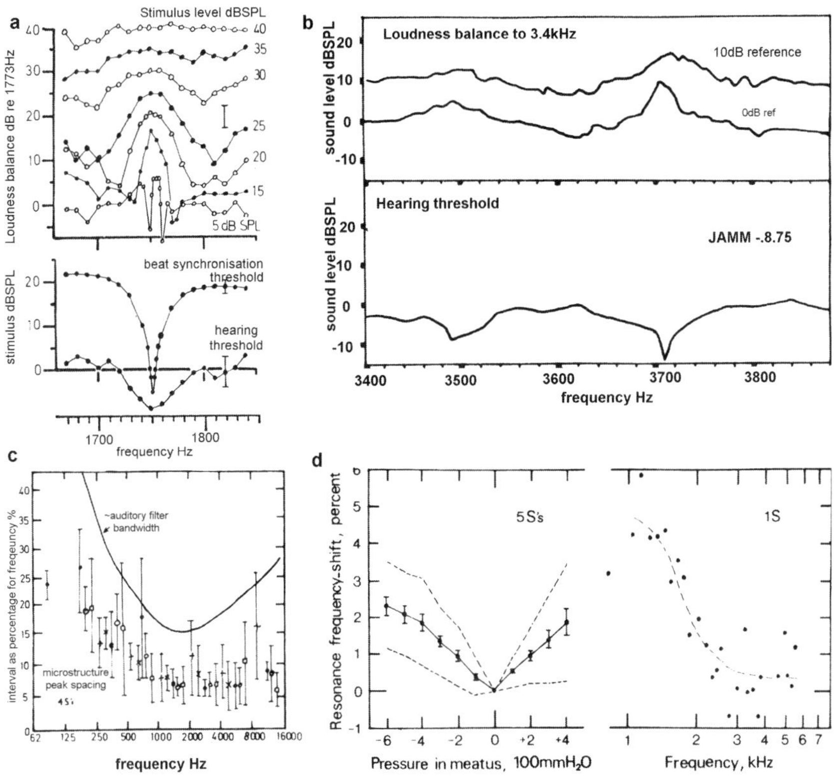

FIGURE 1.4. Psychoacoustic pointers to otoacoustic emissions, all from Kemp (1979). (**a**) A detailed study of loudness and threshold in a narrow frequency region surrounding a frequency of intermittent tonal tinnitus. Loudness and sensitivity are enhanced in this region. The 3dB bandwidth of the enhancement peak at 20dBSPL stimulation is 25Hz, giving a Q3dB of 70. The Q3dB increased for lower stimulus levels and the enhancement is insignificant for stimulus levels above 30dBSPL. The V-shaped 'synchronization threshold' boundary shown separates perception of the stimulus as a pure tone above the boundary from blurred perception due to beats below the boundary (after Wegel 1931; Flottorp 1953; Ward 1955.) (**b**) An example of localized enhancements of loudness and correlated enhancements in threshold sensitivity, which occur across frequency to varying degrees, in healthy ears at roughly regular intervals, usually without the 'beat' phenomena present in (a). (after Elliot 1958; Kemp and Martin 1976) (**c**) Data on inter-loudness-peak intervals across frequency from four subjects showing a high degree of uniformity and roughly following the contour of the auditory filter bandwidth (after Thomas 1975). This is evidence for a cochlear origin of OAE. The rough regularity of loudness enhancements suggested a standing wave/reflection explanation. (**d**) The key evidence for the physical nature of the auditory microstructure and a standing wave origin. Left side: Loudness enhancement peaks shift to slightly higher frequencies when the ear drum is made more stiff with either positive or negative air pressure. Right: The percentage SOAE frequency change in response to 600mm H20. This is greatest below 2kHz. Middle ear motion is strongly affected by tympanic stiffness change around and below its resonant frequency of 1kHz. See also Kemp (1981).

den Brink noted a relationship between threshold, pitch, and loudness variations. Thomas (1975) reexamined Elliot's ripple, referring to it for the first time as the "microstructure" of the auditory threshold, and noted a relationship between the peak-to-peak frequency interval and the auditory bandwidth (Fig. 1.4c). Following up this work, Kemp and Martin (1976) confirmed a strong relationship between fluctuations of hearing threshold as a function of frequency on the one hand, and fluctuations in the loudness of near-threshold tones and binaural pitch anomalies on the other. They drew attention to the quasiperiodic nature of these fluctuations, having a period of about 150 Hz at around 2 kHz. Using very detailed loudness balance measurements with frequency, they discovered that the sharpest peaks in hearing sensitivity resembled resonances with Q_{3dB} of around 100. Furthermore, it was found that the "beats," double sounds, and "idiotone" phenomena reported by Wegel, Ward, and Flottorp respectively occurred primarily around the frequencies of the deepest threshold and strongest loudness enhancements (Fig. 1.4a). This implied a direct link between the apparently rare phenomenon of "phantom" internal oscillations, which caused beats, and the more common microstructure of hearing threshold reported by Elliot and Thomas.

Crucially, Kemp and Martin deduced that all these phenomena were *physical* and not neural. This was based on their finding that the frequency at which these phenomena occurred could be altered slightly by the application of excess air pressure to the ear canal (Fig. 1.4d), which stiffened the middle air. Their initial conclusion presented at the XIIIth International Congress of Audiology, held in Firenze in 1976, was that "a [regular] framework of highly tuned resonant systems assisted in the reception of near threshold sounds. The mechanism must be active and must be at least partly peripheral" (Kemp and Martin 1976; this abstract is reproduced in Kemp 1998 and discussed in Kemp 2003).

2.3 The Final Clue to OAEs: A Standing-Wave Model of the Auditory Microstructure

Kemp and Martin (1976) explanation of the auditory microstructure as due to discrete oscillators was highly speculative and justifiable only by the need to understand their strange experimental findings, which were clearly outside the scope of contemporary auditory theory. Although their initial "multiple resonators" explanation was wrong, it did mark the first attempt to find a mechanistic explanation of auditory microstructure phenomena. Their work continued with further detailed psychoacoustic exploration into the form of the loudness microstructure and its middle ear pressure dependency. The parallel between the shape of the notches in loudness and physical standing wave interference (Fig. 1.4b) emerged as the final clue to a new, testable hypothesis, which led to the first recorded OAEs. The psychoacoustic data collected between 1975 and 1977 (Kemp 1979) were initially rejected for publication and accept only after the publication of work on otoacoustic emissions.

The concept that led to the first recording of OAEs was that the auditory microstructure was due to physical standing waves occurring in the cochlea (Kemp 1979, 1980, 1998). Whenever a propagating wave is contained within boundaries, the possibility of waves reflecting from the boundaries exists and there is the opportunity for standing waves to develop. In a room, for example, sound waves are reflected from the walls and are recirculated. Sound levels are enhanced in specific locations in the room at specific frequencies where the circulation time is a whole number of wave cycles. This is because repeated circulations remain in synchrony and so intensity accumulates. Intensity is depressed at other frequencies and places where repeated circulations arrive in antiphase. Peaks of sound pressure occur at specific locations at frequency intervals of $1/T$ Hz where T is the time for one circuit of the wave. This is the standing-wave phenomenon. The periodicity with respect to frequency and, in particular, the shape of the peaks and valleys in the auditory microstructure indicated that they could be due to standing-wave phenomena (Kemp 1980). If so, then the $\approx$ 100-Hz interloudness peak interval would translate into a wave circulation time of $\approx$ 10 ms, i.e., 10 cycles for 1000 Hz. This was far too long for *acoustic* propagation within the middle ear cavity or the cochlea. In air, sound would require a 3-m path length for this delay. But in the case of the much more slowly *transverse* traveling wave propagation, which takes place in the cochlea, a 10-ms delay was perhaps possible.

The hypothesis developed by 1977 was that the normal, apically traveling cochlear wave was somehow being reflected back to the middle ear and then reflecting again back off the middle ear into the cochlea in an endless circuit of diminishing amplitude, resulting in reverberation, standing waves, and the frequency periodic enhancements of cochlear excitation seen in the threshold and loudness microstructure. Although a single reflection would cause periodic interference notches in the ear canal (see Zweig and Shera 1995), this alone would not cause periodic enhancement of sensitivity and loudness without a second reflection back from the middle ear. Such a second reflection at the middle ear is, however, almost inevitable due to mechanical impedance inequality. The standing-wave model also explained the observed influence of ear canal pressure on the auditory microstructure seen by Kemp and Martin (1976), because changing the tension in the middle ear would slightly change the phase of reflection, causing a shift in microstructure frequencies. Finally, standing waves would modify the shape of the traveling wave and hence periodically influence pitch, as reported by van den Brink (1970).

The time delay predicted by auditory microstructure periodicity was of the order of 10 ms at 1 kHz, i.e., 5 ms in, 5 ms out. This seemed just plausible in this frequency range, given that it was already known that the cochlear traveling wave took several cycles to reach its peak along the basilar membrane. But in addition to needing reflection in the cochlea, which had never been observed, the standing-wave model also needed cochlear damping to be much less than had ever been previously supposed in order to allow standing-wave resonances to develop.

The proposed cochlear reflector needed to have nonlinear, level-dependent properties to explain the observed "saturation" of the loudness microstructure with increased stimulus levels (Fig. 1.4a). As a bonus, this provided a way to explain the generation of aural combination tone production (e.g., $2f_1 - f_2$), which had already been shown to interact subjectively with the auditory microstructure phenomenon (Ward 1955).

In 1977, the standing-wave model was satisfyingly consistent with a variety of unexplained psychoacoustic experimental data, but was in stark conflict with contemporary understanding of the cochlea. First, there was no prior evidence suggesting that the traveling wave could ever be reversed or reflected in the cochlea. Unidirectional propagation and the absence of traveling-wave reflection were cornerstones of cochlear mechanics at that time (von Békésy 1960; Zwislocki 1983), denying all possibility of standing waves. Furthermore, the psychophysical evidence of more than 12-dB periodic peak-to-valley enhancements and depressions of cochlear excitation (the "standing-wave ratio") required round-trip attenuation with two reflections to be less than 40 %. All experimental observations of the traveling wave after von Békésy had confirmed high levels of damping and energy loss, so that even if partial cochlear reflection were contemplated, a passive standing-wave model for the auditory microstructure was not viable. It could not reproduce the high standing-wave ratios required by the psychophysics.

The counterindications were so strong that the cochlear standing-wave hypothesis would probably never have been experimentally tested were it not for evidence in the reports by Wegel, Ward, and Flottorp of what seemed to be "internal oscillation" and which, if physically present, clearly required an energy source—an "active process." Gold's proposal for active damping reduction was based on theoretical considerations to explain the extremely sharp mechanical frequency tuning Gold believed the cochlea had to possess. The standing-wave model of auditory microstructure also needed low damping but for the much more modest purpose of allowing standing-wave resonance to occur. It was not intended as a speculative new model of cochlear function, but simply as an attempt to explain a set of actual psychoacoustical phenomena. It is a curious coincidence that the resonant Q values of ≈ 100 that Gold wrongly believed were needed for hearing did actually appear in the auditory microstructure because of whole cochlear feedback oscillation!

Gold's model predicted sharply tuned mechanics in the healthy living cochlea. This prediction had been tested by several workers, but up to 1977, only negative findings had been reported (e.g., Evans and Wilson 1975). Gold's model had also predicted tonal spontaneous acoustic emissions as a result of *abnormal* function, which he failed to confirm, and in 1971, Glanville et al. (1971) did not consider a cochlear origin in their report of objective tonal tinnitus phenomena.

In contrast, the standing-wave model predicted a variety of new measurable external acoustic phenomena during *normal* audition. The standing-wave model predicted that (as with a reverberant room) it should be possible to demonstrate physically that reverberation was occurring by means of external acoustic measurement in the absence of self-sustained oscillation. In the case of a reverberant room, for example, the acoustic impedance at the windows and doors is modified slightly by the sound pressure from internal resonances. This comparison led to the idea that a microphone sealed in a stimulated ear canal should be able to record the periodic peaks and valleys in sound pressure with frequency matching the subjective auditory microstructure. Furthermore, any sustained self-oscillation that depended on standing-wave resonance would also be detectable in the ear canal without stimulation because the middle ear was a partially reflecting boundary. Any distortion product generated in the cochlea should also be detectable. Finally, if reflections were occurring inside the cochlea, then there should be a detectable "echo" of a short sound stimulus after a time given by the periodicity of Elliot's threshold "ripples" i.e., $\approx$ 10 ms.

The standing-wave model for the auditory microstructure was not arrived at by the author either fortuitously or by pure deduction from the psychophysics, nor from his experience in industrial noise control. Like Gold, the author constructed primitive radio receivers as a boy and observed positive feedback together with the frequency selectivity and the oscillations this created. His familiarity with standing-wave concepts came from his PhD research experience in geophysics. Due to low ionospheric absorption, global radio wave propagation is possible below 60 Hz, and radio emissions from lightning storms continuously excite resonances in the earth – ionosphere cavity at 8 Hz and other higher resonant modes. These resonances are detectable in the background radio noise spectrum on the ground and also as wave impedance variations with frequency. In this case, no reflector is needed other than the two spherical boundaries of the earth and the ionosphere, as the transmission path is spherical. No active process is needed, and there are (fortunately) no self-sustaining oscillations of the earth's electric field to explain (see Kemp 1971, 1998)!

2.4 The First Observations of OAEs and Early Concepts of Origin

Experiments to test the standing-wave resonance model of the auditory microstructure took place between July and August 1977 in the basement of the Nuffield Hearing and Speech Center of the Royal National Throat Nose and Ear Hospital, London. Using ear-impression putty, a miniature hearing aid microphone was sealed into the ear canal of a healthy human ear that exhibited strong auditory microstructure, some monaural diplacusis, and related occasional transient tonal tinnitus. This was initially the author's own ear with results confirmed on colleagues. Sound stimulation was applied via a headphone placed over the sealed ear, with the external stimulus level raised as required to compensate for the attenuation of the sealing plug. The internal stimulus level was monitored via the ear canal microphone. First an attempt was made to record the combination tone heard when an applied tone interacted with an internal tone (after Ward). A single stimulus tone was applied at 2 kHz, 40 dB SPL. The psychoacoustic measurements had indicated that an internal oscillation was occurring at 1764 Hz and that an aural combination tone would be produced at 1518 Hz, consistent with a $2f_1 - f_2$ intermodulation distortion being generated by the applied 2-kHz stimulus (f_2) and the presumed *internal* 1764-Hz oscillation (f_1). Using a 10-Hz-bandwidth heterodyne analyzer to examine the ear canal microphone acoustic signal, clear acoustic signals were detected at both 1518 Hz (the distortion product) and also at 1764 Hz (the internal oscillation). This was the first objective observation of distortion product (DPOAE) and spontaneous (SOAE) emission. It confirmed the physical presence of the internal oscillation and the combination tone suspected by Ward as being due to "a basic instability of some part of the auditory system." But this experiment was not sufficient to confirm the standing-wave hypothesis.

In a second experiment, the tunable narrow-band analyzer was motorized and swept slowly from 1 to 2 kHz. Using the beat frequency sinusoidal output of the analyzer as the stimulus (which was automatically locked to the analysis frequency), the ear canal sound level was continuously recorded on a chart recorder as the stimulus frequency was swept. Since the stimulus was supplied externally through the sealed ear canal, the sound source impedance was very high. This allowed the sound pressure developed in the ear canal to fluctuate with frequency as the acoustic impedance at the eardrum fluctuated due to cochlear reflections arriving in or out of phase in parallel with the internal (perceived) excitation fluctuations. Such a physical fluctuation would imply that the middle ear was coupled to the system (i.e., the cochlea) in which the standing waves were occurring and that the middle ear was most likely a reflecting component in the standing-wave system. Strictly speaking, the external ear canal sound-pressure level would have varied with frequency due to interference between the stimulus and only a single cochlear reflection, but secondary (and multiple) reflections back from the base and standing waves would be essential to explain the *internal* periodic excitation fluctuations of the auditory microstructure seen psychoacoustically. The data obtained from this experiment showed clear evidence of periodic eardrum impedance fluctuations correlating with the individual's auditory microstructure, thereby supporting the standing-wave hypothesis. This led to the first stimulus frequency emission (SFOAE) recordings (see Fig. 1.5 and also Kemp 2003 and Kemp 2007). It paralleled much earlier recordings of unexplained tympanic impedance fine structure made by Moller (1960). It is quite possible that the fine structure that Møller observed in eardrum impedance was due to cochlear reflections, but this possibility was not recognized at the time.

These two new experimental results supported the hypothesis that a standing-wave mechanism was responsible for the psychoacoustic microstructure and that oscillations and nonlinearity existed in the ear, but they did not prove that the phenomena were associated with the hearing process. Proof of a *cochlear* origin for emissions was obtained only by a third experiment, designed to associate the ear canal phenomena with the frequency selectivity that is the defining characteristic of the cochlea. The heterodyne analyzer was converted so that it could be electronically swept repeatedly over a small frequency range ($\approx$ 100 Hz) that contained a clear acoustic impedance undulation (i.e., an ear-canal

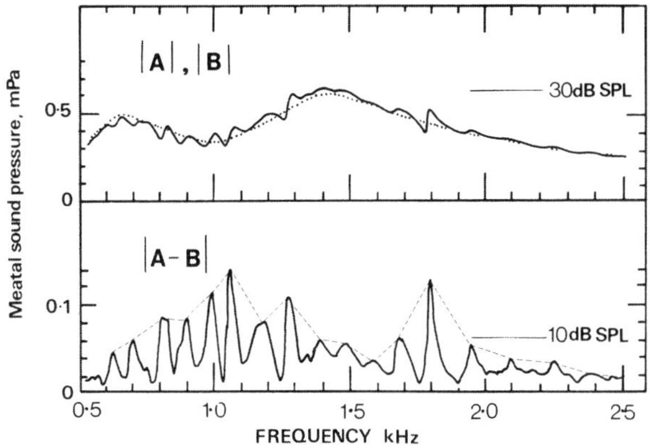

FIGURE 1.5. Early observations of stimulus frequency otoacoustic emissions from Kemp 1980. The upper panel shows raw ear canal sound-pressure measurements as a function of stimulus frequency in an ear exhibiting strong auditory microstructure. Note that the ordinate is sound pressure, not decibels. A 30-dB SPL calibration line is provided. The broad changes in ear canal sound levels can be attributed to transducer response factors, but the narrow periodic undulations in the solid line (A) are due to acoustic interference in the ear canal between the stimulus frequency OAEs and the stimulus at 28 dB SPL (average). The dotted line (B) is data from a 68-dB SPL sweep scaled down by 40 dB (i.e., 1/100). These data show no significant interference ripples, and this is explained by the saturation of the emission reflection mechanism at high levels. The lower panel shows the vector subtraction of A and B, which takes into account phase and amplitude differences between the two measurements. This difference signal represents all level-dependent changes in ear canal sounds, which are attributed to the cochlea. These difference signals are known as stimulus frequency otoacoustic emissions (SFOAEs). In this example, the SFOAE has a strong periodic structure at approximately 100 Hz. This is attributed to standing-wave resonances occurring in the cochlea between the middle ear and the source(s) of reflection. The round-trip delay time can be deduced for the interpeak interval, e.g., at around 700 Hz it is 80 Hz, giving a delay of 12 ms; and around 2000 Hz, the interval is 150 Hz, giving a 7.5-ms delay.

sound-pressure-level inflection) as found in the previous experiment. A second stimulus tone was then introduced at a fixed frequency and its level increased until the impedance undulation leveled out, indicating that the source of "reflection" had been "suppressed." Plotting the level of suppressor tone needed to suppress the reflection at all frequencies of suppressor revealed a tuning curve as sharp as any psychoacoustic tuning curve, and much sharper than the accepted tuning for cochlear mechanics at that time (Kemp and Chum 1980b); see Figure 1.6. This was the first cochlear mechanical tuning curve to show the full frequency selectivity of the cochlea and its nonlinearity, and it was conducted on human ears, only later to be confirmed using OAE in laboratory animals (Brown and Kemp 1984).

The fourth experiment was carried out to verify directly that a true "reflection" was occurring by recording the cochlear echo of a transient stimulus. Connecting

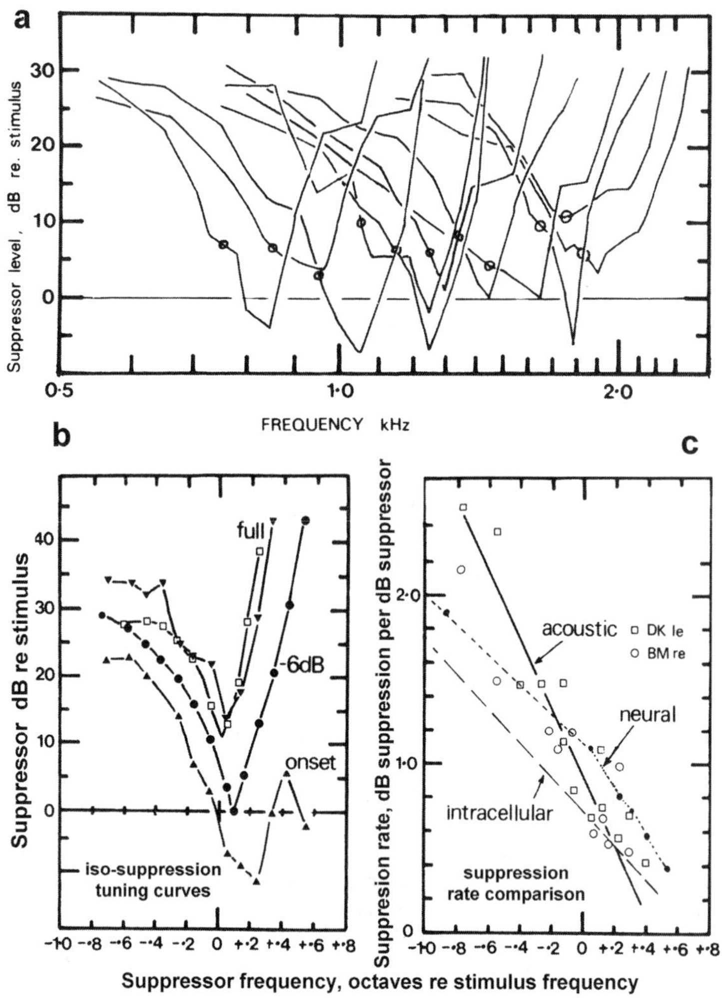

FIGURE 1.6. From Kemp and Chum 1980b; this figure documents the discovery of the close parallels between stimulus frequency emission, cochlear tuning, and nonlinearity respectively. In these stimulus frequency emission experiments, the stimulus was kept constant and the level and frequency of a suppressor level tone were altered to obtain a specific degree of suppression. (**a**) *Top*: 50 % (−6 dB) isosuppression tuning curves for OAEs at several frequencies from one human ear. The stimulus frequency is marked with a circle. If is significant that maximum suppressibility was obtained with suppressors ≈90 Hz (i.e., ≈8 % or ≈1/8 octave) *above* the stimulus frequency. The average Q_{10}dB was 5.9 ±2.7, which can be compared to auditory nerve tuning curve tips seen in laboratory animals. (**b**) *Bottom left*: Stimulus frequency-emission isosuppression curves for various degrees of suppression from first detectable to maximum obtainable. The stimulus was 40 dB SPL, 1.1 kHz. Note that the frequency of maximum suppression moves from the stimulus frequency to higher frequencies for decreasing suppression criteria and that higher-frequency suppressors cause a small suppression at very low levels. (**c**) *Bottom right*: The rate of suppression at the 50 % suppression point expressed as dB OAE

the ear canal microphone this time to a clinical auditory evoked-response averaging system and synchronizing the averager to a repeated acoustic click stimulus, the sound level in the ear canal was repeatedly recorded for 31 ms after each stimulus. The expectation was to see multiple "echoes" of the applied click stimulus with an interecho interval of about 10 ms, as anticipated from the interfrequency peak interval of the auditory microstructure. The transient evoked emissions actually obtained were prolonged and complex, and were unique to each individual ear. However, no discrete "click" echoes could be identified. This result should have been anticipated. The cochlear traveling-wave propagation was known to be frequency dispersive, and the auditory microstructure was itself known to be a highly individual phenomenon with complex frequency fine structure. Frequency-specific stimulation with tone pips demonstrated that the response was indeed a true "reproduction" or reflection of the applied stimulus (Fig. 1.2). The reflection of the stimulus is grossly distorted by frequency dispersion (i.e., the spreading out of the frequency components in time) and a little by cochlear nonlinearity (i.e., intermodulation), as demonstrated much later by Yates and Withnel (1999). This fourth experiment to observe individual discrete cochlear "reflections" was not strictly necessary to prove the standing-wave hypothesis, since the frequency periodicity observed in ear canal impedance adequately proved the existence of reflection. However, it was perceived as being necessary to demonstrate in unambiguous terms to understandably skeptical auditory researchers that an unknown process existed in the cochlea.

Finally, to establish that the new process was somehow linked to an *essential* auditory function, transient acoustic emission recordings were attempted on ears with hearing impairment. No such emissions were found at hearing loss frequencies in human ears (see Fig. 1.2 and Kemp 1978). The influence of deafness on human DPOAE was demonstrated later, after laboratory studies (Kemp et al. 1986).

From this whole series of experiments, the cochlear "echo" experiment was selected for first publication, as the logic was the easiest to describe. Convincing reviewers that there was a previously unknown acoustic "auditory response" seemed to be less of a challenge than trying to convince them that an obscure "modulation" of middle ear impedance was unambiguous evidence of a new biological process! The paper was submitted for publication in the *Journal of the Acoustical Society of America* (Kemp 1978). This was after first being rejected in 1977 by reviewers of the journal *Nature* on the grounds that it had no theoretical basis and that the observations had not been substantiated in animal studies. The

FIGURE 1.6. suppression per decibel suppressor increase. This falls from a steep 3 dB/dB rate for suppressors one octave below the stimulus to a very shallow 0.2 dB/dB a half octave above. This is very characteristic of the auditory filter, as illustrated by contemporary data from single nerve fibers in the cat from Abbas and Sachs (1976) and from guinea pig inner hair cell receptor potentials from Sellick and Russell (1979).

term "acoustic emissions" used to describe the new "response" was borrowed from materials science, where it is applied to sounds created, for example, in welded materials due to the spontaneous release of internal stresses: not really a good analogy in retrospect. More recently, the term has been applied to the acoustic observations of molecular biological processes (Cooper et al. 2001). Evidence for a new factor in cochlear mechanics was first reported publicly to a very skeptical meeting of the British Society of Audiology at Keele University in April 1978.

2.5 Early Confirmation and Questions

Following publication, confirmation of these new and controversial experimental findings by other workers was rapid. Prior to 1978, Wilson had been researching the auditory microstructure at the Department of Human Communications, Keele, and was interested in its possible association with tinnitus. He quickly confirmed that the auditory microstructure was reflected in ear canal acoustics and that spontaneous emission of sound from ears was associated with the auditory microstructure (Wilson 1980b). The close association with cochlear function was demonstrated in primates, where the emissions were appropriately depressed by noise and by ototoxic drugs (Anderson and Kemp 1979). Like Wilson, Zurek was researching tinnitus at that time at the Central Institute for the Deaf, in St. Louis, and was able to confirm spontaneous emissions (Zurek 1981). Zurek must also be credited with being the first to use the term *oto*acoustic emission. Kim et al. (1979) and Kim (1980a) confirmed the of existence acoustic distortion products emissions in the ear canals of cats. These emissions were remarkably more robust than in human ears. Wit and Ritsma (1979, 1980), Rutten (1980), and Wilson (1980b) each confirmed the transient-evoked cochlear "echo" phenomenon.

Despite this rapid confirmation and acceptance of otoacoustic phenomena, acceptance of the fundamental challenge that they posed to cochlear theory was slow. What was this challenge? The challenge was that if OAEs and the associated psychoacoustic phenomena were to be explained by nonlinear reflections, standing waves, and oscillations in a cochlea with low damping and sharp mechanical tuning, then the contemporary view of the cochlea as a passive, heavily damped receptor was wrong. This concept was far too radical for most auditory researchers to accept at that time. Proof that sound emissions from the cochlea existed was just not sufficient to overturn decades of laboratory findings apparently supporting the von Békésy view. To be even tempted to embrace the opposite view would have been judged scientifically irresponsible by most leading figures at the time.

Those most committed to the von Békésy model supported the second filter hypothesis (see Cooper et al., Chapter 2) and maintained that cochlear mechanics had been adequately proved to be linear and only broadly tuned (e.g., Evans and Wilson 1975). Several alternative explanations of OAEs were proposed that avoided an association between OAEs and basilar membrane characteristics. Evans, for example, proposed that the sound emissions were due to stapedius muscle reflex activity (Evans et al. 1981). Wilson (1980c) proposed that the emissions were generated in the cochlea by cellular swelling and transmitted to the middle ear by a pressure wave, and therefore not directly coupled to basilar membrane motion. These proposals did not survive experimental testing. Emissions survived sectioning of the stapedius muscle. Emissions showed very little adaptation to greatly increased stimulation rates, eliminating both myogenic and neural involvement. Volumetric outer hair motility was not found.

For an overview of the debate about the relevance of OAEs to cochlear function at that time, see the proceedings of the 1979 London symposium on "Nonlinear and Active Mechanical Processes in the Cochlea" (Kemp and Anderson 1980), especially the discussion session chaired by Kim (1980b). The London meeting was called to discuss the implications of otoacoustic emissions and other emerging auditory phenomena that suggested an active interaction between physical and biological systems. Use of the term "active process" in this context may well stem from the title of that symposium.

Those who already believed that a major revision of cochlear theories was needed (e.g., Kim, Molnar, and Rhode) were in the minority at that time, but the tide was already beginning to turn against the conservative view. At that same London meeting, LePage and Johnstone (1980) announced the first observations of nonlinear basilar membrane motion in the guinea pig, a characteristic previously seen only by Rhode (1971) in the squirrel monkey. Fettiplace and Crawford (1980) reported on the motility they had found in the hair bundle of turtle cochlear hair cells, and Flock (1980) presented compelling histological evidence for outer hair cell motile capability. All these new observations, together with the OAEs phenomena, had the effect of renewing interest in Gold's hypothesis of active feedback in the cochlea. Zwicker (1979) published an outline of an active cochlear model citing the outer hair cell as an "amplifier" and Kim et al. (1980) reported the functionally useful effect of incorporating negative damping in models of the cochlear partition. When sharp nonlinear cochlear mechanical tuning was finally directly demonstrated in 1982 by Sellick et al. (1982) and by Khanna and Leonard (1982), the balance of evidence finally tipped in favor of a nonlinear sharply tuned basilar membrane. It was the revered Hallowell Davis who led the general switch in opinion from passive to active cochlear mechanics in his 1983 paper (Davis 1983) by embracing the concepts of the "active process" and "cochlear amplifier," terms that have come to denote the mechanism whereby cellular electromotility enhances the traveling wave. Significantly, Hallowell Davis was, according to Gold, one of those who rejected the cochlear amplifier concept in 1948, and Gold greatly appreciated a letter from Hallowell Davis in 1983 acknowledging his error and eventual conversion! Most remaining doubts about the new view of the cochlea were swept away by the discovery of fast mammalian outer hair cell motility (Brownell et al. 1985; Kashar et al. 1986). Thereafter, models of the cochlea that incorporated amplification into basilar membrane motion began to be developed in earnest. Neely (1986) published a model incorporated traveling wave amplification to enhance cochlear sensitivity and frequency selectivity, a process rather different from the positive feedback model Gold had proposed. This generated a conclusion that still stands: that the "cochlear amplifier" must inject its energy into the traveling wave at some distance *before* the frequency place peak in order to create a stable wave peak of a shape matching that seen in basilar membrane motion.

So what is the answer to the question, Why were OAEs not identified before 1977? It cannot be claimed that suitable measurement technology was entirely absent, or that there were no indications in the literature, or that the idea of cochlear electromotility and amplification had not been thought of. The truth is probably that the required concepts contradicted established knowledge so starkly that it inhibited their consideration. But three decades on from the start of these dramatic developments, we still do not have a truly comprehensive model of cochlear mechanical function. This is understandable, given the complexity of the cochlea and the fact that so many physical and physiological cochlear parameters have yet to be determined. But in addition, the computational task of modeling the true physical response of the cochlea to acoustic stimulation in three dimensions—including the impact of the organ of Corti with its embedded electromotile elements—is immense.

Given this complexity and the fact that OAEs are only a by-product of normal function, it is not surprising that there is as yet no realistic global cochlear model that accurately predicts the detailed behavior of otoacoustic emissions. Nevertheless, essential parts of the theory of OAEs have been worked out, and in a sense these wait to be incorporated into a realistic cochlear model. It is not the case that the complex structure and behavior of OAEs needs to be understood before the cochlea can be fully modeled. It is the case that the lack of understanding of the complex structure of OAEs interferes with our ability to use them to best advantage to extract useful information about the cochlea and the "active process."

3. Concepts of OAE Generation

3.1 The Source of OAE

For a brief period between 1978 and the confirmation of hair cell motility in 1983, OAEs were the only direct evidence of an active and nonlinear mechanical process in the cochlea. While cochlear nonlinearity was quickly accepted, it was not at all clear during that period whether the "active process" demonstrated by SOAEs served to enhance normal cochlear function or that it was even relevant. The great majority of auditory researchers now accept that a nonlinear and active process enhances the sensitivity, selectivity, and dynamic range of the cochlea (as discussed in this book), and also that OAEs are a by-product of that same process. But why is such an important cochlear mechanism so "leaky" as to produce emissions? Why is stimulus-related energy wasted at all? It is counterintuitive that an organ like the cochlea, in which there is clearly a great investment in sensitivity, should allow this to happen. Are OAEs somehow inevitable because of the design of the cochlea? Are they even essential? To begin to answer these questions, we need to examine the source(s) of the OAEs we find in the ear canal.

The concept of cochlear traveling-wave "reflection" as a source of OAEs arose from psychoacoustic observations of the auditory microstructure, and the first stimulus-frequency and transient evoked otoacoustic emission measurements supported this notion. The concept of cochlear standing waves explained how human auditory microstructure arose. Together with amplification it was possible to explain how spontaneous emissions could be created by sustained standing-wave oscillations when the round-trip reflection gain exceeded unity. Interestingly, the role played by standing waves in determining the precise frequency of spontaneous emissions was not universally accepted, and other explanations were explored (see van Dijk and Wit 1990). Only recently has the standing-wave concept been independently proved (Shera 2003).

What is the nature of the cochlear reflector? Irregularity or discontinuity in a propagating medium is the typical cause of wave reflection, and this logically translates into mechanical irregularities on the basilar membrane. Irregularity cannot be considered as essential to cochlear function, but might be considered inevitable. Given the anatomical evidence for slight hair cell disorder in all mammals and other animals (e.g., Manley 1983) together with active gain, some scattering of wave energy is inevitable. Early OAE measurements appeared to require the reflector to be nonlinear with respect to stimulus intensity. At this point in 1977, the logical and economical assumption was made that the cochlear reflector was nonlinear (Kemp 1978). This explained the appearance of distortion products when either two tones were reflected from the same region or when a stimulus-tone frequency came near to the frequency of a spontaneous emission. The false assumption that the reflector was the sole cause of OAE nonlinearity (as opposed to the reflected signal having passed through a nonlinear cochlear amplifier) persisted through the 1980s and early 1990s despite evidence that some properties of distortion-product OAEs could not be explained by a reflector fixed to specific places on the basilar membrane (e.g., Kemp et al. 1986).

From the beginning of OAE research, the nature of the "cochlear reflector," i.e., the process that reverses the forward flow of energy in the cochlea, has been explored through measurements of OAEs *phase* in relation to the stimulating sound. The rate of change of phase with frequency can, under common circumstances, indicate the inherent delay in the entire circuit, i.e., the time between stimulation "in" and emission "out" or the "group delay." This delay in seconds is given by dividing "cycles of decrease in phase" by "hertz increase in frequency," and this interpretation makes the assessment of the inherent delay of a continuous stimulus frequency emission possible. It is even possible to assess the delay of distortion product emission by this method even though the output is of a different frequency from that of the stimuli. To do so, the phase of the distortion product is measured relative to that of the same distortion component synthesized directly from the stimuli present in the ear canal before any delay (e.g., Mahoney and Kemp 1995).

The “phase gradient” method of delay measurement is, however, “unsafe” in frequency dispersive systems and especially when, as in the cochlea, it is possible for the source of the returned signal (i.e., the OAEs) to move its location as the frequency is changed. If, for example, the returned signal came from the peak of the cochlear traveling wave, then the frequency-phase gradient appropriate to its true physical delay would be hidden by an almost equal and opposite change in phase due to the intrinsic change in time delay of the traveling wave to its new peak. This is because the peak would move along the basilar membrane, and due to the near logarithmic frequency scaling of the cochlea, the phase of an OAE transmitted from the peak of a traveling wave would change only very slowly with frequency change. The phase gradient method would (wrongly) indicate very little delay. OAEs emanating from a reflector at a fixed place on the basilar membrane would not suffer this erosion of phase gradient and would therefore show a greater and truer group delay.

Early observations of human stimulus frequency OAE phase gradients by Wilson (1980b) and Kemp and Chum (1980a) revealed substantial delays of around 10 ms, which agreed with the direct determination of OAE delay by transient excitation (Fig. 1.2) and were consistent with the periodicity in the auditory microstructure. It implied that the cochlear traveling-wave delay (taken to be half the OAE delay) was around 5 ms in humans, much longer than had been determined by electrophysiological and other methods at the time (e.g., Ruggero 1980). The magnitude of OAE delay was taken by some, including Wilson (1980c), to be evidence against the traveling-wave reflection hypothesis. Wilson and others believed that most OAE delay arose from resonant delay (tuned system ringing) and not through wave-propagation travel time. Interestingly, the debate about whether OAE delays truly indicate traveling-wave delays continues (Ruggero and Temchin 2005, 2006), but most agree that stimulus frequency and transient emissions arise mostly by reflection from some irregularities in the traveling-wave propagating medium and that SFOAE and TEOAE delay is (at least) a function of traveling-wave delay.

Milliseconds of observed delay supported a place-fixed reflector, but it was puzzling as to why SFOAEs were seen with relatively constant delay and intensity over frequency ranges larger than that anticipated from the width of the auditory filter. Traveling-wave morphology dictates that the intensity of the signal reflected from a single place should rise and fall with frequency as the traveling-wave peak crosses over the reflector place, with the phase gradient progressively increasing with frequency. But in practice, extended frequency ranges of relatively constant emission intensity and delay were found (e.g., Wilson 1980b). Strube (1989) offered a solution to this problem by proposing a spatially regular distribution of reflectors spaced at one-half the wavelength of the traveling wave at its peak, which is relatively constant with frequency. This ensured a consistent reflection as frequency is changed. However, without any anatomical evidence for such a structural feature, this solution was little more justified than, though different from, the periodically placed resonators first suggested by Kemp and Martin (1976).

A rigorously derived and elegant explanation was eventually developed by Zweig and Shera (1995; see Shera and Guinnan, Chapter 9). They showed that a dense random distribution of irregularities could result in substantial traveling-wave reflection with a contiguous steep and relatively constant phase gradient, provided that the traveling wave shape conformed to a "tall and broad" profile. In this model, the traveling wave effectively "selects" those irregularities occurring over a limited length of basilar membrane that have optimum spatial reflector periodicity. Only reflections from this subset of reflectors with spatial periodicity appropriate to the peak traveling-wave wavelength will travel back to the base coherently to create a significant OAE. All other reflectors and reflections will be self-canceling and therefore irrelevant. Hence the term *coherent reflection filtering*.

Returning now to the historical development of OAE concepts, Kemp (1978) had presumed that the OAE reflector was inherently nonlinear and hence it would be responsible for the generation and not only the reflection of distortion-product OAE (DPOAE). There was some evidence for this, in that distortion products from close tone stimuli ($f_2/f_1 < 1.1$) showed large delays similar to those of SFOAEs and TEOAEs and this fitted the "place-fixed" reflector model. But the strongest $2f_1 - f_2$ DPOAEs were produced by stimulus frequencies about 20 % apart in frequency. These emissions invariably showed the shallow-phase gradients and hence a small apparent group delay indicative of an origin fixed to the envelope of the traveling wave (wave fixed) and not to the basilar membrane (place fixed) (Kemp et al. 1986). At that time, Kemp interpreted wave-fixed emission as a nonlinear "reflection" phenomenon due to the intensity of excitation at the peak of the traveling wave somehow locally modulating wave propagation characteristics and resulting in a temporary mechanical irregularity. Non linear reflection of the traveling wave at the peak would necessarily include distortion products generated by the same nonlinearity. But Kemp's hypothesis left many questions unanswered: Why, for example, did distortion product emissions switch from a place-fixed "reflection" source to a wave-fixed source at $f_2/f_1 \approx 1.1$ (Kemp et al. 1986)? Why, if local nonlinear "saturation" of cochlear mechanics resulted in a wave-fixed reflector of distortion, did it not also produce wave-fixed (low apparent delay) stimulus frequency emissions? On the latter crucial point, such emissions had been reported in gerbils (Kemp and Brown 1983a,b), but they had never been seen in human ears.

The puzzle left by Kemp in 1986 was solved step by step. Brown et al. (1996) demonstrated that there could be two different sources of the same DP frequency. Shera and Guinnan (1999) clarified the matter still further by rejecting Kemp's presumption that mechanical nonlinearity was a factor in the reflection of traveling-wave energy to form emissions. Although the traveling wave inherited nonlinearity from the cochlear amplifier, they emphasized that a linearly behaving reflector would give rise to both stimulus-frequency emissions and also (provided distortion products generated elsewhere created apical traveling waves that passed the reflector) distortion-product emissions. They renamed place-fixed emissions "reflection emissions" to underline the fact that they believed linear reflection was responsible for the retransmission of traveling-wave energy back to the base of the cochlea and not mechanical nonlinearity. The mechanism proposed for these emissions, coherent reflection from random irregularities, has already been described. The nonlinearity evident in reflection emissions, they argued, is solely due to the nonlinearity that is inherent in the cochlear amplifier that delivers the amplified wave that is being reflected. Shera and Guinnan described distortion-product emissions that had low-delay "wave-fixed" properties (i.e., $2f_1 - f_2$ for $f_2/f_1 > 1.1$) as "distortion emissions" to emphasize that their transmission back to the base of the cochlea had nothing to do with a reflection but was inherent in cochlear nonlinearity.

Kemp and Knight (1999; see also Kemp 2006) considered why the level of distortion emission was frequency-ratio dependent, peaking at $f_2/f_1 \approx 1.2$, and why distortion-products components with frequencies higher than f_2 were always seen to exhibit place-fixed characteristics and never wave-fixed characteristics for any frequency ratio (Knight and Kemp 2001). They concluded that an optimum output stimulus ratio (f_2/f_1) can occur only for distortion components below f_1 (i.e., lower side bands $Nf_1 - (N-1)f_2, f_2 > f_1$) because for this group of distortions the *spatial* phase gradients within the stimuli's traveling waves counteracted each other in a unique way. This allowed the spatial phase distribution of simultaneously generated elemental (hair cell) DP vibrations related to *lower* sideband intermodulation distortion to have a much longer "wavelength" along the basilar membrane than either primary frequency. It can even become negative, i.e., pointing *basal-ward* toward the base. Kemp and Knight (1999) introduced the term "virtual reflector" to describe this phenomenon. If the DP's distributed source "wavelength" is not only negative but also appropriate to a *reverse* traveling wave of the DP frequency, then such a spatial distribution will optimally excite distortion product waves traveling coherently toward the base. Thus, although the distortion product itself is generated by hair cell nonlinearity all along the length of the overlapping stimulus traveling waves, it is only over a limited length and in a restricted stimulus-frequency ratio range that *coherent wave* transmission of DP to the base becomes optimal. This concept is inherent in previous treatments (e.g., Kanis and Boer 1997).

In contrast, at close primary ratios, the DP source wavelength is always positive and so will transmit energy more efficiently apically. It will be emitted as a DPOAE only if this forward traveling wave meets a reflector (Kemp and Tooman 2006). For distortion-product frequencies above f_2 (e.g., DP $2f_2 - f_1$), the "virtual reflector" phenomenon cannot occur. One way of conceptualizing this process is to first consider the stimulus wavelengths in the region where distortion is mainly generated: under the f_2 traveling-wave envelope. Cochlear mechanics ensures that the wavelength of the f_1 wave is always longer than that of the higher-frequency f_2 at any place. However, the wavelength of the notional frequency $2f_1$ is half that of f_1 and therefore can be comparable to that of f_2. In the creation of "$2f_1$" – "f_2" the DP phase can (by subtraction of phase gradients and depending on f_2/f_1) have a spatial phase gradient that is positive or negative (or zero). The same can never true for $2f_2 - f_1$.

The coherent-wave virtual-reflector explanation of DP optimum ratio appears to conflicts with an earlier hypothesis that a local mechanical resonance offset to a lower frequency than the place frequency influences DP emission and so determines the optimum frequency ratio for lower sideband DPOAEs (Brown et al. 1992; Allen and Fahey 1993). However, the participation of a second resonance in the DP emission process is not excluded by the virtual reflector hypothesis.

4. OAEs in Auditory Reseach

The concepts of OAE generation introduced above are developed more fully in other chapters of this book. However, place-fixed coherent reflection emissions, wave-fixed coherent wave transmission of distortion emissions, plus the concept of internal reflection of emission energy by the middle ear and standing waves, together provide a foundation from which the main features of OAEs can be derived. Even so, serious debate continues about the nature of the "reflection" process (e.g., Siegel et al. 2005) and of the retransmission process (Ren 2004).

Despite these fundamental questions, the characteristic fine-frequency structure of OAEs can be largely understood. This structure is quite prominent in human ears and closely parallels the psychophysical microstructure that originally provided the clue to recording OAEs. It is due to interference between multiple reflections of the emission signals inside the cochlea combined with interference between multiple OAE sources in the case of DPOAE. While very interesting phenomenologically, this aspect of OAEs is really secondary in comparison with the deeper issues of cochlear function. Microstructure has little to tell us directly about the underlying processes that are important for hearing. Individual fine-structure features can, however, sometimes serve to magnify changes in the status of the cochlea, because by their nature, cancellation notches, resonant peaks, and spontaneous oscillations depend critically on cochlear status. Thus the effects of, for example, noise exposure or activation of the medial olivary cochlear reflex are most easily seen at these "critical" stimulation frequencies. But equally, a small change in middle ear status can shift this fine structure in frequency, change the OAE intensity, or start and stop a spontaneous OAE because the middle ear is an essential part of the OAE transmission system and of cochlear standing-wave phenomena. This can undermine interpretation.

A major challenge in OAE research is to extract fundamentally important information about hair-cell and cochlear status from OAE data. A tantalizing fact is that given the input–output characteristics and the operating point of a hair cell, the profile of distortion-product components it will produce locally to any given stimulus is completely determinable by calculation (e.g., Russell and Lukashkin 2005). Unfortunately, the profile of emitted distortion-product components (see Fig. 1.3) does not allow the unambiguous determination of hair cell operating characteristics, not least because the relative DP proportions are grossly affected by the emissions mechanism and the integration of multiple cellular sources. The situation will be further complicated if it becomes necessary to account for the nonlinear motility of both the hair cell body and the hair bundle (Fettiplace and Hackney 2006; see also Martin, Chapter 4).

The way to obtain greater value from otoacoustic emission measurements is by devising experiments that yield data of a nature that is largely independent of the emissions escape mechanism and of the effects of interference. Certain OAE observations are by their nature less affected by the complexity of the OAE mechanism than others, and these have enhanced value in auditory research. Phase gradients (group delays) sustained across a range of frequencies are among these. Also (provided middle ear status is controlled for), a general depression or enhancement of OAE intensity over frequency ranges broader than the fine scale of interference effects is more likely to be the result of changes to outer hair cell input–output characteristics or operating point than to a spurious change in the summation of OAE components.

From the start of OAE research, its has been known that such general changes in OAE can occur as a result of ototoxic drugs (e.g., Anderson and Kemp 1979; Long and Tubis 1982), during temporary mechanical biasing of the basilar membrane with low-frequency pressure fluctuations (Zwicker 1981; Zwicker and Manley 1981), after overstimulation (Kemp 1982), with direct efferent stimulation (Siegel and Kim 1982), or during contralateral stimulation with fluctuating acoustic stimulation (e.g., noise) (Collet et al. 1990). These effects are crucially important in understanding functional cochlear homeostasis. The magnitude and time course of OAEs changes resulting from challenges that perturb the resting cochlear state are very unlikely to be determined by spurious characteristic of the OAE generator. They are most likely to reflect characteristics of the homeostatic mechanisms associated with the cochlea. The most dramatic of these may be the oscillatory sequence of changes in cochlear sensitivity occurring with a 2-minute period following brief overstimulation by a low-frequency (e.g., 150 Hz) tone. This oscillation suggests a second-order control system capable of enhancing OHC activity (Kemp et al. 1986; Kirk et al. 1997; O'Beirne 2002; Kemp and Brill 2007).

OAEs therefore have a special role to play in the characterization of individual cochleae with respect to their susceptibility (e.g., Maison and Liberman 2000). The analysis of cochlear perturbation following an acoustic challenge using OAEs as the monitor may provide important clues as to the nature of the differences between individuals' peripheral auditory systems, differences that might contribute to premature presbyacusis, noise-induced hearing loss, or adverse reactions to oto-challenging drugs. OAEs may also serve to demonstrate the benefits of oto-protective drugs and so contribute to their development and ultimately to a reduction in the prevalence of acquired hearing loss.

Appendix: Modes of OAE observation

The commonly used modes of OAE observation are listed below, each denoted by its usual acronym and followed by a summary of:

- the nature of the stimulus applied;
- the usual method of analysis used to separate it from other signals with any additional procedures needed;
- the nature of the OAE signal;
- the means of confirming its cochlear origin.

SOAE: Spontaneous otoacoustic emission. No stimulus is applied. Narrow-band frequency analysis is used for signal recovery. Activity is identified statistically by its level above the background noise. SOAEs in humans usually consist of a stable tone with a bandwidth (frequency stability) of less than 1 Hz. A cochlear origin is confirmable by frequency-selective suppression of the emission by an applied tone, which reveals a sharp tuning curve.

SFOAE: Stimulus frequency otoacoustic emission. A single-tone stimulus is applied. Narrow-band frequency analysis is used to remove noise. Separation of the SFOAE from the stimulus is on the basis of the OAE nonlinear growth properties and is achieved by scaled vector subtraction linked to stimulus-level manipulations or, alternatively, the introduction of a second suppressor tone of a different frequency. SFOAEs consist of a pure tone at the stimulus frequency. A cochlear origin is confirmable by frequency-selective suppression by an applied tone, which reveals a sharp tuning curve.

TEOAE, CEOAE: Transient or click-evoked otoacoustic emission. A short-duration stimulus (< 3 ms) is applied repetitively, either a wideband or frequency-limited stimulus. Synchronous averaging is used for signal recovery from noise and is followed by frequency analysis. Separation from the stimulus and the ear-canal-plus-middle-ear acoustic responses is usually by time gating and/or waveform subtractions linked to stimulus-level manipulations exploiting the nonlinear growth of OAEs. TEOAEs consist of a complex frequency-dispersed oscillation consisting mainly of the frequencies present in the stimulus. True TEOAE exhibit both time- and frequency-dependent suppression in response to additional stimuli (Kemp and Chum 1980a).

DPOAE: Distortion product otoacoustic emission. Two stimulus tones are applied, normally less than a half octave apart. Narrow-band frequency analysis is used for signal recovery from noise. The response is identified statistically by level above the background noise at neighboring frequencies. DP frequency is precisely related to the stimulus frequencies f_1 and f_2 by the formulas $f_1 - \mathrm{N}(f_2 - f_1)$ for the lower band and $f_2 + \mathrm{N}(f_2 - f_1)$ for the upper side band (N an integer greater than 1). DPOAE are often thought of as pure-tone emissions, but this is true only when a single component is separated by processing. Composite DPOAEs (taking all components) are complex modulated sine wave oscillations pulsing at the stimulus difference frequency (see Fig. 1.3 and Kemp et al. 1986). Similar distortion-product generation is common in electromechanical systems. A cochlear origin is confirmable by frequency-selective suppression by an applied tone, which reveals a tuning curve, and also by determination of the delay.

Useful though this classification is for practical purposes as evidenced by its presence in most texts on the subject of the concept of OAEs, the use of the term "types" distracts from the subject of this chapter, which concerns the origins of OAEs in the cochlear process.

References

Abbas PJ, Sachs MB (1976) Two tone suppression in auditory nerve fibres: Extension of a stimulus-response relationship. J Acoust Soc Am 59:112–122.

Allen JB, Fahey PF (1993) A second cochlear-frequency map that correlates distortion product and neural tuning measurements. J Acoust Soc Am 94:809–8016.

Anderson SD, Kemp DT (1979) The evoked cochlear mechanical response in laboratory primates. A preliminary report. Arch Oto Laryngol 224:47–54.

Boege P, Janssen TH (2002) Pure tone threshold estimation from extrapolated distortion product otoacoustic emission I/O-functions in normal and cochlear hearing loss ears. J Acoust Soc Am 111:1810–1818.

Brown AM, Kemp DT (1984) Suppressibility of the 2f1−f2 stimulated acoustic emissions in gerbil and man. Hear Res 13:29–37.

Brown AM, Gaskill SA, Williams DM (1992) Mechanical filtering of sound in the inner ear. Proc Biol Sci 250:29–34.

Brown AM, Harris FP, Beveridge HA (1996) Two sources of acoustic distortion products from the human cochlea. J Acoust Soc Am 100:3260–3267.

Brownell WE (1983) Observations on a motile response in isolated outer hair cells. In: Webster DB and Aitkin LM (eds) Neural Mechanisms of Hearing. Clayton, Australia: Monash U.P., pp. 5–10.

Brownell WE, Bader CR, Bertrand D, de Ribaupierre Y (1985) Evoked mechanical responses of isolated cochlear hair cells. Science 227:194–196.

Collet L, Kemp DT, Veuillet E, Duclaux R, Moulin A, Morgon A (1990) Effect of contralateral auditory stimuli on active cochlear micro-mechanical properties in human subjects. Hear Res 43:251–261.

Cooper MA, Dultsev FN, Minson T, Ostanin VP, Abell C, Klenerman D (2001) Direct and sensitive detection of a human virus by rupture event scanning. Nat Biotechnol 19:833–837

Dallos PJ (1966) On the generation of odd-fractional subharmonics. J Acoust Soc Am 40:1381–1391.

Davis H (1983) An active process in cochlear mechanics. Hear Res 9:79–90.

Elliot E (1958) A ripple effect on the audiogram. Nature (Lond) 181:1076.

Evans EF, Wilson JP (1975) Cochlear tuning properties: concurrent basilar membrane and single nerve fiber measurements. Science 190:1218–1221.

Evans EF, Wilson JP, Borerwe TA (1981) Animal models of tinnitus. In: Tinnitus. Evered D, Lawrenson G (eds) Ciba Foundation Symposium 85. London: Pitman Books, pp. 108–138.

Fettiplace R, Crawford, AC (1980) The origin of tuning in turtle cochlear hair cells. Hear Res 2:447–454.
Fettiplace R, Hackney CM (2006) The sensory and motor roles of auditory hair cells. Nat Rev Neurosci 7:19–29.
Flock A (1980) Contractile proteins in hair cells. Hear Res 2:411–412.
Flottorp G (1953) Pure-tone tinnitus evoked by acoustic stimulation: the idiotone effect. Acta Oto Laryngol 43:396–415.
Glanville JD, Coles RR, Sullivan BM (1971) A family with high-tonal objective tinnitus. J Laryngol Otol 85:1–10.
Gold T (1948) Hearing II. The physical basis of the action of the cochlea. Proc R Soc Lond B135:492–498.
Gold T (1980) Letter to D. J. Pye on recent developments in auditory science and Gold's positive feedback theory. Unpublished private correspondence.
Gold T (1989) Historical background to the proposal 40 years ago of an active model for cochlear frequency analysis. In: Wilson JP, Kemp DT (eds) Cochlear Mechanisms: Structure, Function and Models. London: Plenum Press, pp. 299–306.
Gold T, Kemp DT (2003) Recorded discussion between Thomas Gold and David Kemp on the topic of Gold's auditory and other technical research in the 1940s and 30s (unpublished).
Gold T, Pumphrey RJ (1948) Hearing I. The cochlea as a frequency analyser. Proc R Soc B135: 462–491.
Kanis LJ, Boer E de (1997) Frequency dependence of acoustic distortion products in a locally active model of the cochlea. J Acoust Soc Am 101:1527–1531.
Kashar B, Brownell WE, Altschuler R, Fex J (1986) Electrokinetic shape changes of cochlear outer hair cells. Nature 322:356–368.
Kemp DT (1971) A new technique for the analysis of transient ELF electromagnetic disturbances within the Earth-ionosphere cavity. J Atmos Terr Pys 33:567–572.
Kemp DT (1978) Stimulated acoustic emissions from within the human auditory system. J Acoust Soc Am 64:1386–1391.
Kemp DT (1979) The evoked cochlear mechanical response and the auditory microstructure—evidence for a new element in cochlear mechanics. Scand Audiol Suppl 9:35–47.
Kemp DT (1980) Towards a model for the origin of cochlear echoes. Hear Res 2: 533–548.
Kemp DT (1981) Physiologically active cochlear micromechanics-one source of tinnitus. In: Tinnitus. Ciba Foundation Symp, D Evered, G Lawrenson (Eds), London: Pitman Books Ltd. pp. 54–81
Kemp DT (1982) Cochlear echoes: Implications for noise-induced hearing loss. In: Hamernik RP, Henderson D, Salvi R (eds) New Perspectives on Noise-Induced Hearing Loss. New York: Raven Press, pp. 189–207.
Kemp DT (1986) Otoacoustic emissions, travelling waves and cochlear mechanisms. Hear Res 22:95–104.
Kemp DT (1998) Otoacoustic emissions:distorted echoes of the cochlea's traveling wave. In: Berlin C (ed), Otoacoustic Emissions—Basic Science and Clinical Applications. San Diego: Singular Press, pp. 1–59.
Kemp DT (2001) Exploring Cochlear Status with OAE—the potential for new clinical applications. In: Robinette M andGlattke TJ (eds), Otoacoustic Emissions: Clinical Applications, 2nd ed. New York: Theime, pp. 1–47.
Kemp DT (2003) The OAE Story. Hatfield, UK: Otodynamics.

Kemp DT (2007) Otoacoustic emissions—the basics, the science and the future potential. In: Robinette M andGlattke TJ (eds) Otoacoustic Emisions: Clinical Applications, 3rd ed. New York: Theime, Chapter 1.

Kemp DT, Anderson SD (eds) (1980) Proceedings of the symposium on Active and Nonlinear Mechanical Processes in the Cochlea. Hear Res 2:533–548.

Kemp DT, Brown AM (1983a) A comparison of mechanical nonlinearities in the cochleae of man and gerbil from ear canal measurements. In: Klinke R, Hartmann R (eds) Hearing—Physiological Bases and Psychophysics. Berlin: Springer-Verlag, pp. 82–88.

Kemp DT, Brown AM (1983b) An integrated view of cochlear mechanical nonlinearities observable from the ear canal. In: Boer E de, Viergever MA (eds), Mechanics of Hearing. Delft: Delft University Press, pp. 75–82.

Kemp DT, Brown AM (1986) Wideband analysis of otoacoustic intermodulation. In: Allen JB, Hall JL, Hubbard A, Neely ST, Tubis A (eds) Peripheral Auditory Mechanisms. New York: Springer-Verlag, pp. 306–313.

Kemp DT, Brill O (2007) Slow oscillatory cochlear adaptation to brief low frequency over stimulation: the human OAE "bounce" effect revisited. Midwinter meeting of the Association for Research in Otolaryngology, Abstract No. 412.

Kemp DT, Chum R (1980a) Properties of the generator of stimulated acoustic emissions. Hear Res 2:213–232.

Kemp DT, Chum R (1980b) Observations on the generator mechanism of stimulus frequency acoustic emissions, two-tone suppression. In: Psychophysical, Physiological and Behavioural Studies in Hearing, van den Brink G, Bilsen FA (eds), Delft: Delft University Press, pp. 34–41.

Kemp DT, Knight RD (1999) Virtual reflectors explains DPOAE wave and place fixed dichotomy. Midwinter meeting of the Association for Research in Otolaryngology, Abstract No. 396.

Kemp DT, Martin JA (1976) Active Resonant Systems in Audition. Abstracts of XIII International Congress of Audiology Firenze 1976. International Society of Audiology, Geneva. pp. 64–65

Kemp DT, Tooman PF (2006) DPOAE Micro and Macrostructure—their origin and significance. In: Nuttall A, Ren, T, Gillespie, P, Grosh, K, de Boer E. (eds) Auditory Mechanisms: Processes and Models. Proceedings of the Ninth International Symposium. Hackensack, NJ: World Scientific, pp. 308–314.

Kemp DT, Bray P, Alexander L, Brown AM (1986) Acoustic emission cochleography-practical aspects. Scand Audiol Suppl 25:71–95.

Khanna SM, Leonard DGB (1982) Basilar membrane tuning in the cat cochlea. Science 215:305–306

Kim DO (1980a) Cochlear mechanics: implications of electrophysiological and acoustical observations. Hear Res 2:297–317.

Kim DO (1980b) Transcript of the mid-symposium discussion of the symposium on nonlinear and active mechanical processes in the cochlea. Hear Res 2:581–587.

Kim DO, Seigel JH, Molnar CE (1977) Cochlear distortion product: Inconsistency with linear motion of the cochlear partition. J Acoust Soc Am 61:S2(A)

Kim DO, Molnar, CE, Matthews, JW (1979) Cochlear mechanics: nonlinear behaviour in two-tone responses as reflected in ear-canal pressure and cochlear-nerve-fiber responses J Acoust Soc Am 65:S28(A)

Kim DO, Neely, ST Molnar CE, Matthews JW (1980) An active cochlear model with negative damping in the partition; comparison with Rhode's ante- and post-mortem observations. In: Brink G van den, Bilsen FA (eds) Psychophysical, Physiological and Behavioural Studies in Hearing. Delft: Delft University Press, pp. 7–14.

Kirk DL, Moleirinho A, Patuzzi RB (1997) Microphonic and DPOAE measurements suggest a micromechanical mechanism for the "bounce" phenomenon following low frequency tones. Hear Res 112:69–86

Knight RD, Kemp DT (2001) Wave and Place fixed DPOAE maps of the human ear. J Acoust Soc Am 109:1513–1525.

LePage EL, Johnstone BM (1980) Nonlinear mechanical behaviour of the basilar membrane in the basal turn of the guinea pig cochlea. Hear Res 2:183–189.

Long G, Tubis A (1982) Modification of spontaneous and evoked otoacoustic emissions and associated psychoacoustic microstructure by aspirin consumption. J Acoust Soc Am 84(4):1343–1353.

Lukashkin AN, Russell IJ (2005) Dependence of the DPOAE amplitude pattern on acoustical biasing of the cochlear partition. Hear Res 203:45–53.

Mahoney CF, Kemp DT (1995) Distortion product otoacoustic emission delay measurement in human ears. J Acoust Soc Am 97:3721–3735.

Maison SF, Liberman MC (2000) Predicting vulnerability to acoustic injury with a noninvasive assay of olivocochlear reflex strength. J Neurosci 20:4701–4707.

Manley GA (1983) Frequency spacing of acoustic emissions: a possible explanation. In: Webster W R, Aitken L M (eds), Mechanisms of Hearing. Melbourne, Australia pp. 36–39.

Manley GA (2001) Evidence for an active process and a cochlear amplifier in nonmammals. J Neurophysiol 86:541–549.

Møller AR (1960) Improved technique for detailed measurement of the middle ear impedance. J.Acoust Soc Am 32:250–258.

Neely ST Kim DO (1986) A model for active elements in cochlear biomechanics. J Acoust Soc Am 79:1472–1480.

O'Beirne GA, Patuzzi RB (2002) Modelling the role of outer hair cells in cochlear regulation and tinnitus. In: Patuzzi R (ed), Proceedings of the Seventh International Tinnitus Seminar Perth: University of Western Australia, pp. 62–67.

Pye JD (1980) The nonlinear spiral. Nature 283:137–138.

Ren T (2004). Reverse propagation of sound in the gerbil cochlea. Nat Neurosci 7: 333–334.

Rhode WS (1971) Observations of the vibration of the basilar membrane in squirrel monkey using the Mossbauer technique. J Acoust Soc Am 49:1218–1231.

Ruggero MA (1980) Systematic errors in the direct estimates of basilar membrane travel times. J Acoust Soc Am 67:707–710.

Ruggero M, Temchin AN (2005) Unexceptional sharpness of frequency tuning in the human cochlea. Proc Natl Acad Sci USA 102:18614–1869.

Ruggero M, Temchin AN (2006) Traveling-wave delays in the human cochlea are not exceptionally long. Midwinter meeting of the Association for Research in Otolaryngology, Abstract No. 1175.

Rutten WLC (1980) Evoked acoustic emissions from within normal and abnormal human ears: comparison with audiometric and electrocochleographic findings. Hear Res 2: 263–271.

Sellick PM, Russel IJ (1979) Two-tone suppression in cochlear hair cells. Hear Res 1:227–236

Sellick PM, Patuzzi RB, Johnstone BM (1982) Measurement of basilar membrane motion in the guinea pig using the Mossbauer technique. J Acoust Soc Am 72(1):131–141.

Shera CA (2003) Mammalian spontaneous otoacoustic emissions are amplitude-stabilized cochlear standing waves. J Acoust Soc Am 114:244–262.

Shera CA, Guinnan JJ (1999) Evoked otoacoustic emissions arise by two fundamentally different mechanisms. J Acoust Soc Am 105:782–798.

Siegel JH, Kim DO (1982) Efferent neural control of cochlear mechanics? Olivocochlear bundle stimulation affects cochlear mechanical nonlinearity. Hear Res 6:171–182.

Siegel JH, Cerka AJ, Recio-Spinosa A, Temchin AN, van Dijk P, Ruggero MA (2005) Delays of stimulus frequency emissions and cochlear vibrations contradict the theory of coherent reflection filtering. J Acoust Soc Am 118:2434–2443.

Strube HW (1989) Evoked otoacoustic emissions as cochlear Bragg reflections. Hear Res 38:35–45.

Thomas LB (1975) Microstructure of the pure tone threshold. J Acoust Soc Am 57:26–27.

van den Brink G (1970) Experiments in binaural diplacusis and tonal perception. In: Plomp R, Smoorenburg GF (eds), Frequency Analysis and Periodicitiy Detection in Hearing. Leiden: A.W. Sijthoff, pp. 362–374.

van Dijk P, Wit HP, (1990) Amplitude and frequency fluctuations of spontaneous otoacoustic emissions. J Acoust Soc Am 88:1779–1793

von Békésy G (1960) Experiments in Hearing. Wever EG (ed). New York: McGraw-Hill.

Ward WD (1955) Tonal monaural diplacusis. J Acoust Soc Am 27:365–372.

Wegel RL (1931) A study of tinnitus. Arch Oto Laryngol 14:158–165.

Wilson JP (1980a) Recording of the Kemp Echo and tinnitus from the ear canal without averaging. J Physiol 298:8–9P.

Wilson JP (1980b) Evidence for a cochlear origin for acoustic re-emissions, threshold fine-structure and tonal tinnitus. Hear Res 2:233–252.

Wilson JP (1980c) Model for cochlear echoes and tinnitus based on an observed electrical correlate. Hear Res 2: 527–532.

Wilson JP, Sutton GJ (1981) Acoustic correlates of tonal tinnitus. In: Evered D, Lawrenson G (eds) Tinnitus. Ciba Foundation Symposium 85. London: Pitman Books, pp. 82–107.

Wit HP, Ritsma RJ (1979) Stimulated acoustic emissions from the human ear. J Acoust Soc Am 66:911–913.

Wit HP, Ritsma RJ (1980) Evoked acoustical responses from the human ear: some experimental results. Hear Res 2:253–261.

Yates GK, Withnel RH (1999) The role of intermodulation distortion in transient-evoked otoacoustic emissions. Hear Res 136:49–64.

Zweig G, Shera CA (1995) The origin of periodicity in the spectrum of evoked otoacoustic emissions. J Acoust Soc Am 98:2018–2047.

Zwicker E (1979) A model describing nonlinearities in hearing by active processes with saturation at 40 dB. Biol Cybern 35:243–250.

Zwicker E (1981) Masking period patterns and cochlear acoustic responses. Hear Res 4(2):195–202.

Zwicker E, Manley G (1981) Acoustic responses and suppression- period patterns in guinea pigs. Hear Res 4:43–52

Zwislocki JJ (1983) Sharp vibration maximum in the cochlea without wave reflection. Hear Res 9:103–111.

Zurek PM (1981) Spontaneous narrowband acoustic signals emitted by human ears. J Acoust Soc Am 69:514–523.

2
Traveling Waves, Second Filters, and Physiological Vulnerability: A Short History of the Discovery of Active Processes in Hearing

NIGEL P. COOPER, JAMES O. PICKLES, AND GEOFFREY A. MANLEY

1. A History of Theories of Active Cochlear Mechanics

Descartes, in the *Treatise of Man* (1662), suggested how nerve stimulation could evoke reflex responses. Descartes was aware of the principle of place coding for touch and for vision, and suggested how the stimulation of different nerves in the sensory input could lead to stimulation of different nerves in the motor output and so to different reflex responses. However, he did not report any idea of frequency analysis (later to be Fourier analysis), and therefore was not able to suggest the place coding of frequency by the ear. Instead, Descartes suggested that neural responses to sound were entirely driven by the amplitude of the individual sound impulses, and that frequency coding in hearing was accomplished by the time pattern of the neural stimulation from the ear to the brain.

By the time of Helmholtz (1863, 1885), Fourier analysis was well established. Also well established was the use of hollow vessels, constructed to resonate in response to sound waves of different frequencies, which could be used as a way of detecting the component frequencies of a complex sound. Helmholtz suggested that the cochlea worked in a similar way; the rods of the arches of Corti would resonate in response to different frequencies of stimulation, and this would be detected by the neural innervation. The perception of sound was therefore entirely dependent on detecting the Fourier components of the stimulation. Temporal information was preserved in the transients of the envelope of the stimulus, but time coding as we now know it, in the sense that frequency information was also conveyed in the time pattern of the neural stimulation, was absent.

One objection to Helmholtz's theory was that the fluid in the ear would damp the vibrations of the fine rods, and do this to such an extent that the degree of frequency resolution of hearing that was observed behaviorally would not be possible. In addition, indirect evidence from patterns of noise damage had suggested that tones produced vibrations along a considerable length of the cochlea, rather than only discrete regions, with high-frequency tones producing vibrations near the base, and low-frequency tones producing vibrations in a longer region that spread toward the apex. This led to suggestions that the frequency of a tone could be determined from the length of basilar membrane that was activated (ter Kuile 1900).

Rather different, though emerging again in a different form in later theories incorporating active processes, was the standing-wave theory of Ewald (1899), where the damping in the cochlear duct was so low that standing waves would be set up, and the frequency of a tone determined from the pattern of peaks and troughs. Békésy realized that the different theories could be interconverted merely by changing the elasticity of the basilar membrane (see Békésy 1961; see also a summary of his research in Békésy 1960). After making some preliminary observations on human inner ears, Békésy (1928) investigated the parameters for the elasticity, and then studied the anticipated pattern of vibration of the basilar membrane by means of mechanical models. Mechanical models, such as ripple tanks, were in common use at that time to investigate complex wave phenomena, either as a way of developing physical insight before an analytical solution was attempted, or because the system was too complex for an analytical solution to be possible.

Békésy showed that stimulation by a single tone produced a broad peak of vibration along the model cochlear duct, and that a high-frequency tone produced a peak near the base, and low-frequency tone a peak nearer the apex (Békésy 1928). Békésy's initial papers gave only limited phase data, and it is not clear from the data presented to what extent his models developed true traveling waves, as they were later known. Clear results on the spatial localization of the vibrations of different frequencies on the cochlear partition, determined with human cadaver cochleae, were published by Békésy in 1943. The amplitude and phase data that finally established the classic traveling wave were published by Békésy in 1947. Soon thereafter, Zwislocki (1948) published an analytical solution to a one-dimensional model of an idealized cochlea, which closely matched the pattern of the traveling wave described by Békésy. By the 1950s, therefore, the dominant view was that the experimental findings and the theory of operation of the cochlea were so well established that attacks on the position were pointless.

Such an approach, however, reckoned without one of the most fertile and original minds to have contributed to auditory theory. Thomas Gold "was known for his deep physical insight and willingness to resort to bold and unconventional theories when conventional ones would not fit the facts. For this reason, Gold proposed a number of substantially incorrect theories" (Weisstein 2004). It should be pointed out that Gold also proposed a number of important and correct theories. In the cochlea, Gold's theory was correct, although it was based on incorrectly interpreted experimental data and was dismissed for 30 years, as well as apparently being treated with hilarity among the few who in fact did remember it.

Gold suggested that the cochlea contained an active mechanical amplifier. The experimental data were presented by Gold and Pumphrey (1948), who had attempted to measure the frequency selectivity of the ear by determining the ability of a human observer to detect the phase relationships between two successive tone pulses. The theory suggested that if the first pulse stimulated a resonator in the ear that was still resonating at the moment when the second pulse was presented, the response to the second pulse would either interfere with or reinforce the response to the first, depending on their phase relationships, and hence alter the percept. The values of Q (in this context, the Q_{3dB}, i.e., the center frequency divided by 3-dB bandwidth) that they found ranged up to 250 at 10 kHz. The fault in Gold and Pumphrey's interpretation was their suggestion that this was related to the ongoing oscillations of an internal resonator; in fact, the effect was due to the ability to detect the slight shift in the dominant frequency of a gated tone pip, the shift depending on the exact phase of the gating (Green et al. 1975). Nonetheless, the high values of Q, assuming them to be real, were far greater than could be produced by a resonator in fluid, where the limit would be at most 20.

Gold's theoretical arguments were presented entirely within the context of Helmholtz's theory of independent resonant elements in the ear. There is no mention of a longitudinally coupled traveling wave in the cochlea. While results from Békésy's (1928) modeling had been presented many years before, it is perhaps not surprising if his physiological work on cochlear traveling waves, which had been published in Germany during the Second Word War, was not known in England so soon after the war had ended. Gold's previous employment was in radar work during World War II, and therefore he would have been very familiar with tuned resonant electronic circuits and the techniques used to produce a high degree of amplification in the face of resistive damping. In his 1948 paper, Gold explicitly used the analogy of an electronic regenerative receiver, in which a fraction of the output signal is fed back to the input, in order to produce a high degree of narrowband amplification of a signal in the face of resistive losses (Gold 1948).

Gold suggested that in order to cancel damping in the cochlea, the feedback would have to be in phase with the velocity of the input. The level of the feedback would have to be controlled by a self-regulating mechanism, so as not to produce a sustained spontaneous oscillation. He gave subjective ringing in the ear as an example of excessive feedback that would persist until the adjusting mechanism had regained control. Gold even noted that in this case, some of the energy should be radiated out of the ear, and could be picked up by microphones of sufficient sensitivity. In addition, it should be possible to produce subjective beats between an external tone and the spontaneous oscillation. As to the mechanism, he suggested that energy might be fed back from the cochlear microphonic; at the time of writing this is still a substantial possibility. He also pointed out the evolutionary advantage of regenerative feedback in the ear; not only would thresholds be lowered because the final transducers would be able to extract more energy from the stimulus than was originally presented, but the thermal noise limit would be set by the noise in the amplification pathway, rather than that of the detector. Since the former is a tuned element, but the detector is not, the noise limit can be lowered substantially. The only important speculation that is missing in Gold's construct is that sensorineural hearing loss of cochlear origin might be due to loss of the feedback amplification.

The matter rested at that point for many years, Gold moving out of hearing to become well known for his theory of the origin of the universe. However, auditory scientists continued to be busy in their own area and produced clues that, with hindsight, were used to support Gold's suggestion. For example, Elliott (1958) showed that the audiogram could have fine ripples that were relatively regularly spaced as a function of frequency. Others found frequency regions where single tones could produce discordant sensory sensations, as though multiple tones were present, and Glanville et al. (1971) reported a family whose ears produced continuous high tones that could be measured by microphones.

David Kemp started to follow up Elliott's observations in 1975, and confirmed the presence of very fine, sharply tuned, and stable irregularities in the audiogram. The notches were so sharply tuned, in fact, that Kemp's initial suggestion was that there must be some "superfibers" in the auditory nerve with Q values of 200 or more. However, Kemp soon realized that the extreme frequency stability of the notches, to within a single hertz over months or more and far more stable than most electronic oscillators of the time, meant that their frequency must be fixed by anatomical and geometrical parameters rather than by a purely physiological process. Moreover, as a tone was moved through the notch, the quality of the sensation changed in such as way as though the externally introduced tone were beating with an internally produced one. In his insights as to the mechanism, Kemp was undoubtedly helped by his previous employment, where he had been a sound engineer for the Central Electricity Generating Board of the United Kingdom. Kemp was charged with devising active sound reduction systems and spent much of his time lowering microphones down power station chimneys (Kemp 2003). In viewing the cochlea as a tiny version of a power station chimney containing an active sound source, Kemp realized that if a small region of the basilar membrane was producing a mechanical oscillation, and if the frequency and position were right, a regenerative standing wave could be set up in the cochlear duct. This wave would be fixed in frequency both by the mechanical properties of the duct and by the anatomical position of the generator in the duct, reminiscent of Ewald's standing-wave theory. If the generator were not intrinsically locked to precisely one frequency, a set of oscillations at regularly spaced frequencies could be generated by one active region, where there were different numbers of standing waves (e.g., 3, 4, 5) to the point of reflection of the waves in the duct (presumably the oval window).

These speculations led Kemp to insert a microphone, connected to a wave analyzer, into his own ear canal. He introduced an external tone (at frequency f_1), which when combined with the suspected internally produced tone (matched to frequency f_2), would give a distortion product, at the frequency $2f_1 - f_2$, to which the analyzer was tuned. The distortion tone was indeed present, showing that a real tone of frequency f_2 was present in the ear. Kemp then tuned the wave analyzer to f_2, and found that the ear was emitting a signal of this frequency (as later reviewed by Kemp 2003). These experiments with tones were followed up by another demonstration of active processes, in which transient emissions were produced in response to a click stimulus applied to the ear (Kemp 1978).

Kemp (1979) explicitly suggested that these active processes could lead to sharpening of frequency selectivity and amplitude sensitivity in the cochlea, as presaged by Gold, and that they could explain the physiological vulnerability and amplitude nonlinearity of cochlear responses. Kemp presented these results at a meeting of the Inner Ear Biology workshop in September 1978, where they came to the attention of auditory electrophysiologists and modelers such as Duck On Kim. Kim et al. (1980b) were the first to incorporate active elements into a formal model of cochlear tuning that produced realistic and sharp tuning. The active region was confined, by unspecified mechanisms, to a region just basal to the peak of the traveling wave. This, in its basic essence, is the model for active cochlear mechanics as we now know it. Passive models at the time included those of Zwislocki (1980), in which the tectorial membrane vibrating radially acted as a second resonator in the cochlea, and a similar resonator was later included in the model of Neely and Kim (1983). This gave a mechanism by which the active area could be confined to a limited region of the cochlea.

2. Early Measurements of the Motion of the Cochlear Partition

The first extensive experimental studies of the mechanical responses of the inner ear to sound were carried out by Georg Békésy, whose early work (1928 to 1958) was made available in the English language by E. Glen Wever (Békésy 1960). Most of Békésy's work was carried out using stroboscopic illumination of Reissner's membrane in the cochleae of various mammals and birds (nearly all of which were dead). Békésy sprinkled tiny crystals of silver over Reissner's membrane to enhance its visibility, and observed its movements using a microscope. The stroboscopic technique limited Békésy's observations to relatively low frequencies, and the use of a visual detection technique limited his observations to high sound pressure levels (120 to 140 dB SPL). Nonetheless, Békésy's approach had one major advantage over many later methods, because it enabled him to observe what was happening over a longer stretch of the hearing organ and not at just a single spot. Békésy established that with careful dissection, "the displacement of the partition for frequencies below 3000 cps showed no phase difference between the movement of Reissner's membrane and that of the basilar membrane" (Békésy 1960, p. 447).

Békésy's experiments established (1) that pure-tone sound induces a traveling-wave displacement of the organ of Corti, whereby the energy of the wave is translated through the surrounding fluids rather than through the sensory tissues themselves; (2) that each site in the cochlea responds best to one particular frequency of stimulation—in later studies, this became known as the site's characteristic frequency (CF); (3) that the traveling waves are tonotopically organized, lower-frequency tones evoking waves that travel farther toward the apex of the cochlea than higher-frequency tones; and (4) that the traveling wave is asymmetrical, with a much steeper cutoff slope on the apical side of its peak than on the basal side. These kinds of data were consistent with "place" theories of audition, in which the character of a sensation is determined by the place of excitation in the hearing organ and thus the unique set of nerve fibers that are stimulated (cf. Helmholtz 1863; see Wever 1949 for a more thorough review). Considering the fact that we now know that the auditory organ's responses are highly nonlinear and physiologically vulnerable, it is remarkable that Békésy's data were sufficiently reasonable as to be immediately accepted and to provoke further studies.

2.1 Problems with Békésy's Findings

One of the most puzzling aspects of Békésy's data was the issue of sensitivity, since the extrapolation of his results to threshold levels of human hearing predicted a displacement amplitude of the cochlear partition of something like 1/500 of the diameter of a hydrogen atom, and thus well below thermal threshold. It was not until 1971, when Rhode (1971, see Section 4) first demonstrated a prominent nonlinearity in the motion of the squirrel monkey's cochlear partition, that it became obvious that linear extrapolations of Békésy's data were untenable.

Another puzzling aspect of Békésy's data was that, compared to information available from early psychoacoustic studies of human hearing, the frequency selectivity displayed by the envelopes of his traveling waves was very poor indeed. This fact occupied Békésy for many years. He became fascinated by early data from Hartline's laboratory, which had studied the visual system of *Limulus*, the horseshoe crab (Hartline and Ratliff 1957). In the lateral complex eyes of this species, the nerve fibers emerging from the individual ommatidial units not only provide the input to the brain, but have collaterals that synapse on nearby fibers and have an inhibitory effect on their activity. The inhibitory effect became known as lateral inhibition, because the fibers inhibited their neighbors by laterally running fiber branches, and this inhibition decreased with distance from the active ommatidium. In whole eyes, lateral inhibition leads to contrast enhancement, emphasizing the difference in illumination intensity across an edge of illumination. Békésy suspected that the inner ear could work in a similar fashion, lateral inhibition between nerve fibers emerging from the tonotopically organized organ, leading to a sharpening of the contours of the frequency responses. At that time, there were no recordings from single nerve fibers in auditory nerves, and a direct approach to the problem was not possible. To study the phenomenon further, therefore, Békésy used mechanoreceptors of the skin of the human lower arm and demonstrated in an extensive study that there, too, lateral-inhibitory phenomena were measurable (Békésy 1970, cited in Fex 1974). This work became rapidly irrelevant to audition, however, since evidence was accumulating that there was no neural basis for lateral inhibition in the cochlea (see Section 3).

3. Early Measurements of Neural Activity

In the field of auditory research, most of the major advances in knowledge have been made as the result of very significant new developments in the available technologies (such as electronics, microelectrodes, the Mössbauer technique, fast digital spectrum analyzers, and laser interferometry). Thus, many major advances in knowledge had to wait until suitable techniques had been developed, often as offshoots of technical advances in other scientific fields.

In the 1940s, microelectrodes for recording from nervous tissue became available, and a small number of researchers began to record from single primary auditory nerve fibers and second-order cells (e.g., Galambos and Davis 1943; Tasaki 1954). These early neural recordings showed some features that were to be expected on the basis of Békésy's mechanical findings. The nerve fibers were frequency selective, and their sensitivity contours were sharper on the high-frequency side, for example. The selectivity and sensitivity of the neurons, especially the primary neural units, was not particularly good, also in agreement with Békésy's mechanical measurements. However, the poor selectivity was actually the result of a problem that has always plagued auditory physiologists: the fact that the inner ear is very sensitive to damage. In the early neural experiments, considerable damage was probably caused by drilling into the cochlea to expose the nerve fibers (e.g., Tasaki 1954). The physiological sensitivity of the cochlea is also reflected in a dramatic change in its characteristics after the death of the animal (recall that Békésy worked mainly on dead tissue).

Over time, physiological recordings from single auditory neurons, and of the cochlear microphonic potentials produced by hair cells (e.g., Tasaki et al. 1952), confirmed many features of Békésy's mechanical findings. Additional support for Békésy's findings was provided by experiments studying the effects of local lesions of the basilar membrane on an animal's audiogram (e.g., Schuknecht 1960). Although it is often assumed that Békésy found poor tuning because he worked only with dead animals, this is incorrect. Békésy also worked on living animals, and stated, "I have found no important difference between the observations done on a live guinea pig and one killed three hours earlier" (Békésy's 1953). It is now clear that all of the early observations on live guinea pigs must have involved severely compromised cochleae, but it took many years of work before measurements from undamaged cochleae were made (e.g., see Section 4).

3.1 Refined Neural Data Initiate Controversy

In the late 1960s and 1970s, several important advances dominated the field. Extensive anatomical studies of the innervation of the cochlear partition clearly established that there was no neural basis for lateral inhibition in the cochlea (e.g., Spoendlin 1969), so that the final frequency selectivity of the auditory periphery must be established at the level of the hair cell afferent synapses. With the passing of time, and the development of refined physiological recording techniques, conditions for studying the auditory nerve improved dramatically (e.g., Kiang et al. 1965; Evans 1972). It soon became clear that the frequency selectivity of nerve fibers leaving the cochlea was directly related to the cochlea's condition (Evans 1972; Robertson and Manley 1974) and that in the best recordings, there was a substantial difference between the frequency selectivity of primary nerve fibers and Békésy's measurements of the displacement of the cochlear partition.

Evans (1972, 1975) suggested that there must be an intervening filtering process between the selectivity of the motion of the sensory elements in the cochlea (the first filter) on the one hand and the responses of the auditory nerve on the other, leading to a concept known as the "second filter." For years, discussion continued on the possible nature of this second filter. Four types of mechanism were developed as possible bases for the second filter (see Dallos 1981 for a review), and the discussion of possible mechanisms continued until well after Sellick et al. (1982) had established that mechanical and neural tuning were the same, since some of the proposed theories involved interactions within the cochlea that would feed back into the mechanical motion.

4. Refined Measurement of Basilar Membrane Motion

Spurred on by the growing divide that was becoming apparent between neural tuning and Békésy's mechanical measurements, new techniques began to be applied to cochlear mechanics in the late 1960s. One of the key advances was made by Brian Johnstone and colleagues at the University of Western Australia, who applied the newly discovered phenomenon of the "Mössbauer effect" (Mossbauer 1962) to cochlear mechanics in 1967 (Johnstone and Boyle 1967). The Mössbauer technique involved placing a tiny source of radioactivity on the basilar membrane, and detecting the variations in the frequency of the emitted radiation as this source rode up and down on the back of the basilar membrane traveling wave. The technique was ideally suited to measurements of high-frequency vibrations, as the key variable was the velocity of the source, not its displacement. Johnstone and Boyle (1967) were therefore able to make measurements in the high-frequency region of the living guinea-pig cochlea (where the CFs were around 18 kHz) at levels much lower than Békésy could envision, and well into the physiologically relevant range (e.g., 60–95 dB SPL).

Along with the technical improvements, and the switch to measuring in the basal turns of the cochlea, came an increase in the apparent sharpness of the mechanical tuning: Q_{3dB}'s increased from around 1.6 (for Békésy's apical turn measurements) to 2.5 (for Johnstone and Boyle's basal turn measurements). The mechanical tuning still fell far short of the neural tuning observed in the neural studies of Kiang et al. (1965), but at least the high-frequency cutoff slopes of the basilar membrane vibrations (around 70 dB/oct. in 1967, and 95 dB/oct.

by 1970) were starting to become comparable with those of the auditory nerve (Johnstone et al. 1970).

Another key development occurred in 1971, when Rhode published his first measurements of basilar membrane motion in the basal turns of the squirrel-monkey cochlea. These were the first measurements to demonstrate a mechanical nonlinearity in the cochlea: the nonlinearity was both compressive, with growth rates of as little as 0.4 dB/dB, and frequency specific, having its greatest effect near the CF of the site being studied (partway round the basal turn in the squirrel monkey cochlea, the CF was around 8 kHz). The net result was that the basilar membrane's tuning appeared sharper as the intensity of a stimulus decreased. The measured mechanical tuning also appeared closer in form to the neural tuning, with speculations rising that a very close correspondence would be possible if the mechanical nonlinearity extended down to neural threshold levels (e.g., see Rhode 1971; Geisler et al. 1974).

Despite substantial supporting evidence from Rhode's own laboratory (Rhode 1974, 1978, 1980; Rhode and Robles 1974; Robles et al. 1976), recordings from other laboratories showed no signs of nonlinearity throughout the 1970s. Little or no nonlinearity was found in either guinea-pig or cat cochleae, whether measured by Mössbauer (Helfenstein 1973; Johnstone and Yates 1974; Rhode and Robles 1974; Rhode 1978), capacitive probe (Wilson and Johnstone 1973, 1975; Evans and Wilson 1975; Wilson and Evans 1983), or optical detection techniques (Kohlloffel 1972). Even regions of the squirrel-monkey cochlea that were either more apical or more basal than the 8-kHz site appeared to vibrate linearly (Rhode and Robles 1974; Rhode 1978, 1980). Two opposing schools of thought began to develop.

5. Basilar Membrane Nonlinearity or a Second Filter?

Throughout the 1970s, most investigators were reluctant to accept Rhode's findings, and favored the idea that a second filter was needed to account for fundamental differences between the mechanical and neural tuning in the cochlea. Numerous mechanisms were suggested to sharpen up the broad tuning imparted by the basilar membrane, including both electrical and mechanical interactions across the width and along the length of the organ of Corti (e.g., Duifhuis 1976; Manley 1977; Zwislocki 1977; Zwislocki and Kletsky 1979). While few of these theories were directly amenable to experimental testing, the definitive experiment to prove the existence of some form of second filter appeared to have been performed in 1975: Evans and Wilson (1975) made interleaved measurements of both the mechanical and neural tuning in individual cat ears, and found exactly what the proponents of a second filter would expect—broad tuning on the basilar membrane, and sharp tuning in the auditory nerve, as illustrated in Figure 2.1A.

Despite the growing evidence to support the existence of a second filter, a few theoreticians (most notably a group working at Washington University, in

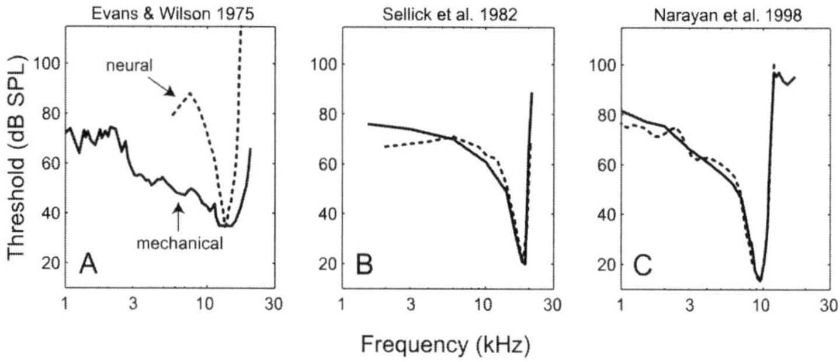

FIGURE 2.1. The rise and fall of the second filter. Comparisons between neural and mechanical tuning curves at three key historical points. Neural data (*dashed lines*) were recorded from individual fibers of the auditory nerve, while mechanical tuning (*solid lines*) was observed on the basilar membrane, and is represented in the form of iso-velocity contours. (**A**) Evans and Wilson found marked discrepancies between neural and mechanical tuning in individual cat cochleae. (Adapted with permission from Figure 2 of Evans and Wilson 1975, with linear extrapolation of the mechanical data to a velocity criterion of 1 μm/s. Copyright 1975 American Association for the Advancement of Science.) (**B**) Sellick et al. observed mechanical tuning that was almost as sharp as the neural tuning observed in normal-hearing guinea pigs. (Adapted with permission from Figures 5 and 10 of Sellick et al. 1982 using a velocity criterion of 90 μm/s. Copyright 1982 Acoustical Society of America.) (**C**) Narayan et al. observed almost identical neural and mechanical tuning in individual chinchilla cochleae. (Reproduced with permission from Figure 1 of Narayan et al. 1998, with a velocity criterion of 164 μm/s. Copyright 1998 American Association for the Advancement of Science.)

Missouri) still clung to the view that Rhode's nonlinearity was essential, and that no second filter was needed (e.g., Kim and Molnar 1975). The idea that the level dependence of the cochlear partition's vibrations might be nonlinear was originally proposed to counter the impossibly small vibration amplitudes that could be predicted at threshold from Békésy's mechanical observations (e.g., see Whitfield 1967). Another factor that strongly supported this idea (in addition to Rhode's direct observations) was the presence of highly tuned distortion products in the ear (Goldstein 1967; Goldstein and Kiang 1968; Smoorenburg 1972). These distortion products were not present in the stimulus being delivered to the ear, but they were perceived as though they were real stimuli, and they affected individual auditory nerve fibers in exactly the same way that an externally applied stimulus would (Goldstein and Kiang 1968; Kim et al. 1980a). At the time that direct evidence of distortion products traveling on the basilar membrane would have been crucial, however, mechanical distortion products could not be observed (Wilson and Johnstone 1973; Rhode 1977). Those who favored the idea of nonlinearity being the key to the second filter debate hence had very little left to cling to, other than some seemingly minor technicalities (e.g., see Greenwood 1977), and the fact that Rhode's nonlinearity was physiologically vulnerable (Rhode 1973, 1978; Rhode and Robles 1974).

5.1 Confirmation of Mechanical Nonlinearity and the Demise of the Second Filter

Two key turning points in the debate between those against and those in favor of a second filter occurred in 1978. First, Ian Russell and Peter Sellick published the first intracellular recordings of sound-evoked responses in living cochlear hair cells (Russell and Sellick 1978). These recordings demonstrated that inner hair cells were just as sharply tuned as auditory nerve fibers, and they hence narrowed down the conceptual gap that a putative second filter could occupy. The second turning point came when David Kemp (1978) provided direct evidence that the cochlea was an active mechanical device, by showing that it was capable of emitting sound as well as absorbing it. Kemp's initial evidence was strongly refuted by many of the second-filter proponents, but his findings were almost universally acknowledged by 1980 when more extensive studies had been carried out (for historical perspectives see Kemp, Chapter 1 and 1998). In essence, Kemp's studies proved not only that the cochlea is mechanically active, but that it is also mechanically nonlinear.

The search to confirm and extend Rhode's original observations was now on, and the initial results looked promising: LePage and Johnstone (1980), working on the guinea-pig basilar membrane, finally found signs of a nonlinearity that looked very similar to those found in the squirrel monkey almost a decade earlier. These guinea-pig experiments did not show sharp tuning, but the growth rates for the responses near the location's CF were as low as 0.5 dB/dB, and there was no reason to preclude a theoretical extrapolation to sharp tuning at low sound pressure levels. LePage and Johnstone's experiments also cast doubt on one of the crucial assumptions that had been used in the earlier experiments of Evans and Wilson (1975). Both of these studies used a capacitive probe to measure the basilar membrane's motion, and this involved draining the scala tympani such that little or no perilymph was left to cover the basilar membrane. While Evans and Wilson had assumed (and to some extent, checked) that this procedure did not influence the operation of the cochlea, LePage and Johnstone found that it could elevate cochlear thresholds within tightly defined frequency regions by up to 70 dB (cf. Robertson 1974; Sellick et al. 1982). The implication (spelled out explicitly in Sellick et al. 1982) was that the experiments of Evans and Wilson had been flawed, because they had not monitored the cochlea's thresholds on a frequency-by-frequency basis. Frequency-specific threshold losses associated with partial drainage of the scala tympani could therefore have gone unnoticed and have been overlooked in the neural part of Evans and Wilson's study. Other experiments in the late 1970s highlighted another problem for Evans and Wilson's study: the CFs of auditory nerve fibers shifted considerably (by up to one-half of an octave) when the cochlea's sensitivity was impaired (e.g., Robertson et al. 1979; Robertson et al. 1979). Evans and Wilson's assumption that their sharply tuned nerve fibers originated from the same point in the cochlea that they had measured mechanically was therefore flawed in two respects. Almost a decade after its original proposal, therefore, it could be argued that there was still no evidence to support the existence of a second filter.

The final nail in the coffin of the original second-filter proposal was put in place in 1982, when two groups of investigators made independent claims that the basilar membrane was almost as sharply tuned as any auditory nerve fiber (Khanna and Leonard 1982; Sellick et al. 1982). Both of these groups argued that the main difference between their observations and those made in all previous studies lay in the physiological condition of their preparations. However, while Khanna and Leonard based their arguments on the use of cochlear microphonic, histological, and acoustic controls that were later viewed as questionable (e.g., see Cooper and Rhode 1992), Sellick et al. relied on a control technique that became the gold standard in subsequent studies: frequency-specific compound action potential recordings. As described below, the studies of Sellick et al. (1982) laid the foundations for all future studies of the basilar membrane, and are therefore now viewed as seminal.

Sellick et al. (1982) extended Rhode's observations in several ways. By applying an improved Mössbauer technique to the basal turn of the guinea-pig cochlea, Sellick et al. were able to observe nonlinearity down to levels within 20 dB of neural threshold (e.g., to 35 dB SPL). Their observations extended to even lower levels for near-CF tones, but the vibrations observed at these levels grew at rates that were close to 1 dB/dB. Sellick et al. thus provided the first direct evidence that cochlear mechanical nonlinearities had a "threshold" of their own, and this finding was to prove instrumental in the development of realistic cochlear models (e.g., see Johnstone et al. 1986). Sellick et al. also extended Rhode's (1973, 1978) findings on physiological vulnerability: they demonstrated that almost all of the vulnerability observed at the level of the auditory nerve could be accounted for by the vulnerability of the basilar membrane's nonlinearity. The findings of Sellick et al. clearly reinforced Békésy's (1953) opinion that "It is probably not a good idea to work on something as complicated and as easy to damage as the organ of Corti," although of course he (and others) continued to do so for many years! Finally, of course, Sellick et al. (1982) demonstrated that the basilar membrane is almost as sharply tuned as the auditory nerve, at least for frequencies in the region of the tuning curve's tip, as illustrated in Figure 2.1B.

6. Early Concepts of the Cochlear Amplifier

The need for a second filter to explain physiological vulnerability and sharp tuning in the auditory nerve had disappeared almost overnight. It had been replaced by the need for a "cochlear amplifier": a mechanism that could enhance both the tuning and the sensitivity of the basilar membrane at low sound pressure levels, in just the same way that the compressive nonlinearities originally discovered by Rhode did. This mechanism had to be biologically active, such that it would be physiologically vulnerable. A mechanism that was also active in a physical sense (i.e., one that was capable of injecting mechanical energy into the cochlea) offered the additional benefit of being able to account for some of the peculiar characteristics that had recently been discovered in the field of otoacoustic emissions. In much the same way that Ted Evans had coined the "second filter" term almost one decade earlier, Hallowell Davis (1983) now summarized the need for a "cochlear amplifier" without a specific mechanism in mind.

The chief protagonists of cochlear amplification were the mathematical modelers, who had been struggling to match physiological tuning curves for over a decade. Prompted by the discovery of otoacoustic emissions in 1978, the modelers had already begun to resort to active processes by 1980 (Zwicker 1979; Kim et al. 1980b). By the mid-1980s, the idea that some form of positive feedback was necessary both to sensitize and to sharpen the basilar membrane's response (as originally proposed by Gold in 1948) was almost universally accepted (de Boer 1983; Neely and Kim 1983, 1986a,b; Johnstone et al. 1986; Zwicker 1986; Patuzzi et al. 1989).

Since the early 1980s, many cochlear models had already been based on the assumption of the existence of some kind of cochlear amplifier, but the authors had used other terms to describe the phenomenon, such as "feedback motor," "negative damping," "active feedback," and "reverse transduction" (Neely and Kim, Chapter 10). The basic idea in all cases was that hair cells can react in a frequency-selective fashion to faint sounds with a response that favorably affects the local motion of the hearing organ. Initially, the molecular basis of the active process was not important for modelers, provided they were fast enough to enhance the motion on a cycle-by-cycle basis at auditory frequencies. Nonetheless, the elucidation of the molecular basis of the active process quickly became a focus of physiological studies. Potential candidates to fill the role of a cochlear amplifier did not take long to materialize and are briefly referred to below, before further expansion in Section 8. The strongest candidates (cf. Kim 1986) were two different forms of hair cell motility: somatic motility (Section 8.2), and hair bundle motility (Section 8.1).

7. Modern Measurements of Basilar Membrane Motion

At the time that cellular investigations into the nature of the cochlear amplifier were progressing rapidly, direct mechanical investigations of intact, living ears were undergoing a second revolution. The era of the Mössbauer technique was drawn to an elegant end by Mario Ruggero, Luis Robles, and colleagues, who convincingly replicated the findings of Sellick et al. in another mammalian species, the chinchilla (Robles et al. 1986). The Mössbauer technique was superseded by heterodyne laser interferometry, an optical technique that offered vastly improved bandwidth, sensitivity, and linearity, as well as the potential to cause less damage to the cochlea (Willemin et al. 1988). The new technique quickly permitted the first demonstration of fully reversible physiological vulnerability on the basilar membrane (Ruggero and Rich 1991), and eventually led to some of the most definitive measurements of the cochlear amplifier at work (Nuttall and Dolan 1996; Ruggero et al. 1997; Rhode and Recio 2000).

Robust confirmations of earlier findings from Mössbauer studies of two-tone suppression (Rhode and Robles 1974; Rhode 1977; Patuzzi et al. 1984b; Robles et al. 1989) and the effects of acoustic trauma (Patuzzi et al. 1984a) were readily provided by numerous laser interferometric studies (Cooper and Rhode 1992; Ruggero et al. 1992; Nuttall and Dolan 1993; Rhode and Cooper 1993; Cooper 1996; Ruggero et al. 1996; Fridberger et al. 2002). The new technique even provided the long-awaited proof that nonlinear distortion products exist on the basilar membrane, with characteristics well suited to explain previous psychophysical studies (Goldstein 1967) and certain features of distortion-product otoacoustic emissions (Robles et al. 1991, 1997; Cooper and Rhode 1997). The most solid rebuttal of the second-filter idea also followed soon thereafter: Narayan et al. (1998) found an extremely close correspondence between concurrent measurements of basilar membrane and neural tuning in individual chinchilla cochleae, as illustrated in Figure 2.1C.

7.1 New Challenges to Traditional Traveling-Wave Concepts

Another advance that was facilitated by the switch to laser interferometry was the opportunity to study the motion of structures other than the basilar membrane: basically, any structure that reflected light could, at least in theory, have its vibrations measured. In practice, however, most studies were restricted to making observations from artificial reflectors that had been placed on the basilar membrane, because the inherent reflectance of a healthy cochlear partition is remarkably low (Khanna et al. 1989a). Only groups using custom-made interferometers were able to record vibrations without using added reflectors. The most intriguing and challenging results to come out of these studies were presented by Shyam Khanna in 1989, who observed substantially tuned responses from the apical surfaces of the hair cells (i.e., the reticular lamina) near the apex of an excised guinea-pig cochlea, despite the virtual absence of a response on the nearby basilar membrane (e.g. see Khanna et al. 1989b). These observations provided a considerable challenge for those who believed that it was the basilar membrane that supported a functional traveling wave. When viewed alongside measurements of the fluid pressure distributions within living guinea-pig cochleae (Dancer and Franke 1980), Khanna's measurements led to a revival of the pre Békésy "resonance" theories of hearing (cf. Wever 1949), in which basilar membrane traveling waves were viewed as inconsequential byproducts of a coupled system of individual resonators (e.g., hair cells; cf. Dancer 1992).

The idea that the hair cells could respond more vigorously than the basilar membrane received additional support from other in vitro studies: Mammano and Ashmore (1993), for example, observed that electrically stimulated outer hair cells moved the reticular lamina between 5 and 10 times more than they moved the basilar membrane. Further measurements of both hair cell and basilar membrane vibrations in the living guinea-pig cochlea (Khanna and Hao 2000) eventually led to the idea that the cochlear amplifier depends on a negative feedback loop, which actively clamps the vibrations of the basilar membrane but amplifies those of the reticular lamina, at least in the apical turns of the cochlea (Khanna 2004). Another negative feedback proposal was made to explain the occurrence of physiologically vulnerable expansive nonlinearities in the apical turns of the guinea-pig cochlea (Zinn et al. 2000). Other studies of apical cochlear mechanics have been more consistent with the idea of positive feedback, however, and have demonstrated both compressive nonlinearity, two-tone suppression, and two-tone distortion in the apical turns of living chinchilla and guinea-pig cochleae (e.g., Cooper and Rhode 1995, 1996, 1997; Rhode and Cooper 1996; Cooper and Dong 2003).

The final, or at least latest, twist in the tale of traveling waves and active processes in the cochlea has come during a search for backward (or "reverse") traveling waves in the cochlea. These waves were originally proposed to carry the energy of an otoacoustic emission away from its site of generation and out of the cochlea (cf. Kemp 1978), and have played a central role in otoacoustic models ever since (e.g., see Zweig and Shera 1995). However, both Ren (2004) and Ruggero (2004) have recently argued that backward basilar membrane traveling waves either do not exist or are unimportant from a functional point of view. Much the same conclusion was reached more than 20 years earlier by Wilson (1980), who proposed that sound-evoked volumetric changes in hair cells might initiate an alternate form of backward wave, traveling almost instantaneously to the stapes via a compressional wave in the cochlea's fluids. If this turns out to be true, it may well provide a crucial test to distinguish between the roles of somatic and hair bundle motility in the active cochlea (see Section 8).

8. The Existence of Active Processes

In mammals, a cochlear amplifier is taken to underlie the high sensitivity to sound, the narrow frequency selectivity (tuning) of hair cells or their afferent nerve fibers, the typical patterns in rate-level functions of primary auditory neurons, the sensitivity of both threshold and frequency selectivity of auditory afferent nerve fibers to hypoxia and other insults, and the presence and characteristics of otoacoustic emissions. All of these phenomena have been found not only in mammalian but also in nonmammalian data (Manley 2001), suggesting that at least some manifestations of active mechanisms arose very early in the evolution of vertebrates.

8.1 An Active Process in the Hair Bundle

In the first model that proposed a specific cellular mechanism for an active process in hearing, Weiss (1982) assumed that a hair cell's stereovillar bundle could move under the influence of membrane-potential changes. Since such changes in potential could result from the transduction of sound stimuli, the motion of the hair bundle would set up a *feedback* mechanism, and if this feedback had an appropriate phase relationship, it could amplify the stimulus signal.

Electrically driven hair-bundle motion was demonstrated early in a semi-intact cochlear preparation from the turtle, *Trachemis scripta elegans* (Crawford and Fettiplace 1985). The stereovillar bundles of some of these cells were observed to oscillate at audio frequencies (30–170 Hz) when stimulated either mechanically, using a flexible glass fiber, or electrically, by injecting current into the hair cell's body. Some of the bundles were even observed to oscillate spontaneously, and clearly delivered mechanical energy into the motion of the glass fiber.

Following Crawford and Fettiplace's (1985) description of active motions in an auditory hair bundle, Howard and Hudspeth (1987) demonstrated the presence of an adaptational motor system in the stereovillar bundles of vestibular (saccular) hair cells from the frog, *Rana catesbeiana*. Also using the frog saccule, Denk and Webb (1989) found a correlation between hair-cell intracellular electrical noise and bundle position that suggested a reverse-transduction process as Weiss (1982) had proposed. Similarly, Rusch and Thurm (1990) described both spontaneous and electrically induced movements of both kinocilia and stereovilli of vestibular ampullary hair cells. Jaramillo et al. (1990) then demonstrated that saccular hair bundles could execute more rapid movements than those observed during adaptation, and Benser et al. (1996) showed bundle movements whose frequency components lay in the same range as those necessary for normal saccular responses. More recently, Hudspeth's group has demonstrated that saccular hair bundles can actually perform mechanical work, and thus can amplify responses to mechanical stimuli (Martin and Hudspeth 1999). Ricci et al. (2000) have also provided further evidence for a calcium-sensitive force generator linked to the gating of the transducer channels in the auditory hair cells of *Trachemis scripta elegans*.

In 2001, Manley et al. described evidence from in vivo experiments using ac electrical current injection into the cochlea of the bobtail lizard, *Tiliqua rugosa*, that can be explained only on the assumption that an active process is present in the hair-cell bundle. Further, Martin and Hudspeth (2001) demonstrated that the saccular hair bundle's active response to mechanical stimulation incorporates a compressive nonlinearity. The localization of both active processes and this nonlinearity to the bundle suggested that both the activity and the nonlinearity of mammalian basilar-membrane responses might be traceable to the characteristics of the sensory hair bundles themselves. Martin (2001, 2003) showed that the active process could also produce spontaneous oscillation of the saccular hair bundles. In the auditory hair bundles of *Trachemis scripta elegans*, Ricci et al. (2002) described the different components of bundle motion that could contribute to active processes.

Although all of the early work on active hair bundle motion was carried out in nonmammalian systems and formed the basis for various models of cochlear amplification (Choe et al. 1998; Camalet et al. 2000; Hudspeth et al. 2000; Vilfan and Duke 2003; see also Martin, Chapter 4), no evidence either for or against such a mechanism in mammals was published until 2003. Nonetheless, an active process in the bundle of the mammalian hair cell was held by some authors to be at least as appropriate as the oft-discussed active process that is sited in the cell membrane (Yates and Kirk 1998; see also Section 8.2). And since 2003, signs of active processes have been found in the hair bundles of neonatal rat outer hair cells (Kennedy et al. 2003, 2005) and in the gerbil cochlea (Chan and Hudspeth 2005). This latest evidence has clearly renewed interest in the hair bundle as a source of cochlear activity (Liberman et al. 2004; Martin, Chapter 4).

8.2 An Active Process in the Soma of the Outer Hair Cell

Another form of hair cell motility was first observed by Bill Brownell and colleagues in 1983. Brownell et al. (Brownell 1983; Brownell et al. 1985) studied the behavior of isolated outer hair cells from the guinea-pig cochlea. When stimulated by either intra- or extracellular currents, or by the application of the neurotransmitter acetylcholine, the cell bodies (or soma) of these hair cells were observed to lengthen and/or contract over time-courses that ranged from a fraction of a second to several minutes. Subsequent studies showed that there were at least two different subtypes of somatic motility: a fast form, which was voltage-dependent, but unaffected by intracellular calcium and/or ATP levels (Ashmore 1987; Santos-Sacchi and Dilger 1988), and a slower form, which depended on both calcium and ATP (Zenner et al. 1985; Flock et al. 1986). Only the fastest form was thought, and subsequently proved, to be fast enough to amplify sounds on a cycle-by-cycle basis (cf. Dallos and Evans 1995; Zheng et al. 2000; Liberman et al. 2002; see Fakler et al., Chapter 7).

The discovery of somatic motility unleashed extensive efforts by many laboratories to describe the phenomenon in detail, to quantify its capabilities, and, of course, to incorporate the idea into cochlear models (e.g., Neely and Kim 1986a,b). Today, the majority of papers describe the active process in mammalian outer hair cells as resulting from very fast configurational changes in densely packed complexes of the protein prestin (Zheng et al. 2000) in the outer hair cell's basolateral plasma membrane.

In parallel with the impressive cellular and molecular discoveries concerning active mechanisms in the cochlea, the manifestation of the activity in the form of otoacoustic emissions has opened up vast new fields of study (Manley and van Dijk, Chapter 6; Shera and Guinan, Chapter 8; Lonsbury-Martin and Martin, Chapter 11) and new opportunities to use OAE as noninvasive tools in clinical studies (Janssen and Muller, Chapter 12).

9. Summary

The history of the study of the mechanical motions, neural responses, and cellular and molecular mechanisms underlying auditory perception illustrates well how science advances. The original idea of a "second filter" in the auditory periphery turned out to be something of a red herring, but took over a decade to be replaced by our present concept of a "cochlear amplifier." Perhaps this was unavoidable, given the extreme and unexpected sensitivity of the cochlea to damage on the one hand, and the time needed to develop adequate recording techniques on the other. The development of modern in vitro and in vivo preparations, cheap but powerful computers, and fast digital analysis techniques has since enabled a series of interlocking discoveries that have made possible enormous advances in both our theoretical and experimental understanding of the active cochlea.

References

Ashmore JF (1987) A fast motile response in guinea-pig outer hair cells: the cellular basis of the cochlear amplifier. J Physiol (Lond) 388:323–347.

Békésy G (1928) Zur Theorie des Hörens; die Schwingungsform der Basilarmembran. Physik Zeits 29:793–810.

Békésy G (1943) Über die Schwingungen der Schneckentrennwand beim Präparat und Ohrenmodell. Akust Zeits 7:173–186.

Békésy G (1947) Über die Resonanzkurve und die Abklingzeit der verschiedenen Stellen der Schneckentrennwand. Akust Zeits 8:66–76.

Békésy G (1953) Description of some mechanical properties of the organ of Corti. J Acoust Soc Am 25:770–785.

Békésy G (1960) Experiments in Hearing, Wever EG (ed) New York: McGraw-Hill.

Békésy G (1961) Concerning the pleasures of observing, and the mechanics of the inner ear. Nobel Lectures, Physiology or Medicine 1942–1962, Amsterdam: Elsevier.

Benser ME, Marquis RE, Hudspeth AJ (1996) Rapid, active hair-bundle movements in hair cells from the bullfrog's sacculus. J Neurosci 16:5629–5643.

Brownell WE (1983) Observations on a motile response in isolated outer hair cells. In: Webster DB, Aitkin LM (eds) Neural Mechanisms of Hearing. Clayton, Australia: Monash U.P., pp. 5–10.

Brownell WE, Bader CR, Bertrand D, de Ribaupierre Y (1985) Evoked mechanical responses of isolated cochlear outer hair cells. Science 227:194–196.

Camalet S, Duke T, Jülicher F, Prost J (2000) Auditory sensitivity provided by self-tuned oscillations of hair cells. Proc Natl Acad Sci USA 97:3183–3188.

Chan DK, Hudspeth AJ (2005) Ca^{2+} current-driven nonlinear amplification by the mammalian cochlea in vitro. Nat Neurosci 8:149–155.

Choe Y, Magnasco M, Hudspeth AJ (1998) A model for amplification of hair-bundle motion by cyclical binding of Ca^{2+} to mechanoelectrical transduction channels. Proc Natl Acad Sci USA 95:15321–15326.

Cooper NP (1996) Two-tone suppression in cochlear mechanics. J Acoust Soc Am 99:3087–3098.

Cooper NP, Dong W (2003) Baseline position shifts and mechanical compression in the apical turns of the cochlea. In: Gummer AW (ed) The Biophysics of the Cochlea: Molecules to Models. Singapore: World Scientific, pp. 261–270.

Cooper NP, Rhode WS (1992) Basilar membrane mechanics in the hook region of cat and guinea-pig cochleae: sharp tuning and nonlinearity in the absence of baseline position shifts. Hear Res 63:163–190.

Cooper NP, Rhode WS (1995) Nonlinear mechanics at the apex of the guinea-pig cochlea. Hear Res 82:225–243.

Cooper NP, Rhode WS (1996) Two-tone suppression in apical cochlear mechanics. Audit Neurosci 3:123–134.

Cooper NP, Rhode WS (1997) Mechanical responses to two-tone distortion products in the apical and basal turns of the mammalian cochlea. J Neurophysiol 78:261–270.

Crawford AC, Fettiplace's R (1985) The mechanical properties of ciliary bundles of turtle cochlear hair cells. J Physiol 364:359–379.

Dallos P (1981) Cochlear physiology. Annu Rev Psychol 32:153–190.

Dallos P, Evans BN (1995) High-frequency motility of outer hair cells and the cochlear amplifier. Science 267:2006–2009.

Dancer A (1992) Experimental look at cochlear mechanics. Audiology 31:301–312.

Dancer A, Franke R (1980) Intracochlear sound pressure measurements in guinea pigs. Hear Res 2:191–205.

Davis H (1983) An active process in cochlear mechanics. Hear Res 9:79–90.

de Boer E (1983) No sharpening? a challenge for cochlear mechanics. J Acoust Soc Am 73:567–573.

Denk W, Webb WW (1989) Simultaneous recording of fluctuations of hair-bundle deflection and intracellular voltage in saccular hair cells. In: Wilson JP, Kemp DT (eds) Cochlear Mechanisms: Structure, Function and Models. New York: Plenum Press, pp. 125–132.

Descartes R (1662) L'Homme. Translated as Treatise of Man, Hall TS. 1972, Cambridge, MA: Harvard University Press.

Duifhuis H (1976) Cochlear nonlinearity and second filter: possible mechanism and implications. J Acoust Soc Am 59:408–423.

Elliott E (1958) A ripple effect in the audiogram. Nature 181:1076.

Evans EF (1972) The frequency response and other properties of single fibres in the guinea-pig cochlear nerve. J Physiol 226:263–287.

Evans EF (1975) The sharpening of cochlear frequency selectivity in the normal and abnormal cochlea. Audiology 14:419–444.

Evans EF, Wilson JP (1975) Cochlear tuning properties: concurrent basilar membrane and single nerve fiber measurements. Science 190:1218–1221.

Ewald JR (1899) Zur Physioloie des Labyrinths. VI. Eine neue Hörtheorie. Pflüg Arch Physiol 76:147–188.

Fex J (1974) Neural excitatory processes of the inner ear. In: Keidel WD, Neff WD (eds), Handbook of Sensory Physiology: Auditory System. V/I. Berlin: Springer-Verlag. pp. 585–646.

Flock A, Flock B, Ulfendahl M (1986) Mechanisms of movement in outer hair cells and a possible structural basis. Arch Otorhinolaryngol 243:83–90.

Fridberger A, Zheng J, Parthasarathi A, Ren T, Nuttall A (2002) Loud sound-induced changes in cochlear mechanics. J Neurophysiol 88:2341–2348.

Galambos R, Davis H (1943) The response of single auditory-nerve fibers to acoustic stimulation. J Neurophysiol 6:39–58.

Geisler CD, Rhode WS, Kennedy DT (1974) Responses to tonal stimuli of single auditory nerve fibers and their relationship to basilar membrane motion in the squirrel monkey. J Neurophysiol 37:1156–1172.

Glanville JD, Coles RR, Sullivan BM. (1971) A family with high-tonal objective tinnitus. J Laryngol Otol 85:1–10.

Gold T (1948) Hearing II. The physical basis of the action of the cochlea. Proc R Soc Lond B 135:492–498.

Gold T, Pumphrey RJ (1948) Hearing. I. The cochlea as a frequency analyser. Proc R Soc B 135:462–491

Goldstein JL (1967) Auditory nonlinearity. J Acoust Soc Am 41:676–689.

Goldstein JL, Kiang NY (1968) Neural correlates of the aural combination tone 2f1−f2. Proc IEEE 56:981–992.

Green DM, Wier CC, Wightman FL (1975) Gold and Pumphrey revisited, again. J Acoust Soc Am 57:935–938

Greenwood DD (1977) Comments on paper by Rhode (Some observations on two-tone interaction measured with the Mössbauer effect). In: Evans EF, Wilson JP (eds) Psychophysics and Physiology of Hearing. London: Academic Press, p. 40.

Hartline HK, Ratliff F (1957) Inhibitory interaction of receptor units in the eye of *Limulus*. J Gen Physiol 40:357–376.

Helfenstein W (1973) Beitrag zur Messung der akustisch bedingten Bewegungen us Identifikation des mechanischen Teils des Innenohrs der Katze. Zurich, Switzerland, Eidgenossische Technische Hochschule.

Helmholtz HLF (1863) Die Lehre den Tonempfindungen als physiologische Grundlage für die Theorie der Musik. Brunswick: Vieweg-Verlag.

Helmholtz HLF (1885) On the Sensations of Tone. New York: Dover.

Howard J, Hudspeth AJ (1987) Mechanical relaxation of the hair bundle mediates adaptation in mechanoelectrical transduction by the bullfrog's saccular hair cell. Proc Natl Acad Sci USA 84:3064–3068.

Hudspeth AJ, Choe Y, Mahte AD, Martin P (2000) Putting ion channels to work: Mechanoelectrical transduction, and amplification by hair cells. Proc Natl Acad Sci USA 97:11765–11772.

Jaramillo F, Howard J, Hudspeth AJ (1990) Calcium ions promote rapid mechanically evoked movements of hair bundles. In: Dallos P, Geisler CD, Matthews JW, Ruggero MA, Steele CR (eds) The Mechanics and Biophysics of Hearing. New York: Springer-Verlag, pp. 26–33.

Johnstone B, Boyle A (1967) Basilar membrane vibration examined with the Mössbauer effect. Science 158:389–390.

Johnstone BM, Yates GK (1974) Basilar membrane tuning curves in the guinea pig. J Acoust Soc Am 55:584–587.

Johnstone BM, Taylor KJ, Boyle AJ (1970) Mechanics of the guinea pig cochlea. J Acoust Soc Am 47:504–509.

Johnstone BM, Patuzzi RB, Yates GK (1986) Basilar membrane measurements and the travelling wave. Hear Res 22:147–153.

Kemp DT (1978) Stimulated acoustic emissions from within the human auditory system. J Acoust Soc Am 64:1386–1391.

Kemp DT (1979) Evidence of mechanical nonlinearity and frequency selective wave amplification in the cochlea. Arch Otorhinolaryngol 224:37–45.

Kemp DT (1998) Otoacoustic emissions: Distorted echoes of the cochlea's traveling wave. In: Berlin CI (ed) Otoacoustic Emissions: Basic Science and Clinical Applications. San Diego: Singular, pp. 1–59.

Kemp DT (2003) 25 years of progress with OAEs: the inside story. Presentation to ASHA conference, Chicago, November 2003: www.asha.org.

Kennedy HJ, Evans MG, Crawford AC, Fettiplace R (2003) Fast adaptation of mechanoelectrical transducer channels in mammalian cochlear hair cells. Nat Neurosci 6:832–836.

Kennedy HJ, Crawford AC, Fettiplace R (2005) Force generation by mammalian hair bundles supports a role in cochlear amplification. Nature 433:880–883.

Khanna SM (2004) The response of the apical turn of cochlea modeled with a tuned amplifier with negative feedback. Hear Res 194:97–108.

Khanna SM, Hao LF (2000) Amplification in the apical turn of the cochlea with negative feedback. Hear Res 149:55–76.

Khanna SM, Leonard DG (1982) Basilar membrane tuning in the cat cochlea. Science 215:305–306.

Khanna SM, Willemin JF, Ulfendahl M (1989a) Measurement of optical reflectivity in cells of the inner ear. Acta Otolaryngol Suppl 467:69–75.

Khanna SM, Flock A, Ulfendahl M (1989b) Comparison of the tuning of outer hair cells and the basilar membrane in the isolated cochlea. Acta Otolaryngol Suppl 467:151–156.

Kiang NYS, Watanabe T, Thomas EC, Clark LF (1965) Discharge Patterns of Single Fibers in the Cat's Auditory Nerve. Cambridge, MA: MIT Press.

Kim DO (1986) Active and nonlinear cochlear biomechanics and the role of outer-hair-cell subsytem in the mammalian auditory system. Hear Res 22:105–114.

Kim DO, Molnar CE (1975) Cochlear mechanics: measurements and models. In: Tower DB (ed) The Nervous System. New York: Raven Press, pp. 57–68.

Kim DO, Molnar CE, Matthews JW (1980a) Cochlear mechanics: nonlinear behavior in two-tone responses as reflected in cochlear-nerve-fiber responses and in ear-canal sound pressure. J Acoust Soc Am 67:1704–1721.

Kim DO, Neely ST, Molnar CE, Matthews JW (1980b) An active cochlear model with negative damping in the partition: comparison with Rhode's ante- and post-mortem observations. In: Van den Brink G, Bilsen FA (eds) Psychophysical, Physiological and Behavioural Studies in Hearing. Delft, the Netherlands: Delft University Press, pp. 7–15.

Kohllöffel LUE (1972) A study of basilar membrane vibrations III: The basilar membrane frequency response curve in the living guinea-pig. Acustica 27:82–89.

Kuile E ter (1900) Die richtige Bewegungsform der Membrana basilaris. Pflüg Arch Physiol 79:484–509.

LePage EL, Johnstone BM (1980) Nonlinear mechanical behaviour of the basilar membrane in the basal turn of the guinea pig cochlea. Hear Res 2:183–189.

Liberman MC, Gao J, He DZZ, Wu X, Jia S, Zuo J (2002) Prestin is required for electromotility of the outer hair cell and for the cochlear amplifier. Nature 419:300–304.

Liberman MC, Zuo J, Guinan JJ, Jr. (2004) Otoacoustic emissions without somatic motility: can stereocilia mechanics drive the mammalian cochlea? J Acoust Soc Am 116:1649–1655.

Mammano F, Ashmore JF (1993) Reverse transduction measured in the isolated cochlea by laser Michelson interferometry. Nature 365:838–841.

Manley GA (1977) Frequency dependent extracellular interaction between hair cells as a possible mechanism for cochlear frequency sharpening. In: Evans EF, Wilson JP (eds), Psychophysics and Physiology of Hearing. London: Academic Press, pp. 139–146.

Manley GA (2001) Evidence for an active process and a cochlear amplifier in nonmammals. J Neurophysiol 86:541–549.

Manley GA, Kirk D, Köppl C, Yates GK (2001) In-vivo evidence for a cochlear amplifier in the hair-cell bundle of lizards. Proc Natl Acad Sci USA 98:2826–2831.

Martin P, Hudspeth AJ (1999) Active bundle movements can amplify a hair cell's response to oscillatory mechanical stimuli. Proc Natl Acad Sci USA 96:14306–14311.

Martin P, Hudspeth AJ (2001) Compressive nonlinearity in the hair bundle's active response to mechanical stimulation. Proc Natl Acad Sci USA 98:14386–14391.

Martin P, Bozovic D, Choe Y, Hudspeth AJ (2003) Spontaneous oscillation by hair bundles of the Bullfrog's sacculus. J Neurosci 23:4533–4548.

Mössbauer RL (1962) Recoilless Nuclear Resonance Absorption. Annu Rev Nuclear Sci 12:123–152.

Narayan SS, Temchin AN, Recio A, Ruggero MA (1998) Frequency tuning of basilar membrane and auditory nerve fibers in the same cochleae. Science 282:1882–1884.

Neely ST, Kim DO (1983) An active cochlear model showing sharp tuning and high sensitivity. Hear Res 9:123–130.

Neely ST, Kim DO (1986a) An active cochlear model showing sharp tuning and high sensitivity. Hear Res 9:123–130.

Neely ST, Kim DO (1986b) A model for active elements in cochlear biomechanics. J Acoust Soc Am 79:1472–1480.

Nuttall AL, Dolan DF (1993) Two-tone suppression of inner hair cell and basilar membrane responses in the guinea pig. J Acoust Soc Am 93:390–400.

Nuttall AL, Dolan DF (1996) Steady-state sinusoidal velocity responses of the basilar membrane in guinea pig. J Acoust Soc Am 99:1556–1565.

Patuzzi RB, Johnstone BM, Sellick PM (1984a) The alteration of the vibration of the basilar membrane produced by loud sound. Hear Res 13:99–100.

Patuzzi RB, Sellick PM, Johnstone BM (1984b) The modulation of the sensitivity of the mammalian cochlea by low frequency tones. III. Basilar membrane motion. Hear Res 13:19–27.

Patuzzi RB, Yates GK, Johnstone BM (1989) Outer hair cell receptor current and sensorineural hearing loss. Hear Res 42:47–72.

Ren T (2004) Reverse propagation of sound in the gerbil cochlea. Nat Neurosci 7: 333–334.

Rhode WS (1971) Observations of the vibration of the basilar membrane in squirrel monkeys using the Mössbauer technique. J Acoust Soc Am 49:1218–1231.

Rhode WS (1973) An investigation of post-mortem cochlear mechanics using the Mössbauer effect. In: Moller A (ed) Basic Hearing Mechanisms. New York: Academic Press, pp. 49–63.

Rhode WS (1974) Measurement of vibration of the basilar membrane in the squirrel monkey. Ann Otol Rhinol Laryngol 83:619–625.

Rhode WS (1977) Some observations on two-tone interaction measured with the Mössbauer effect. In: Evans EF, Wilson JP (eds) Psychophysics and Physiology of Hearing. London: Academic Press, pp. 27–38.

Rhode WS (1978) Some observations on cochlear mechanics. J Acoust Soc Am 64: 158–176.

Rhode WS (1980) Cochlear partition vibration—recent views. J Acoust Soc Am 67: 1696–1703.

Rhode WS, Cooper NP (1993) Two-tone suppression and distortion production on the basilar membrane in the hook region of cat and guinea pig cochleae. Hear Res 66:31–45.

Rhode WS, Cooper NP (1996) Nonlinear mechanics in the apical turn of the chinchilla cochlea in vivo. Audit Neurosci 3:101–121.

Rhode WS, Recio A (2000) Study of mechanical motions in the basal region of the chinchilla cochlea. J Acoust Soc Am 107:3317–3332.

Rhode WS, Robles L (1974) Evidence from Mössbauer experiments for nonlinear vibration in the cochlea. J Acoust Soc Am 55:588–596.

Ricci AJ, Crawford AC, Fettiplace R (2000) Active hair bundle motion linked to fast transducer adaptation in auditory hair cells. J Neurosci 20:7131–7142.

Ricci AJ, Crawford AC, Fettiplace R (2002) Mechanisms of active bundle motion in auditory hair cells. J Neurosci 22:44–52.

Robertson D (1974) Cochlear neurons: frequency selectivity altered by perilymph removal. Science 186:153–155.

Robertson DR, Johnstone BM (1979) Aberrant tonotopic organisation in the inner ear damaged by kanamycin. J Acoust Soc Am 66:466–469.

Robertson DR, Manley GA (1974) Manipulation of frequency analysis in the cochlear ganglion of the guinea pig. J Comp Physiol 91:363–375.

Robertson DR, Cody AR, Bredberg G, Johnstone BM (1979) Response properties of spiral ganglion neurons in cochleas damaged by direct mechanical trauma. J Acoust Soc Am 67:1295–1303.

Robles L, Rhode WS, Geisler CD (1976) Transient response of the basilar membrane measured in squirrel monkeys using the Mössbauer effect. J Acoust Soc Am 59: 926–939.

Robles L, Ruggero MA, Rich NC (1986) Basilar membrane mechanics at the base of the chinchilla cochlea. I. Input-output functions, tuning curves, and response phases. J Acoust Soc Am 80:1364–1374.

Robles L, Ruggero MA, Rich NC (1989) Nonlinear interactions in the mechanical response of the cochlea to two-tone stimuli. In: Wilson JP Kemp DT (eds) Cochlear Mechanisms: Structure, Function, and Models. New York: Plenum, Press, pp. 369–375.

Robles L, Ruggero MA, Rich NC (1991) Two-tone distortion in the basilar membrane of the cochlea. Nature 349:413–414.

Robles L, Ruggero MA, Rich NC (1997) Two-tone distortion on the basilar membrane of the chinchilla cochlea. J Neurophysiol 77:2385–2399.

Ruggero MA (2004) Comparison of group delays of 2f1−f2 distortion product otoacoustic emissions and cochlear travel times. Acoust Res Lett Online 5:143–147.

Ruggero MA, Rich NC (1991) Furosemide alters organ of Corti mechanics: evidence for feedback of outer hair cells upon the basilar membrane. J Neurosci 11:1057–1067.

Ruggero MA, Robles L, Rich NC (1992) Two-tone suppression in the basilar membrane of the cochlea: mechanical basis of auditory-nerve rate suppression. J Neurophysiol 68:1087–1099.

Ruggero MA, Rich NC, Recio A (1996) The effect of intense acoustic stimulation on basilar-membrane vibrations. Audit Neurosci 2:329–345.

Ruggero MA, Rich NC, Recio A, Narayan SS, Robles L (1997) Basilar-membrane responses to tones at the base of the chinchilla cochlea. J Acoust Soc Am 101: 2151–2163.

Rüsch A, Thurm U (1990) Spontaneous and electrically induced movements of ampullary kinocilia and stereovilli. Hear Res 48: 247–264.

Russell IJ, Sellick PM (1978) Intracellular studies of hair cells in the mammalian cochlea. J Physiol (Lond) 284:261–290.

Santos-Sacchi J, Dilger JP (1988) Whole cell currents and mechanical responses of isolated outer hair cells. Hear Res 35:143–150.

Schuknecht HF (1960) Neuroanatomical correlates of auditory sensitivity and pitch discrimination in the cat. In: Rasmussen GL, Windle W (eds) Neural Mechanisms of the Auditory and Vestibular Systems. Springfield, IL: Charles C Thomas, pp. 105–115.

Sellick PM, Patuzzi RB, Johnstone BM (1982) Measurement of basilar membrane motion in the guinea pig using the Mössbauer technique. J Acoust Soc Am 72:131–141.

Smoorenburg GF (1972) Combination tones and their origin. J Acoust Soc Am 52: 615–632.
Spoendlin H (1969) Innervation patterns in the organ of Corti of the cat. Acta Otolaryngol 67:239–254.
Tasaki I (1954) Nerve impulses in individual auditory nerve fibers of guinea pig. J Neurophysiol 17:97–122.
Tasaki I, Davis H, Legouix JP (1952) The space-time pattern of the cochlear microphonics (guinea pig) as recorded by differential electrodes. J Acoust Soc Am 24:502–519.
Vilfan A, Duke, T (2003) Two adaptation processes in auditory hair cells together can provide an active amplifier. Biophys J 85:191–203.
Weiss TF (1982) Bidirectional transduction in vertebrate hair cells: A mechanism for coupling mechanical and electrical processes. Hear Res 7:353–360.
Weisstein EW (2004) Gold, Thomas. http://scienceworld.wolfram.com/biography/Gold.html
Wever EG (1949) Theory of Hearing. New York: John Wiley & Sons.
Whitfield IC (1967) The Auditory Pathway. Baltimore: Williams & Wilkins.
Willemin JF, Dandliker R, Khanna SM (1988) Heterodyne interferometer for submicroscopic vibration measurements in the inner ear. J Acoust Soc Am 83:787–795.
Wilson JP (1980) Model for cochlear echoes and tinnitus based on an observed electrical correlate. Hear Res 2:527–532.
Wilson JP, Evans EF (1983) Some observations on the "passive" mechanics of cat basilar membrane. In: Webster WR and Aitkin LM (eds), Mechanisms of Hearing. Clayton: Monash University Press, pp. 30–35.
Wilson JP, Johnstone JR (1973) Basilar membrane correlates of the combination tone 2f1−f2. Nature 241:206–207.
Wilson JP, Johnstone JR (1975) Basilar membrane and middle-ear vibration in guinea pig measured by capacitive probe. J Acoust Soc Am 57:705–723.
Yates GK, Kirk DL (1998) Cochlear electrically evoked emissions modulated by mechanical transduction channels. J Neurosci 18:1996–2003.
Zenner HP, Zimmermann U, Schmitt U (1985) Reversible contraction of isolated mammalian cochlear hair cells. Hear Res 18:127–133.
Zheng J, Shen W, He DZ, Long KB, Madison LD, Dallos P (2000) Prestin is the motor protein of cochlear outer hair cells. Nature 405:149–155.
Zinn C, Maier H, Zenner H, Gummer AW (2000) Evidence for active, nonlinear, negative feedback in the vibration response of the apical region of the in-vivo guinea-pig cochlea. Hear Res 142:159–183.
Zweig G, Shera CA (1995) The origin of periodicity in the spectrum of evoked otoacoustic emissions. J Acoust Soc Am 98:2018–2047.
Zwicker E (1979) A model describing nonlinearities in hearing by active processes with saturation at 40 dB. Biol Cybern 35:243–250.
Zwicker E (1986) A hardware cochlear nonlinear preprocessing model with active feedback. J Acoust Soc Am 80:146–153.
Zwislocki JJ (1948) Theorie der Schneckenmechanik. Acta Otolayngol Suppl 72:1–76.
Zwislocki JJ (1977) Further indirect evidence for interaction between cochlear inner and outer hair cells. In: Evans EF, Wilson JP (eds) Psychophysics and Physiology of Hearing. London: Academic Press, pp. 125–135.
Zwislocki JJ (1980) Theory of cochlear mechanics. Hear Res 2:171–182.
Zwislocki JJ, Kletsky EJ (1979) Tectorial membrane: a possible effect on frequency analysis in the cochlea. Science 204:639–641.

3
Critical Oscillators as Active Elements in Hearing

Thomas A.J. Duke and Frank Jülicher

1. Introduction

Active processes play a crucial role in the hearing organs of many animals, where they amplify sound stimuli in a level-dependent and frequency-selective manner (Dallos 1992; Hudspeth 1997; Manley et al. 2001). The most readily observable signature of active processes—otoacoustic emissions—have been found in all classes of land vertebrates (Kemp, Chapter 1; Manley and van Dijk, Chapter 6; Probst 1990; Köppl 1995; Manley and Köppl 1998; Manley 2000), and spontaneous movements have even been reported in the hearing apparatus of some insects (Göpfert, Chapter 5; Göpfert et al. 2005). In different animal species, the physiology of the hearing organs and the molecular basis of active processes varies in significant ways (Manley et al. 2004). Despite this biological diversity, vertebrate ears share many features that characterize the active amplifier. These include otoacoustic emissions, a compressive nonlinearity, high-frequency selectivity, and the physiological vulnerability of all these phenomena. The generality of this observed behavior suggests that a common physical principle may underlie active amplification in all auditory systems (Camalet et al. 2000; Eguiluz et al. 2000). We propose that the physics of dynamical systems provides an appropriate framework for identifying this principle and argue that the basic element of the active amplifier is a "critical oscillator"—an active dynamical system that is on the verge of an oscillatory instability. In this chapter, we discuss how the concept of critical oscillators provides a unifying description of the general characteristics of the active process in hearing. Different physiological realizations of critical oscillators in different species, and the various ways in which these oscillators can be coupled together, can lead to diverse properties of certain aspects of auditory systems, while the generic features are conserved.

The importance of active processes in the ear was first noted by Gold (Gold 1948; see also Kemp, Chapter 1; Cooper, Pickles, and Manley, Chapter 2). He estimated the viscous damping of vibrating structures within the basilar membrane, which limits the sharpness of passive resonances, and concluded that a purely passive system could not provide the frequency discrimination observed in the auditory system. He formulated his regenerative hypothesis: If active processes could overcome the viscous friction, resonances of high quality could be achieved. He thus proposed an amplifier based on positive feedback. Because such an active system might become self-oscillatory, he realized that a mechanism of self-regulation would be required for it to operate reliably, and predicted the occurrence of otoacoustic emissions. All these ideas are important in the framework of critical oscillators.

In order to cover a large dynamic range, it is important to amplify weak signals but not strong ones. This dependence of amplification on signal level corresponds to a compressive nonlinearity. Such a nonlinearity is well known to exist in the cochlear response, where it permits operation over a dynamic range of 120 dB of sound pressure levels (Rhode 1971; Ruggero et al. 1997; Robles and Ruggero 2001). Goldstein pointed out that this nonlinearity is "essential": nonlinear effects do not disappear for small stimuli (Goldstein 1967). He also discussed the properties of combination tones that occur in the presence of two stimulus frequencies, which are an intrinsically nonlinear phenomenon. He argued that combination tones correspond to mechanical vibrations on the basilar membrane and suggested that they result from nonlinearities in the basilar membrane mechanics. Consequently, he suggested that the basilar membrane itself possesses an essential nonlinearity. These ideas are also closely related to the concept of critical oscillators. While nonlinearities can become important at large vibration amplitudes in a passive mechanical system, passive systems cannot be essentially nonlinear in response to periodic stimulation. Essential nonlinearities are thus a signature of active processes and can be directly linked to the cochlear amplifier.

This chapter is organized as follows: In Section 2, we discuss general aspects of sound detection and amplification by nonlinear dynamic oscillators. This framework allows us to describe the characteristic properties of a broad class of active systems that respond to vibrations. We can formulate the ideas of Gold and Goldstein in a common framework that allows us to distinguish general properties from properties that are related to the specific architecture of a given auditory system. In Section 3, various different ways to build active oscillators are discussed. We also show how a mechanism of self-regulation can maintain the active oscillators on the verge of an oscillatory instability. In Section 4, we weigh the evidence for the actual physical basis of critical oscillators in the auditory systems of insects, nonmammalian vertebrates, and mammals. The mammalian cochlea is distinguished by the propagation of a traveling wave along the basilar membrane. In Section 5, we present a description of the cochlear traveling wave, based on the idea that critical oscillators with a range of characteristic frequencies drive the basilar membrane, and are coupled together through the cochlear fluid.

2. Generic Response of Critical Oscillators

2.1 Active and Passive Oscillators

It is important to distinguish active oscillators from passive harmonic or anharmonic oscillators. Active oscillators are dynamical systems that are powered by a source of energy. Consequently, they can generate spontaneous oscillations that persist even in the presence of damping; they settle to a steady state of oscillation in which the heat dissipated by the damping is compensated by energy input from the source (Strogatz 2001). This contrasts with passive oscillators, for which the amplitude of oscillation always eventually decays. The amplitude of spontaneous oscillation of an active oscillator evidently depends on the internal state of the dynamical system, which can usually be summarized by a parameter C known as the "control parameter." For one range of values of the control parameter, oscillations are absent and the system remains quiescent; in another region of parameter space, the system oscillates with a particular amplitude and a particular frequency, both of which vary smoothly as C varies. There is one particular value of the control parameter—the critical value C_c—which marks the boundary between these two regions (see Fig. 3.1). This is the precise point at which the system becomes unstable, and spontaneous oscillations of vanishing amplitude, but with a well-defined characteristic frequency ω_c, are generated. It corresponds mathematically to a "supercritical Hopf bifurcation" of the dynamical system, i.e., the bifurcation of a fixed point to a limit cycle (Strogatz 2001). We call a dynamical system that is poised exactly at this critical point a "critical oscillator."

In practice, spontaneous oscillations usually occur in complex systems consisting of many components. In this case, the variations of a large number of rapidly changing degrees of freedom result in fluctuations, which can be described as background noise acting on the more slowly oscillating displacement of the system (van Kampen 1992). The oscillation frequency is then not perfectly fixed, because the noise causes the phase of the oscillation to wander over time. Indeed, there is typically a finite coherence time, after which the phase of oscillations is randomized (Stratonovich 1981). But in the limit case in which the number of interacting degrees of freedom is very large, the coherence time can become infinite. Then the system has a persistent, perfectly periodic output. Such coherent oscillations are characterized by a power spectrum that exhibits delta peaks (infinitely sharp peaks) at the frequency of coherent spontaneous oscillations, together with a broader spectrum reflecting fluctuations and noise. Note that coherent oscillations cannot occur in a passive system, but only in active

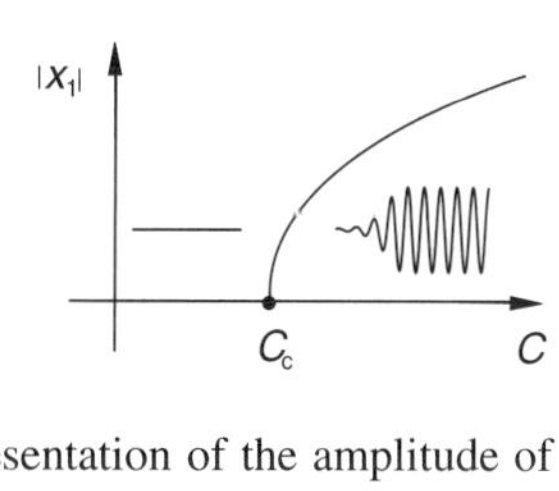

FIGURE 3.1. Schematic representation of the amplitude of spontaneous limit cycle oscillation $|x_1|$ as a function of a control parameter C near a supercritical Hopf bifurcation. At a critical value $C = C_c$, a quiescent, nonoscillating state becomes unstable. For $C > C_c$, spontaneous oscillations appear with an amplitude and frequency that depend on C. The oscillation amplitude grows continuously as $|x_1| \sim (C - C_c)^{1/2}$.

systems. The occurrence of phase coherence in a multicomponent dynamical system is related to the appearance of order in large thermodynamic systems, where collective effects give rise to abrupt transitions in behavior. The Hopf bifurcation of a complex dynamical system is thus a generalization of the concept of equilibrium phase transitions, applied in this case to an active, nonequilibrium system (Risler et al. 2004, 2005).

Extensive studies of thermodynamic phase transitions have revealed one remarkable feature: when a system is close to a critical point, the way that it responds to external influences is universal and independent of the many microscopic details of the system (Ma 2000). How a liquid turns into a gas with rising temperature is the same for all materials, for example, and the mathematics that describes this transition also accounts for the onset of magnetism in ferromagnetic substances. The same universality of behavior applies to complex dynamical systems close to a Hopf bifurcation. Thus many of the characteristic properties of critical oscillators are generic. In particular, their response to an oscillatory stimulus can be discussed quite generally, irrespective of the specific components from which they are built.

2.2 Nonlinear Response Function

In the context of hearing, we are interested in knowing the displacement $x(t)$ of an active oscillator in response to a stimulus that is periodic in time, with angular frequency ω. The external force acting on the system may then be written as a Fourier series, $f(t) = \sum f_n e^{in\omega t}$, where the f_n denote complex Fourier coefficients with $f_n = f^*_{-n}$.

We first consider a simpler passive system that does not generate spontaneous motion, so that $x = 0$ in the absence of external forcing. The response contains the same Fourier modes as the stimulus and is expressed as $x(t) = \sum x_n e^{in\omega t}$, where the $x_n = x^*_{-n}$ are the complex Fourier components of the displacements. If fluctuations are present, $x(t)$ is defined as the average displacement observed after time t when the stimulus is repeated many times. For sufficiently small stimulus amplitudes we can express the response to a periodic stimulus as an expansion:

$$x_n = G^{(1)}_n f_n + \sum_k G^{(2)}_{nk} f_{n-k} f_k + \sum_{kl} G^{(3)}_{nkl} f_{n-k-l} f_k f_l + O(f^4). \tag{3.1}$$

This expression is the most general form of the nonlinear response that satisfies time-translation invariance, ensuring that the properties of the system are independent of its history. Note that in the absence of a stimulus ($f_n = 0, \forall n$), the response vanishes, $x_n = 0$, as expected for a passive system. In Eq. (3.1), the complex quantity $G^{(1)}_n(\omega)$ is the frequency-dependent linear response function. The effects of nonlinearities are captured by the higher-order coefficients $G^{(k)}$ with $k > 1$, which are also complex numbers.

Consider now an active system that is capable of generating spontaneous oscillations. In the region of phase-space where the system is quiescent, Eq. (3.1) remains valid. But what happens if the control parameter C is varied so that the system moves toward the Hopf bifurcation? As the oscillating instability is approached, the linear response function $G_1^{(1)}(\omega)$ increases. Right at the critical point, $C = C_c$, where the system becomes unstable, $G_1^{(1)}(\omega)$ formally diverges for the specific frequency $\omega = \omega_c$. Here, ω_c is the characteristic frequency of the spontaneous oscillations that emerge in the absence of a stimulus. As a consequence of this divergence, the validity of Eq. (3.1) is restricted to smaller and smaller amplitudes of force as the Hopf bifurcation is approached: it can be used only as long as $|G_1^{(1)} f_1|$ remains finite.

There is, however, a related description of the response that is valid near the instability. Before reaching the bifurcation point, we can formally invert Eq. (3.1) for small $|f_k|$ within the radius of convergence of the expansion, and reexpress it as an expansion of f_n in terms of x_n,

$$f_n = F_n^{(1)} x_n + \sum_k F_{nk}^{(2)} x_{n-k} x_k + \sum_{kl} F_{nkl}^{(3)} x_{n-k-l} x_k x_l + O(x^4), \tag{3.2}$$

where x_n are assumed to be small. All coefficients $F^{(n)}$ can be related to the coefficients $G^{(n)}$ of Eq. (3.1). In particular, $F_n^{(1)} = 1/G_n^{(1)}$ is simply the inverse of the linear response function, while the nonlinear terms are related in a more complicated way. As the Hopf bifurcation is approached from the quiescent side, $G_1^{(1)}(\omega)$ diverges for the characteristic frequency, $\omega = \omega_c$, which implies that $F_1^{(1)}(\omega)$ becomes zero at this frequency. Therefore Eq. (3.2) is the appropriate expansion, valid for small amplitudes, in the vicinity of the instability and can even be used once the instabilty has been traversed and spontaneous oscillations occur. Note that all expansion coefficients $F^{(n)}$ are, in general, functions of frequency ω.

2.3 Hopf Bifurcation

We shall now examine the response of a critical oscillator to a simple sinusoidal stimulus of small amplitude f_1, with $f_k = 0$ for $k > 1$. The response is principally at the same frequency, but higher harmonics x_k with $k > 1$ are also present and have amplitudes $x_k = O(x_1^k)$. Focusing on x_1 and x_2, we find (Jülicher and Prost 1997; Camalet et al. 2000) that

$$f_1 \simeq F_1^{(1)} x_1 + (F_{1,2}^{(2)} + F_{1,-1}^{(2)}) x_{-1} x_2 + (F_{1,1,1}^{(3)} + 2F_{1,-1,1}^{(3)}) x_1^2 x_{-1}, \tag{3.3}$$

$$0 \simeq F_2^{(1)} x_2 + F_{2,1}^{(2)} x_1^2, \tag{3.4}$$

where we have assumed, without loss of generality, that $F_{nkl}^{(3)} = F_{nlk}^{(3)}$. Equation (3.4) determines $x_2 \simeq -(F_{2,1}^{(2)}/F_2^{(1)}) x_1^2$ as a function of x_1. Inserting this relation in Eq. (3.4), we obtain a simple expression for the response amplitude x_1 in the presence of a sinusoidal stimulus:

$$f_1 \simeq \mathcal{A}x_1 + \mathcal{B}|x_1|^2 x_1. \tag{3.5}$$

Here, we have defined the effective linear and cubic response functions $\mathcal{A} \equiv F_1^{(1)}$ and $\mathcal{B} \equiv -(F_{2,1}^{(2)}/F_2^{(1)})(F_{1,2}^{(2)} + F_{1,-1}^{(2)}) + (F_{1,1,1}^{(3)} + 2F_{1,-1,1}^{(3)})$. Note that while the linear coefficient $\mathcal{A}$ is directly given by the linear response function, the structure of nonlinearities is more subtle. Significantly, Eq. (3.5) does not contain quadratic terms; the dominant nonlinearity is cubic. However, the coefficient $\mathcal{B}$ of the cubic term contains linear, quadratic, and cubic coefficients of the expansion (2). This derivation is valid for any complex system near a Hopf bifurcation, so Eq. (3.5) is the generic form of the response of a critical oscillator, independent of the physical details of the system.

The above description also accounts for spontaneous oscillations in the absence of a stimulus. In this case, Eq. (3.5) with $f_1 = 0$ has a nontrivial solution with amplitude

$$|x_1|^2 \simeq -\frac{\mathcal{A}}{\mathcal{B}}. \tag{3.6}$$

The frequency ω_c of these oscillations is determined by the criterion that $-\mathcal{A}/\mathcal{B}$ is real and positive. The critical point occurs for the particular case $\mathcal{A}(\omega_c) = 0$ for the critical frequency. As discussed in the preceding text, the distance of the system from the critical point is controlled by the parameter C. We can therefore express the linear coefficient $\mathcal{A}$ for small frequency difference $\omega - \omega_c$ and small deviation of the control parameter from its critical value $C - C_c$ as

$$\mathcal{A}(\omega, C) \simeq \alpha(\omega - \omega_c) + \beta(C - C_c), \tag{3.7}$$

where α and β are complex coefficients, the values of which depend on the particular system. Equations (3.5) and (3.7) together characterize the generic behavior of a critical oscillator, describing both how oscillations are generated and how the system responds to a stimulus.

2.4 Generic Mechanical Response

Although the frequency dependence of the coefficients $\mathcal{A}$ and $\mathcal{B}$ in Eq. (3.5) varies from one system to another, Eq. (3.7) holds for any system that is close to a Hopf bifurcation (Camalet et al. 2000). Consequently, the response of a critical oscillator exhibits some very general features. Most importantly, if the oscillator is poised exactly at the critical point with $C = C_c$, and is stimulated at its characteristic frequency, $\omega = \omega_c$, the linear coefficient $\mathcal{A}$ vanishes and the amplitude of the response varies as

$$|x_1| \simeq |\mathcal{B}|^{-1/3}|f_1|^{1/3}. \tag{3.8}$$

This power law is formally satisfied for arbitrarily small f_1, so the response displays an "essential nonlinearity."

It is easy to see why an active oscillator is ideally suited to act as a mechanosensor in the auditory system. First, the compressive nonlinearity permits the detection of a wide range of signals. Forces that vary by a factor of one million cause displacements of the detection apparatus that vary by a factor of only one hundred—a physically reasonable range. Second, the critical oscillator acts as an active amplifier that boosts faint signals. Defining the signal gain as the ratio $|x_1|/|f_1|$ of output and stimulus amplitudes, Eq. (3.8) implies that this gain diverges with a power law

$$\frac{|x_1|}{|f_1|} \sim |f_1|^{-2/3}, \tag{3.9}$$

so the weakest stimuli are infinitely amplified. Third, the response is highly frequency selective. If the critical oscillator is stimulated at a frequency that differs from the characteristic frequency, $\omega \neq \omega_c$, the nonlinear response can be lost: Eqs. (3.5) and (3.7) indicate that the response becomes linear if $|\omega - \omega_c| \gg \Delta\omega_a$, where

$$\Delta\omega_a = \frac{|\mathcal{B}^{1/3}|}{|\alpha|}|f_1^{2/3}|. \tag{3.10}$$

This therefore defines an "active bandwidth," $\Delta\omega_a$, within which active amplification occurs. Note that the bandwidth depends on the stimulus amplitude, and the tuning becomes very sharp at low levels.

To profit best from the active response, the system should be poised exactly at the Hopf bifurcation. As we discuss later, adjustment of mechanosensors to the proximity of the critical point is likely to be automatically achieved by a feedback control mechanism. The adjustment cannot be perfect, however. When the system is close to, but not exactly at, the critical point, $C \neq C_c$, the linear term $\mathcal{A}$ is nonzero for all frequencies. Therefore, the power-law response is apparent only for $|f_1| \gg |\mathcal{A}/\mathcal{B}|^{1/2}$ when the nonlinearity dominates; for smaller $|f_1|$, the response is linear. The generic response of a dynamic oscillator to periodic stimulation is displayed in Figure 3.2.

The concept of critical oscillators provides a natural explanation for the mechanical responses observed in auditory systems, where there is considerable evidence of an essential compressive nonlinearity. The above analysis shows that an essential nonlinearity is possible only in an active system. Indeed, the vanishing of $\mathcal{A}(\omega_c)$ for nonzero ω_c is impossible in a passive system, because the unavoidable damping necessarily leads to a nonzero imaginary part of $\mathcal{A}$. Conversely, the vanishing of $\mathcal{A}$, which leads to an essential singularity, is necessarily linked to the existence of spontaneously oscillating solutions at neighboring parameter values, since the functions $\mathcal{A}$ and $\mathcal{B}$ are continuous and differentiable on general physical grounds. Thus an essential nonlinearity in the response of a complex system to a periodic stimulus provides direct evidence that the system is on the verge of an oscillating instability.

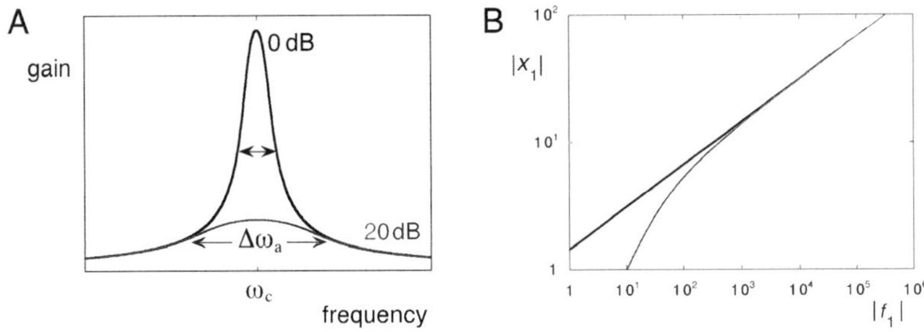

FIGURE 3.2. Typical response of an active oscillator, poised at a Hopf bifurcation, to mechanical stimulation at fixed frequency. (**A**) As the stimulus level decreases, the gain increases and the bandwidth $\Delta\omega_a$ of active amplification decreases. (**B**) A truly critical oscillator, stimulated at its characteristic frequency, displays an essential singularity, Eq. (3.8) (*black curve*). If a critical oscillator is stimulated at a frequency that differs from its characteristic frequency, or if a dynamic oscillator is not precisely critical, its response is linear at low levels (*gray curve*).

2.5 Normal Form

The preceding generic description of a critical oscillator is closely related to the so-called normal form of a dynamical system in the proximity of a Hopf bifurcation (Strogatz 2001; Wiggins 2003). The normal form describes the long-time relaxation of an active oscillator to its limit cycle. Usually, the normal form is defined in the absence of an external stimulus and in the absence of fluctuations. A nonlinear variable transformation is used to define a complex variable Z. Expressed in this variable, the limit cycle corresponds to motion along a circle in the complex plane. The normal form can be linked to the generic expansion (3.5) together with the expression (3.7). We define the complex variable $Z(t) = x_1(t)e^{i\omega t}$ and describe the relaxation to the limit cycle as a slow variation of the amplitude x_1 of the unstable mode with time. This implies that near the bifurcation, $x(t) \simeq x_1(t)e^{i\omega t} + x_{-1}(t)e^{-i\omega t}$, and that $dZ/dt = (dx_1/dt + i\omega x_1)e^{i\omega t} \simeq i\omega x_1 e^{i\omega t}$. Here, we assume that relaxation is slow compared to the oscillation frequency. Rapid variations of $x(t)$ due to the presence of higher harmonics are ignored; they simply lead to nonlinear corrections to the linear relation $x \simeq 2\mathrm{Re}(Z)$. Equation (3.5) together with Eq. (3.7), in the time domain, can be expressed by the dynamic equation

$$\frac{dZ}{dt} = -(r + i\omega_0)Z + B|Z|^2 Z + \frac{e^{i\theta}}{\Lambda} f(t). \tag{3.11}$$

This is the normal form of a system close to a Hopf bifurcation, in the presence of a force $f(t)$. The real parameter $r = (C - C_c)\mathrm{Im}(\beta/\alpha)$ characterizes the proximity to the bifurcation, which is situated at $r = 0$. For $r < 0$ there is a limit-cycle solution, representing spontaneous oscillations at frequency $\omega_0 = \omega_c + (C - C_c)\mathrm{Re}(\beta/\alpha)$, whose amplitude is governed by the complex nonlinear coefficient $B = -\mathcal{B}/\alpha$. The external force is multiplied by a complex factor $e^{i\theta}/\Lambda = i/\alpha$. The real coefficient Λ is an effective friction, and the phase θ characterizes the phase shift at which the external force acts on the variable Z.

The essential nonlinearity of a critical oscillator is clear from Eq. (3.11). When the system is precisely at the bifurcation, $r = 0$, and $f(t)$ is a sinusoidal force at the characteristic frequency $\omega = \omega_c$, the solution is $Z = x_1 e^{i\omega t}$ with $|x_1| \sim |f|^{1/3}/|\Lambda B|^{1/3}$. The normal form is also convenient for characterizing the linear response that occurs at small forces in the case that the active oscillator is not exactly critical. For a real stimulus force with $f_{-1} = f_1^*$, and for small $\omega - \omega_0$, the response is $x_1 \simeq \chi(\omega) f_1$, where the linear response function $\chi = \mathcal{A}^{-1}$ is

$$\chi \simeq \frac{1}{\Lambda}\left[\frac{e^{i\theta}}{i(\omega_0 - \omega) + r} + \frac{e^{-i\theta}}{-i(\omega_0 + \omega) + r}\right]. \tag{3.12}$$

2.6 *Effects of Fluctuations*

From the point of view of complex systems, a Hopf bifurcation is an idealization that can strictly be realized only in the thermodynamic limit: only an infinite number of interacting degrees of freedom can generate spontaneous oscillations that are truly coherent (Risler et al. 2004, 2005). In small systems, such as single cells, an oscillatory state is subject to fluctuations caused by the irregular movements of many molecular or microscopic components. These fluctuations destroy the precision of the oscillation period. As a result, the phase coherence of the oscillations is limited to a coherence time τ (Stratonovich 1981). Fluctuations therefore conceal the Hopf bifurcation that occurs in the equivalent deterministic system.

Even so, a noisy active oscillator can function effectively as a detector of perodic mechanical stimuli. Suppose that an active dynamical system is close to the point where a Hopf bifurcation would occur in the absence of fluctations. It would then spontaneously oscillate at low amplitude and in an irregular fashion, with a short coherence time. If the system is then stimulated by a sinusoidal force $f(t) = f_1 e^{i\omega t}+$ c.c. where c.c denotes the complex conjugate at its characteristic frequency $\omega = \omega_c$, the response at that frequency x_1 varies with the level of the stimulus according to Eq. (3.5). The augmentation of this Fourier component of the displacement is equivalent to an increase of phase coherence in the response (see Fig. 3.3), even though the overall amplitude of the noisy oscillations does not necessarily increase (Camalet et al. 2000). This entrainment, or "phase-locking," of the dynamical system can be detected to infer the presence of the stimulus. In the case of hearing, low-level sounds could be detected by analyzing the timings of spike trains elicited in the auditory nerve.

The effect of noise on a dynamical system that is close to a Hopf bifurcation is most readily discussed using the normal form. Fluctuations can be modeled by adding to Eq. (3.11) a stochastic force $\eta(t)$ with zero average $\langle\eta\rangle = 0$. Assuming for simplicity Gaussian white noise, we use $\langle\eta(t)\eta(0)\rangle = 2D(t)$ and

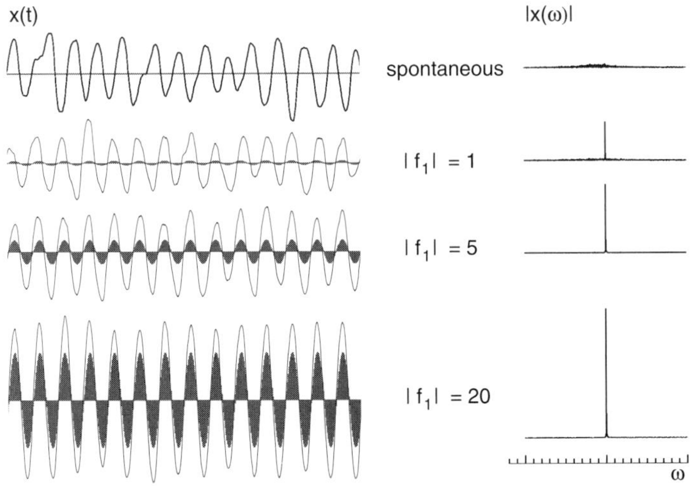

FIGURE 3.3. Role of fluctuations in the response of a critical oscillator. Spontaneous oscillations (*top line*) are noisy and lack phase coherence. A periodic stimulus at the characteristic frequency of the oscillator (with level indicated by the *shaded curve*) causes phase locking, although the overall amplitude of the response does not necessarily increase. Corresponding to this enhancement of the phase coherence, a peak emerges in the Fourier transform of the response (*right column*).

$\langle \eta(t)\eta^*(0)\rangle = 0$. The spectral density $\tilde{C}(\omega)$, which is the Fourier transform of the autocorrelation function $C(t) = \langle x(\tau+t)x(\tau)\rangle$, is of the form (Martin et al. 2001)

$$C(\omega) \simeq \frac{1}{\Lambda^2}\left[\frac{\tilde{D}(\omega)}{r^2+(\omega-\omega_0)^2}+\frac{\tilde{D}(-\omega)}{r^2+(\omega+\omega_0)^2}\right], \tag{3.13}$$

where the effect of nonlinearities has been neglected. This spectrum, which exhibits a peak at the characteristic frequency ω_0, is characteristic of noisy active oscillators. The coherence time $\tau \simeq r^{-1}$ characterizes the width of the peak.

The observation of noisy oscillations with a finite coherence time τ is insufficient to determine whether a given system is a passive damped oscillator subject to thermal fluctuations, or whether it is an active nonlinear system that oscillates spontaneously. Comparison of the spectral density with the response to mechanical stimuli, however, provides key information about the passive or active nature of the system. In the case of a passive system in thermodynamic equilibrium at temperature T, fluctuations result from Brownian motion and satisfy a fluctuation-dissipation theorem. This relates the autocorrelation function to the linear response function $\tilde{\chi}(\omega) = \tilde{\chi}'(\omega) + i\tilde{\chi}''(\omega)$, where $\tilde{\chi}'$ and $\tilde{\chi}''$ denote the real and the imaginary part, respectively. It can be written as (Forster 1990)

$$\tilde{C}(\omega) = \frac{2k_BT}{\omega}\tilde{\chi}(\omega)''. \tag{3.14}$$

Therefore, for a passive system the linear response function is completely determined by a knowledge of the autocorrelation function $C(t)$. For an active oscillator, on the other hand, the fluctuation-dissipation theorem does not apply, and the linear response function is not so constrained. So by measuring both the autocorrelation function and the linear response function of a noisy oscillator, and seeing whether the fluctuation-dissipation theorem is obeyed or violated, it is possible to determine whether a system is passive or active (Martin et al. 2001).

2.7 Response to Multiple Frequencies

While a critical oscillator provides the benefit of active amplification, the nonlinearity of its response means that different frequencies interfere with one another, so that the spectrum of the response differs from the spectrum of the stimulus. The nonlinearities naturally lead to the generation of new frequencies or "distortion products," and also to "two-tone suppression."

To illustrate the basic interference phenomena that arise from nonlinearities, we consider how a critical oscillator responds to a signal containing two frequency components ω_1 and ω_2,

$$f(t) = f_{10}e^{i\omega_1 t} + f_{01}e^{i\omega_2 t} + c.c., \tag{3.15}$$

where $c.c.$ denotes the complex conjugate and f_{10} and f_{01} are the Fourier amplitudes of frequency components ω_1 and ω_2, respectively. For a dynamical system in the vicinity of a Hopf bifurcation, we can generalize the arguments that lead to Eq. (3.2) (Jülicher et al. 2001; Stoop and Kern 2004). The nonlinearities in the system lead to many more frequency components in the response, which can be generally written as

$$x(t) = \sum_{kl} x_{kl}e^{i(k\omega_1 + l\omega_2)t}. \tag{3.16}$$

In the vicinity of a Hopf bifurcation, we can generalize Eq. (3.5) and write

$$f_{10} \simeq \mathcal{A}(\omega_1)x_{10} + \mathcal{B}|x_{10}|^2 x_{10} + \bar{\mathcal{B}}|x_{01}|^2 x_{10}, \tag{3.17}$$

$$f_{01} \simeq \mathcal{A}(\omega_2)x_{01} + \mathcal{B}|x_{01}|^2 x_{01} + \bar{\mathcal{B}}|x_{10}|^2 x_{01}. \tag{3.18}$$

This expression reveals the basic features of the response to two tones. We assume that ω_1 is the critical frequency of the oscillator. If a stimulus component at this frequency is present, $f_{10} \neq 0$, but the second frequency component is absent, $f_{01} = 0$, then $x_{01} = 0$ and we recover the general response to a single stimulus, Eq. (3.5). But if the second frequency component is introduced, $f_{01} \neq 0$, a corresponding frequency component is generated in the response, $x_{01} \neq 0$. This frequency component x_{01} changes the response to the first stimulus, which now satisfies

$$f_{10} \simeq \mathcal{A}' x_{10} + \mathcal{B}|x_{10}|x_{10}, \tag{3.19}$$

where $\mathcal{A}' = \mathcal{A} + \bar{\mathcal{B}}|x_{01}|^2$. The presence of the second frequency component thus changes the linear response. The critical oscillator consequently responds to the stimulus at the characteristic frequency as if it were not at the critical point. This implies that the nonlinear amplification is reduced in the presence of a second frequency, and the threshold of detection is increased (Ruggero et al. 1992). This feature is thus similar to the general phenomenon of two-tone suppression in hearing—reduced sensitivity to a sound stimulus when a second nearby frequency is simultaneously present (Zwicker and Fastl 1999). This two-tone suppression is most important when the frequency difference $\Delta\omega = \omega_1 - \omega_2$ is small, because $x_{01} \sim f_{01}/\Delta\omega$ then becomes large.

In addition to two-tone suppression, new frequencies are generated by nonlinearities in the response (see Fig. 3.4). In general, and according to Eq. (3.16), these frequency components are given by

$$\omega_{kl} = k\omega_1 + l\omega_2, \tag{3.20}$$

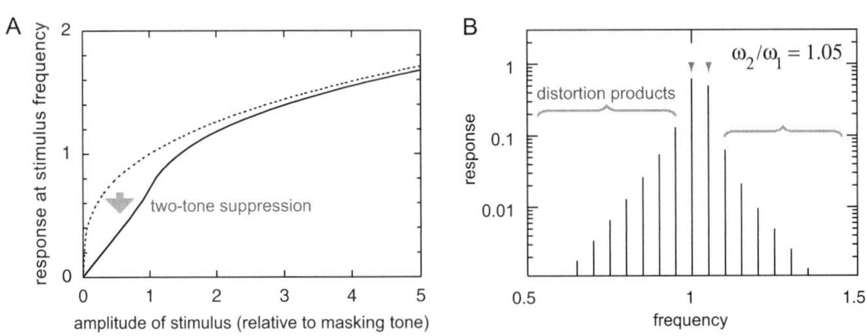

FIGURE 3.4. Two-tone suppression and distortion-product generation in the response of a critical oscillator. (**A**) The response of a critical oscillator to a stimulus at its characteristic frequency (*dashed line*) is suppressed (*full line*) in the presence of a second, "masking" frequency. (**B**) The response of a critical oscillator to a stimulus containing two frequencies ω_1 and ω_2 (arrow) displays a characteristic spectrum of distortion products. New components appear at frequencies $n\omega_1 + m\omega_2$, where n and m are integers, because of nonlinear couplings. Most notable are the distortion-product frequencies $2\omega_1 - \omega_2$ and $2\omega_2 - \omega_1$, but many other components with frequencies $\omega_1 + n\Delta\omega$, where $\Delta\omega = \omega_1 - \omega_2$, are important.

where k and l are integers. Particular examples are the frequencies $\omega_{2,-1} = 2\omega_1 - \omega_2$ and $\omega_{-1,2} = 2\omega_2 - \omega_1$, which are the most prominent frequencies of distortion products in hearing. For small $\Delta\omega$ there is a particularly strong response at these frequencies, because they are directly generated by the two frequency components in the stimulus. These new frequencies mix with the original ones to create further distortion products, so a whole hierarchy of frequencies is generated:

$$\omega_{k+1,-k} = \omega_1 + (k-1)\Delta\omega. \tag{3.21}$$

The amplitude of these distortion products decays exponentially as

$$x_{k+1,-k} \sim e^{-\lambda|k-3/2|}. \tag{3.22}$$

The number of strongly excited distortion products is given by $\lambda^{-1} \sim \Delta\omega^{-1}$, and so the response contains many components when the two stimuli are close in frequency.

This exponential decay of the spectrum of distortion products is characteristic of dynamic oscillators that are close to a Hopf bifurcation. A general nonlinear system (or a system far from a Hopf bifurcation) does not generate this response (Jülicher et al. 2001). The observation of an exponential spectrum in the two-tone response of the basilar membrane (Robles et al. 1997) therefore indicates that critical oscillators might be an important part of the active process in the mammalian cochlea.

3. Building Critical Oscillators

3.1 Physical Realizations of Active Mechanical Oscillators

Any active mechanical system that is close to a supercritical Hopf bifurcation behaves in the general way described in the previous section. Because the supercritical Hopf bifurcation is the simplest type of transition from a quiescent to an oscillatory regime, such dynamical systems are easy to manufacture. Here, we consider a variety of physical realizations of critical oscillators that might be relevant in hearing.

Inertial oscillator with negative damping: One of the simplest active oscillators to envisage is a damped inertial oscillator, supplemented by an internal driving force that counteracts the friction. The equation of motion, when the system is under the influence of an external force f, is

$$m\ddot{x} = -\lambda\dot{x} - kx + f_{\mathrm{a}} + f, \tag{3.23}$$

where m is a mass, λ is a coefficient of damping, k is the modulus of an elastic spring, and f_{a} is the internally generated active force. If the active force is in phase with the velocity of the oscillator, it acts as a negative friction, and if it overcomes the damping, spontaneous oscillations are generated. Some kind of nonlinearity is essential to stabilize the oscillations. One way of doing this is for the active force to diminish with amplitude, $f_a = (C - Bx^2)\dot{x}$, so that the total friction becomes positive for large oscillations. The system is then known as a van der Pol oscillator (van der Pol 1926; Strogatz 2001), and obeys

$$m\ddot{x} = -(\lambda - C + Bx^2)\dot{x} - kx + f. \tag{3.24}$$

If the driving force is periodic with frequency ω, the response at the same frequency satisfies Eq. (3.5) with

$$\mathcal{A} = k - m\omega^2 + i\omega(\lambda - C), \tag{3.25}$$

$$\mathcal{B} = i\omega B. \tag{3.26}$$

The coefficient C plays the role of a control parameter, and the Hopf bifurcation occurs at $C = C_c = \lambda$. The system then generates spontaneous oscillations with characteristic frequency $\omega_c = (k/m)^{1/2}$. Note that it is necessary for $\mathcal{B}$ to be purely imaginary to stabilize the system. It is also important that the nonlinearity be in the damping; the combination of an active force that is simply proportional to the velocity, and a nonlinear spring $k(x) = k'x + B'x^3$, for example, would not create a stable oscillator. Alternative stabilizing nonlinearities in the damping are possible, however, such as an active force that decreases with velocity, $f_a = (C - B\dot{x}^2)\dot{x}$. In this case $\mathcal{A}$ is unchanged, but $\mathcal{B} = 3i\omega^3 B$. This oscillator thus has the same critical point and characteristic frequency, but the frequency-dependence of the oscillation amplitude is different.

Time-delayed feedback: It is easy to understand how the above kind of active undamping works, but it is less obvious how a feedback mechanism could be constructed to ensure that the active force is precisely in phase with the velocity. An alternative is to use a feedback proportional to the displacement, but with a time delay τ, so that $f_a = Cx(t - \tau)$ (Boukas and Liu 2002). In the case that the mechanical system has a simple nonlinear drag and a nonlinear elasticity, its equation of motion is then

$$m\ddot{x} = -(\lambda + Bx^2)\dot{x} + (k + B'x^2)x + Cx(t - \tau) + f. \tag{3.27}$$

The coefficients of Eq. (3.5) for this system are

$$\mathcal{A} = k - m\omega^2 + i\omega\lambda + Ce^{-i\omega\tau}, \tag{3.28}$$

$$\mathcal{B} = 3B' + i\omega B. \tag{3.29}$$

The strength of the feedback C acts as the control parameter. The critical point C_c and characteristic frequency ω_c are given by the solutions of the two equations $k - m\omega_c^2 = -C_c \cos(\omega_c \tau)$ and $\omega_c \lambda = C_c \sin(\omega_c \tau)$. If the feedback is rapid compared to the period of oscillation, these expressions reduce to $C_c \approx \lambda/\tau$ and $\omega_c^2 \approx k/m + \lambda/\tau m$.

Two first-order systems: On the cellular scale, inertial effects are negligible in most cases, and the dynamics of an active mechanical system is described by a first-order differential equation

$$\lambda \dot{x} = -k(x)x + f_a + f, \tag{3.30}$$

where $k(x)$ is the nonlinear stiffness of the system. In this case, oscillations can be generated if the active force f_a also evolves in time according to a first-order differential equation, which is coupled to Eq. (3.30). The simplest conceivable situation is akin to "stretch-activation" in muscle, i.e., we consider an active process that generates a restoring force when the system is displaced, but which relaxes with a time constant τ when the system is stationary (Martin et al. 2001; Vilfan and Duke 2003b). To obtain oscillations, it is also important for the mechanical system to display a regime of negative elasticity, $k(x) = k - C + Bx^2$; here k is the bare stiffness and C is a control parameter characterizing a reduction of the stiffness. When $C > k$, the stiffness becomes negative at small displacements.

The combination of negative stiffness and an internal, dynamically regulated active force is represented by the pair of equations

$$\lambda \dot{x} = -(k - C + Bx^2)x + f_a + f, \tag{3.31}$$

$$\tau \dot{f}_a = -f_a - \bar{k}x. \tag{3.32}$$

The coefficients of Eq. (3.5) for this system are

$$\mathcal{A} = \frac{k + \bar{k} - C - \tau\lambda\omega^2 + i\omega(\lambda + \tau(k - C))}{1 + i\omega\tau}, \tag{3.33}$$

$$\mathcal{B} = 3B\frac{1 + 3i\omega\tau}{1 + i\omega\tau}. \tag{3.34}$$

Inspection of Eq. (3.33) reveals that the Hopf bifurcation occurs at $C = C_c = k + \lambda/\tau$ and that the characteristic frequency is given by $\omega_c^2 = \bar{k}/\lambda\tau - 1/\tau^2$. Thus in this case, the frequency of oscillation is determined by the properties of the active force-generating system, rather than by the mechanical properties of the system.

3.2 Self-Adjustment to the Critical Point

While the properties of critical oscillators can account for many general features of sound detection in hearing, it remains to be explained how a dynamical system could operate sufficiently close to a Hopf bifurcation for the generic behavior of critical oscillators to be apparent. The necessary fine-tuning of parameters raises doubts as to whether a living system can profit from the special properties at a critical point in a reliable way. These concerns have been allayed, however, by the proposition of a mechanism that enables an oscillator to adjust automatically to its critical point.

Feedback regulation of the control parameter (Coullet et al. 1989; Camalet et al. 2000; Moreau et al. 2003) is a simple and general mechanism to maintain a dynamical system at a point of operation close to the bifurcation point. This kind of automatic self-adjustment ensures that when the system is quiescent, the control parameter gradually changes so that the system approaches the instability; but as soon as oscillations are detected, the control parameter is stabilized. Self-adjustment works best in the absence of external stimuli, when the highest sensitivity is needed, by adapting the control parameter according to the detected amplitude of vibrations.

Consider a system for which the control parameter can vary in the range from zero to C_{max}, and which oscillates for $C > C_{\mathrm{c}}$. Self-adjustment can be described by the dynamic equation for the control parameter

$$\frac{dC}{dt} = \left(1 - \frac{x^2}{\delta^2}\right)(C_{\mathrm{max}} - C)/\tau_{\mathrm{a}}. \tag{3.35}$$

Here δ characterizes the threshold amplitude at which the feedback becomes effective. If the system is quiescent ($x = 0$), the control parameter rises toward its maximum value with a characteristic adaptation time τ_{a}. When C exceeds C_{c}, the system starts to oscillate and the amplitude near the critical point varies as

$$|x_1|^2 \simeq \Delta^2(C - C_{\mathrm{c}})/C_{\mathrm{c}}, \tag{3.36}$$

where Δ denotes a characteristic amplitude of spontaneous oscillations. The feedback control, Eq. (3.35), thus brings the system to an operating point $C = C_0$, close to the critical point, for which $x_1^2 \simeq \delta^2$ and (Camalet et al. 2000)

$$\frac{C_0 - C_{\mathrm{c}}}{C_{\mathrm{c}}} \simeq \frac{\delta^2}{\Delta^2}. \tag{3.37}$$

This quantity describes the relative distance to the critical point achieved by self-regulation. Because the threshold of detection δ can be much less than the typical amplitude of spontaneous oscillations Δ, the self-regulation can ensure that the system operates sufficiently close to the critical point for the generic nonlinear behavior to dominate.

Thus in the absence of a stimulus, a self-adjusted active oscillator will be poised very close to the point where it is most sensitive. When stimulated by a weak signal, such that the amplitude of the response only slightly exceeds δ, Eq. (3.35) indicates that the control parameter is not significantly perturbed by the stimulus; thus the oscillator remains in the immediate vicinity of the critical point, and active amplification can be maintained. If the stimulus is stronger, however, and elicits a large response, Eq. (3.35) indicates that the control parameter gradually decreases, so that the system drifts away from the critical point, into the quiescent regime. As a consequence, the amount of active amplification diminishes with time—a feature that does not pose a problem because a strong stimulus can be detected without the aid of amplification. After the termination of a sustained, high-level stimulus, it takes a typical time τ_a for the system to readjust and return to the vicinity of the critical point, where it is most sensitive. Thus this mechanism of detection naturally accounts for a period of adaptation, following a strong stimulus, before the weakest signals can be detected. Note that the adaptation time τ_a must be long compared to the characteristic period of the oscillator, so that the variations in the control parameter caused by the feedback mechanism do not interfere with the limit cycle.

4. Critical Oscillators in the Auditory System

4.1 Variations on a Theme

The main signature of active amplification in the auditory system is otoacoustic emissions. Such emissions have been detected in all classes of land vertebrates (Manley and van Dijk, Chapter 6; Probst 1990; Köppl 1995; Manley and Köppl 1998; Manley 2000), and more recently an active process has been discovered in the hearing of some insects (Göpfert, Chapter 5; Göpfert and Robert 2003; Göpfert et al. 2005). The wide range of animal species that benefit from active audition suggests that a variety of realizations of active oscillators may have evolved. The way that oscillators of different frequencies are coupled to one another in the auditory system might also be expected to vary according to the inner-ear physiology.

While the existence of otoacoustic emissions is well established, the identification of the cellular processes that are at the origin of these emissions has proved difficult. There thus remains controversy about the nature of the active elements in hearing and their relation to nonlinear oscillators. Evidence from many sources links active amplification to the biophysics of mechanosensory cells and strongly suggests that it is these cells themselves that host the active process. In vertebrates, hair cells are thus the prime candidates for the generation of mechanical oscillations (Dallos 1992; Hudspeth 1997). Active oscillations of the hair bundle have been directly observed in amphibians and reptiles (Crawford and Fettiplace 1985; Howard and Hudspeth 1987; Denk and Webb 1992; Benser et al. 1996; Martin and Hudspeth 1999; Martin et al. 2000). In the mammalian cochlea, inner and outer hair cells can be distinguished (Dallos et al. 1996). Because of their electromotility, outer hair cells have been the focus of the search for active elements (Brownell et al. 1985; Kachar et al. 1986; Ashmore 1987). However, the hair bundles of both inner and outer hair cells could act as active amplifiers, as they do in other vertebrates.

From an evolutionary standpoint, it seems probable that the first active mechanosensors would have been based on a motile system that naturally produces vibrations. An oscillating axoneme, coupled to a mechanosensitive channel, is a possible candidate. Natural selection could have refined a simple feedback system to provide a self-adjustment mechanism that keeps the oscillator close to its critical point. The ions that enter through the channel might bind to the motor proteins that drive the axoneme, for example, thereby affecting the quality of the oscillations. The auditory system of some insects, which is based on mechanosensory neurons, has been shown to employ dynamic oscillators (Göpfert, Chapter 5). The anatomy of their sensory apparatus (Eberl 1999) suggests that a structure related to an axoneme might be the mechanical basis of the active oscillator.

4.2 Vertebrate Hair Bundles

The hair bundles of vertebrate hair cells consist of about 50 stereocilia: stiff, rodlike extensions of the cell with a length of several microns and a diameter of about 300 nm. The stereocilia are arranged in rows of increasing height, and each stereocilium is connected, at its tip, to a stereocilium in the adjacent row. The fine filament that forms this bridge is known as a "tip link" (Pickles et al. 1984; Kachar et al. 2000). Shear deformation of the hair bundle stretches the tip links and triggers the opening of mechanosensitive transduction channels to which they are joined (Hudspeth 1989). This leads to a transduction current: an influx of K^+ and Ca^{2+} ions into the cell. Active internal forces, generated internally within the stereocilia by myosin motor proteins, can also modulate the tension in the tip links and thereby affect the transduction current (Hudspeth and Gillespie 1994).

The micromechanical properties of hair bundles in living hair cells range from adaptive movements in response to abrupt force steps (which have both fast and slow components [Howard and Hudspeth 1988; Benser et al. 1996; Ricci et al. 2000]), to spontaneous oscillations (Crawford and Fettiplace 1985; Howard and Hudspeth 1987; Denk and Webb 1992; Benser et al. 1996; Martin and Hudspeth 1999; Martin et al. 2000, 2003). These are discussed in detail in Martin, Chapter 4. Micromanipulation experiments performed on spontaneously oscillating hair bundles have revealed a frequency selectivity and a compressive nonlinear response that are compatible with the general properties of critical oscillators (Martin and Hudspeth 1999, 2001; Martin et al. 2001). In the case of hair bundles in the bullfrog sacculus, the violation of the fluctuation-dissipation theorem, Eq. (3.14), has been shown experimentally (Martin et al. 2001). Thus Brownian motion has been ruled out as the source of fluctuations, and the hair bundle has been proved to be a noisy, active oscillator. Direct evidence of active amplification by hair bundles in the intact reptilian cochlea has also been presented (Manley et al. 2001).

Spontaneous oscillations of the hair bundle could be generated by several different mechanisms, which can be characterized by two or more coupled first-order equations, as described in Section 3. In the absence of active processes, the hair bundle mechanics are well characterized by Eq. (3.30) with $f_a = 0$, where x denotes the hair bundle displacement and f the applied external force. The hair bundle stiffness $k(x)$ is nonlinear and depends on displacement because the transduction channels contribute a gating compliance to the bundle stiffness (Howard and Hudspeth 1988). Measurements of the stiffness of hair bundles from the bullfrog sacculus show that $k(x)$ can become negative (Martin 1999; Kennedy et al. 2005).

The generation of spontaneous oscillations requires an active force f_a, generated inside the hair bundle. The observed properties of hair bundle oscillations in the bullfrog sacculus are best described by assuming that the active force f_a is generated by myosin-Ic adaptation motors in the stereo-cilia, which are themselves regulated by the intracellular Calcium concentration (Hudspeth and Gillespie 1994; Holt et al. 2002; Kros et al. 2002). Using the hair bundle deflection x, the motor force f_a, and the Calcium concentration c, this dynamical system is described by three coupled first-order systems (Martin et al. 2003; Nadrowski et al. 2004). For low-frequency oscillations such as those observed in the bullfrog, the Calcium concentration changes almost instantaneously compared to the other variables, and the dynamical system then reduces to two coupled first-order systems. It can be represented by equations of the form (3.30) and (3.32). Here Eq. (3.32) captures the combined effects of the force–velocity relationship of the motors and the regulation of motors by a changing Calcium concentration, in a simplified linear description. The existence of spontaneous oscillations requires both negative hair bundle stiffness ($C > \bar{k}$) and sufficiently strong Calcium feedback of the motor force $\bar{k} > \lambda/\tau$. The full state diagram of the system is discussed in Nadrowski et al. 2004 .

The interplay of nonlinear hair-bundle stiffness, myosin motors, and calcium dynamics in the hair bundle can also lead to spontaneous oscillations by other mechanisms (Choe et al. 1998; Vilfan and Duke 2003a). If the reclosure of transduction channels can be triggered by an increase of the intracellular calcium concentration (Howard and Hudspeth 1988; Benser et al. 1996; Wu et al. 1999; Ricci et al. 2000), this also generates an active force that depends on the calcium concentration and can also be described by Eqs. (3.30) and (3.32). In such a scenario, the dynamics of adaptation motors is slow and could be used for self-regulation to the critical regime (Vilfan and Duke 2003a).

Finally, a sufficiently large number of motor proteins that operate collectively could become self-oscillatory, even without being coupled to the calcium dynamics. This possibility is based on a general physical mechanism that could operate over a large frequency range (Jülicher and Prost 1997). Such motor-induced mechanical oscillations might be relevant to axonemal vibrations and beating (Camalet et al. 2000).

4.3 Mammalian Cochlea

The physical basis of the active amplifier in the mammalian cochlea remains controversial. There is much evidence that the specialized outer hair cells can pump energy into the basilar membrane (Chapter 7; Dallos et al. 1996). The somatic electromotility of outer hair cells can certainly produce large, rapidly changing forces, but there is currently no indication that it can generate spontaneous oscillations. It is possible that it acts as a linear amplifier that boosts nonlinear oscillations generated by another source. Indeed, the hair bundles of outer hair cells have recently been found to generate active forces that could contribute to the amplifier (Kennedy et al. 2005), and an in vitro preparation has provided direct evidence for a characteristic resonance in the mammalian cochlea, linked to a compressively nonlinear amplifier that depends on the flow of Ca^{2+} ions through hair-bundle transduction channels (Chan and Hudspeth 2005).

Despite the uncertainty about the physical basis of the mechanism, there have been numerous propositions that active oscillators play a crucial role in the mammalian cochlea. Undamping of an inertial oscillator is the active mechanism that Gold proposed to exist in the mammalian cochlea (Gold 1948). He reasoned that the motion of individual elements of the basilar membrane is damped by the cochlear fluid, and that a resonant response could occur only if an internally generated active force, in phase with the velocity, counteracted the damping. Following this lead, nonlinear active models of the cochlear response have been constructed (Neely and Kim, Chapter 10; duifhuis 1985; Kolston et al. 1990; de Boer 1996; see also Cooper et al., Chapter 2). Based on an analysis of the basilar membrane response, Zweig has argued that the active amplifier consists of a set of time-delayed feedback oscillators (Zweig 1991). In the following section, we will discuss a more general model of the cochlear response based on the generic response of critical oscillators (Duke and Jülicher 2003; Magnasco 2003). We argue that this model should be valid whatever the underlying physical basis of the mammalian cochlear amplifier, and that it provides a framework for discussing energy flow and nonlinear effects in cochlear traveling waves.

5. Traveling Waves Powered by Critical Oscillators

5.1 Traveling Waves in the Cochlea

The basic physics of cochlear traveling waves may succinctly be described using a simplified one-dimensional model (Zwislocki 1948; Zweig 1976; de Boer 1980; Lighthill 1981; Zweig 1991). In this approach, the propagation of a pressure wave in the cochlear fluid and the propagation of an associated disturbance of the basilar membrane are analogous to the propagation of voltage and current in an electromagnetic transmission line.

The cochlear duct is separated into two channels by the basilar membrane, but these are connected at the apex of the cochlea by a small aperture, the helicotrema (see Fig. 3.5). A sound stimulus impinging on the oval window, at the base of the cochlea, causes changes in the pressures $P_1(x, t)$ and $P_2(x, t)$ in both channels. Here t is the time and x is the position along the cochlea, with the oval window at $x = 0$ and the helicotrema at $x = L$. The pressure gradients induce longitudinal currents $J_1(x, t)$ and $J_2(x, t)$, which flow in opposite directions in the two

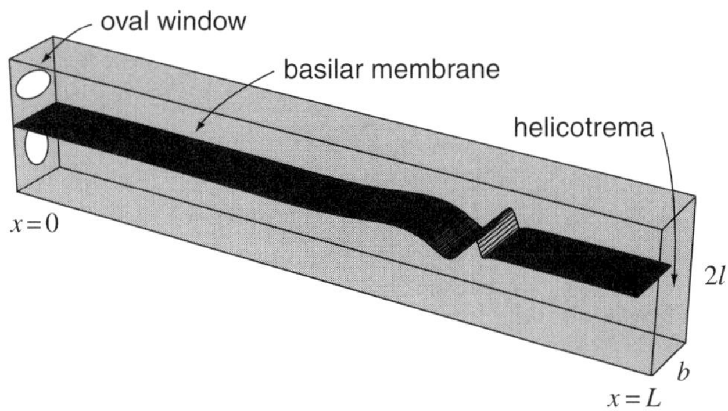

FIGURE 3.5. Schematic representation of the cochlea. The fluid-filled duct is divided into two channels by the basilar membrane. Sound enters at the base and travels toward the apex, where the two channels are connected by the helicotrema. The system behaves as a transmission line and couples basilar membrane vibrations hydrodynamically. We assume that critical oscillators operate along the cochlear partition as local active elements, and that their characteristic frequency decreases continuously from base to apex.

channels. We define the relative current $j \equiv J_1 - J_2$ and the pressure difference $p \equiv P_1 - P_2$. The balance of pressure gradients with inertial and viscous forces in the fluid, together with the incompressibility of the fluid, leads to the equation

$$2\rho b \partial_t^2 h + \eta \partial_t h = \partial_x \left[bl \partial_x p \right], \tag{3.38}$$

which relates the height profile $h(x, t)$ of the basilar membrane to the pressure difference across it. Here, b and l denote the width and height of the cochlear channels, respectively. The damping coefficient η is proportional to the fluid viscosity, and ρ is the fluid density.

The pressure difference p acts to deform the basilar membrane, and the resultant wave propagation depends crucially on the way the cochlear partition responds. If the partition is simply a passive element (e.g., in the dead cochlea), the response relation close to the basal end, where the stiffness $K(x)$ of the basilar membrane is high, takes the simple form $p = K(x)h$. Then Eq. (3.38) becomes a damped linear wave equation, the speed of propagation of the membrane disturbance is $c = (K(x)l/2\rho)^{1/2}$, and the distance that the wave travels before it gradually gets dissipated depends on a balance between damping and elastic forces. If the cochlear partition is an active system, however, the wave can be regenerated as it progresses. We will see that this provides a way of transporting the incident energy to a localized region of the cochlea.

5.2 Active Traveling Waves

We describe the amplifier in the active cochlea as a set of critical oscillators that are located on the cochlear partition and that are positioned in such a way that they can drive the motion of the basilar membrane. Motivated by the observed variation of the characteristic frequency with place in the cochlea (von Békésy 1960; Greenwood 1990), we describe the position-dependence of characteristic frequencies as

$$\omega_r(x) = \omega_0 e^{-x/d}, \tag{3.39}$$

so that the oscillators near the base have the highest frequencies, and those near the apex the lowest.

When a sound stimulus with a single frequency component enters the ear, the local pressure changes periodically in time, $p(t) = \tilde{p}e^{-i\omega t} + c.c.$, where $c.c.$ denotes the complex conjugate. If the response of the cochlear partition is governed by the critical oscillators, the local displacement at frequency ω can be written

$$\tilde{p} \simeq \mathcal{A}(\omega)\tilde{h} + \mathcal{B}|\tilde{h}|^2\tilde{h}. \tag{3.40}$$

Here $\tilde{h}$ is the Fourier amplitude of the resulting periodic vibration $h(x,t) \simeq \tilde{h}(x)e^{-i\omega t} + c.c.$, and $\mathcal{A}$ and $\mathcal{B}$ denote the complex coefficients introduced in (5).

Combining Eq. (3.38) describing the wave mechanics with Eq. (3.40) for the local amplifier, we obtain a nonlinear wave equation for the deformation of the basilar membrane. In frequency representation, it reads (Duke and Jülicher 2003):

$$-2\rho b\omega^2\tilde{h} - i\omega\eta\tilde{h} = \partial_x\left[bl\partial_x\left(\mathcal{A}(x,\omega)\tilde{h} + \mathcal{B}|\tilde{h}|^2\tilde{h}\right)\right]. \tag{3.41}$$

Here the complex linear response coefficient $\mathcal{A}(x,\omega)$ depends on both position and frequency. There are, however, a number of constraints on its functional form. Most importantly, because each of the oscillators is assumed to be critical, the linear response vanishes at the local characteristic frequency $\omega_r(x)$: $\mathcal{A}(x,\omega) \simeq \alpha(\omega - \omega_r(x))$ for small $\omega - \omega_r$. In addition, decomposing $\mathcal{A} = A' + iA''$ into real and imaginary parts, we note that $A'(\omega)$ is an even function, while $A''(\omega)$ is odd. Both vanish at the characteristic frequency ω_r. Finally, the response of the cochlear partition at zero frequency should reflect a simple, passive elastic response, so that $\mathcal{A}(x,0) \simeq K(x)$. The typical functional form of $\mathcal{A}(\omega)$ is shown in Figure 3.6.

Whatever the precise choice of $\mathcal{A}(\omega)$, Eq. (3.41) can be solved for the amplitude $\tilde{h}(x)$ of the basilar membrane displacement. The appropriate boundary conditions are $|\tilde{p}(0)| = p_{\text{in}}$, where p_{in} is the amplitude of the incoming pressure, and $\tilde{p}(L) = 0$ because the channels are connected at the helicotrema. The traveling wave patterns that correspond to these solutions are $h(x,t) = \text{Re}(\tilde{h}(x)e^{i\omega t})$. Examples of these waveforms are shown in Figure 3.7.

There is a characteristic place $x = x_r$ in the cochlea where the frequency of the wave is matched by the frequency of the critical oscillators, $\omega = \omega_r(x_r)$. Far from

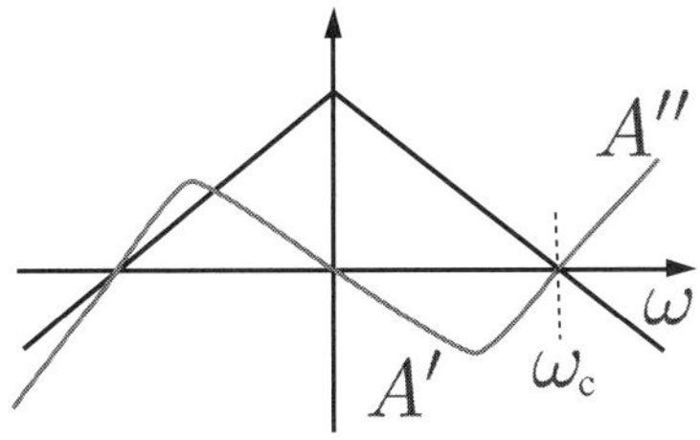

FIGURE 3.6. Schematic representation of the linear response function $\mathcal{A} = A' + iA''$ of a critical oscillator. The real (reversible) part A' is an even function of ω, while the imaginary (irreversible) part is odd. Both functions vanish for $\omega = \pm\omega_c$, where ω_c is the oscillation frequency at the critical point.

this resonance point, the wave equation Eq. (3.41) describes traveling waves that are linear for small vibration amplitudes. A wave that enters at $x = 0$ first encounters active oscillators that have a high characteristic frequency compared to the wave frequency, $\omega < \omega_r$. At this point, the response is dominated by the real part of the linear response A', which differs little from the passive stiffness of the membrane. But the imaginary part is negative, $A''(\omega) < 0$, indicating that energy is pumped into the wave by the critical oscillators. This pumping can negate, or even overcome, the effects of viscous friction and thus maintain or enhance the amount of energy carried by the wave. As the wave continues, it passes regions with smaller values of A'', and its speed of propagation $c \approx (A'l/\rho)^{1/2}$ declines. Because the energy flux is undiminished (or has even increased, due to the active pumping), the fall in wave velocity implies an increase in the amplitude of the wave. The displacement of the basilar membrane therefore grows as the wave approaches the place of resonance. In the immediate vicinity of the characteristic place, $|A|$ becomes small while $\tilde{h}$ increases. Thus the cubic term in Eq. (3.41)

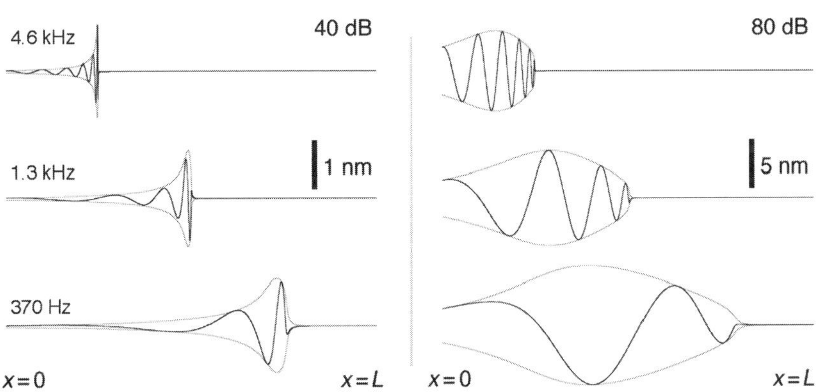

FIGURE 3.7. Active traveling waves resulting from the hydrodynamic coupling of critical oscillators located on the cochlear partition, with characteristic frequencies decreasing from base to apex. These waves are solutions to the nonlinear wave equation, Eq. (3.41).

rapidly becomes more important than the linear term. This leads to a strongly nonlinear response of the basilar membrane in this region. The wave slows to a halt as it approaches the characteristic place, and as it does so, the energy that it carries is absorbed by the active oscillators (the imaginary part of the cubic term in the response is positive). At positions beyond the characteristic place, $x > x_r$, A' becomes negative, and consequently the wave number $q \sim \omega/(A')^{1/2}$ becomes imaginary, indicating the breakdown of wave propagation. Any small amount of energy that remains in the wave is thus reflected from the characteristic place, and the displacement of the basilar membrane decays very sharply for $x > x_r$.

A cochlear partition activated by critical oscillators thus provides a natural mechanism for the phenomenon known as "critical layer absorption," whereby a wave comes to a stop at a particular place, but takes sufficiently long to do so that practically all of the energy that it carries can be absorbed there (Lighthill 1981). The traveling wave attains a peak amplitude before it reaches the resonant point, at a location $x = x_p < x_r$ that is very close to the characteristic place at low amplitudes but that moves toward the base as the level of the stimulus increases. The peak response displays the characteristic nonlinearity of critical oscillators, $\tilde{h}(x_p) \sim \tilde{p}(x_p)^{1/3}$. However, the vibration amplitude as a function of sound pressure level at a fixed position can exhibit responses that are not simple power laws. Direct obervations of the motion of the basilar membrane (Rhode 1971; Sellick et al. 1982; Ruggero et al. 1997; Robles and Ruggero 2001) reveal many features of the cochlear response that are consistent with the general properties of an active traveling wave that is driven by critical oscillators (Duke and Jülicher 2003).

5.3 Two-Tone Interference in Active Traveling Waves

In Section 2, we discussed the nonlinear interference effects that appear in the response of an individual critical oscillator when it is stimulated by multiple frequencies. These have their counterparts in an active traveling wave that is powered by a set of critical oscillators. When a sound stimulus containing two frequency components ω_1 and ω_2 (with $\omega_1 < \omega_2$) enters the cochlea, the critical oscillators on the cochlear partition respond at both those frequencies, but also at the distortion product frequencies. As a result, the traveling wave contains a whole spectrum of components (as has been observed experimentally [Robles et al. 1991; Rhode and Cooper 1993; Robles et al. 1997]), and each of these components reaches a peak amplitude at a different place. In particular, the distortion products that have a lower frequency than the stimulus tones travel farther than the latter, and peak at their characteristic place, where the principal components ω_1 and ω_2 have already decayed. This localized oscillation of the membrane at distortion-product frequencies explains why these lower frequencies, such as $2\omega_1 - \omega_2$ and $3\omega_1 - 2\omega_2$, can be heard (Helmholtz 1954; Zwicker and Fastl 1999). At the same time, the peak response to ω_1 is diminished by the presence of ω_2, and vice versa. Such two-tone suppression has been directly observed in the response of the basilar membrane (Ruggero et al. 1992;

Rhode and Cooper 1993). For stimuli with components of equal level, the suppressive effect of the lower-frequency component ω_1 is greater, because this component travels through the characteristic place of the other component ω_2 and interferes with the response of the critical oscillators there. The response of a cochlea in which the active elements are critical oscillators thus accounts for the main features of the psychoacoustic phenomenon of two-tone suppression (Zwicker and Fastl 1999).

6. Discussion

Traditionally, the cochlear amplifier has been modeled using an approach that starts with a linear harmonic description and adds a positive feedback. As the feedback strength increases, the gain of the system grows and the system eventually becomes unstable. A description of the cochlear amplifier based on critical oscillators is not inconsistent with such an approach, but it has several advantages. First, because critical oscillators are inherently nonlinear systems, the description underlines that nonlinearities are closely linked to the active mechanism and are, in fact, an unavoidable byproduct of active amplification. Second, the concept of the active amplifier as a self-regulated system that automatically adjusts itself to the vicinity of a critical point solves a difficulty of traditional approaches, namely that unstable behavior must be avoided, but at the same time positive feedback is required to produce an active amplifier. Third, our analysis indicates that the characteristic frequency of oscillation depends on the properties of the underlying nonlinear dynamical system; instead of depending on the stiffness and mass of passive harmonic oscillator elements alone, the characteristic frequency is determined, to a large degree, by the typical time scales of the active cellular processes that drive the cochlear amplifier.

Previous models of the cochlear traveling wave that have been based on coupled nonlinear oscillators have encountered the problem that such a system can become chaotic. The description presented in Section 5 neatly side-steps this issue, by basing the analysis on the fundamental Fourier modes, together with the higher harmonics resulting from nonlinearities. Deterministic chaotic dynamics is irrelevant in the cochlea because stochastic fluctuations are important at the cellular scale on which the active amplifier operates. These fluctuations conceal any chaotic behavior that could result from the coupling of several nonlinear oscillators, but emphasize the generic behavior in the critical regime. Even though the Hopf bifurcation is hidden by fluctuations, the general signatures of critical oscillators are typically apparent over a wider range of parameters for noisy systems than for deterministic ones.

The noise that occurs in complex dynamical systems due to the rapid, irregular dynamics of many degrees of freedom might also provide an explanation of otoacoustic emissions, the prime indicator of active processes in the cochlea. The spontaneous noisy motion of a critical oscillator, in the absence of any external stimulus, can exhibit spectral peaks at the oscillation frequency, because the noise is amplified by the active dynamical system. In the cochlea, the noisy motion of many critical oscillators, vibrating at different frequencies, would lead to the generation of many wavelike excitations that travel along the basilar membrane and interact via nonlinearities. In such a situation, certain spectral components could be selected as eigenmodes of the cochlear cavity, and these frequencies would appear in the spectrum of spontaneous emissions (Shera and Guinan, Chapter 9; Shera 2003).

In conclusion, the concept of self-regulated critical oscillators provides a unifying description of the principal features of sound detection in auditory systems. The essential singularity in the response of an individual critical oscillator provides the frequency selectivity, high sensitivity, and a wide dynamic range required by hearing organs. Clear experimental evidence that hair bundles in the bullfrog sacculus operate as critical oscillators has been presented. As yet, there is no such evidence at the cellular level for critical oscillators in the mammalian cochlea. But we have shown that critical oscillators that cover a range of frequencies, that are located within the cochlea partition, and whose dynamics is coupled through the fluid can account for the transport of energy to a particular frequency-dependent location in the cochlea, the nonlinearity of the response at that place, the generation of distortion products, and the existence of spontaneous otoacoustic emissions. The close correspondence with the measured features of the cochlear response provides a compelling argument that critical oscillators are the key element of the active cochlear amplifier.

References

Ashmore J (1987) A fast motile response in guinea-pig outer hair cells: The cellular basis of the cochlear amplifier. J Physiol 388:323–347.

Benser ME, Marquis RE, Hudspeth AJ (1996) Rapid active hair bundle movements in hair cells from the bullfrog's sacculus. J Neurosci 16:5629–5643.

Boukas EK, Liu ZK (2002) Deterministic and Stochastic Time-Delay Systems. Boston: Birkhäuser.

Brownell WE, Bader CR, Bertrand D, de Ribaupierre Y (1985) Evoked mechanical responses of isolated cochlear outer hair cells. Science 227:194–196.

Camalet S, Duke T, Jülicher F, Prost J (2000) Auditory sensitivity provided by self-tuned critical oscillations of hair cells. Proc Natl Acad Sci USA 97:3138–3188.

Chan DK, Hudspeth AJ (2005) Ca^{2+} current-driven nonlinear amplification by the mammalian cochlea in vitro. Nat Neurosci 8:149–155.

Choe Y, Magnasco MO, Hudspeth AJ (1998) A model for amplification of hair-bundle motion by cyclical binding of Ca^{2+} to mechanoelectrical-transduction channels. Proc Natl Acad Sci USA 95:15321–15326.

Coullet P, Goldstein RE, Gunaratne GH (1989) Parity-breaking transitions of modulated patterns in hydrodynamic systems. Phys Rev Lett 63:1954–1957.

Crawford AC, Fettiplace R (1985) The mechanical properties of ciliary bundles of turtle cochlear hair cells. J Physiol 364: 359–379.

Dallos P (1992) The active cochlea. J Neurosci 12:4575–4585.

Dallos P, Popper AN, Fay RR (eds) (1996) The Cochlea. New York: Springer-Verlag.
de Boer E (1980) Auditory physics. Physical principles in hearing theory. Phys Rev 62:87–174.
de Boer E (1996) Mechanics of the cochlea. In: Dallos P, Popper AN, Fay RR (eds.), The Cochlea. New York: Springer-Verlag, pp. 258–317.
Denk W, Webb WW (1992) Forward and reverse transduction at the limit of sensitivity studied by correlating electrical and mechanical fluctuations in frog saccular hair cells. Hear Res 60:89–102.
Duifuis H, Hoogstraten HW, van Netten SM, Diependaal RJ, Bialek W (1985) Modelling the cochlear partition with coupled van der Pol oscillators. In: Allen JB (ed) Peripheral Auditory Mechanisms. New York: Springer-Verlag, pp. 290–297.
Duke T, Jülicher F (2003) Active traveling wave in the cochlea. Phys Rev Lett 90: 158101(1–4).
Eberl DF (1999) Feeling the vibes: chordotonal mechanisms in insect hearing. Curr Opin Neurobiol 9:389–393.
Eguiluz V, Ospeck M, Choe Y, Hudspeth AJ, Magnasco MO (2000) Essential nonlinearities in hearing. Phys Rev Lett 84:5232–5235.
Forster D. (1990) Hydrodynamic Fluctuations, Broken Symmetry and Correlation Functions. Redwood City, CA: Addison-Wesley.
Gold T (1948) The physical basis of the action of the cochlea. Proc R Soc B 135:492–498.
Goldstein JL (1967) Auditory nonlinearity. J Acoust Soc A 41:676–689.
Göpfert MC, Robert D (2003) Motion generation by *Drosophila* mechanosensory neurons. Proc Natl Acad Sci USA 100:5514–5519.
Göpfert MC Humphris ADL, Albert JT, Robert D Hendrich O (2005) Power gain exhibited by motile mechanosensory neurons in *Drosophila* ears. Proc Natl Acad Sci USA 102:325–330.
Greenwood DD (1990) A cochlear frequency-position function for several species—29 years later. J Acoust Soc Am 87:2592–2605.
Helmholtz H (1954) On the Sensation of Tone. New York: Dover.
Holt JR, Gillespie SK, Provance DW, Shah K, Shokat KM, Corey DP, Mercer JA, Gillespie PG (2002) A chemical-genetic strategy implicates myosin-1c in adaptation by hair cells. Cell 108: 371–381.
Howard J, Hudspeth AJ (1987) Mechanical relaxation of the hair bundle mediates adaptation in mechanoelectrical transduction by the bullfrog's saccular hair cell. Proc Natl Acad Sci USA 84:3064–3068.
Howard J, Hudspeth AJ (1988) Compliance of the hair bundle associated with gating of mechanoelectrical transduction channels in the bullfrog's saccular hair cell. Neuron 1:189–199.
Hudspeth AJ (1989) How the ear's works work. Nature 341:397–404.
Hudspeth AJ (1997) Mechanical amplification of stimuli by hair cells. Curr Opin Neurobiol 7: 480–486.
Hudspeth AJ, Gillespie PG (1994) Pulling springs to tune transduction: adaptation by hair cells. Neuron 12:1–9.
Jülicher F, Prost J (1997) Spontaneous oscillations of collective molecular motors. Phys Rev Lett 78:4510–4513.
Jülicher F, Andor D, Duke T (2001) Physical basis of two-tone interference in hearing. Proc Natl Acad Sci USA 98:9080–9085.
Kachar B, Brownell WE, Altschuler R, Fex J (1986) Electrokinetic shape changes of cochlear outer hair-cells. Nature 322:365–368.

Kachar B, Parakkal M, Kurc M, Zhao Y, Gillespie PG (2000) High-resolution structure of hair-cell tip links. Proc Natl Acad Sci USA 97:13336–13341.

Kennedy HJ, Crawford AC, Fettiplace R (2005) Force generation by mammalian hair bundles supports a role in cochlear amplification. Nature 433:880–883.

Kolston PJ, de Boer E., Viergever MA, Smoorenburg GF (1990) Realistic mechanical tuning in a micromechanical cochlear model. J Acoust Soc Am 88:1794–1801.

Köppl C (1995) Otoacoustic emissions as an indicator for active cochlear mechanics: a primitive property of vertebrate auditory organs. In: Manley GA, Klump GM, Kppl C, Fastl H, Oeckinghaus H (eds) Advances in Hearing Research, Singapore: World Scientific, pp. 207–218.

Kros CJ, Marcotti W, van Netten SM, Self TJ, Libby RT, Brown SD, Richardson GP, Steel KP (2002) Reduced climbing and increased slipping adaptation in cochlear hair cells of mice with Myo7a mutations Nat Neurosci 5:41–47.

Lighthill J (1981) Energy-flow in the cochlea. J Fluid Mech 106:149–213.

Ma SK (2000) Modern Theory of Critical Phenomena. New York: Perseus Books.

Magnasco MO (2003) A wave traveling over a Hopf instability shapes the cochlear tuning curve. Phys Rev Lett 90:058101-1–058104-4.

Manley GA (2000) Cochlear mechanisms from a phylogenetic viewpoint. Proc Natl Acad Sci USA 97:1173–11743.

Manley GA (2001) Evidence for an active process and a cochlear amplifier in nonmammals. J Neurophysiol 86:541–549.

Manley GA, Köppl C (1998) Phylogenetic development of the cochlea and its innervation. Curr Opin Neurobiol 8:468–474.

Manley GA, Kirk DL, Köppl C, Yates GK (2001) In vivo evidence for a cochlear amplifier in the hair-cell bundle of lizards. Proc Natl Acad Sci USA 98:2826–2831.

Manley GA, Popper AN, Fay RR, eds (2004) Evolution of the Vertebrate Auditory System, Springer Handbook of Auditory Research. New York: Springer-Verlag.

Martin P, Hudspeth AJ (1999) Active hair-bundle movements can amplify a hair cell's response to oscillatory mechanical stimuli. Proc Natl Acad Sci USA 96:14306–14311.

Martin P, Hudspeth AJ (2001) Compressive nonlinearity in the hair bundle's active response to mechanical stimulation. Proc Acad Natl Sci USA 98:14386–14391.

Martin P, Metha AD, Hudspeth AJ (2000) Negative hair-bundle stiffness betrays a mechanism for mechanical amplification by the hair cell. Proc Natl Acad Sci USA 97:12026–12031.

Martin P, Hudspeth AJ, Jülicher F (2001) Comparison of a hair bundle's spontaneous oscillations with its response to mechanical stimulation reveals the underlying active process. Proc Natl Acad Sci USA 98:14380–14385.

Martin P, Bozovic D, Choe Y, Hudspeth AJ (2003) Spontaneous oscillation by hair bundles of the bullfrog's sacculus. J Neurosci 23:4533–4548.

Moreau L, Sontag E, Arcat M (2003) Feedback tuning of bifurcations. Syst Contr Lett 50:229–239.

Nadrowski B, Martin P Jülicher F (2004) Active hair-bundle motility harnesses noise to operate near an optimum of mechanosensitivity. Proc Natl Acad Sci USA 101: 12195–12200.

Nilsen KE, Russell IJ (2000) The spatial and temporal representation of a tone on the guinea pig basilar membrane. Proc Natl Acad Sci USA 97:11751–11758.

Pickles JO, Comis SD, Osborne MP (1984) Cross-links between sterocilia in the guinea-pig organ of Corti, and their possible relation to sensory transduction. Hear Res 15: 103–112.

Probst R (1990) Otoacoustic emissions: an overview. Adv Otorhinolaryngol 44:1–91.
Rhode WS (1971) Observations of the vibration of the basilar membrane in squirrel monkeys using the Mossbauer technique. J Acoust Soc Am 49:1218–1231.
Rhode WS, Cooper NP (1993) Two-tone suppression and distortion production on the basilar membrane in the hook region of the cat and guinea pig cochleae. Hear Res 66:31–45.
Ricci AJ, Crawford AC Fettiplace R. (2000) Active hair bundle motion linked to fast transducer adaptation in auditory hair cells. J Neurosci 20:7131–7142.
Risler T, Prost J, Jülicher F (2004) Universal critical behavior of noisy coupled oscillators. Phys Rev Lett 93:175702–175704.
Risler T, Prost J, Jülicher F(2005) Universal critical behavior of noisy coupled oscillators: a renormalization group study. Phys Rev E 72:016130–1 to 016130–18.
Robles L, Ruggero M (2001) Mechanics of the mammalian cochlea. Physiol Rev 81: 1305–1352.
Robles L, Ruggero MA, Rich NC (1991) Two-tone distortion in the basilar membrane of the cochlea. Nature 349:413–414.
Robles L, Ruggero MA, Rich NC (1997) Two-tone distortion on the basilar membrane of the chinchilla cochlea. J Neurophysiol 77:2385–2399.
Ruggero MA, Robles L, Rich NC (1992) Two-tone suppression in the basilar membrane of the cochlea: mechanical basis of auditory-nerve rate suppression. J Neurophysiol 68:1087–1099.
Ruggero MA, Rich NC, Recio A, Narayan SS, Robles L (1997) Basilar-membrane responses to tones at the base of the chinchilla cochlea. J Acoust Soc Am 101: 2151–2163.
Sellick PM, Patuzzi R, Johnstone BM (1982) Measurement of basilar membrane motion in the guinea pig using the Mössbauer technique. J Acoust Soc Am 72: 131–141.
Shera CA (2003) Mammalian spontaneous otoacoustic emissions are amplitude-stabilized cochlear standing waves. J Acoust Soc Am 114:244–262.
Stoop R, Kern A (2004) Two-tone suppression and combination tone generation as computations performed by the Hopf cochlea. Phys Rev Lett 93:268103–268104.
Stratonovich RL (1981) Topics in the Theory of Random Noise, vol. II. New York: Gordon and Breach.
Strogatz SH (2001) Nonlinear Dynamics and Chaos. New York: Perseus Books.
van der Pol B (Nov. 1926) On relaxation oscillations. Philos Mag 2:978–992.
van Kampen NG (1992) Stochastic Processes in Physics and Chemistry. Amsterdan: North-Holland.
Vilfan A, Duke T (2003a) Instabilities in the transient response of muscle. Biophys J 85:818–827.
Vilfan A, Duke T (2003b) Two adaptation processes in auditory hair cells together can provide an active amplifier. Biophys J 85:191–203.
von Ɓekésy G (1960) Experiments in Hearing. New York: McGraw-Hill.
Wiggins S (2003) Introduction to Applied Nonlinear Dynamics and Chaos. New York: Springer.
Wu YJ, Ricci AJ, Fettiplace R (1999) Two components of transducer adaptation in auditory hair cells. J Neurophysiol 82:2171–2181.
Zweig G (1976) Basilar membrane motion. Cold Spring Harbor Symp Quant Biol 40: 619–633.

Zweig G (1991) Finding the impedance of the organ of Corti. J Acoust Soc Am 89: 1229–1254.

Zwicker E, Fastl H. (1999) Psychoacoustics. Berlin: Springer-Verlag.

Zwislocki J (1948) Theorie der Schneckenmechanik: Qualitative und quantitative Analyse Acta Otolaryngol Suppl 72:1–76.

4
Active Hair-Bundle Motility of the Hair Cells of Vestibular and Auditory Organs

Pascal Martin

1. Introduction

The hair bundle is the mechanosensitive organelle that protrudes from the hair cell's apical surface. Sound in hearing organs, head acceleration in vestibular organs and water flow in the lateral line of fish and amphibians ultimately evoke hair-bundle deflections, to which the hair cell responds by generating an electrical signal. Mechanical-to-electrical transduction, a process that is fast enough to enable the hair cell to respond at auditory frequencies up to $\approx$100 kHz, most probably results from direct mechanical gating of mechanosensitive ion channels. The hair bundle is more than a passive receptor, a sort of mechanical antenna; this organelle can also convert energy of biochemical origin into mechanical work and thus act as a small engine. The present chapter focuses on active hair-bundle motility, as revealed by experiments in in vitro preparations of excised sensory epithelia from the vertebrate ear and on the relevance of motility for the detection of minute oscillatory stimuli by the internal ear.

2. The Architecture of Mechanoelectrical Transduction

2.1 Hair-Bundle Structure

The hair bundle is a tuft of a few tens to a few hundreds of cylindrical processes, called *stereocilia*, which project from the hair cell's apical surface (Fig. 4.1). Each stereocilium,[1] whose length ranges typically from one to ten microns and

[1]In cell biology, a true "cilium" is based on microtubules and designates a structure that can sometimes beat. An example is given by the flagellum of sperm cells, but ciliated cells are also found in many other epithelial tissues. The denomination "stereocilium" for an actin-rich structure is thus misleading. Stereocilia are similar to the actin-based microvilli of intestinal epithelial cells, and it has been proposed that they be called "stereovilli." Here the most common, though imperfect, usage "stereocilia" is followed.

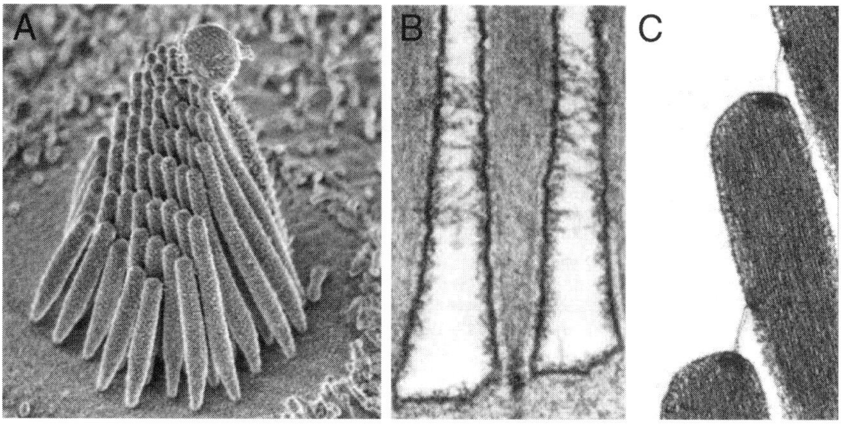

FIGURE 4.1. Hair-bundle structure. (**A**) Staircase pattern of the actin-based stereocilia that protrude from the apical surface of a hair cell from the bullfrog's saccule. The bulb of the microtubule-based kinocilium stands at a distance of 6.7 μm from the cell's surface (scanning electron micrograph provided by Dr. Peter Gillespie, Oregon Science and Health University, Portland, OR). (**B**) Stereocilia taper near their point of insertion in the cuticular plate and are interconnected by lateral links. These links are severed by the enzymatic treatment (subtilisin) that is often used to disconnect the hair bundle from overlying structures and are not essential to mechanoelectrical transduction. About one-tenth of the actin filaments that fill each stereocilium insert in the cuticular plate, where they form a rootlet. The diameter of an unfixed stereocilium is ≈450 nm (bullfrog's saccule; from Jacobs and Hudspeth 1990). (**C**) Tip links in the bullfrog's saccule. Tip links can be quickly severed by application of tetracarboxylate chelators of Ca^{2+}, for instance 1,2-Bis(2-aminophenoxy)ethane-N,N,N′,N′-tetra-acetic acid (BAPTA) (Assad et al. 1991), but mechanoelectrical transduction is not affected by lowering the calcium concentration down to a few tens of nanomolars with antipyrylazo III or other nontetracarboxylate chelators of Ca^{2+} (Marquis and Hudspeth 1997). (Transmission electron micrograph provided by Dr. A.J. Hudspeth, Howard Hugues Medical Institute at the Rockefeller University, New York, NY.)

whose width is in the range of one hundred to a few hundreds nanometers (Tilney and Saunders 1983), contains a dense para-crystalline array of cross-linked actin filaments oriented in parallel to the long axis of the stereocilium (Tilney et al. 1980), sheathed in the hair cell's plasma membrane like a finger in a glove. A stereocilium tapers near its base. Only a few tens out of a few hundred actin filaments penetrate the actin-rich cuticular plate at the hair cell's apical surface, where they form a rootlet. Arranged in rows of increasing height, the stereocilia form a staircase pattern with a vertical plane of bilateral symmetry. Stereocilia are interconnected by numerous lateral links (Pickles et al. 1984; Jacobs and Hudspeth 1990; Goodyear and Richardson 1992, 1999, 2003). In particular, the tip of each stereocilium is attached to the flank of the nearest taller neighbor by a single fine filament called the tip link (Pickles et al. 1989; Nagel et al.

1991), which plays a critical role in mechano-electrical transduction (Pickles et al. 1984).

In front of the tallest row of stereocilia, some hair bundles also display one true cilium—called the kinocilium—that contains an axoneme, i.e., 9 + 1 doublets of microtubules. The kinocilium is not essential to mechano-electrical transduction (Jacobs and Hudspeth 1990); its role is obscure. Although kinocilia are present during development, they recede and then disappear as the hair cells mature in mammalian and many avian auditory hair cells.

2.2 General Features of Bundle Motion

Upon deflection, the stereocilia do not bend or buckle (Fig. 4.2). Instead, each stereocilium pivots about its point of insertion in the cuticular plate (Flock et al. 1977; Hudspeth 1983a; Howard 2001). This movement thus induces shearing between adjacent stereocilia, which presumably slide along one another in their region of approximation. The hair bundle appears to move as a unit; high-resolution video microscopy in the cochlea of the turtle *Pseudemys scripta elegans* (Crawford et al. 1989) as well as interferometric recordings of the movements of the stereocilia at the opposite edges of single hair bundles from the saccule of the bullfrog *Rana catesbeiana* (Kozlov et al. 2006) have failed to reveal splaying or other internal modes of movement within the hair bundle.

That a force applied at the bundle's top evokes a movement of the whole cluster of stereocilia demonstrates that the stimulus propagates across the hair bundle. Two ultrastructural properties of the hair bundle might explain this observation. First, interciliary links could distribute mechanical stress across stereocilia. If appreciably elastic, however, these links would allow the bundle to splay apart when a force is applied only at the bundle's top. Second, at least for the hair cells of the saccule of the bullfrog, the cuticular

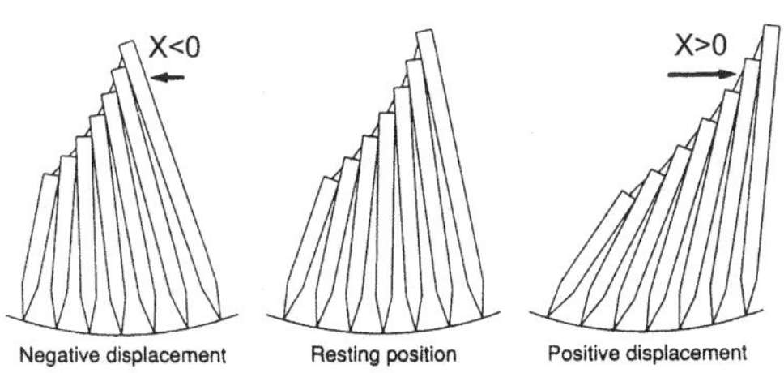

FIGURE 4.2. Schematic representation of hair-bundle motion in the bullfrog's saccule. The positions of seven stereocilia are represented in the plane of bilateral symmetry of a hair bundle. In response to a horizontally directed deflection (→) of amplitude X, each stereocilium pivots about its insertion point in the cuticular plate and the hair bundle appears to move as a unit. This movement induces shearing between adjacent stereocilia, which, assuming that the stereocilia are rigid rods, affects the lengths of the tip links: a positive stimulus extends the tip links, whereas a negative stimulus has the opposite effect. Note that the cuticular plate is convex and that the stereocilia are in close approximation with one another near the hair bundle's top. (From Jacobs and Hudspeth 1990.)

plate is convex (Hudspeth 1983b; Jacobs and Hudspeth 1990). Because each stereocilium is inserted orthogonally to this curved plate, the stereocilia must push against each other near their tips and thus be spring-loaded (Fig. 4.2). Although of unrecognized significance in other organs, this geometrical constraint forces stereocilia into contact and prevents splaying (Kozlov et al. 2006).

2.3 Mechanoelectrical Transduction

The hair bundle is a mechanical antenna, which, when deflected by an external force, generates an electrical signal (Hudspeth and Corey 1977; Hudspeth 1989). The transduction currents elicited by a series of step displacements of varying magnitudes describe a sigmoidal relation. The range of mechanosensitivity spans only a few hundreds of nanometers of bundle displacements, which amounts to a couple of degrees of bundle angular deflection. In the following sections, some essential features of the mechanoelectrical process are introduced, as well as a theoretical framework to describe how transduction happens.

2.3.1 The Ionic Basis of Transduction

Transduction is mediated by mechanosensitive ion channels that are embedded in the plasma membrane near the tip of the stereocilia (Hudspeth 1982; Jaramillo and Hudspeth 1991; Denk et al. 1995; Lumpkin and Hudspeth 1995). These transduction channels pass a variety of small monovalent and divalent cations with poor selectivity (Corey and Hudspeth 1979a; Ohmori 1985; Howard et al. 1988), except for Ca^{2+}, which acts as a permeant blocker of the channel's pore (Crawford et al. 1991; Lumpkin et al. 1997; Ricci et al. 1998). By finding circumstances under which one and only one transduction channel in a hair bundle retained its ability to gate, it became possible to directly measure a unitary conductance of $\approx$ 80–300 pS (Crawford et al. 1991; Géléoc et al. 1997; Ricci et al. 2003). Because the unitary conductance is large, a few tens of open transduction channels per hair cell would be enough to produce significant receptor potentials. Noise analysis of transduction currents by hair cells from the bullfrog's saccule indeed indicates that there are fewer than four operational channels per stereocilium (Holton and Hudspeth 1986). The greatest transduction conductances reported for voltage-clamped hair cells of the turtle's basilar papilla correspond to about one operational channel per stereocilium (Crawford et al. 1989). Although damage inflicted upon the delicate hair bundles during dissection and stimulation must reduce the number of operational channels in vitro, each stereocilium is probably endowed with at most only a few transduction channels in vivo, possibly a single one.

In vivo, hair cells are inserted in an epithelium that separates fluids of differing ionic compositions. The hair-cell somata are immersed in Na^+-rich perilymph, whereas the cells' apical surfaces, where hair bundles are situated, are bathed by K^+-rich, Na^+-poor endolymph that is also generally characterized by a reduced Ca^{2+} concentration ($[Ca^{2+}] \approx$ 30–300 μM) compared to other extracellular fluids

($[Ca^{2+}] \approx 1$ mM). Probably because Ca^{2+} obstructs the way for the more abundant K^+ ions, lower Ca^{2+} concentrations in the ionic milieu bathing the hair cells' apical surfaces yield higher hair-bundle conductances. In response to saturating bundle deflections, the transduction current is two to three times larger when hair bundles are immersed in artificial endolymph with $[Ca^{2+}] \approx 0.1$ mM than when they are exposed to artificial perilymph with $[Ca^{2+}] \approx 1$ mM (Corey and Hudspeth 1983b; Crawford et al. 1991). Correspondingly, the conductance of a single transduction channel increases from a value $\approx$100 pS when measured in $[Ca^{2+}] \approx 1$ mM (Crawford et al. 1991; Géléoc et al. 1997) to about twice that value in artificial endolymph (Ricci et al. 2003).

2.3.2 Transduction Kinetics

As required by auditory sensitivities extending above 100 kHz in animals such as whales and bats (Sales and Pye 1974), the mechanoelectrical process is very fast compared to other sensory modalities. In the bullfrog's saccule, abrupt step deflection of an ensemble of a few hundred hair cells evokes a transduction current after a delay of only 25 μs at 28°C (Corey and Hudspeth 1979b, 1983a). The time constant τ_C of the approach to steady state depends on the magnitude of the hair-bundle displacement and temperature (Corey and Hudspeth 1983a). At 4°C, a temperature low enough that transduction kinetics are limited by the channels' gating and not by the speed of stimulation, this time constant ranges from 100 μs for saturating deflections of $\approx 1\,\mu$m to 500 μs for small deflections of $\approx 0.1\,\mu$m. The dependence on temperature of the channel's relaxation kinetics in response to a step from $-0.4\,\mu$m to $+0.2\,\mu$m is well described by an Arrhenius equation

$$\tau_C(T) = \tau_0 \exp\left(\frac{\Delta G_b}{k_B T}\right), \tag{4.1}$$

in which $1/\tau_0 \approx 6 \cdot 10^{12}\,\text{s}^{-1}$ is an intrinsic frequency factor and $\Delta G_b \approx 80$ zJ $(= 20 k_B T$ at room temperature) is an energy barrier that depends on the stimulus magnitude but not, or weakly, on temperature. This result implies that τ_C is roughly halved when the temperature increases by 10°C:

$$Q_{10} = \frac{\tau_C(T+10)}{\tau_C(T)} \approx 0.5. \tag{4.2}$$

The transduction current from single hair cells from the turtle's cochlea shows very similar kinetic properties (Crawford et al. 1989; Ricci et al. 2003, 2005), but that from the mammalian cochlea is significantly faster with $\tau_C < 50$ μs for small displacements at room temperature with 1.5 mM Ca^{2+} in the artificial endolymph (Kennedy et al. 2003; Ricci et al. 2005). In addition, activation kinetics of the transduction channels depends on Ca^{2+}. In the turtle's cochlea, τ_C increases by twofold from 0.37 ms to 0.75 ms for small bundle displacements when the Ca^{2+} concentration in endolymph is lowered from 2.8 mM to 0.05 mM (Ricci et al. 2005).

2.3.3 The Gating-Spring Hypothesis

That transduction channels respond so quickly to hair-bundle displacements suggests a direct mechanical activation of the transduction apparatus without the intervention of an intermediate enzyme or second messenger (Corey and Hudspeth 1979b, 1983a). In this scheme, mechanical stimuli affect the extension of elastic mechanical linkages—the "gating springs"—attached to, or part of, the transduction channels (Fig. 4.3). In the simplest formulation of the gating-spring model (Corey and Hudspeth 1983a; Howard et al. 1988; Howard et al. 1988), the channel is either open or closed. The free-energy difference $\Delta G = G^{\mathrm{OPEN}} - G^{\mathrm{CLOSED}}$ between closed and open states of one transduction channel together with its gating spring is the sum of two contributions,

$$\Delta G = \Delta\mu + \Delta G_{\mathrm{el}}(X) \tag{4.3}$$

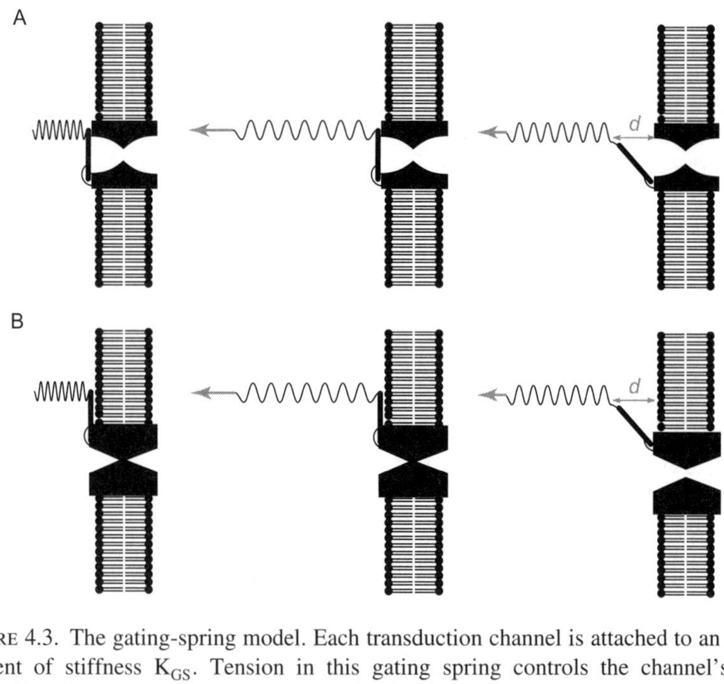

FIGURE 4.3. The gating-spring model. Each transduction channel is attached to an elastic element of stiffness K_{GS}. Tension in this gating spring controls the channel's open probability. If tension in the spring is low (**left**), the channel spends most of its time in a closed state. Pulling on the spring (*gray arrow*) raises the channel's open probability. A gating spring shortens by a distance d when a channel opens, thereby producing a gating force $K_{GS}d$ that reduces tension in the spring (Section 5.1). Although the gating swing can be depicted as a movement of a gate that obstructs the channel's pore (**A**), its exact molecular nature is unknown. For instance, this conformational change could be mediated by the movement of a molecular lever associated with channel gating and thus be of much larger magnitude than that corresponding to the opening of the channel's pore (**B**).

where $\Delta\mu$ is the energy difference between open and closed states in the absence of a gating spring and ΔG_{el} accounts for the elastic energy stored in the gating spring. We denote by X the hair-bundle deflection evoked by an external force acting on the hair bundle along its horizontal axis of mechanosensitivity in the bundle's plane of bilateral symmetry (Shotwell et al. 1981a); positive displacements ($X > 0$) are here defined as excitatory and thus correspond to opening of the transduction channels. As required for a displacement-sensitive apparatus, channel gating must be associated with a conformational change that affects the extension of the gating spring. Assuming that a gating spring shortens by the distance d when a channel opens and that the stiffness $\mathrm{K_{GS}}$ of a gating spring is the same in the open and in the closed states,

$$\begin{aligned} G_{\mathrm{el}}^{\mathrm{CLOSED}} &= \tfrac{1}{2}\mathrm{K}_{GS}\left(x_r + x + \tfrac{d}{2}\right)^2, \\ G_{\mathrm{el}}^{\mathrm{OPEN}} &= \tfrac{1}{2}\mathrm{K}_{GS}\left(x_r + x - \tfrac{d}{2}\right)^2, \end{aligned} \tag{4.4}$$

with x_r the gating-spring extension at rest ($X = 0$) when the channel is midway between open and closed and $x = \gamma X$ the gating-spring extension evoked by the external force. The geometric projection factor γ relates the extension of the gating spring to the hair-bundle displacement X. It follows that the free-energy difference between the open and the closed states is a linear function of hair-bundle displacement,

$$\Delta G(X) = -ZX + \Delta G(0), \tag{4.5}$$

in which the magnitude of the gating force $Z = \gamma \mathrm{K_{GS}} d$ characterizes the bundle's sensitivity to displacement. Assuming that the transduction channels equilibrate fast enough to be at equilibrium, the open probability P_{o} of a transduction channel is given by a Boltzmann distribution

$$P_o = \frac{1}{1 + \exp\left(\frac{\Delta G}{k_B T}\right)} = \frac{1}{1 + \exp\left(-\frac{Z(X - X_0)}{k_B T}\right)}, \tag{4.6}$$

with $X_0 = \frac{\Delta\mu}{Z} - \frac{x_r}{\gamma}$ the position at which half the channels are open. This simple model thus indicates that the average conductance of a hair bundle is a sigmoidal function of displacement. The farther the bundle is deflected in the positive direction, the greater the tension in the gating spring and the lower the energy of the open state with respect to the closed; a positive stimulus thus biases the channel to spend more time open. A key feature of all hair cells is that small bundle deflections X can elicit significant transduction currents. This is possible only if the gating spring bears sufficient tension at rest to open a fraction of the transduction channels. For a resting open probability $P_o = 0.5$, Eq. (4.6) yields the condition $x_r = \frac{\Delta\mu - k_B T}{\mathrm{K_{GS}} d} \approx \frac{\Delta\mu}{\mathrm{K_{GS}} d}$, with the reasonable assumption that $\Delta\mu >> k_{\mathrm{B}} T$ (Hudspeth 1992). The virtual mechanical work to extend a gating spring to its resting extension against a constant force given by the gating force of a single channel, $(\mathrm{K_{GS}} d)\, x_{\mathrm{r}}$, must be comparable to the intrinsic energy difference $\Delta\mu$ between open and closed states of the channel.

Because the height of the energy barrier between closed and open states contains a fraction δ_{OC} of the energy difference ΔG, transduction kinetics also depend on displacement. From Eyring's theory of transition rates (Hänggi et al. 1990), the transition rate k_{CO} from closed to open states (respectively k_{OC} from open to closed states) is given by

$$\begin{aligned} k_{CO} &= k_0 \exp\left(-\frac{G_b^{\varnothing}+\delta_{CO}\Delta G(X)}{k_B T}\right), \\ k_{CO} &= k_0 \exp\left(-\frac{G_b^{\varnothing}-(1-\delta_{CO})\Delta G(X)}{k_B T}\right), \end{aligned} \tag{4.7}$$

in which k_0 is an attempt rate and $G_b^{\varnothing}$ is the intrinsic energy barrier. The time constant τ_C for equilibration of the transduction channels' open probability after the hair bundle has been abruptly deflected is given by $\tau_C = \frac{1}{(k_{CO}+k_{OC})}$. Activation kinetics of the transduction channels thus impose first-order low-pass filtering on mechanoelectrical transduction with a cutoff frequency $(2\pi\tau_C)^{-1}$ (Ricci et al. 2005). For large negative displacements, the time constant of equilibration of the transduction current no longer depends on bundle displacement (Corey and Hudspeth 1983a). This finding suggests that the gating spring works in extension but not in compression, which would be the case if the gating spring became slack for sufficiently large negative displacements of the hair bundle ($X \approx -100$ nm).

Although a two-state model suffices to account for the basic features of mechanoelectrical transduction, it does not predict the sigmoidal onset of the transduction channels' activation kinetics and the detailed shape of the curve that relates transduction currents to the hair-bundle displacements. Experimental current-displacement curves often saturate more gradually for positive than for negative displacements, whereas Eq. (4.6) produces a symmetrical sigmoid. To accommodate this observation, the two-state model can be expanded by adding a second closed state (Markin and Hudspeth 1995).

2.4 The Molecules of Transduction

2.4.1 The Transduction Channel

Perhaps because the vertebrate ear is endowed with only a few tens of thousands of hair cells, each containing a small number of transduction channels, the molecular identity of the channel has remained mysterious. Putative candidates have emerged from recent progress in invertebrates, after one ion channel from the transient–receptor–potential (TRP) family was found to mediate, at least in part, mechanosensation in the bristle of the fly with similar kinetics and sensitivity as in the hair cell (Walker et al. 2000; Strassmaier and Gillespie 2002). Following this breakthrough, expression of candidate genes from the TRP family was tested in the hair cell from mice, and one member, called *TRPA1*, was found to have the right properties to be a component of the transduction channel (Corey et al. 2004). In particular, the onset of expression coincides developmentally with that of mechanosensitivity and antibodies against *TRA1* label the hair bundles, especially at their tips. Moreover, inhibition of *TRPA1* expression in zebrafish results in a significant diminution of the microphonic potentials elicited by mechanical stimulation. Knockout mice, in which the essential exons for proper function of the *TRPA1* gene are missing, have recently been generated (Kwan et al. 2006). Unfortunately, these modified mice display normal hearing and no detectable impairment of the transduction process. The molecular nature of the transduction channel is thus still elusive.

2.4.2 The Gating Spring

Three lines of evidence implicate the tip link as a component of the gating spring. First, these oblique filaments are oriented along an axis that coincides with the hair cell's axis of maximal sensitivity (Shotwell et al. 1981; Pickles et al. 1984). Second, their points of insertion are near the sites of entry of the transduction current into the hair cell (Hudspeth 1982; Lumpkin and Hudspeth 1995). Third, chemical disruption of these links results in a rapid and complete loss of mechanoelectrical transduction (Assad et al. 1991), which can be restored if the tip links are allowed to regenerate (Zhao 1996). Although the tip link is ideally placed to constitute the gating spring, electron micrographs reveal a structure that is probably too stiff and inextensible (Kachar et al. 2000; Tsuprun et al. 2004). The recent proposal that cadherin 23 might be a component of the tip link (Siemens et al. 2004; Sollner et al. 2004), together with molecular dynamics simulations of the response of cadherin 23 filaments to an applied force, buttresses the conclusion that the tip link is not compliant enough to be the gating spring (Sotomayor et al. 2005). If the tip link is not the gating spring, a compliant linkage would have to be connected in series to the tip link. This linkage could be a part of the transduction channel itself, as was proposed for a channel (TRPN1, previously called NOMP-C) that mediates mechanosensation in the bristle sensilla of *Drosophila* (Walker et al. 2000). TRPN1 is indeed endowed with an unusual intracellular domain composed of 29 ankyrin repeats at its N terminus. These 29 repeats are expected to form almost exactly one turn of a helix, a structure expected to display a similar stiffness to that of the gating spring (Corey et al. 2004; Howard and Bechstedt 2004). Finally, because they must anchor the transduction apparatus to the actin filaments of a stereocilium (Hudspeth and Gillespie 1994), the motor molecules that mediate adaptation (see Section 3.4.1) might also contribute to the gating spring. In the bullfrog, the adaptation motor is composed of a few dozens of myosin Ic molecules (Gillespie et al. 1993). For this nonprocessive motor, typically $\approx 5\% - 10\%$ of the myosins are strongly bound to actin at any given time, and each bound myosin should form an elastic link with a stiffness of ≈ 0.5 mN·m^{-1} (Batters et al. 2004b). The motor assembly would thus be expected to contribute a combined stiffness of 1.5 to 3 mN m^{-1}, which is comparable to the gating-spring stiffness (see Section 5.).

2.5 Geometrical Relationship Between Hair-Bundle Displacement and Gating-Spring Extension

Identifying the tip link (or a link in series) with the gating spring permits an estimate of the geometrical factor γ that relates the horizontal hair-bundle deflection X at the bundle's top with gating-spring extension x (Section 2.3.3). A tip link is so positioned that its extension nearly equals the shear between adjacent stereocilia. One can approximate

$$x = \gamma x \approx \frac{s}{h}, \tag{4.8}$$

where s is the mean interstereociliary spacing and h the height at the point of shear (Fig. 4.4). Computer simulations that account for the geometry of hair bundles in the bullfrog's saccule (Jacobs and Hudspeth 1990) indicate that over a physiological range of horizontal deflections in the bundle's plane of mirror symmetry, each tip link's extension is proportional to the bundle's displacement X. Moreover, despite the differing heights of the stereocilia, all tip links across the hair bundle undergo very nearly the same elongation for a given hair-bundle deflection, and thus receive the same stimulus: although morphologically in series, tip links operate as if they were in parallel. In the bullfrog's saccule, $\gamma = 0.14$ (Jacobs and Hudspeth 1990); for turtle auditory hair cells in the mid-frequency range, $\gamma = 0.06$ (Hackney et al. 1993); in the mouse utricle, $\gamma = 0.05$ (Stauffer et al. 2005). Because the interstereociliary distance $s \approx 1\,\mu\text{m}$ is relatively invariant within hair cells from the same organ and there is at most a fivefold variation in bundle height in a given cochlea (Fettiplace and Fuchs 1999), the geometrical gain factor γ is expected to increase by about fivefold from the low-frequency region to the high-frequency region of auditory organs.

3. Adaptation to Sustained Stimuli

Like other sensory cells, hair cells remain sensitive to small, time-varying stimuli even in the presence of a constant background that threatens to saturate the transduction process (Eatock 2000). In response to a prolonged step displacement of the hair bundle in the positive direction ($X > 0$), a hair cell generates a

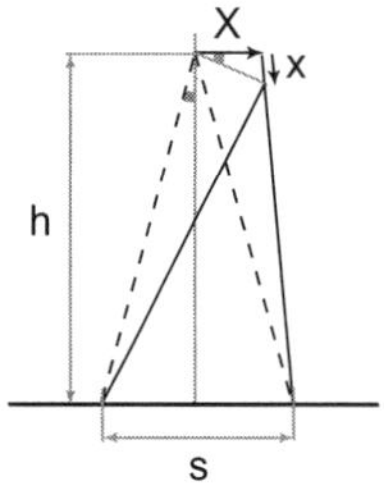

FIGURE 4.4. Geometrical relation between hair-bundle displacement X and shearing x between adjacent stereocilia. With an interstereociliary distance s, a height h at the point of shear and assuming $s << h$ (not drawn to scale for clarity), one finds similar triangles (marked by *gray bands*) that yield $x/X \approx s/h$ ($<< 1$) in the small-angle approximation.

transduction current that first increases in magnitude, reflecting the opening of transduction channels, but then declines with time (Fig. 4.5A). Conversely, a negative bundle displacement elicits first a decrease of the current flowing into the hair cell, but at later times, although the bundle position is maintained, the current increases. During the adaptation process, a brief stimulus superimposed on the first evokes the maximum transduction current again, thus demonstrating that adaptation does not result from inactivation of the transduction channels. Instead, channels reclose when they are opened by a positive step stimulus and conversely reopen for a sustained stimulus of opposite polarity. Because adaptation has now been reported in frog (Eatock et al. 1987), turtle (Crawford et al. 1989), mouse (Russell et al. 1989; Kros et al. 1992) and rat (Kennedy et al. 2003), this process is likely to be ubiquitous across vertebrate hair cells.

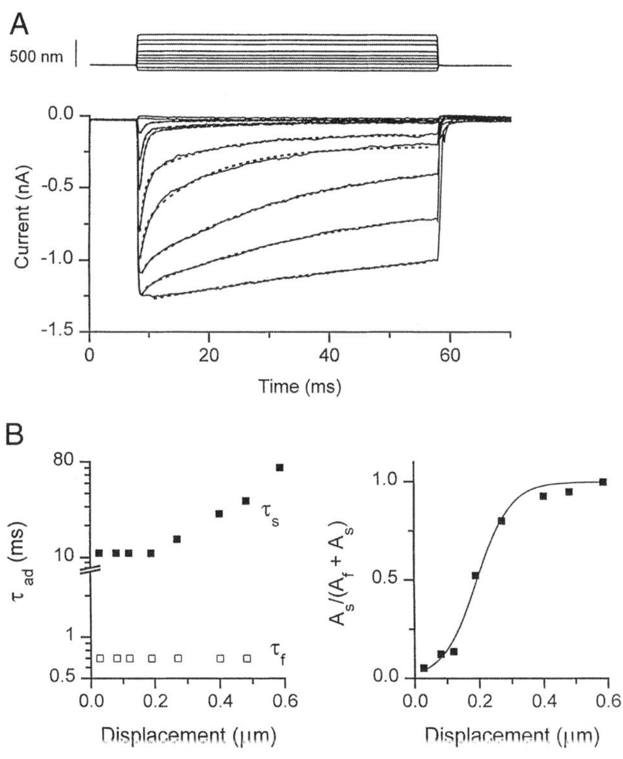

FIGURE 4.5. Adaptation. (**A**) Transduction currents (bottom) plotted as a function of time in response to step displacements (*top*) of a hair bundle from the turtle's cochlea. During a protracted stimulus, the transduction current declines over time toward its resting level. Adaptation is more rapid and complete in response to small than to large stimuli. (**B**) The current traces plotted in (**A**) were fitted by the sum of two exponentials, yielding slow (indexed "s") and fast (indexed "f") components of adaptation. The time constant (*left*) and the relative amplitude of the slow component (right) are plotted as a function of bundle displacement. (From Wu et al. 1999.)

3.1 A Biasing Process

By measuring the transduction currents evoked by a family of brief bundle movements of varying sizes (Fig. 4.6), it was shown that adaptation is associated with a shift along the displacement axis of the cell's instantaneous current-displacement relation (Corey and Hudspeth 1983b; Shepherd and Corey 1994). Adaptation thus brings the channels' open probability back toward its resting value and resets the bundle's sensitivity to a new range of displacements. For a sustained stimulus that opens a large fraction of the transduction channels, however, the transduction current fails to return fully to its resting value, and adaptation is incomplete. The current-displacement relation shifts a distance ≈80 % of an applied hair-bundle displacement of 200 nm or more (Crawford et al. 1989; Hacohen 1989). In contrast, adaptation can be nearly complete for stimuli producing less than 10 % of the maximum response (Wu et al. 1999; Kennedy et al. 2003).

3.2 Adaptation Kinetics: Fast and Slow Components of Adaptation

Adaptation kinetics are complex (Fig. 4.5). For small bundle deflections, adaptation is fast; its time course can be described by a single exponential with a time constant as short as 40 μs in outer hair cells from the rat cochlea with 1.5 mM Ca^{2+} in the artificial endolymph (Ricci et al. 2005) but usually of

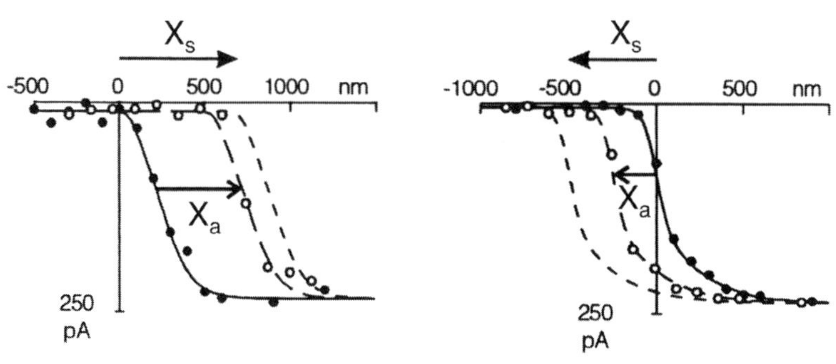

FIGURE 4.6. Adaptive shift of the current–displacement relation $I(X)$. A burst of test deflections X was applied to a hair bundle from the bullfrog's saccule, and the peak transduction currents were measured to obtain the $I(X)$ curve at the resting position (•). A static deflection $X = X_s$ was then applied to the hair bundle, and the $I(X)$ curve was measured again (o) after the bundle was allowed enough time to reach its adapted state, here 2 s. A positive deflection $X_s = +700$ nm (left) evoked a positive shift of the $I(X)$ curve by $X_a = +500$ nm, whereas a negative deflection of $X_s = -500$ nm evoked a negative shift by $X_a = -245$ nm. Adaptation is thus incomplete. A broken line shows the $I(X)$ curve that would be expected if adaptation were to be complete. (Adapted from Shepherd and Corey 1994.)

$\approx$0.3–4 ms in other species (Howard and Hudspeth 1987; Wu et al. 1999). In the turtle's cochlea, the time constant for fast adaptation is inversely related to the hair cell's characteristic frequency of operation (Fettiplace and Ricci 2003). Correspondingly, auditory hair cells generally present faster adaptation than vestibular hair cells. This suggests that the high-pass filtering provided by adaptation participates in setting the characteristic frequency of the hair cell. Activation of the transduction channels (Section 2.3.2) together with adaptation provides a (broad) band-pass filter whose center frequency corresponds to the characteristic frequency of the hair cell (Ricci et al. 2005). For larger stimuli, a single exponential is not sufficient to describe the decline of the current, and more, progressively slower, exponential components with time constants of $\approx$10–1,000 ms must be added (Hacohen 1989; Wu et al. 1999). In the displacement range of 0–800 nm, with hair cells from the bullfrog's saccule and the mouse utricule, the time course of the adaptive shift along the displacement axis of a current-displacement relation is well described by the sum of two exponential functions, thereby (artificially) separating fast and slow components of adaptation (Stauffer et al. 2005). The fast component of adaptation dominates for small positive bundle displacements ($<$ 200 nm) that open less than 60 % of the transduction channels at the peak of the transduction current, but is supplanted by the slow component as the stimulus becomes more intense and transduction nearly saturates. Fast and slow components combine linearly to produce the total extent of adaptation. A two-component decomposition of the adaptation process (Fig. 4.5B and C) was performed earlier for the turtle's cochlea by analyzing the time course of the transduction current (Wu et al. 1999).

3.3 Regulation of Adaptation

3.3.1 By Calcium

Calcium affects the kinetics and the extent of the adaptation process, as well as the open probability of the transduction channels at rest (Fig. 4.7). Reducing the Ca^{2+} concentration in the proximity of the channels' pores by lowering the extracellular Ca^{2+} concentration (Corey and Hudspeth 1983b; Eatock et al. 1987; Hacohen 1989; Holt et al. 1997), depolarizing the hair cell (Assad et al. 1989), or introducing Ca^{2+} chelators (BAPTA or EGTA) into the cell soma (Crawford et al. 1989, 1991; Ricci et al. 1998; Wu et al. 1999) slows the adaptation process and restricts its extent. Because adaptation can be abolished by depolarizing the hair cell at large positive potentials, a Ca^{2+} influx appears to be necessary for adaptation to proceed (Assad et al. 1989; Crawford et al. 1991). These findings suggest that at normal membrane potentials (≈ -60 mV), the Ca^{2+} component of the transduction current promotes the adaptive shift of the current-displacement relation in response to protracted hair-bundle deflections. At rest, the fraction of open transduction channels depends on the extracellular Ca^{2+} concentration (Corey and Hudspeth 1983b) or on the concentration of Ca^{2+} buffer in the hair cell (Ricci et al. 1998): lower Ca^{2+} concentrations result in higher open

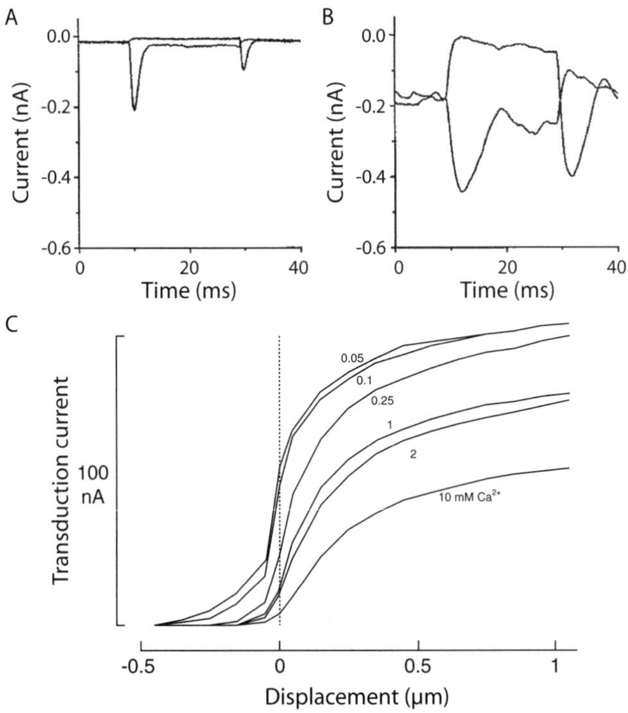

FIGURE 4.7. The effects of Ca^{2+} on adaptation. In the turtle's cochlea, reducing the calcium concentration from 2.8 mM (**A**) to 0.05 mM (**B**) in the artificial endolymph that bathes the hair bundles slows down the adaptive decline of the inward transduction current that is produced by +200 nm displacement steps of a hair bundle. A single-exponential fit to the current decay yields here characteristic time constants for the adaptation process of 1.1 ms (**A**) and 3.5 ms (**B**). Comparing the response to positive and negative steps reveals that the fraction of transduction channels that are open at rest as well as the magnitude of the maximal transduction current increase with decreasing extracellular calcium concentration. Shown here in the bullfrog's saccule (**C**), the current–displacement curve shifts positively when the Ca^{2+} concentration is increased from 50 μM to 10 mM in artificial endolymph. Conversely, the proportion of peak current that is activated at rest decreases from 52 % in low calcium to 8 % or less in high calcium. This recording was obtained by stimulating a few hundred hair bundles en masse and measuring the total transduction current that flows through the sensory epithelium. (Graphs **A** and **B** are adapted from Ricci et al. 1998; **C** is reproduced from Corey and Hudspeth 1983b.)

probabilities. Adaptive shifts of the current-displacement relation are probably controlled by the Ca^{2+} concentration at an intracellular site (Assad et al. 1989; Ricci et al. 1998). If so, this intracellular site must reside within ≈50 nm of the channels' pores to account for fast adaptation kinetics (Ricci et al. 1998). The introduction of a Ca^{2+} buffer into the hair cell affects both the channels' open probability and adaptation kinetics. These two parameters, however, can be manipulated independently.

3.3.2 By the cAMP Second-Messenger Pathway

When an analogue of cAMP is introduced into a hair cell from the turtle's cochlea, the current-displacement relation shifts along the displacement axis toward positive displacements by ≈260 nm on average, and the fraction of open transduction channels increases (Ricci and Fettiplace 1997). Adaptation kinetics, however, are unaffected. The mechanism of action of cAMP remains obscure. In many cells, Ca^{2+} interacts with calmodulin to activate adenylyl cyclase, producing cAMP that stimulates protein kinase A. This kinase then phosphorylates target proteins, possibly the channel itself or a protein attached to it. The bullfrog's myosin Ic, a molecular motor that is implicated in the adaptation process (Section 3.4.2), displays a consensus phosphorylation site that might impede the motor's action when phosphorylated (Martin et al. 2003).

3.4 Mechanisms of Adaptation

The gating-spring model provides useful insight into possible mechanisms of adaptation (Assad and Corey 1992; Hudspeth and Gillespie 1994; Holt and Corey 2000). The adaptation process buttresses negative feedback on the fraction of open transduction channels. Because gating-spring tension and intrinsic energy balance between open and closed forms of a transduction channel control the channel's open probability (Eq. [4.6]), adaptation must regulate one or both of these parameters.

3.4.1 The Active Motor Model

If a positive displacement of the hair bundle tenses elastic elements that pull the transduction channels open, then the channels would reclose if tension in these gating springs somehow declined. Conversely, if negative stimulation slackened the gating springs and thus initially closed the transduction channels, channels would reopen if the tension in the gating springs were restored. Adjustment of the tension in the gating springs could be mediated by a longitudinal movement of the anchoring point of the transduction apparatus along the actin core of a stereocilium (Hudspeth and Gillespie 1994). Such a process requires the activity of a molecular motor. Adaptation to positive stimuli might correspond to passive slippage if the increased tension in the gating springs pulled the transduction apparatus toward the stereociliary pivots, thereby reducing gating-spring tension (Fig.4.8). Positive adaptation, however, requires an upward movement of the transduction apparatus that increases tension in the gating springs as the movement proceeds; energy of biochemical origin must be used to perform mechanical work. In this model (Fig. 4.8), the adaptation motor constitutes an intracellular link in series with the channel and the gating spring (Fig. 4.9). Destined to climb up the stereocilia, the adaptation motor builds tension in the gating springs until this tension is great enough to stall the motor. There, the force f_{STALL} produced by the motor in each stereocilium along the gating-spring axis must be equal and opposite to the tension in this spring: $f_{STALL} \approx K_{GS} x_r$,

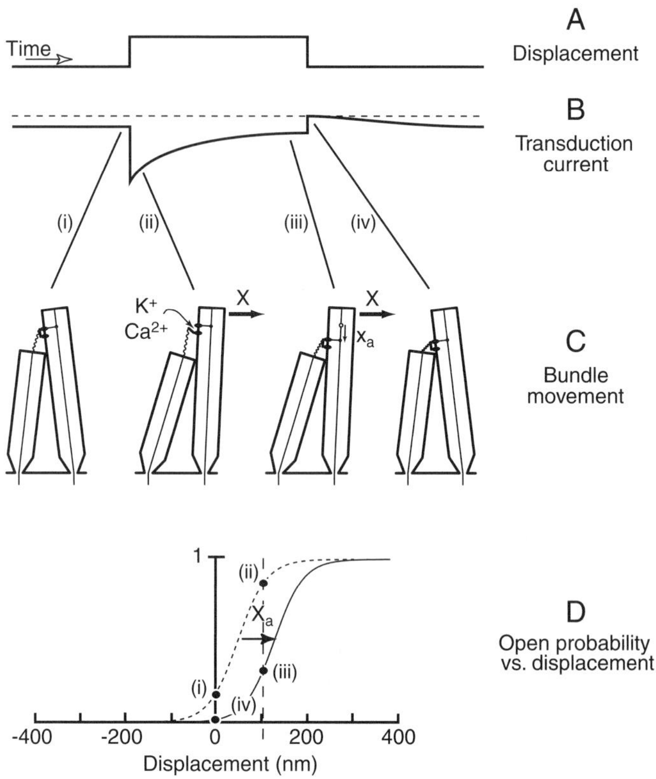

FIGURE 4.8. The active-motor model of adaptation. (**A**) A positive deflection $X = 100$ nm is applied to a hair bundle for 100 ms. (**B**) This mechanical stimulus evokes an inward transduction current, which, as adaptation proceeds, declines toward a plateau. (**C**) The hair bundle is represented by a pair of stereocilia in which actin filaments are pictured by a single line. A transduction channel embedded in the membrane of the taller stereocilium is connected to the actin cytoskeleton of the same stereocilium by an intracellular link that comprises the adaptation motor (myosin molecules) and to the tip of the shorter neighboring stereocilium by the extracellular tip link. The tip link embodies here the gating spring, but any more compliant linkage in series with it could also play this role with no effect on the description. Positive stimulation increases gating-spring tension and thus fosters opening of the transduction channels. The entry of Ca^{2+} ions in the cytoplasm, together with the increased tension of the gating spring, provokes slipping of the myosin molecules (*vertical arrow*) by a distance x_a. This motor movement in turn decreases gating-spring tension, thereby allowing the transduction channel to reclose. (**D**) The effect of adaptation on the instantaneous relation between the channel's open probability and bundle displacement. Assuming that the transduction channels carry about 15 % of their maximal current at rest (point *i*), the abrupt stimulus increases the open probability to a high value (point *ii*). As a result of adaptation and with the bundle clamped at a fixed position of +100 nm, transduction channels reclose (point *iii*). The ensuing decline of the transduction current is due to a shift of the open probability–displacement relation along the displacement axis in the direction of bundle movement by a distance $X_a = x_a/\gamma$, in which γ is the geometrical factor that relates hair-bundle deflection to gating-spring

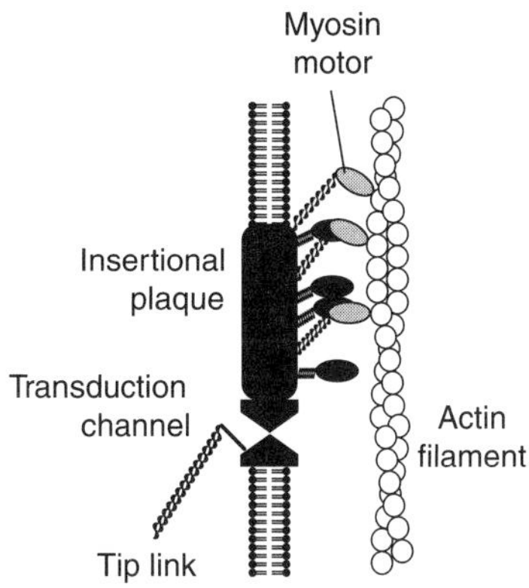

FIGURE 4.9. Functional view of the transduction complex. Each transduction channel is mechanically coupled to a neighboring stereocilium by a tip link and to the actin core of the stereocilium to which it belongs by a collection of myosin molecules. Active force production by ATP-driven power strokes of these motor molecules is directed toward the tip of the stereocilium (top of the figure), where the barbed ends of the actin filaments are located, thereby exerting tension on the tip link and thus on the transduction channel that is attached in between. Because myosin is a nonprocessive motor, only a small fraction of the motor collection is attached to actin at any given time (*gray heads*). The channel's open probability depends on motor strength: the stronger the motor, the larger the open probability. (Adapted from Hudspeth and Gillespie 1994.)

if $x_r >> d/2$, with notation of Section 2.3.3. In this scheme, the resting tension in the gating springs and thus the fraction of the channels open at rest (see Eq. [4.6]) are controlled by the motor strength. The stronger the motor, the larger the gating-spring extension at rest x_r and thus the greater the open probability will be.

When a hair bundle is deflected at its top by a horizontal distance X from its resting position, the adaptation motor will move along the longitudinal axis of the stereocilia by a distance x_a (Fig. 4.8). In the following, the sign convention is such that motor movement in the positive direction ($x_a > 0$) corresponds to a motor sliding toward the stereociliary base; such a movement is expected to be evoked by positive stimuli. Because adaptation develops on a much longer time

FIGURE 4.8. extension (Section 2.5). At the stimulus offset, the channel's open probability returns to a lower value than at rest (point *iv*). Similarly, adaptation to negative hair-bundle displacements (not shown) is interpreted as the result of a climbing motion of the adaptation motor along the actin core of the stereocilia, which evokes a negative shift of the open probability–displacement relation along the displacement axis. (Adapted from Hudspeth and Gillespie 1994.)

scale than channel gating, one can assume that the transduction channels obey equilibrium statistics and thus that the channels' open probability still satisfies (6). The bundle position X_0 at which half the transduction channels are open, depends on motor position,

$$X_0 = \frac{\Delta\mu}{Z} - \frac{x_r}{\gamma} + \frac{x_a}{\gamma}, \tag{4.9}$$

with the parameter definitions of Section 2.3.3. In this formulation of the active-motor model (Assad and Corey 1992), adaptation thus results in a simple shift of the probability-displacement relation along the displacement axis (Fig. 4.8).

3.4.2 The Implication of Myosin Molecules in Adaptation

Because each stereocilium is endowed with an actin core, myosin molecules are natural candidates to embody the adaptation motor (Hudspeth and Gillespie 1994). The actin filaments have their barbed ends located at the top of the stereocilia and are thus properly oriented to support upward movements of most members of the myosin family. Five unconventional myosins have so far been localized in hair cells: myosin Ic, IIIa, VI, VIIa, and XVa (Hasson et al. 1997). Among these isosymes, myosin Ic (Gillespie et al. 1993; Holt et al. 2002) and myosin VIIa (Kros et al. 2002) have so far been implicated in the adaptation process.

In hair cells from the bullfrog's saccule, a few dozen myosin Ic molecules per stereocilium appear to be appropriately concentrated near the tips of the stereocilia (Gillespie et al. 1993). The implication of myosin Ic in the adaptation process has been carefully tested in the mouse, because the gene encoding for mouse myosin Ic can be mutated. Engineering of an enlarged nucleotide binding pocket in myosin Ic made it possible to specifically block this motor with large ADP analogues (Gillespie et al. 1999). This strategy provided strong evidence that myosin Ic is essential to both fast and slow components of the adaptation process, at least in the vestibular system (Holt et al. 2002; Stauffer et al. 2005).

Myosin VIIa is the other myosin isosyme implicated in adaptation. Mutations in myosin VIIa cause hereditary deafness in mice and humans. Although cochlear hair bundles of mutant mice are disorganized, which suggests that myosin VIIa is implicated in the maintenance of the bundle's architecture (Boeda et al. 2002), their transduction apparatus works (Kros et al. 2002). The transduction channels, however, are shut at rest and the hair bundles need to be displaced well beyond their physiological operating range ($X > +150$ nm) for transduction channels to open significantly. This result suggests that the gating springs are slack when myosin VIIa is inactive. Conversely, active myosin VIIa must pull on the gating springs to open a fraction of the transduction channels and thus set the operating point of the transduction apparatus in a region of mechanosensitivity. Is myosin VIIa the adaptation motor in hair cells from the mouse cochlea? Even if deficient in myosin VIIa, a hair cell adapts with both fast and slow components to positive step stimuli with magnitudes in the bundle's (abnormal) sensitivity range. Moreover, adaptation in mutant mice is more complete than in heterozygous control mice. It is unclear at the moment whether positive adaptation is mediated by another myosin isosyme or simply results from passive slippage of the transduction apparatus. In particular, there is no evidence for or against a functional role of myosin Ic in cochlear hair cells. Because adaptation to positive stimuli requires myosin Ic in the vestibular system of mice, however, it would be curious if positive adaptation in the cochlea were entirely passive.

Inhibiting the ATPase activity of myosin molecules with vanadate or butanedione monoxime provides supporting evidence for the operation of two mechanisms of adaptation (Wu et al.1999). When a hair cell is exposed to the inhibitor and abruptly deflected in the positive direction, the slow component of adaptation vanishes but the initial fast component of the current decline survives. Two mechanisms of adaptation might thus cooperate in the hair cell (Howard et al. 1988; 1991 1991): one ATP-dependent mechanism that relies on the active attachment–detachment cycle of myosin molecules along the actin core of the stereocilia and a second mechanism that does not rely on the completion of an actin–myosin ATPase cycle. As detailed below (Section 3.4.3), this second mechanism might still involve myosin, for instance if myosin were able to rock back and forth without detaching from actin (Stauffer et al. 2005).

3.4.3 Ca^{2+}-Reclosure Model

What is the mechanism that underlies fast adaptation? Because of its importance in the adaptation process (see Section 3.3), Ca^{2+} probably interacts with some component of the transduction apparatus to foster channel closure. Calcium might interact directly with the intracellular side of the channel's pore, or with an associated protein, to stabilize the closed form of the channel (Howard et al. 1988; Crawford et al. 1989). In this formulation of the Ca^{2+}-reclosure model, Ca^{2+} shuts the channel directly and thus increases gating-spring tension. In a simple two-state model of transduction (Section 2.3.3), this mechanism corresponds to a free-energy difference $\Delta\mu$ between the open and the closed states of a transduction channel that depends on the intracellular calcium concentration $[Ca^{2+}]$ at the binding site, with $\frac{d\mu}{d[Ca^{2+}]} > 0$. In support of this model, a recent analysis of the mechanical correlate of fast adaptation—the "twitch" (Section 4.1)—suggests that a Ca^{2+}-bound channel requires $\approx$3 pN more force to open (Cheung and Corey 2005).

Calcium might instead allow the transduction channels to close by interacting with myosin Ic to decrease its stiffness (Howard and Spudich 1996; Bozovic and Hudspeth 2003; Martin et al. 2003), reduce the myosin's duty ratio and thus the average number of bound myosin molecules at any given time (Hudspeth and Gillespie 1994; Nadrowski et al. 2004), or elicit a conformational release (Benser et al. 1996; Batters et al. 2004a). In preparations where the shape of the current-displacement relation often does not change significantly as positive adaptation proceeds (Hacohen 1989; Shepherd and Corey 1994; Holt et al. 1997; Vollrath and Eatock 2003), a softening of the actin–myosin crossbridge is less likely than a conformational release by the motor (Stauffer et al. 2005).

What could be the origin of the release? Mechanical measurements of force–displacement relations on a single myosin molecule (Batters et al. 2004a) as well as earlier measurements on muscle fibers (Huxley and Simmons 1971) suggest that an adversary force of sufficiently large magnitude might be capable of reversing the conformational swing of the myosin's neck that is normally powered by ATP hydrolysis (Howard 2001). Positive fast adaptation might thus correspond to a reversal of the myosin-Ic power stroke for those molecules that are strongly bound to actin and initially poised (Batters et al. 2004b) in a post-stroke conformation. Even if this were the case, however, the detailed relation between gating of the transduction channels and the mechanical properties of myosin Ic would still remain to be elucidated, in particular to clarify exactly how Ca^{2+} binding would facilitate reversal of the power stroke.

Calmodulin is a cytoplasmic receptor for Ca^{2+} that is concentrated near the tips of the stereocilia (Walker et al.1993). Because calmodulin antagonists block the adaptation process (Walker and Hudspeth1996), calmodulin probably controls the Ca^{2+} dependence of adaptation. Although of unknown significance in hair cells, calmodulin may constitute a subunit of the transduction channel (Saimi and Kung2002) and thus affect channel gating. Alternatively, because it binds to the neck of myosin Ic molecules, calmodulin might regulate the motor activity.

4. Active Hair-Bundle Movements

4.1 Mechanical Correlates of Adaptation: Hair-Bundle Movements in Response to Steps

The gating-spring model intimately associates the fraction of transduction channels that are open with tension in the gating spring (Eq. [4.6]). Within this framework, the adaptive current decline that is observed in response to a positive step force (Section 3.) must involve changes in gating-spring tension and thus internal forces. Because these forces act on the stereociliary pivots, the hair bundle should move as adaptation proceeds. Three lines of evidence support this inference (Benser et al.1996; Ricci et al.2000a). First, the time course of the hair-bundle motion often mirrors that of the cell's electrical response during adaptation (Fig. 4.10). Second, the bundle movement evoked by a step stimulus is profoundly affected when the transduction channels are blocked with aminoglycosides or when the bundle is stimulated along a direction orthogonal to the bundle's axis of mechanosensitivity, thereby demonstrating that part of the motion is associated with channel gating. Finally, lowering Ca^{2+} slows both adaptation of the electrical response and the associated hair-bundle movement, suggesting that the Ca^{2+} component of the transduction current evokes mechanical rearrangements that drive the bundle movements. However, although the transduction current often declines monotonically from an initial peak in response to a positive stimulus, the mechanical correlate of adaptation

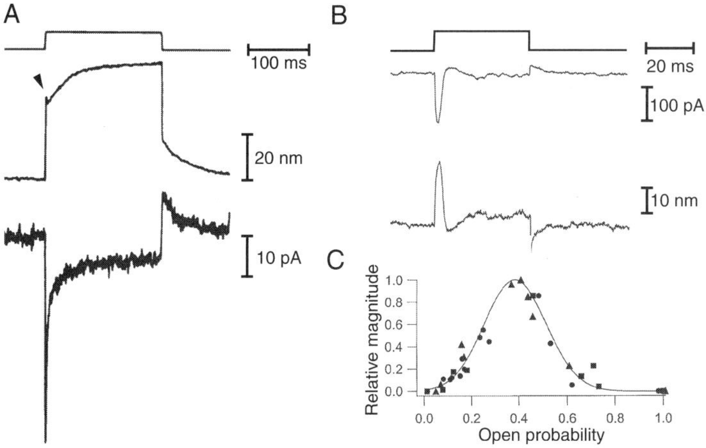

FIGURE 4.10. Active hair-bundle movements in response to positive force steps. (**A**) A hair bundle from the bullfrog's saccule displays a fast recoil (*arrowhead, middle*), also termed "twitch," against the stimulus (top, ≈63 pN). This transient movement is associated with fast reclosure of the transduction channels (*bottom*). A slower relaxation of opposite polarity follows the twitch, with a similar time course as the concomitant decline of the transduction current. (Adapted from Howard and Hudspeth 1987.) (**B**) In the turtle's cochlea, the time course of hair-bundle movement (*bottom*) resembles that of the transduction current (*top*) for a force step of ≈37 pN (*top*). Although the bundle's response is composed of two components of opposite polarities and distinct kinetics, the fast component dominates here the overall process. (**C**) By simultaneously measuring hair-bundle movements and transduction currents in response to a family of force steps, one can determine how the magnitude of the twitch varies with the fraction of the transduction channels that is opened at the peak of the transduction current. The twitch magnitude is here plotted with respect to its maximum value (*squares*, 32 nm; *circles*, 16 nm; *triangles*, 5.2 nm) and the smooth curve is a Gaussian. (Graphs **B** and **C** are adapted from Ricci 2000.)

can be nonmonotonic (Fig. 4.10A); fast and slow components of adaptation are in fact associated with movements of opposite polarities. When an abrupt stimulus is applied in the positive direction, the hair bundle first moves in the direction of the applied force but then displays a transient movement in the opposite direction called the "twitch" (Fig. 4.10A and B). This fast recoil happens on a millisecond (or less) time scale and coincides with the initial, fast reclosure of the transduction channels. The magnitude and duration of the twitch depend on the bundle's resting position and on the magnitude of the stimulus. The relation between twitch magnitude and bundle displacement is bell-shaped, with a maximum magnitude in the 20–40 nm range. The twitch is visible within the narrow region of bundle displacements that elicit opening of the transduction channels and is most prominent when fewer than 50% of the transduction channels are opened by the stimulus (Fig. 4.10C). For the twitch to be visible, however, the rise time of hair-bundle motion must be significantly shorter than the time scale of the current decline. In the opposite case, the hair bundle adapts while it is being deflected, and the hair-bundle position correspondingly relaxes slowly in the direction of the stimulus. Recent experiments with outer hair cells from the rat cochlea revealed fast adaptation of the transduction current in response to positive, abrupt step stimuli (Kennedy et al. 2003) but failed to show the associated twitch. It is likely that when flexible fibers are used to measure force production by the hair bundle, the rise time of the stimulus (150–200 μs) is not short enough to elicit a twitch. In agreement with this interpretation, the current-displacement relation displays an unusually large width of ≈500 nm when measured with a flexible fiber but a normal width of ≈300 nm when measured with a rigid fiber (stimulus rise time in the range 50–100 μs). Because adaptation is associated with a shift of the current-displacement relation along the displacement axis (see Section 3.1), slow stimuli result in a measurement that integrates the instantaneous relation as it shifts; gating will appear to happen at larger bundle displacements than would be observed with fast stimuli.

4.2 Electrically Evoked Movements of Quiescent Hair Bundles

By patch-clamping a single hair cell or passing an electrical current trough the sensory epithelium of an entire organ, one can affect the membrane potential of the hair cells' apical surface and therefore the electrical field that drives Ca^{2+} ions into the hair cell through open transduction channels. Because the Ca^{2+} component of the transduction current affects the channels' open probability (Section 3.3), a change of the cell's membrane potential will evoke channel gating and thus hair-bundle movements. Depolarization of a hair cell from the bullfrog saccule by a voltage step evokes a hair-bundle movement in the negative direction with a time course (≈30–200 ms) that resembles that of the slow component of adaptation (Assad et al. 1989; Assad and Corey 1992). This behavior accords with the active-motor model of adaptation (Section 3.4.1): reduction of the intracellular Ca^{2+} concentration promotes climbing of the adaptation motor up the stereocilia, in turn increasing gating-spring tension and thus evoking a movement of the hair bundle in the negative direction. Similar depolarization of a hair cell from the cochlea of the turtle (Fig. 4.11), however, produces a bundle movement of opposite polarity and with a time course (≈1 ms) that matches that of fast adaptation (Ricci et al. 2000b). This behavior was interpreted as evidence for a second mechanism of adaptation, whereby Ca^{2+} impoverishment near the intracellular side of a transduction channel's pore promotes fast channel opening (Section 3.4.3). If channel opening in turn reduces gating-spring tension, one indeed expects a relaxation of gating-spring tension and thus a bundle movement of the observed (positive) polarity. Surprisingly (Fig. 4.11), the polarity of the electrically evoked movement in the turtle can be reversed if a sufficiently large positive offset is applied to the hair bundle (Ricci et al. 2002). This polarity

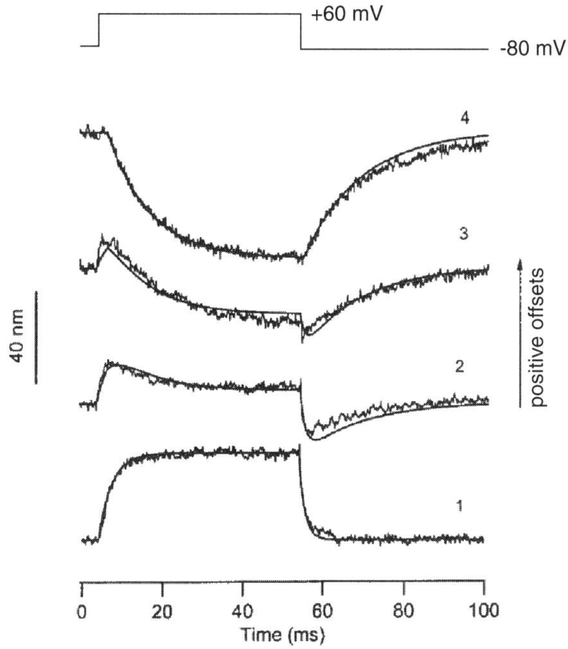

FIGURE 4.11. Hair-bundle movements in response to depolarizing voltage steps. A hair cell from the turtle's cochlea is initially voltage clamped at −80 mV and then abruptly depolarized at +60 mV for 50 ms. In addition, a flexible fiber is used to impose positive offsets to the hair-bundle position. When no offset is applied (*trace 1*), the hair bundle displays a movement in the positive direction, with a time course that can be fitted by a single exponential of time constant 3.3 ms. Large positive offsets can reverse the polarity of the response (*trace 4*), which then resembles the responses measured with hair cells from the saccule of the bullfrog when no offset is applied. The reversed response is also characterized by monoexponential kinetics, but with a larger time constant of 10.3 ms. For intermediate offsets (*traces 2* and *3*), the response is biphasic and can be described as a linear weighted sum of normal (fast and positive) and reversed (slow and negative) components. (Adapted from Ricci et al. 2002.)

reversal concurs with slower kinetics, producing a behavior that resembles that observed in the frog. At intermediate magnitudes of the offset, the response is biphasic: the hair bundle initially moves quickly in the positive direction, reaches a peak, and then moves slowly in the negative direction toward a plateau. Biphasic responses have also been reported for hair cells from the frontal ampulla of the young eel *Anguilla anguilla*, but with the opposite sequence: depolarization of the cell's apical surface first evokes a negative deflection and then a movement toward the positive direction (Rusch and Thurm 1990). Both positive (fast) and negative (slow) electrically evoked movements can thus be observed with the same hair cell. Recent experiments (Bozovic and Hudspeth 2003) revealed that both polarities of electrically evoked movements can be observed in the bullfrog's saccule as well. Electrically evoked biphasic movements are similar in polarity and kinetics to those observed in response to abrupt force steps (the "twitch"; see Section 4.1); they probably result from the same underlying mechanism. What could be the effect of an offset? Because adaptation is incomplete for bundle deflections of sufficiently large magnitudes, applying an offset affects the open probability at the rest of the transduction channels (Section 3.1). These experiments thus suggest that polarity and kinetics of electrically evoked movements depend on the operating point of the hair bundle (Tinevez et al. 2007).

4.3 Spontaneous Hair-Bundle Oscillations

Spontaneous oscillations of the hair bundle have been observed in the cochlea of the turtle (Crawford and Fettiplace 1985), the frontal ampulla of the eel (Rusch and Thurm 1990), and the saccule of the bullfrog (Howard and Hudspeth 1987; Denk and Webb 1992; Martin and Hudspeth 1999; Martin et al. 2003). Should hair-bundle oscillations happen in vivo, a rhythmic activity of the primary afferent fibers that innervate the hair cells would be expected. Preferred intervals have been measured in the spontaneous activity of afferent fibers in turtle, lizard, and birds (Manley 1990).

4.3.1 Properties of Spontaneous Oscillations from the Bullfrog's Saccule

By placing the sensory epithelium in a two-compartment chamber, it is possible to mimic the ionic environment that hair cells experience in vivo, in particular the relatively low Ca^{2+} concentration in the artificial endolymph that bathes the hair bundles (Martin and Hudspeth 1999). Under such ionic circumstances, hair bundles from the bullfrog's saccule routinely display spontaneous oscillations at 5–50 Hz with peak-to-peak magnitudes as large as 100 nm (Fig. 4.12A). The bundle's motion is not sinusoidal but instead resembles the relaxation oscillations observed in various dynamical systems (Strogatz 1997): in each half-cycle, a slow excursion in one direction is interrupted by an abrupt jump in the opposite direction. As a consequence, the probability distribution of the bundle's position is bimodal, with a local minimum near $X = 0$ (Fig. 4.12B). This distribution resembles that observed for sound pressure at the frequency of a spontaneous otoacoustic emission from the human ear (Bialek and Wit 1984), the frog ear (van Dijk et al. 1989), and lizard and bird ears (van Dijk et al. 1996).

Oscillations are noisy. The spectral density of each oscillation displays a well-defined peak centered at a frequency ν_0 that indicates the characteristic frequency of the oscillation (Fig. 4.12C). Frequency fluctuations around ν_0 are characterized by the quality factor $Q = \frac{\nu_0}{\Delta\nu}$ in the range of about 1 to 3, in which $\Delta\nu$ is the width of the spectrum at half its maximal value. The hair-bundle movement thus loses phase coherence after a couple of cycles of an oscillation.

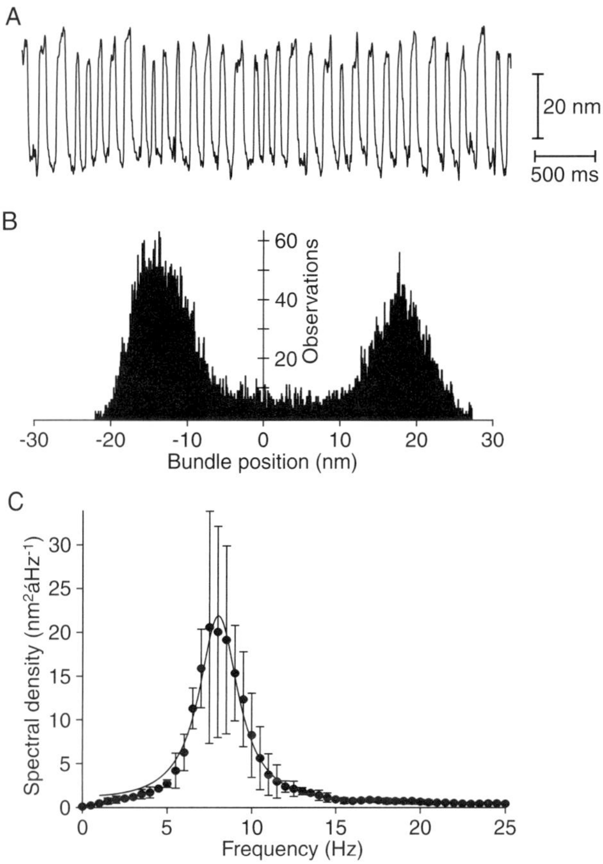

FIGURE 4.12. Spontaneous oscillations of a hair bundle from the saccule of the bullfrog. (**A**) The position of a glass fiber attached to the hair bundle's top reports the time course of the oscillation, which had a root-mean-squared magnitude of 28 nm. (**B**) The probability distribution of bundle positions is bimodal. (**C**) The signal's spectrum displays a broad peak at ≈8 Hz with a width at half its maximum value of 2.8 Hz. The smooth line is a fit by Eq. (4.35) from Chapter 3, by Duke and Jülicher.

4.3.2 Effect of Ca^{2+} on Spontaneous Oscillation

In the bullfrog's saccule, when the Ca^{2+} concentration in the artificial endolymph that bathes the hair bundles exceeds ≈1 mM or falls below ≈100 μM, the oscillations disappear (Martin et al. 2003). Reimposing a Ca^{2+} concentration near that found in physiological endolymph (250 μM for this species) restores the oscillatory behavior. Thus Ca^{2+} is a control parameter of a bundle's dynamic behavior. Transiently elevating the Ca^{2+} concentration with a Ca^{2+}-containing electrode using iontophoresis has three consistent and reversible effects: a net movement in the negative direction, a mild (<20 %) decrease of the oscillation magnitude, and an increase of the oscillation frequency. The increase in oscillation frequency involves principally a shortening of the slow component of bundle movement in the positive direction. Reducing the Ca^{2+} concentration through the iontophoretic application of a Ca^{2+} chelator produces effects opposite to those of Ca^{2+} iontophoresis. As detailed in Chapter 6 by Manley and van Dijk, Ca^{2+} modulates the frequency and magnitude of spontaneous otoacoustic emissions from lizards' ears (Manley and Kirk 2002, 2005; Manley et al. 2004). Otoacoustic emissions in these animals are most probably produced by active hair-bundle movements (Manley et al. 2001) and behave similarly to in vitro spontaneous hair-bundle oscillations in the frog with respect to Ca^{2+} changes: increased Ca^{2+} levels produce faster and generally smaller emissions. Although lizard emissions occur at much higher frequencies than the spontaneous hair-bundle oscillations observed in the bullfrog, active oscillatory movements in the two systems might be produced by a similar Ca^{2+}-regulated mechanism.

5. Force–Displacement Relation of the Hair Bundle

Mechanical measurements are complicated by two important features. First, as a result of adaptation (Section 4.1), the hair-bundle deflection in response to a force step is often nonstationary (Howard and Hudspeth1987; Russell et al.1992) and can even be nonmonotonic (Benser et al.1996; Ricci et al.2000a). Second, as detailed below, the relation between hair-bundle deflection and external force is nonlinear (Howard et al.1988; Russell et al.1992; Martin et al.2000; Ricci et al.2002): hair bundles appear to be softer in the narrow region of bundle displacements where transduction channels open or close in response to an external force, a phenomenon termed "gating compliance" (Howard et al.1988). The hair bundle is thus a complex nonlinear dynamical system whose responsiveness to external forces is frequency and amplitude-dependent (Martin and Hudspeth2001; Martin et al.2001).

5.1 Gating Compliance, Negative Stiffness

Hair-bundle stiffness can be determined by applying abrupt step displacements Δ at the base of a flexible fiber of stiffness K_F whose tip is attached to the bundle's tip and monitoring the ensuing bundle movement X with a photometric system (Flock and Strelioff 1984; Crawford and Fettiplace 1985; Howard and Hudspeth 1987). The bundle's elastic properties are usually assayed by doing the measurement as quickly as possible after the bundle has been abruptly deflected, in practice a fraction of one millisecond (Ricci et al. 2002; Kennedy et al. 2005) to a few milliseconds after the stimulus onset (Howard et al. 1988; Martin et al. 2000). This delay is long enough that transduction channels have time to fully respond to the stimulus (the equilibration time with the bundle's position is $\tau_C \approx 100\mu s$; see Section 2.3.2) and that the viscous hydrodynamic drag that necessarily impedes abrupt bundle motion has vanished, but generally short enough to avoid significant development of the adaptation process. When displaced extensively either in the positive or in the negative direction ($P_o = 0$ or $P_o = 1$ respectively), the hair bundle behaves as a linear spring of constant stiffness. For intermediate displacements ($0 < P_0 < 1$), however, the force–displacement relation is nonlinear (Fig. 4.13A): the hair bundle appears to be softer in the narrow region of displacements that elicit channel gating. Nonlinear force–displacement relations have been measured in the saccule of the bullfrog (Howard et al. 1988; Martin et al. 2000), the cochlea of the turtle (Ricci et al. 2002), the vestibule and the cochlea of the mouse (Russell et al. 1992; Géléoc et al. 1997), and the cochlea of the rat (Kennedy et al. 2005).

This reduction in stiffness upon channel gating can readily be interpreted with the gating-spring model. Mechanical activation of the transduction channels indeed dictates a reciprocal relationship between gating of the transduction channels and tension in the gating springs (see Section 2.3.3). An external force in the positive direction will evoke an increase of the channels' open probability; channel opening reduces the extension of the gating spring, thereby generating

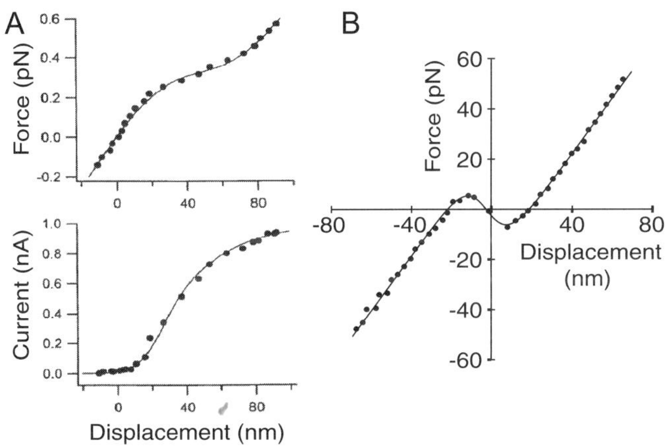

FIGURE 4.13. Nonlinear hair-bundle mechanics (**A**) Here measured in the cochlea of the turtle, the curve that describes the relationship between the external force applied at a hair bundle and the evoked bundle displacement (*top curve*) is steep and linear for large negative or positive bundle displacements but displays a flat nonlinear region in a range of displacements for which a receptor current is measured (*bottom curve*). In this experiment, the preparation was immersed in standard saline containing a Ca^{2+} concentration of 2.8 mM. Bundle position and force were recorded 0.5 to 1 ms after the onset of stimulation. (Adapted from Ricci et al. 2002.) (**B**) In the saccule of the bullfrog, the force-displacement relation can display a region of negative slope. The hair bundle was here immersed in artificial endolymph with a Ca^{2+} concentration of 250 μM, and the force–displacement relation was measured under displacement-clamp circumstances. The origin of the graph, which corresponds to the position of the hair bundle at steady state, was here unstable: the hair bundle oscillated spontaneously. (From Martin et al. 2001.) In both **A** and **B**, the smooth lines correspond to fits with the gating-spring model (Eqs. [4.13] and [4.6]).

an effective "gating force" in the positive direction. Conversely, a negative force closes channels; channel closing in turn produces a force in the negative direction. Along the bundle's horizontal axis of mechanosensitivity, gating of a single channel produces a gating force $Z = \gamma \mathrm{K}_{\mathrm{GS}} d$ (Section 2.3.3). Because the gating force provides an extra give in the direction of the applied force, the stiffness of the transduction apparatus will be lower than that of the gating spring alone, a process termed "gating compliance" (Howard et al. 1988). It is accordingly expected that the stiffness of the hair bundle depends on the open probability of the transduction channels (Eq. [4.6]) and thus on hair-bundle displacement.

Assuming that adaptation is not allowed sufficient time to develop significantly after a step stimulus changes gating-spring extension by x (Section 5.2), the time-averaged force $\langle f_1 \rangle$ exerted by a single transduction element is

$$\langle f_1 \rangle = P_{\mathrm{o}} \frac{\partial G_{el}^{\mathrm{OPEN}}}{\partial x} + (1 - P_{\mathrm{o}}) \frac{\partial G_{\mathrm{el}}^{\mathrm{CLOSED}}}{\partial x} = \mathrm{K}_{\mathrm{GS}} \left(x + x_r + \frac{d}{2} - dP_{\mathrm{o}} \right), \tag{4.10}$$

in which the elastic energies $G_{\mathrm{el}}^{\mathrm{OPEN}}$ and $G_{\mathrm{el}}^{\mathrm{CLOSED}}$ as well as other notations are defined in Section 2.3.3. An external force applied at the bundle's top must be balanced by the elastic restoring forces provided by the combination of stereociliary pivots and gating springs. With the important assumption that N identical transduction apparatuses (channels and gating springs) operate in parallel (Iwasa and Ehrenstein 2002), the mean external force F that must be applied to a hair-bundle top to impose a deflection X is given by

$$F = \mathrm{K}_{\mathrm{SP}} (X - X_{\mathrm{SP}}) + N\gamma \langle f_1 \rangle, \tag{4.11}$$

in which K_{SP} is the combined stiffness of the stereociliary pivots and X_{SP} is the position that the hair bundle would assume in the absence of gating springs. In the saccule of the bullfrog, severing the tip links results in a forward motion of the hair bundle by a distance $X_{SP} = +133$ nm (Assad et al. 1991), thereby demonstrating that the gating springs bear significant tension at rest ($F = 0$; $X = 0$). With the combined gating-spring stiffness $\mathrm{K}_{GS} = N\gamma^2 \mathrm{K}_{GS}$, $X_r = x_r/\gamma$ and $D = d/\gamma$, Eq. (4.11) can be rewritten

$$F = \mathrm{K}_{\mathrm{SP}} (X - X_{\mathrm{SP}}) + \mathrm{K}_{\mathrm{GS}} \left(X + X_{\mathrm{r}} + \frac{D}{2} - DP_{\mathrm{o}} \right), \tag{4.12}$$

or, with the overall hair-bundle stiffness when the channels are either all open or all closed, $\mathrm{K}_{\infty} = \mathrm{K}_{\mathrm{GS}} + \mathrm{K}_{\mathrm{SP}}$, and the constant force $F_o = \mathrm{K}_{\mathrm{GS}} \left(X_{\mathrm{r}} + \frac{D}{2} \right) - \mathrm{K}_{\mathrm{SP}} X_{\mathrm{S}}$,

$$F = \mathrm{K}_{\infty} X - NZP_{\mathrm{o}} + F_{\mathrm{o}}. \tag{4.13}$$

Because the probability for a transduction channel to be open is a sigmoidal function of hair-bundle displacement (Eq. [4.6]), the force–displacement relation is nonlinear. By obstructing the channel's pore, aminoglycoside antibiotics prevent channel gating (Kroese et al. 1989). In accordance with the gating-spring model, these drugs reversibly abolish gating compliance (Howard et al.

1988; Martin et al. 2003). This nonlinear behavior in hair-bundle mechanics is analogous to that observed in muscle under isometric conditions when one measures the tension approached during a quick recovery phase after an abrupt length change as a function of the magnitude of the length step (Huxley and Simmons 1971). In the hair bundle, the nonlinearity depends on the state of the transduction channels, whereas in the muscle it might result from a change of conformation of the myosin molecules. Differentiating equation (13) with respect to X yields the slope stiffness of the hair bundle:

$$\mathrm{K}_{\mathrm{HB}} = \mathrm{K}_{\infty} - \frac{NZ^2}{k_{\mathrm{B}}T} P_{\mathrm{o}}(1 - P_{\mathrm{o}}). \tag{4.14}$$

The hair-bundle stiffness is minimal when half the transduction channels are open. Fitting force–displacements with Eq. (13) yields numerical estimates for the parameters K_{∞}, N, and Z. If the stiffness K_{SP} contributed by the pivots is known (see later), one gets in addition the stiffness $\mathrm{K}_{\mathrm{GS}} = \frac{\mathrm{K}_{\infty} - \mathrm{K}_{SP}}{N\gamma^2}$ of a single gating spring and in turn the magnitude $d = \frac{Z}{\gamma \mathrm{K}_{GS}}$ of the gating swing.

Under ionic circumstances that mimic in vivo conditions with two separate ionic compartments (Martin et al. 2000), the force–displacement relation of a hair bundle from the bullfrog's saccule can display a region of negative stiffness (Fig. 4.13B). Gating compliance can thus be great enough to dominate the bundle's overall stiffness. According to the gating-spring model (Eq. [4.14]), the condition for negative stiffness is

$$Z > \sqrt{\frac{4\mathrm{K}_B T K_{\infty}}{N}}. \tag{4.15}$$

Numerically (Martin et al. 2000), $Z = 0.7 \pm 0.3$ pN, $\mathrm{K}_{\infty} = 680 \pm 360 \mu\mathrm{N} \cdot \mathrm{m}^{-1}$, and $N = 47 \pm 31$ (mean ± standard deviation, 23 measurements). Because successive treatments with the protease subtilisin and BAPTA sever most of the interciliary links, it is possible to estimate the contribution to hair-bundle stiffness that results from bending the actin fascicle at the pivots. Under such enzymatic circumstances with hair cells from the saccule of the bullfrog (Jaramillo and Hudspeth 1993; Marquis and Hudspeth 1997), the translational stiffness measured at 6.7 μm above the cell's apical surface is $\mathrm{K}_{\mathrm{SP}} \approx 200 \mu\mathrm{N} \cdot \mathrm{m}^{-1}$. Assuming that the stimulus recruited all of the $N \approx 60$ stereocilia in a hair bundle, this translational stiffness yields a rotational stiffness per pivot of $\mathrm{K}_{\mathrm{R}} \approx 1.5 \cdot 10^{-16} \mathrm{N} \cdot \mathrm{m} \cdot \mathrm{rad}^{-1}$. With $\mathrm{K}_{\mathrm{SP}} \approx 200 \mu\mathrm{N} \cdot \mathrm{m}^{-1}$, one additionally infers $\mathrm{K}_{\mathrm{GS}} = 570 \pm 250 \mu\mathrm{N} \cdot \mathrm{m}^{-1}$ and $d = 8.2 \pm 1.5$ nm (Martin et al. 2000).

Positions of negative stiffness are unstable. A free or weakly loaded bundle cannot settle there, for any mechanical perturbation will eject the hair bundle to a position of force balance in a flanking region of positive stiffness. Because of the intimate relation between hair-bundle mechanics and channel gating (Eq. [4.14]), there is thus a whole range of open probabilities at which the hair bundle cannot be at steady state. Negative stiffness reflects a thermodynamic instability of the hair bundle analogous to a first-order phase transition. It results from the concerted gating of transduction channels operating in parallel (Iwasa and Ehrenstein 2002, Kozlov et al. 2006) and does not necessitate energy expenditure by the hair bundle. As detailed below, however, this instability can be harnessed by an active motor to generate spontaneous oscillations.

5.2 The Adaptation Motor: A Relaxation Element

Because of adaptation, the apparent stiffness of a hair bundle after it has been abruptly deflected varies over time (Martin et al. 2000; Kennedy et al. 2005). How does adaptation affect a force–displacement relation? Within the framework of the active-motor model (Section 3.4.1), adaptation is a dynamic process that relaxes stimulus-evoked tension in the gating springs (Section 3); after the adaptation motor has moved a distance x_a along the actin core of the stereocilia, tension in a gating spring can be inferred from Eq. (4.1.) by formally substituting x by $x - x_a$:

$$\langle f_1 \rangle = \mathrm{K_{GS}} \left(x - x_\mathrm{a} + x_\mathrm{r} + \frac{d}{2} - dP_\mathrm{o} \right). \tag{4.16}$$

The open probability P_o is itself a function of $x - x_\mathrm{a}$ (combine Eq. [4.6] with Eq. [4.9]). The force–displacement relation of a hair bundle is thus given by

$$F = \mathrm{K_{SP}}\,(X - X_\mathrm{SP}) + \mathrm{K_{GS}} \left(X - X_a + x_r + \frac{D}{2} - DP_o \right), \tag{4.17}$$

in which $X_\mathrm{a} = x_\mathrm{a}/\gamma$. Note that under displacement-clamp circumstances (X constant), adaptation affects tension in the gating springs but not that in the stereociliary pivots. As a consequence, different steady-state positions of the adaptation motors yield force–displacement relations of the same shape but shifted along a line of slope $\mathrm{K_{SP}}$ (Martin et al. 2000).

6. Mechanical Amplification Provided by Active Hair-Bundle Oscillations

A hair cell is capable of producing active movements of its hair bundle (Section 4). Does active hair-bundle motility enhance the cell's performances as a mechanosensor? Mechanical amplification of a hair cell's response to sinusoidal stimuli has indeed been demonstrated experimentally, but only in the case of spontaneously oscillating hair bundles from the bullfrog's saccule (Martin and Hudspeth 1999). In this section, we thus restrict ourselves to this situation.

6.1 Linear Response Function: Demonstration of Activity

In response to an external force $F(t)$, the hair bundle assumes a mean position $\langle X(t)\rangle \neq 0$. For small-enough forces, the relation between bundle position and external force is linear, and $\tilde{\chi}(\omega) = \frac{\langle \tilde{X}(\omega)\rangle}{\tilde{F}(\omega)}$ defines the frequency-dependent response function of the hair bundle. Here, tildes denote Fourier transforms of the corresponding variables and ω is the angular frequency. By applying small sinusoidal mechanical stimuli with a flexible glass fiber, $\tilde{\chi}(\omega) = |\tilde{\chi}(\omega)|\, e^{i\varphi(\omega)}$ can be measured as a function of angular frequency ω (Martin et al. 2001). The modulus $|\tilde{\chi}(\omega)|$ of the response function defines the bundle's sensitivity to external stimulation. The sensitivity of an oscillatory hair bundle is tuned (Fig. 4.14A) and is about threefold greater near its characteristic frequency ν_0 than at higher frequencies. The phase relation $\varphi(\omega)$ between stimulus and response changes sign near ν_0 (Fig. 4.14B). At high frequencies, the bundle movement lags the stimulus. Stimulation at or near ν_0 yields an in-phase response. Finally, at low frequencies, the response displays a phase lead. As $-90° < \varphi < 90°$, the dissipative part of the response function, defined by $|\tilde{\chi}| \sin(\varphi)$, is thus of negative sign at frequencies below ν_0. Such behavior can happen only in an active system, because it indicates that energy is on average withdrawn from the hair bundle by the stimulus fiber to evoke the observed movement. Under such circumstances, a conservative estimate of the active work $\bar{W}_A$ produced by the hair cell during one cycle of stimulation yields $\bar{W}_A \approx 80$ zJ, corresponding to one or two molecules of ATP hydrolyzed per cycle of stimulation (Martin and Hudspeth 1999). Because hair-bundle friction may be an order of magnitude larger than first anticipated (Nadrowski et al. 2004), this measurement clearly represents an underestimate. By comparing the bundle's spontaneous oscillations to its response to small mechanical stimuli, one can demonstrate that an oscillatory hair bundle violates a fundamental principle of systems at thermal equilibrium: the fluctuation–dissipation theorem (Callen and Welton 1951). A hair bundle's spontaneous movements as well as its responsiveness to sinusoidal stimuli at frequencies near that of the spontaneous oscillation are thus produced by energy-consuming elements within the hair cell (Martin et al. 2001).

The behavior of the frequency-dependent linear function is not generic and thus provides useful information about the class of active oscillators that could account for the bundle's mechanical activity. In particular, the van der Pol oscillator, a standard model that generates spontaneous oscillations by introducing negative friction, behaves differently: in this case, the dissipative part of the response function does not change sign. This observation rules out the possibility (Gold 1948) that the active process generates a force proportional to velocity, which negates friction.

6.2 Entrainment of a Spontaneous Oscillation

A conspicuous feature of any spontaneous hair-bundle oscillation, at least in the bullfrog's saccule, is its noisy character. Because of fluctuations inherent

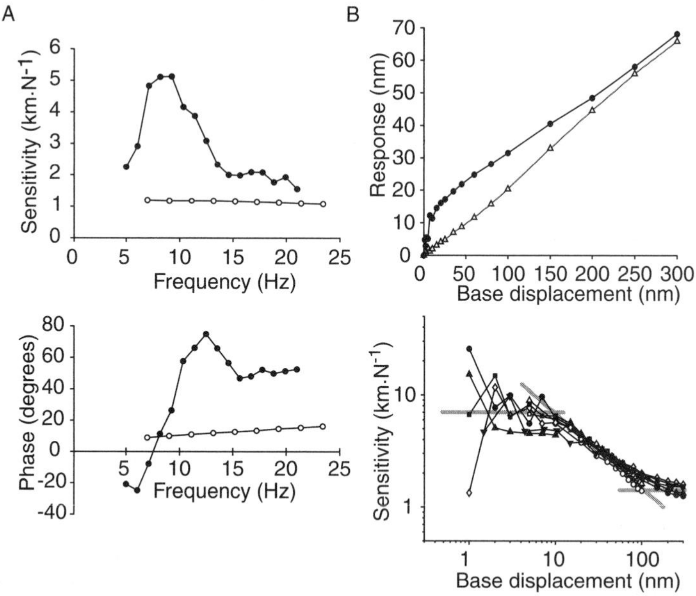

FIGURE 4.14. Response functions in the bullfrog's saccule. (**A**) The modulus (*top curves*) and phase (*bottom curves*) of the linear response function of an active hair bundle that oscillated spontaneously at ≈8 Hz (•) and of a control, passive hair bundle that did not oscillate (∘) are plotted as a function of the stimulus frequency. In contrast to the control, the oscillatory hair bundle displays frequency tuning and enhanced sensitivity near the bundle's characteristic frequency of spontaneous oscillation. (**B**) Stimulated at a frequency of 150 Hz (△), the same oscillatory hair bundle as in **A** displays a phase-locked response whose root-mean-squared magnitude increases linearly with the displacement amplitude of the sinusoidal movement applied to the stimulus fiber's base. In contrast, stimulation near the bundle's characteristic frequency of spontaneous oscillation (•) produces an amplified, nonlinear response (*top curves*). When plotted in a doubly logarithmic form (*bottom curves*), the sensitivity of various hair bundles displays a compressive nonlinearity when the bundles are stimulated near their characteristic frequency. Three regimes of responsiveness are observed: a linear regime of high sensitivity for small stimuli (*top horizontal gray line*), a nonlinear regime in which the sensitivity decreases approximately like the negative two-thirds power of the stimulus amplitude (*oblique gray line*) and finally a linear regime of low sensitivity for intense stimuli (*bottom horizontal gray line*). (Adapted from Martin et al. 2001.)

to a hair bundle, the spectral density of bundle motion is broadly distributed about the oscillation's characteristic frequency ν_0 (Section 4.3). Application of a sinusoidal stimulus, however, renders the hair-bundle movements more regular, by a process called "entrainment" or "phase-locking." For small stimuli, phase-locking is partial. A degree of phase locking can be defined as the fraction of hair-bundle movement that occurs at the stimulus frequency (Martin et al. 2001). Phase-locking is most efficient when the hair bundle is stimulated near its characteristic frequency. In this case, the degree of phase-locking increases steeply with the stimulus intensity, showing a 9 % increase per nanometer displacement. Nanometer displacements of the stimulus fiber's base, corresponding to forces as small as a fraction of one piconewton applied at the bundle's top, suffice to produce a narrow peak at the stimulus frequency in the spectrum of bundle motion. When the stimulus gets more intense, the bundle's response at the stimulus frequency grows at the expense of other frequency components of bundle motion: the power produced by the hair cell to drive its hair-bundle movements is progressively funneled at the frequency of stimulation. At a frequency near the bundle's characteristic frequency of spontaneous oscillation, a ±10-nm stimulus, corresponding to a ±3-pN force, produces near complete entrainment. In contrast, when a hair bundle is stimulated at a frequency for which the spectrum of spontaneous bundle motion displays negligible power, a ±10-nm stimulus does not evoke any significant entrainment, and complete entrainment requires stimuli with amplitudes greater than 100 nm. In this case, the degree of phase locking increases only by 1 % per nanometer of displacement. In both cases, the overall magnitude of hair-bundle motion does not increase significantly until the stimulus is intense enough to completely entrain the bundle's movements.

Partial phase-locking of hair-bundle movements is expected to evoke a more regular release of synaptic transmitter by the hair cell and in turn an afferent-nerve activity partially locked to the stimulus. In birds (Gleich and Narins 1988; Köppl 1997), amphibians (Hillery and Narins 1984), and fish (Fay and Coombs 1983) the threshold for phase-locking of spontaneous neural activity in vivo is 10–20 dB lower than that for increased firing of auditory nerve fibers. This behavior, which resembles that of noisy hair-bundle movements, could be explained if hair bundles in these species oscillated spontaneously and were increasingly entrained by stimuli of growing magnitudes. In the goldfish, the behavioral threshold is not associated with a change in the regularity of activity in saccular neurons (Fay and Coombs 1983), but this observation remains to be confirmed in other species.

6.3 Compressive Nonlinearity of the Bundle's Response to Stimuli of Increasing Magnitudes: Gain of the Amplificatory Process

When an oscillatory hair bundle is subjected to stimuli of increasing magnitudes near its characteristic frequency of spontaneous oscillation, the root-mean-squared magnitude of bundle motion $X_{\mathrm{RMS},\omega_S}$ at the stimulus frequency ω_S exhibits a compressive nonlinearity (Martin et al. 2001). The bundle's response displays three regimes as a function of the magnitude of motion of the stimulus fiber's base (Fig. 4.14B). For small stimuli, in the range 1–10 nm, the response is linear. Correspondingly, the bundle's sensitivity saturates to a constant value in the range 5–20 nm · pN^{-1}. For moderate to intense stimuli, in the range 15–150 nm, the response displays a compressive nonlinearity. The dependence of response magnitude $X_{\mathrm{RMS},\omega_S}$ on the stimulus magnitude Δ is well described by a power law $X_{RMS,\omega_S} \propto \Delta^{\frac{1}{3}}$. The hair-bundle sensitivity correspondingly varies as $\Delta^{-\frac{2}{3}}$. Finally, for stimuli exceeding ±150 nm, the response becomes linear and the sensitivity reaches a plateau at $\approx 1\ \text{nm} \cdot \text{pN}^{-1}$.

In contrast, stimulation at frequencies much higher than the bundle's characteristic frequency of spontaneous oscillation evokes linear responses of low and constant sensitivity $\approx 1\ \text{nm} \cdot \text{pN}^{-1}$ throughout the range of stimulus amplitudes (Fig. 4.14B). This sensitivity equals that of the bundle when stimulated at intense levels near its characteristic frequency of spontaneous oscillation and that of passive cells that do not display spontaneous hair-bundle oscillations. This suggests that stimulation at high frequencies or at intense levels evokes passive responses of an oscillatory hair bundle. By comparing the passive and active behaviors of the hair bundle, one can assess the benefit of active hair-bundle motility. The gain of the amplificatory process can be defined by the ratio of the sensitivity at resonance to small stimuli and that to intense or high-frequency stimuli. By this criterion, an oscillatory hair bundle from the bullfrog's saccule amplifies its phase-locked response to minute stimuli by approximately tenfold. In vivo, the frequency-dependent compressive nonlinearity that has been described here at the level of single hair-bundle mechanics should be reflected in the relation between the discharge rate in afferent fibers innervating the hair cell and stimulus intensity. In the barn owl, this relation is sigmoidal for sounds at frequencies well above or below the characteristic frequency of the nerve fiber but gets distorted near that frequency; this behavior is suggestive of a frequency-dependent nonlinearity in the transduction pathway (Köppl and Yates 1999).

7. A Minimal Theoretical Description of Spontaneous Hair-Bundle Oscillations Driven by Myosin-Based Adaptation Motors

Spontaneous oscillations of the hair bundle can be generated by at least three active mechanisms. First, the collection of myosin-based adaptation motors could display a region of negative slope in its force–velocity relation and become self oscillatory as a result of the elastic coupling provided by the gating springs (Jülicher and Prost 1995). Second, coupling Ca^{2+}-mediated reclosure of the transduction channels (Section 3.4.3) with gating kinetics (Choe et al. 1998) or intracellular Ca^{2+} dynamics (Vilfan and Duke 2003) can yield hair-bundle oscillations. Finally, the interplay between negative hair-bundle stiffness and the Ca^{2+}-dependent activity of the adaptation motors can generate oscillations. This third mechanism, which provides a quantitative description of the hair-bundle oscillations observed in the bullfrog's sacculus (Martin et al. 2003; Nadrowski et al. 2004), is described in the following.

7.1 Active Hair-Bundle Mechanics

We describe here the dynamic interplay between nonlinear hair-bundle mechanics, a direct consequence of the gating-spring theory of mechanoelectrical transduction, and the Ca^{2+}-regulated activity of the adaptation motor (Assad and Corey 1992; Martin et al. 2003; Nadrowski et al. 2004). It is useful to introduce new variables $\bar{X} = X - X_{SP}$ and $\bar{X}_a = X_a - X_{SP} - X_r - \frac{D}{2}$, for which the origin of hair-bundle and motor positions, respectively, is such that the gating springs bear no tension. The dynamic behavior of a hair bundle can be described by the following two coupled equations:

$$\lambda \frac{d\bar{X}}{dt} = -\mathrm{K}_{\mathrm{GS}} \left(\bar{X} - \bar{X}_{\mathrm{a}} - DP_{\mathrm{o}}\right) - \mathrm{K}_{\mathrm{SP}}\bar{X} + F + \eta, \tag{4.18}$$

$$\lambda_a \frac{d\bar{X}_a}{dt} = \mathrm{K}_{\mathrm{GS}} \left(\bar{X} - \bar{X}_{\mathrm{a}} - DP_{\mathrm{o}}\right) - \mathrm{K}_{\mathrm{ES}} \left(\bar{X}_{\mathrm{a}} - \bar{X}_{\mathrm{ES}}\right) - F_{\mathrm{a}} + \eta_{\mathrm{a}}, \tag{4.19}$$

Here, the open probability P_{o} of the transduction channels obeys Eqs. (4.6) and (4.9):

$$P_o = \left(1 + A \exp\left(-\frac{Z\left(\bar{X} - \bar{X}_a\right)}{k_B T}\right)\right)^{-1}, \tag{4.20}$$

with

$$A = \exp\left(\frac{\Delta\mu + \mathrm{K}_{\mathrm{GS}} D^2 / 2N}{k_{\mathrm{B}} T}\right).$$

Equation (4.18) describes the dynamics of the hair-bundle displacement $\bar{X}$. At steady state ($\mathrm{d}\bar{X}/\mathrm{d}t = 0$), the external force F is balanced by the elastic restoring forces provided by the parallel combination of stereociliary pivots and gating springs (Eq. [4.12]); otherwise, the hair bundle moves at a velocity that is inversely proportional to the friction coefficient λ. The noise term η accounts for fluctuations that arise from the Brownian motion of fluid molecules and the stochastic gating of transduction channels (Nadrowski et al. 2004). Noise strength is characterized by the autocorrelation function $\langle \eta(t)\eta(0)\rangle = 2k_{\mathrm{B}} T\lambda\delta(t)$ of time t, in which $\delta(t)$ is the Dirac distribution. Under circumstances in which hydrodynamic friction dominates, $\lambda \approx 130\ \mathrm{nN \cdot s \cdot m^{-1}}$ for a hair bundle of the bullfrog's saccule (Denk et al. 1989). With strong gating forces and $P_o \approx 0.5$, however, the contribution to bundle friction owing to channel clatter can dominate hydrodynamic friction, and λ may be tenfold larger (Nadrowski et al. 2004).

Active hair-bundle movements result here from an active force that is exerted by the collection of molecular motors that mediate adaptation (Sections 3.4.1 and 3.4.2); F_a denotes the projection of the motor stall force on the bundle's horizontal axis of mechanosensitivity. The motors' ascent along a stereocilium is opposed by the downward pull exerted by the force

$$F_{\text{mot}} = K_{\text{GS}}\left(\bar{X} - \bar{X}_{\text{a}} - DP_{\text{o}}\right) - K_{\text{ES}}\left(\bar{X}_{\text{a}} - \bar{X}_{\text{ES}}\right) \tag{4.21}$$

that results from the combined tensions in gating and extent springs, respectively. Equation (4.19) describes the dynamic behavior of these motors by a linear expansion in $F_{\text{mot}} - F_{\text{a}}$ of a general relation between motor velocity $d\bar{X}_{\text{a}}/dt$ and F_{mot} near stall force (Assad and Corey 1992; Martin et al. 2003; Nadrowski et al. 2004)

$$\frac{d\bar{X}_{\text{a}}}{dt} = \frac{1}{\lambda_{\text{a}}}\left(F_{\text{mot}} - F_{\text{a}}\right). \tag{4.22}$$

Here $1/\lambda_a$ is the slope of the force–velocity relation. In the bullfrog, measurements of the initial adaptation rate evoked by positive step displacements of the hair bundle as a function of the step magnitude demonstrate a linear increase of the slipping rate (Hacohen et al. 1989). From these experiments, one can estimate that the coefficient λ_a lies in the range 7–25 $\mu\text{N} \cdot \text{s} \cdot \text{m}^{-1}$. To account for the incompleteness of adaptation, an extent spring of stiffness K_{ES} anchors the transduction apparatus to the stereociliary cytoskeleton (Shepherd and Corey 1994). The extent spring bears no tension when $\bar{X}_{\text{a}} = \bar{X}_{\text{ES}}$. At stall, F_{a} is balanced by F_{mot}. Active force production by the motors corresponds to motors climbing up the actin core of the stereocilia, i.e., $d\bar{X}_{\text{a}}/dt < 0$, which tends to increase the extension of the gating springs and thus open the transduction channels. The stochastic activity of the molecular motors generates a stochastic force η_a. The strength of this active noise term can be written as $\langle \eta_{\text{a}}(t)\eta_{\text{a}}(0)\rangle = 2k_{\text{B}}T_{\text{a}}\lambda_{\text{a}}\delta(t)$, in which $T_a \approx 1.5T$ is an effective temperature that characterizes the motor's activity in the bullfrog's saccule (Nadrowski et al. 2004).

To account for the regulation of adaptation by Ca^{2+} (Section 3.3), it is assumed that the force produced by the motors depends on the Ca^{2+} concentration, denoted by $[Ca^{2+}]$, at the motor site

$$F_{\text{a}} \approx F_{\text{max}} + \left.\frac{dF_a}{d[Ca^{2+}]}\right|_{\text{REF}} \left[Ca^{2+}\right]. \tag{4.23}$$

Equation (4.23) ignores nonlinearities in $F_{\text{a}}([Ca^{2+}])$; this equation represents the truncation to linear order of a general expansion in powers of $[Ca^{2+}]$ with respect to a reference Ca^{2+} concentration. Imposing $\left.\frac{dF_{\text{a}}}{d\left[Ca^{2+}\right]}\right|_{\text{REF}} < 0$ ensures that increased Ca^{2+} levels at the motor site reduce active force generation at the motor site and thus result in lower open probabilities for the transduction channels (Section 3.3). Assuming that relaxation of the intracellular Ca^{2+} concentration near the motors occurs on a shorter time scale than hair-bundle movements (Lumpkin and Hudspeth 1998), the Ca^{2+} concentration at the motor site is approximately determined by the open probability of the transduction channels:

$$[\mathrm{Ca}^{2+}] \approx [\mathrm{Ca}^{2+}]_0 + [\mathrm{Ca}^{2+}]_\mathrm{M} P_\mathrm{o}. \tag{4.24}$$

In Eq. (4.24), the maximal Ca^{2+} concentration at the motor site $[Ca^{2+}]_M >> [Ca^{2+}]_0$ depends on the Ca^{2+} concentration in endolymph and can thus be varied experimentally (Martin et al. 2003). By combining, Eq. (4.23) with Eq. (4.24), one gets

$$F_\mathrm{a} \approx F_\mathrm{max} (1 - S_\mathrm{Ca} P_\mathrm{o}), \tag{4.25}$$

in which the dimensionless parameter

$$S_\mathrm{Ca} = -\frac{[Ca^{2+}]_\mathrm{M}}{F_\mathrm{max}} \left. \frac{dF_a}{d[Ca^{2+}]} \right|_\mathrm{REF}$$

defines the strength of the Ca^{2+} feedback on the motor force, and F_{max} is the maximal force that the motors can generate.

7.2 State Diagram in the Absence of Noise: Hopf Bifurcation

Most parameters in Eqs. (4.18) and (4.19) are constrained by measurements of the bundle's force–displacement relations, which give numerical estimates for K_{GS}, K_{SP}, and D (Section 5), and of adaptation kinetics and extent, which yield estimates for λ_a and K_{ES}, respectively. The bundle's friction coefficient λ can also be estimated (Denk and Webb 1989; Nadrowski et al. 2004). Assuming that these parameters are fixed, the dynamic behavior of a hair bundle depends only on two control parameters: the maximal force F_{max} that the motors can produce and the Ca^{2+}-feedback strength S_{Ca}. Steady states satisfy $d\bar{X}/dt = 0$ and $d\bar{X}_a/dt = 0$. First ignoring fluctuations and with $F = 0$, linear-stability analysis of these steady states reveals that a hair bundle can become oscillatory, provided that the condition for negative stiffness is satisfied (Eq. [4.15]). In this case, the state diagram of a hair bundle displays different regions (Fig. 4.15), including an oscillatory region for intermediate values of F_{max} and S_{Ca} (Nadrowski et al. 2004). Note that in this description, the activity of the adaptation motor must be regulated by Ca^{2+} to obtain oscillations: there are no oscillations at $S_{Ca} = 0$. Because the Ca^{2+}-feedback strength S_{Ca} depends on the Ca^{2+} concentration in endolymph, a hair bundle is expected to oscillate only when this concentration remains within a certain range (Section 4.3.2).

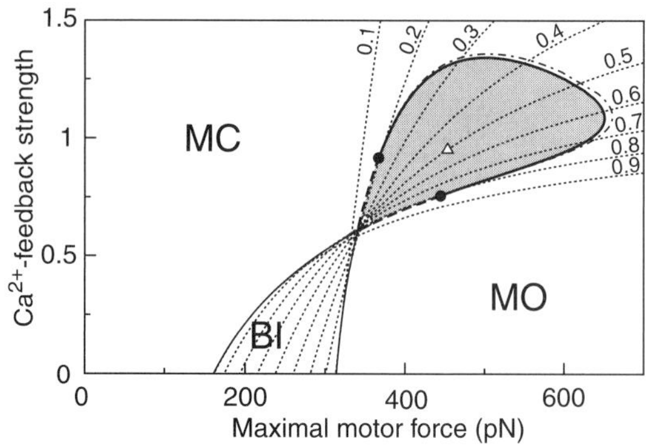

FIGURE 4.15. State diagram of the system described by Eqs. (4.18) and (4.19) as a function of the maximal motor force $\gamma F_{\max}$ that the adaptation motor can produce along the long axis of the stereocilia and of the Ca^{2+}-feedback strength S_{Ca} on motor activity. Lines of equal probability P_o of the transduction channels (*dotted lines*) are superimposed and each indexed by the corresponding value for P_o. The diagram is parsed into regions in which the hair bundle can be monostable with its transduction channels mostly closed (MO), monostable with its transduction channels mostly open (MO), bistable (BI), or oscillatory (*gray*). Hopf bifurcations occur along the line separating oscillating from stable states. For $F_{\max}\gamma = 352$ pN and $S = 0.65$ (∘), stochastic simulations of the system closely match the behavior of a hair bundle that was studied experimentally (Nadrowski et al. 2004). For $F_{\max}\gamma = 439$ pN and $P_o = 0.5$, stochastic simulations indicate that the system displays a global maximum of sensitivity (△) in the presence of noise. Parameters: $\lambda = 2.8\mu N \cdot s \cdot m^{-1}$, $\lambda_a = 10\mu N \cdot s \cdot m^{-1}$, $K_{GS} = 750\mu N \cdot m^{-1}$, $K_{SP} = 200\mu N \cdot m^{-1}$, $K_{ES} = 0\mu N \cdot m^{-1}$, $K_F = 400\mu N \cdot m^{-1}$, $N = 50$, and $D = 60.9$ nm. (Adapted from Nadrowski et al. 2004.)

What makes a hair bundle oscillate? There is a one-to-one correspondence between a bundle's force–displacement and open probability–displacement relations. A region of negative stiffness in a force–displacement relation thus disqualifies a whole range of open probabilities at which a hair bundle would settle stably. Because the channels' open probability is set by the Ca^{2+}-regulated force that the adaptation motor can exert on the gating springs, a stable steady state will be out of reach if the force at which the motor would be at stall corresponded to an unstable open probability of the transduction channels. In this case, the adaptation motor oscillates (Fig. 4.16), perpetually frustrated in its attempt to reach a thermodynamically unstable position (Martin et al. 2000; Le Goff et al. 2005).

As detailed in Chapter 3, by Duke and Jülicher, the boundary between a stable and an oscillatory state corresponds to an oscillatory instability, called a Hopf bifurcation. For the system described by Eqs. (4.18) and (4.19) in the absence of noise, the bifurcation can be approached arbitrarily closely by adjusting the maximal motor force $F_{\max}$ or the Ca^{2+}-feedback strength S_{Ca}. Operating near a Hopf bifurcation is ideal for detecting oscillatory stimuli, for proximity to the bifurcation ensures high sensitivity and frequency selectivity (Choe et al. 1998; Camalet et al. 2000; Eguiluz et al. 2000). Note, however, that these properties

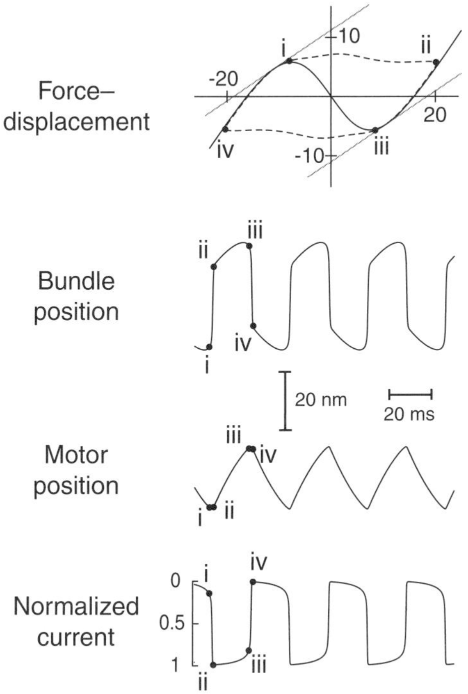

FIGURE 4.16. Simulations of spontaneous oscillations. The first figure from the top (continuous line) describes the relation between the external force F (ordinate, in piconewtons), applied to the combined system of a hair bundle with an attached stimulus fiber, and the resulting deflection X of the hair bundle's top (abscissa, in nanometers) from the system's steady-state position at rest. This relation was calculated under circumstances for which the adaptation motor remained immobile at a steady-state position that was determined in the absence of an external force. The origin of the graph defines the system's operating point. With no external force applied, the system's position varies with time according to the trajectory superimposed as a dashed line on the force–displacement relation. This trajectory represents the force $F - \mathrm{K_{SP}}X_a$ as a function of the relative position $X - X_a$, where the positions are here calculated with respect to those obtained at steady state with $F = 0$. As indicated on the left side, the remaining figures represent, from top to bottom respectively, the position of the hair bundle, that of the adaptation motor, and the normalized transduction current, all plotted as functions of time. The system displays spontaneous oscillations of the bundle's position at a frequency of 27 Hz and with a root-mean-squared magnitude of 15.8 nm. The fast components of bundle motion, during which the adaptation motor remains almost stationary, is associated with horizontal jumps across the unstable region of the force–displacement relation and abrupt changes of the channels' open probability (points *i* to *ii* and *iii* to *iv*). The slow components of bundle motion amounts to adaptive movements along the stable branches of the relation (points *ii* to *iii* and *iii* to *iv*) and are produced by a sawtooth oscillation of the adaptation motor. The transduction current reflects the bundle oscillations: in each half-cycle, a fast movement in the positive direction is associated with a fast opening of a large fraction of the transduction channels, whereas the slow positive movement that follows corresponds to slow adaptive reclosure of the channels. Parameters are the same as in Figure 4.15 with $S = 0.85$ and $F_{\mathrm{max}} = 56$ pN.

are generic of any dynamical system on the verge of an oscillatory instability: they do not pinpoint the specific mechanism that underlies the amplificatory process.

7.3 Fluctuations and Maximal Sensitivity of Oscillatory Hair Bundles

The responsiveness of an oscillatory hair bundle to sinusoidal stimuli (Section 6.3) displays power-law behaviors that resemble those associated with noiseless active systems operating close to a Hopf bifurcation (Section 2.4 in Chapter 3, by Duke and Jülicher). However, the bundle's sensitivity to small stimuli saturates, whereas that of a critical oscillator can be arbitrarily large. Is the hair bundle at an optimal operating point or could its sensitivity be greater than has so far been observed? Hair-bundle oscillations are noisy. Stochastic simulations of Eqs. (4.18) and (4.19) can produce spontaneous oscillations and response functions in quantitative agreement with experiments for a particular choice of the control parameters F_{max} and S_{Ca} (Nadrowski et al. 2004). By exploring all possible dynamic states that the hair bundle can endorse, it was shown that a hair bundle from the bullfrog's saccule operates near an optimum of mechanosensitivity and frequency selectivity. Fluctuations thus restrict the system's sensitivity to oscillatory stimuli as well as the range of stimulus magnitudes over which the compressive nonlinearity of the bundle's response occurs. Although fluctuations conceal the phase coherence of spontaneous oscillations and the bifurcation between stable and oscillatory states, a single hair bundle amplifies (about tenfold) its response to small stimuli, and correspondingly, the characteristic compressive nonlinearity that arises near a Hopf bifurcation remains.

7.4 "Excitable" Regimes: The Twitch Revisited

A hair bundle that operates on the stable side of a Hopf bifurcation does not oscillate spontaneously but can nevertheless display active behaviors in response to stimulation (Tinevez et al. 2007). In the case of a monostable hair bundle with most of the transduction channels closed, a positive step stimulus can evoke a nonmonotonic movement with a recoil that resembles the twitch described in Section 4.1. Such "positive twitch" is favored by large Ca^{2+}-feedback strengths S_{Ca} and thus high extracellular Ca^{2+} concentrations. Conversely, a quiescent hair bundle with most of its channels open, an instance promoted by low values of S_{Ca}, displays a recoil in response to negatively directed stimuli; this "negative twitch" resembles in waveform one half-cycle of a spontaneous oscillation (Fig. 4.17A). The two types of excitable behaviors differ by their kinetics; the positive twitch is much faster than its negative counterpart (Tinevez et al. 2006).

Interestingly, the occurrence of negative stiffness is not mandatory for the observation of a recoil in the bundle's movement evoked by a step stimulus. In the case that the force–displacement relation is everywhere of positive slope,

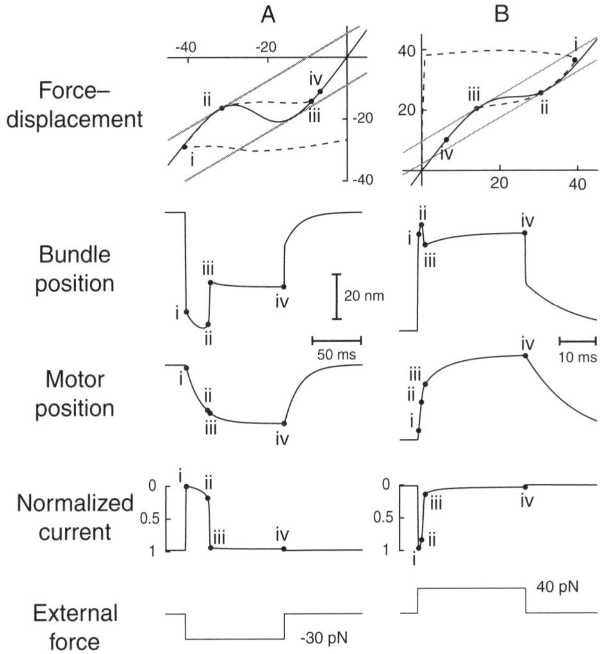

FIGURE 4.17. Twitches: (**A**) Negative twitch. The force–displacement relation contains here a negative-stiffness region that is centered at a position of -16 nm with respect to the position at rest. The system is stable, but a negatively directed force step of -30 pN elicits a twitch of 20-nm magnitude and 25-ms duration that qualitatively resembles one half-cycle of the oscillation shown in Figure 4.12 (points *i* to *iii*). The recoil (points *ii* to *iii*) is associated with a fast jump of the hair bundle across the unstable region of negative stiffness and a massive opening of the transduction channels. This complex, triphasic hair-bundle movement is associated with a monotonic movement of the adaptation motor. Parameters: same as in Figure 4.15 but $\lambda = 0.28 \mu\text{N} \cdot \text{s} \cdot \text{m}^{-1}$, $K_{ES} = 250 \mu\text{N} \cdot \text{m}^{-1}$, $D = 45$ nm, $S = 0.3$, and $F_{max} = 63$ pN. (**B**) Positive twitch. The force–displacement relation is everywhere of positive slope but displays a compliant nonlinear region delimited by two oblique straight lines (*gray*) in which the slope stiffness is lower than the stiffness K_{SP}. This region is centered at a positive position of 18 nm with respect to the position at rest. The brisk application of a $+40$ pN force first evokes a movement in the direction of the applied force that is interrupted by a recoil (points *ii* to *iii*) within the first few milliseconds of motion, during which the system traverses the compliant region of the force–displacement relation. This complex, triphasic movement is produced by a monotonic movement of the adaptation motor toward the stereociliary base. The recoil is associated with a small movement of ≈ 1 nm of the adaptation motor along the long axis of the stereocilia and with reclosure of 65 % of the transduction channels on a millisecond time scale. The late positive component of motion is slower and associated with a component of channel reclosure of similar time course. For smaller stimuli, the system is directly brought into the flat region of its force–displacement relation (not shown). In this case, the first slow component of adaptive motion (points *i* to *ii*) is not observed, and the twitch thus appears as a sharp spikes. Here $\lambda = 0.28 \mu\text{N} \cdot \text{s} \cdot \text{m}^{-1}$, $\lambda_a = 10 \mu\text{N} \cdot \text{s} \cdot \text{m}^{-1}$, $K_{GS} = 1000 \mu\text{N} \cdot \text{m}^{-1}$, $K_{SP} = 200 \mu\text{N} \cdot \text{m}^{-1}$, $K_{ES} = 250 \mu\text{N} \cdot \text{m}^{-1}$, $K_F = 560 \mu\text{N} \cdot \text{m}^{-1}$, $N = 50$, $D = 37.1$ nm, $S = 3.8$, and $F_{max} = 49$ pN.

Equation (4.19) indicates that a small movement of the adaptation motor δX_{a} evokes a hair-bundle displacement δX that can be approximated by

$$\delta X \approx \frac{\mathrm{K}_{\mathrm{GS}}}{\mathrm{K}_{\mathrm{GS}} + \mathrm{K}_{\mathrm{SP}}} \delta X_a, \tag{4.26}$$

where

$$\mathrm{K}_{\mathrm{GS}} = \mathrm{K}_{\mathrm{GS}} \left(1 - \frac{ZD}{k_{\mathrm{B}}T} P_o (1 - P_o) \right).$$

According to our assumption, the slope $\mathrm{K}_{\mathrm{GS}} + \mathrm{K}_{\mathrm{SP}}$ of the force–displacement relation is positive for all positions X. A positive adaptive movement δX_a of the motor, for instance provoked by a positive step stimulus, can thus evoke a negative deflection δX of the hair bundle (a recoil), provided that $\mathrm{K}_{\mathrm{GS}} < 0$ and thus that

$$P_{\mathrm{o}} (1 - P_{\mathrm{o}}) > \frac{k_{\mathrm{B}}T}{ZD}. \tag{4.27}$$

This condition can be fulfilled only if there is a region of the force–displacement curve where the local slope is smaller than the stiffness K_{SP} of the pivots: gating compliance must be strong enough that $ZD > 4k_BT$. This condition is less restrictive than the condition for negative stiffness (Eq. [4.15]). In accordance with recent experiments that implicate myosin Ic in both fast and slow components of adaptation (Stauffer et al. 2005), simulations indicate that myosin-based adaptation, as described by Eqs. (4.19) and (4.25), can account for twitch and the associated fast and slow components of the transduction current (Fig. 4.17B).

How can a fast component be produced by myosin-based adaptation? In the case that X is clamped, adaptation is governed by the relaxation of X_{a} only. In this simple case, adaptation is associated with a shift of the channels' open probability–displacement relation along the displacement axis that is driven by a movement of the adaptation motor. In response to a step displacement, Eq. (4.19) indicates that X_{a} relaxes to the steady state with a time course that can be described at long times by a single exponential with the characteristic time scale

$$\tau_a = \frac{\lambda_a}{\mathrm{K}_{\mathrm{GS}} \left(1 - \frac{D}{\delta} \left(1 - S \frac{F_{\max}}{\mathrm{K}_{\mathrm{GS}} D} \right) P_{\mathrm{o}} (1 - P_{\mathrm{o}}) \right) + \mathrm{K}_{\mathrm{ES}}}, \tag{4.28}$$

in which $\delta = k_{\mathrm{B}}T/Z$ and P_{o} is the open probability at the steady state. Adaptation kinetics thus depend on the bundle's operating point and the Ca^{2+}-feedback strength. For an operating point in the linear part of the force–displacement relation, where P_{o} is near either 1 or 0, adaptation is slow with $\tau_{\mathrm{a}} = \lambda_{\mathrm{a}}/\left(\mathrm{K}_{\mathrm{GS}} + \mathrm{K}_{\mathrm{ES}}\right) = 24$ ms with the parameters of Figure 4.17B and $\lambda = 30 \mu\mathrm{N} \cdot \mathrm{s} \cdot \mathrm{m}^{-1}$. In the nonlinear part of the force–displacement region with $P_o \approx$

0.5, adaptation can be fast if $S > K_{GS}D/F_{max}$. Because S_{Ca} increases with the external Ca^{2+} concentration, this condition is favored by immersing the hair bundles in a high-Ca^{2+} medium. In this case, the gating spring is effectively stiffer, and $\tau_a = 3.7$ ms. These values accord with the fast and slow components of adaptation, respectively, that have been measured in the bullfrog's saccule (Stauffer et al. 2005).

8. Summary

Micromanipulation of single hair bundles allows characterization of a bundle's mechanical properties and responsiveness to external forcing. The dynamic behavior of a hair bundle can be rich, and three classes of active hair-bundle movements have been identified. First, in response to abrupt step stimuli, a quiescent hair bundle can monotonically relax toward a new steady-state position. Second, for small step stimuli, the initial response in the direction of the applied force can be interrupted by a fast recoil, also termed twitch, in opposition to the external stimulus, thereby demonstrating a form of mechanical excitability. Finally, a hair bundle can become unstable and oscillate spontaneously. These various regimes of hair-bundle motility are intimately related to adaptation, the active process that provides negative feedback on the open probability of the transduction channels, and to gating compliance, the mechanical correlate of channel gating. By regulating the adaptation process, inflowing Ca^{2+} ions through open transduction channels provide electrical-to-mechanical feedback. Slow and fast components of adaptation have been identified and associated to hair-bundle movements of similar kinetics but opposite polarities. On the basis of this, it has been proposed that two separate mechanisms are at work, each prompted by the Ca^{2+} component of the transduction current but with conflicting effects on hair-bundle movements. At least in frog and mouse vestibular hair cells, however, myosin-1c motor molecules play an essential role in both fast and slow adaptation. An ATP-dependent movement of the myosin-based adaptation motor along the actin core of the stereocilia probably powers slow adaptation and the associated hair-bundle movement in the direction of the applied force. At a fixed open probability of the transduction channels, the motor indeed works to cancel any change in gating-spring tension that is evoked by external stimulation. However, the means by which myosin intervenes in fast adaptation to foster movements in opposition to the stimulus is still unclear. Attached myosin molecules might rock in place to provide negative feedback on gating-spring tension and thus promote fast gating of the transduction complex. Alternatively, a direct interaction of Ca^{2+} with the transduction channels might stabilize the closed state of the channel, rendering it more difficult to open. In any case, hair-bundle movements in opposition to the stimulus are thought to be produced by internal forces that result from conformational changes associated to channel gating during adaptation. If the position of the adaptation motor is fixed and a positive step stimulus is applied, Ca^{2+}-mediated reclosure of the transduction channels further increases the tension in the gating springs and thus promotes negative hair-bundle movements. Conversely, reopening of the transduction channels that were closed by a negative stimulus yields a decrease of gating-spring tension and fosters positive bundle movements. In contrast to movements of the adaptation motor, channel gating associated with adaptation thus affords positive feedback on gating-spring tension. A nonlinear and dynamic tradeoff between opposing forces that are produced by movements of the adaptation motor on the one end and those exerted by channel gating on the other end probably accounts for the complexity of active hair-bundle motility.

The gating-spring model provides an unsurpassed framework to describe mechano-electrical transduction, adaptation, and active hair-bundle movements. This success is all the more impressive if we consider how little we know about the molecular constituents on which the model rests. In particular, the molecular identity of the conformational change associated with channel gating is still unknown. The distance by which the gating spring shortens as a transduction channel opens is too large to simply reflect the movement of a channel's pore-obstructing gate but approximates the power-stroke size of myosin Ic molecules (Batters et al. 2004b). Is the gating swing the reflection of a quick conformational change of the myosin molecules that mediate adaptation? One would then have to explain how channel gating influences the state of the motor. Interpreting the gating swing at a molecular level is clearly an important challenge for the future. In frogs, this gating swing appears to be as large as $\approx$10 nm, in turn producing gating forces up to $\approx$10 pN per transduction complex, when measured along the axis of the gating spring (Martin et al. 2000). As the total gating force that a hair bundle experiences is proportional to the open probability of the transduction channels, large gating forces result in nonlinear force–displacement characteristics and, if large enough, in negative slope stiffness.

Negative stiffness can destabilize the myosin-based adaptation motor, which results in spontaneous hair-bundle oscillations. It was demonstrated experimentally in frogs that the hair cell can harness hair-bundle oscillations to amplify its response to sinusoidal stimuli, thereby enhancing its sensitivity and frequency selectivity to minute stimuli. Only two ingredients are required in this amplificatory process: Ca^{2+}-regulated adaptation motors and strong gating compliance. Because these ingredients are found in distantly related species such as the turtle (Ricci et al. 2002) and the rat (Kennedy et al. 2005), it is plausible but remains to be demonstrated that the mechanism unveiled in frogs is relevant to other species as well. One concern about myosin-based adaptation is its ability to drive spontaneous oscillations at auditory frequencies of a few kilohertz or more (Manley and Gallo 1997). The ATPase cycle of a single myosin can be rather long, as exemplified by myosin-1c, which can hydrolyze in vitro as little as $\approx$0.6 ATP molecule per second (Gillespie et al. 1999). Because myosin molecules work in groups, however, only a small fraction of the molecules needs to be involved in one cycle of bundle movement. A collection of motor molecules can thus generate much faster movements than the overall biochemical cycle of a single molecule would suggest (Hudspeth and Gillespie 1994), being limited theoretically only by the shortest time scale in the force-producing process (Jülicher and Prost 1997). In addition, the relevant time scale of force production also depends on the rigidity and friction of the elements to which the motor collection is attached: if the myosins are coupled to stiff enough elements, oscillations of a structure like the hair bundle can happen at high frequencies (Camalet et al. 2000). It remains possible, however, that a hair bundle makes use of other or additional elements than molecular motors to produce fast hair-bundle oscillations (Choe et al. 1998; Vilfan and Duke 2003). As described in Chapter 3, by Duke and Jülicher, any dynamical system that operates near an oscillatory instability—a Hopf bifurcation—is endowed with generic properties that are ideally suited for the detection of oscillatory stimuli. Whatever mechanism creates such instability, each hair cell might have found its way to the advantageous proximity of a Hopf bifurcation.

References

Assad JA, Corey DP (1992) An active motor model for adaptation by vertebrate hair cells. J Neurosci 12:3291–3309.

Assad JA, Hacohen N, Corey DP (1989) Voltage dependence of adaptation and active bundle movement in bullfrog saccular hair cells. Proc Natl Acad Sci USA 86:2918–2922.

Assad JA, Shepherd GM, Corey DP (1991) Tip-link integrity and mechanical transduction in vertebrate hair cells. Neuron 7:985–994.

Batters C, Wallace MI, Coluccio LM, Molloy JE (2004a) A model of stereocilia adaptation based on single molecule mechanical studies of myosin I. Philos Trans R Soc Lond B Biol Sci 359:1895–1905.

Batters C, Arthur CP, Lin A, Porter J, Geeves MA, Milligan RA, Molloy JE, Coluccio LM (2004b) Myo1c is designed for the adaptation response in the inner ear. EMBO J 23:1433–1440.

Benser ME, Marquis RE, Hudspeth AJ (1996) Rapid, active hair bundle movements in hair cells from the bullfrog's sacculus. J Neurosci 16:5629–5643.

Bialek W, Wit HP (1984) Quantum limits to oscillator stability: theory and experiments on acoustic emissions from the human ear. Phys Lett A 104:173–178.

Boeda B, El-Amraoui A, Bahloul A, Goodyear R, Daviet L, Blanchard S, Perfettini I, Fath KR, Shorte S, Reiners J, Houdusse A, Legrain P, Wolfrum U, Richardson G, Petit C (2002) Myosin VIIa, harmonin and cadherin 23, three Usher I gene products that cooperate to shape the sensory hair cell bundle. EMBO J 21:6689–6699.

Bozovic D, Hudspeth AJ (2003) Hair-bundle movements elicited by transepithelial electrical stimulation of hair cells in the sacculus of the bullfrog. Proc Natl Acad Sci USA 100:958–963.

Callen HB, Welton TA (1951) Irreversibility and generalized noise. Phys Rev 83:34–40.

Camalet S, Duke T, Jülicher F, Prost J (2000) Auditory sensitivity provided by self-tuned critical oscillations of hair cells. Proc Natl Acad Sci USA 97:3183–3188.

Cheung ELM, Corey DP (2006) Ca2+ Changes the force sensitivity of the hair-cell transduction channel. Biophys J 90:124–139.

Choe Y, Magnasco MO, Hudspeth AJ (1998) A model for amplification of hair-bundle motion by cyclical binding of Ca^{2+} to mechanoelectrical-transduction channels. Proc Natl Acad Sci USA 95:15321–15326.
Corey DP, Hudspeth AJ (1979a) Ionic basis of the receptor potential in a vertebrate hair cell. Nature 281:675–677.
Corey DP, Hudspeth AJ (1979b) Response latency of vertebrate hair cells. Biophys J 26:499–506.
Corey DP, Hudspeth AJ (1983a) Kinetics of the receptor current in bullfrog saccular hair cells. J Neurosci 3:962–976.
Corey DP, Hudspeth AJ (1983b) Analysis of the microphonic potential of the bullfrog's sacculus. J Neurosci 3:942–961.
Corey DP, Sotomayor M (2004) Hearing: tightrope act. Nature 428:901–903.
Corey DP, Garcia-Anoveros J, Holt JR, Kwan KY, Lin SY, Vollrath MA, Amalfitano A, Cheung EL, Derfler BH, Duggan A, Geleoc GS, Gray PA, Hoffman MP, Rehm HL, Tamasauskas D, Zhang DS (2004) TRPA1 is a candidate for the mechanosensitive transduction channel of vertebrate hair cells. Nature 432:723–730.
Crawford AC, Fettiplace R (1985) The mechanical properties of ciliary bundles of turtle cochlear hair cells. J Physiol 364:359–379.
Crawford AC, Evans MG, Fettiplace R (1989) Activation and adaptation of transducer currents in turtle hair cells. J Physiol 419:405–434.
Crawford AC, Evans MG, Fettiplace R (1991) The actions of calcium on the mechano-electrical transducer current of turtle hair cells. J Physiol 434:369–398.
Denk W, Webb WW (1989) Thermal-noise-limited transduction observed in mechanosensory receptors of the inner ear. Phys Rev Lett 63:207–210.
Denk W, Webb WW (1992) Forward and reverse transduction at the limit of sensitivity studied by correlating electrical and mechanical fluctuations in frog saccular hair cells. Hear Res 60:89–102.
Denk W, Webb WW, Hudspeth AJ (1989) Mechanical properties of sensory hair bundles are reflected in their Brownian motion measured with a laser differential interferometer. Proc Natl Acad Sci USA 86:5371–5375.
Denk W, Holt JR, Shepherd GM, Corey DP (1995) Calcium imaging of single stereocilia in hair cells: localization of transduction channels at both ends of tip links. Neuron 15:1311–1321.
Eatock RA (2000) Adaptation in hair cells. Annu Rev Neurosci 23:285–314.
Eatock RA, Corey DP, Hudspeth AJ (1987) Adaptation of mechanoelectrical transduction in hair cells of the bullfrog's sacculus. J Neurosci 7:2821–2836.
Eguiluz VM, Ospeck M, Choe Y, Hudspeth AJ, Magnasco MO (2000) Essential nonlinearities in hearing. Phys Rev Lett 84:5232–5235.
Fay RR, Coombs S (1983) Neural mechanisms in sound detection and temporal summation. Hear Res 10:69–92.
Fettiplace R, Fuchs PA (1999) Mechanisms of hair cell tuning. Annu Rev Physiol 61: 809–834.
Fettiplace R, Ricci AJ (2003) Adaptation in auditory hair cells. Curr Opin Neurobiol 13:446–451.
Flock A, Strelioff D (1984) Graded and nonlinear mechanical properties of sensory hairs in the mammalian hearing organ. Nature 310:597–599.
Flock A, Flock B, Murray E (1977) Studies on the sensory hairs of receptor cells in the inner ear. Acta Otolaryngol 83:85–91.

Géléoc GS, Lennan GW, Richardson GP, Kros CJ (1997) A quantitative comparison of mechanoelectrical transduction in vestibular and auditory hair cells of neonatal mice. Proc R Soc Lond B 264:611–621.

Gillespie PG, Wagner MC, Hudspeth AJ (1993) Identification of a 120 kd hair-bundle myosin located near stereociliary tips. Neuron 11:581–594.

Gillespie PG, Gillespie SK, Mercer JA, Shah K, Shokat KM (1999) Engineering of the myosin-Iβ nucleotide-binding pocket to create selective sensitivity to N(6)-modified ADP analogs. J Biol Chem 274:31373–31381.

Gleich O, Narins PM (1988) The phase response of primary auditory afferents in a songbird (Sturnus vulgaris L.). Hear Res 32:81–91.

Gold T (1948) Hearing. II. the physical basis of the action of the cochlea. Proc R Soc Lond B 135:492–498.

Goodyear R, Richardson G (1992) Distribution of the 275 kD hair cell antigen and cell surface specialisations on auditory and vestibular hair bundles in the chicken inner ear. J Comp Neurol 325:243–256.

Goodyear R, Richardson G (1999) The ankle-link antigen: an epitope sensitive to calcium chelation associated with the hair-cell surface and the calycal processes of photoreceptors. J Neurosci 19:3761–3772.

Goodyear R, Richardson G (2003) A novel antigen sensitive to calcium chelation that is associated with the tip links and kinocilial links of sensory hair bundles. J Neurosci 23:4878–4887.

Hackney CM, Fettiplace R, Furness DN (1993) The functional morphology of stereociliary bundles on turtle cochlear hair cells. Hear Res 69:163–175.

Hacohen N, Assad JA, Smith WJ, Corey DP (1989) Regulation of tension on hair-cell transduction channels: displacement and calcium dependence. J Neurosci 9:3988–3997.

Hänggi P, Talkner P, Borkovec M (1990) Reaction-rate theory: fifty years after Kramers. Rev Mod Phys 62:251–341.

Hasson T, Gillespie PG, Garcia JA, MacDonald RB, Zhao Y, Yee AG, Mooseker MS, Corey DP (1997) Unconventional myosins in inner-ear sensory epithelia. J Cell Biol 137:1287–1307.

Hillery CM, Narins PM (1984) Neurophysiological evidence for a traveling wave in the amphibian inner ear. Science 225:1037–1039.

Holt JR, Corey DP (2000) Two mechanisms for transducer adaptation in vertebrate hair cells. Proc Natl Acad Sci USA 97:11730–11735.

Holt JR, Corey DP, Eatock RA (1997) Mechanoelectrical transduction and adaptation in hair cells of the mouse utricle, a low-frequency vestibular organ. J Neurosci 17:8739–8748.

Holt JR, Gillespie SK, Provance DW, Shah K, Shokat KM, Corey DP, Mercer JA, Gillespie PG (2002) A chemical-genetic strategy implicates myosin-1c in adaptation by hair cells. Cell 108:371–381.

Holton T, Hudspeth AJ (1986) The transduction channel of hair cells from the bull-frog characterized by noise analysis. J Physiol 375:195–227.

Howard J and Bechstedt S (2004) Hypothesis: a helix of ankyrin repeats of the NOMPC-TRP ion channel is the gating spring of mechanoreceptors. Curr Biol 14:R224–226.

Howard J, Hudspeth AJ (1987) Mechanical relaxation of the hair bundle mediates adaptation in mechanoelectrical transduction by the bullfrog's saccular hair cell. Proc Natl Acad Sci USA 84:3064–3068.

Howard J, Hudspeth AJ (1988) Compliance of the hair bundle associated with gating of mechanoelectrical transduction channels in the bullfrog's saccular hair cell. Neuron 1:189–199.

Howard J, Spudich JA (1996) Is the lever arm of myosin a molecular elastic element? Proc Natl Acad Sci USA 93:4462–4464.

Howard J, Roberts WM, Hudspeth AJ (1988) Mechanoelectrical transduction by hair cells. Ann Rev Biophys Chem 17:99–124.

Howard J (2001) Mechanics of Motor Proteins and the Cytoskeleton. Sunderland, MA: Sinauer Associates.

Hudspeth AJ (1982) Extracellular current flow and the site of transduction by vertebrate hair cells. J Neurosci 2:1–10.

Hudspeth AJ (1983a) The hair cells of the inner ear. Sci Am 248:54–64.

Hudspeth AJ (1983b) Mechanoelectrical transduction by hair cells in the acousticolateralis sensory system. Annu Rev Neurosci 6:187–215.

Hudspeth AJ (1989) How the ear's works work. Nature 341:397–404.

Hudspeth AJ (1992) Hair-bundle mechanics and a model for mechanoelectrical transduction by hair cells. Soc Gen Physiol Ser 47:357–370.

Hudspeth AJ, Corey DP (1977) Sensitivity, polarity, and conductance change in the response of vertebrate hair cells to controlled mechanical stimuli. Proc Natl Acad Sci USA 74:2407–2411.

Hudspeth AJ, Gillespie PG (1994) Pulling springs to tune transduction: adaptation by hair cells. Neuron 12:1–9.

Huxley AF, Simmons RM (1971) Proposed mechanism of force generation in striated muscle. Nature 233:533–538.

Iwasa KH, Ehrenstein G (2002) Cooperative interaction as the physical basis of the negative stiffness in hair cell stereocilia. J Acoust Soc Am 111:2208–2212.

Jacobs RA, Hudspeth AJ (1990) Ultrastructural correlates of mechanoelectrical transduction in hair cells of the bullfrog's internal ear. Cold Spring Harb Symp Quant Biol 55:547–561.

Jaramillo F, Hudspeth AJ (1991) Localization of the hair cell's transduction channels at the hair bundle's top by iontophoretic application of a channel blocker. Neuron 7:409–420.

Jaramillo F, Hudspeth AJ (1993) Displacement-clamp measurement of the forces exerted by gating springs in the hair bundle. Proc Natl Acad Sci USA 90:1330–1334.

Jülicher F, Prost J (1995) Cooperative molecular motors. Phys Rev Lett 75:2618–2621.

Jülicher F, Prost J (1997) Spontaneous oscillations of collective molecular motors. Phys Rev Lett 78:4510–4513.

Kachar B, Parakkal M, Kurc M, Zhao Y, Gillespie PG (2000) High-resolution structure of hair-cell tip links. Proc Natl Acad Sci USA 97:13336–13341.

Kennedy HJ, Evans MG, Crawford AC, Fettiplace R (2003) Fast adaptation of mechanoelectrical transducer channels in mammalian cochlear hair cells. Nat Neurosci 6: 832–836.

Kennedy HJ, Crawford AC, Fettiplace R (2005) Force generation by mammalian hair bundles supports a role in cochlear amplification. Nature 433:880–883.

Köppl C (1997) Phase locking to high frequencies in the auditory nerve and cochlear nucleus magnocellularis of the barn owl, *Tyto alba*. J Neurosci 17:3312–3321.

Köppl C, Yates G (1999) Coding of sound pressure level in the barn owl's auditory nerve. J Neurosci 19:9674–9686.

Kozlov AS, Riesler T and Hudspeth AJ (2006) Coherent motion of stereocilia assures the concerted gating of hair-cell transduction channels. Nat Neurosci 10:87–92

Kroese AB, Das A, Hudspeth AJ (1989) Blockage of the transduction channels of hair cells in the bullfrog's sacculus by aminoglycoside antibiotics. Hear Res 37:203–217.

Kros CJ, Rüsch A, Richardson GP (1992) Mechano-electrical transducer currents in hair cells of the cultured neonatal mouse cochlea. Proc R Soc Lond B 249:185–193.

Kros CJ, Marcotti W, van Netten SM, Self TJ, Libby RT, Brown SD, Richardson GP, Steel KP (2002) Reduced climbing and increased slipping adaptation in cochlear hair cells of mice with Myo7a mutations. Nat Neurosci 5:41–47.

Kwan KY, Allchorne AJ, Vollrath MA, Christensen AP, Zhang D-S, Woolf CJ, Corey DP (2006) TRPA1 Contributes to cold, mechanical, and chemical nociception but is not essential for hair-cell transduction. Neuron 50:277–289.

Le Goff L, Bozovic D, Hudspeth AJ (2005) Adaptive shift in the domain of negative stiffness during spontaneous oscillation by hair bundles from the internal ear. PNAS 102:16996–17001.

Lumpkin EA, Hudspeth AJ (1995) Detection of Ca^{2+} entry through mechanosensitive channels localizes the site of mechanoelectrical transduction in hair cells. Proc Natl Acad Sci USA 92:10297–10301.

Lumpkin EA, Hudspeth AJ (1998) Regulation of free Ca^{2+} concentration in hair-cell stereocilia. J Neurosci 18:6300–6318.

Lumpkin EA, Marquis RE, Hudspeth AJ (1997) The selectivity of the hair cell's mechanoelectrical-transduction channel promotes Ca^{2+} flux at low Ca^{2+} concentrations. Proc Natl Acad Sci USA 94:10997–11002.

Manley GA (1990) Peripheral Hearing Mechanisms in Reptiles and Birds. Berlin: Springer-Verlag.

Manley GA, Gallo L (1997) Otoacoustic emissions, hair cells, and myosin motors. J Acoust Soc Am 102:1049–1055.

Manley GA, Kirk DL (2002) The influence of injected AC and DC currents on spontaneous otoacoustic emissions in the bobtail lizard. J Assoc Res Otolaryngol 3:200–208.

Manley GA, Kirk DL, Köppl C, Yates GK (2001) In vivo evidence for a cochlear amplifier in the hair-cell bundle of lizards. Proc Natl Acad Sci USA 98:2826–2831.

Manley GA, Kirk DL (2005) BAPTA induces frequency shifts in vivo of spontaneous otoacoustic emissions of the bobtail lizard. Audiol Neurootol 10:248–257.

Manley GA, Sienknecht U, Köppl C (2004) Calcium modulates the frequency and amplitude of spontaneous otoacoustic emissions in the bobtail skink. J Neurophysiol 92:2685–2693.

Markin VS, Hudspeth AJ (1995) Gating-spring models of mechanoelectrical transduction by hair cells of the internal ear. Annu Rev Biophys Biomol Struct 24:59–83.

Marquis RE, Hudspeth AJ (1997) Effects of extracellular Ca^{2+} concentration on hair-bundle stiffness and gating-spring integrity in hair cells. Proc Natl Acad Sci USA 94:11923–11928.

Martin P, Hudspeth AJ (1999) Active hair-bundle movements can amplify a hair cell's response to oscillatory mechanical stimuli. Proc Natl Acad Sci USA 96:14306–14311.

Martin P, Hudspeth AJ (2001) Compressive nonlinearity in the hair bundle's active response to mechanical stimulation. Proc Natl Acad Sci USA 98:14386–14391.

Martin P, Mehta AD, Hudspeth AJ (2000) Negative hair-bundle stiffness betrays a mechanism for mechanical amplification by the hair cell. Proc Natl Acad Sci USA 97:12026–12031.

Martin P, Hudspeth AJ, Jülicher F (2001) Comparison of a hair bundle's spontaneous oscillations with its response to mechanical stimulation reveals the underlying active process. Proc Natl Acad Sci USA 98:14380–14385.

Martin P, Bozovic D, Choe Y, Hudspeth AJ (2003) Spontaneous oscillation by hair bundles of the bullfrog's sacculus. J Neurosci 23:4533–4548.

Nadrowski B, Martin P, Julicher F (2004) Active hair-bundle motility harnesses noise to operate near an optimum of mechanosensitivity. Proc Natl Acad Sci USA 101: 12195–12200.

Nagel G, Neugebauer D-C, Schmidt B, Thurm U (1991) Structures transmitting stimulatory force to the sensory hairs of vestibular ampulae of fishes and frogs. Cell Tissue Res 265:567–578.

Ohmori H (1985) Mechano-electrical transduction currents in isolated vestibular hair cells of the chick. J Physiol (Lond) 359:189–217.

Pickles JO, Comis SD, Osborne MP (1984) Cross-links between stereocilia in the guinea pig organ of Corti, and their possible relation to sensory transduction. Hear Res 15: 103–112.

Pickles JO, Brix J, Comis SD, Gleich O, Köppl C, Manley GA, Osborne MP (1989) The organization of tip links and stereocilia on hair cells of bird and lizard basilar papillae. Hear Res 41:31–41.

Ricci AJ, Fettiplace R (1997) The effects of calcium buffering and cyclic AMP on mechanoelectrical transduction in turtle auditory hair cells. J Physiol 501:111–124.

Ricci AJ, Fettiplace R (1998) Calcium permeation of the turtle hair cell mechanotransducer channel and its relation to the composition of endolymph. J Physiol 506:159–173.

Ricci AJ, Wu YC, Fettiplace R (1998) The endogenous calcium buffer and the time course of transducer adaptation in auditory hair cells. J Neurosci 18:8261–8277.

Ricci AJ, Crawford AC, Fettiplace R (2000a) Active hair bundle motion linked to fast transducer adaptation in auditory hair cells. J Neurosci 20:7131–7142.

Ricci AJ, Gray-Keller M, Fettiplace R (2000b) Tonotopic variations of calcium signalling in turtle auditory hair cells. J Physiol 524 Pt 2:423–436.

Ricci AJ, Crawford AC, Fettiplace R (2002) Mechanisms of active hair bundle motion in auditory hair cells. J Neurosci 22:44–52.

Ricci AJ, Crawford AC, Fettiplace R (2003) Tonotopic variation in the conductance of the hair cell mechanotransducer channel. Neuron 40:983–990.

Ricci AJ, Kennedy HJ, Crawford AC, Fettiplace R (2005) The transduction channel filter in auditory hair cells. J Neurosci 25:7831–7839.

Rusch A, Thurm U (1990) Spontaneous and electrically induced movements of ampullary kinocilia and stereovilli. Hear Res 48:247–263.

Russell IJ, Richardson GP, Kossl M (1989) The responses of cochlear hair cells to tonic displacements of the sensory hair bundle. Hear Res 43:55–69.

Russell IJ, Kossl M, Richardson GP (1992) Nonlinear mechanical responses of mouse cochlear hair bundles. Proc R Soc Lond B 250:217–227.

Saimi Y, Kung C (2002) Calmodulin as an ion channel subunit. Annu Rev Physiol 64:289–311.

Sales G, Pye D (1974) Ultrasonic Communication by Animals. London:Chapmann and Hall

Shepherd G, Corey D (1994) The extent of adaptation in bullfrog saccular hair cells. J Neurosci 14:6217–6229.

Shotwell SL, Jacobs R, Hudspeth AJ (1981a) Directional sensitivity of individual vertebrate hair cells to controlled deflection of their hair bundles. Ann NY Acad Sci 374:1–10.

Siemens J, Lillo C, Dumont RA, Reynolds A, Williams DS, Gillespie PG, Muller U (2004) Cadherin 23 is a component of the tip link in hair-cell stereocilia. Nature 428:950–955.

Sollner C, Rauch GJ, Siemens J, Geisler R, Schuster SC, Muller U, Nicolson T (2004) Mutations in cadherin 23 affect tip links in zebrafish sensory hair cells. Nature 428: 955–959.

Sotomayor M, Corey DP, Schulten K (2005) In search of the hair-cell gating spring elastic properties of ankyrin and cadherin repeats. Structure (Camb) 13:669–682.

Stauffer EA, Scarborough JD, Hirono M, Miller ED, Shah K, Mercer JA, Holt JR, Gillespie PG (2005) Fast adaptation in vestibular hair cells requires myosin-1c activity. Neuron 47:541–553.

Strassmaier M, Gillespie PG (2002) The hair cell's transduction channel. Curr Opin Neurobiol 12:380–386.

Strogatz ST (1997) Nonlinear Dynamics and Chaos, 7th edition. Reading, MA: Addison-Wesley.

Tilney LG, Saunders JC (1983) Actin filaments, stereocilia, and hair cells of the bird cochlea. I. Length, number, width, and distribution of stereocilia of each hair cell are related to the position of the hair cell on the cochlea. J Cell Biol 96:807–821.

Tilney L, Derosier D, Mulroy M (1980) The organization of actin filaments in the stereocilia of cochlear hair cells. J Cell Biol 86:244–259.

Tinevez J-Y, Jülicher F, Martin P (2007) Unifying the various incarnations of active hair-bundle motility by the vertebrate hair cell. Biophys J in press.

Tsuprun V, Goodyear RJ, Richardson GP (2004) The structure of tip links and kinocilial links in avian sensory hair bundles. Biophys J 87:4106–4112.

Van Dijk P, Wit HP, Segenhout JM (1989) Spontaneous otoacoustic emissions in the European edible frog. Hear Res 42:273–282.

Van Dijk P, Manley GA, Gallo L, Pavusa A, Taschenberger G (1996) Statistical properties of spontaneous otoacoustic emissions in one bird and three lizard species. J Acoust Soc Am 100:2220–2227.

Vilfan A, Duke T (2003) Two adaptation processes in auditory hair cells together can provide an active amplifier. Biophys J 85:191–203.

Vollrath MA, Eatock RA (2003) Time course and extent of mechanotransducer adaptation in mouse utricular hair cells: comparison with frog saccular hair cells. J Neurophysiol 90:2676–2689.

Walker RG, Hudspeth AJ (1996) Calmodulin controls adaptation of mechanoelectrical transduction by hair cells of the bullfrog's sacculus. Proc Natl Acad Sci USA 93: 2203–2207.

Walker RG, Hudspeth AJ, Gillespie PG (1993) Calmodulin and calmodulin-binding proteins in hair bundles. Proc Natl Acad Sci USA 90:2807–2811.

Walker RG, Willingham AT, Zuker CS (2000) A *Drosophila* mechanosensory transduction channel. Science 287:2229–2234.

Wu YC, Ricci AJ, Fettiplace R (1999) Two components of transducer adaptation in auditory hair cells. J Neurophysiol 82:2171–2181.

Zhao Y, Yamoah EN, Gillespie PG (1996) Regeneration of broken tip links and restoration of mechanical transduction in hair cells. Proc Natl Acad Sci USA 93:15469–15474.

5 The Morphological Specializations and Electromotility of the Mammalian Outer Hair Cell

RICHARD HALLWORTH AND HEATHER JENSEN-SMITH

1. Introduction

More than 20 years have elapsed since the discovery of induced and electrically driven length changes in isolated outer hair cells (OHCs) (Brownell 1984; Brownell et al. 1985). At the time, the phenomenon was embraced as the potential solution to two problems: the paradoxical nature of the OHC and the origin of the frequency- and level-specific mechanical energy input to the cochlea that we call the cochlear amplifier. Today, while we know much more about the physiology and electromechanical activity of the OHC, we are no closer to understanding its role in cochlear transduction. Indeed, as we shall see, there are good reasons to believe that its role in hearing is something far distant from that originally envisaged.

To understand the paradox, recall that OHCs are an unusual hair cell in many respects. As we describe here, the OHC is highly specialized and has characteristics possessed by no other vertebrate hair cell. Foremost among these is the electromotility of the OHC, its ability to extend and contract at acoustic frequencies in response to membrane potential change. OHCs are clearly the most vulnerable cell type in the cochlea. Major hearing pathologies such as aminoglycoside ototoxicity, noise trauma, and aging strike primarily at OHCs. And yet, while OHCs are four times as numerous as inner hair cells (IHCs), they receive one-twentieth of the afferent innervation.

The consequence of OHC loss appears to be loss of the cochlear amplifier, that is, a profound loss of sensitivity and tuning that is altogether incongruous when compared to the sparse and diffuse innervation of OHCs. What could OHCs contribute that was so important, if not afferent activity? Arguing for the need for mechanical amplification in the vibration of the basilar membrane (BM), investigators seized on the newly discovered electromechanical activity of OHCs as the solution to both problems.

For a mechanism such as OHC electromotility to be considered as an explanation of the cochlear amplifier, it must fulfill at least the following criteria: (1) it must add energy to the vibration of the cochlear partition (Robles and Ruggero 2001) (2) it must be frequency specific (that is, local to a best frequency place) (Russell and Nilsen 1997), (3) it must add nonlinearity to the vibration pattern of the BM even at low stimulus levels (Robles et al. 1997), and, since the cochlea amplifier loses effectiveness with increasing stimulus amplitude, (4) it must turn off at moderate to high sound stimulus levels (Robles and Ruggero 2001). For good measure, it should also be metabolically vulnerable (Rhode and Robles 1974; Ruggero and Rich 1991). Significant problems with this idea remain, as discussed below.

2. Morphological Specializations of the OHC

In this first part of the chapter, the unusual, even exotic, morphology and molecular structure of the OHC is considered and related to the current state of our understanding of its function and its role in the cochlear amplifier.

2.1 Anatomical Distinctiveness

The OHC has a most unusual morphology compared to other hair cells, one that is not, in our opinion, sufficiently appreciated. The cylindrical shape of the OHC, with constant diameter but varying length as a function of best frequency location, taken together with its eccentrically placed nucleus, are all hallmarks of a radically different kind of hair cell (Fig. 5.1). Also noteworthy are the lack of intercellular attachments (except at the apex and Deiters cell cup) which, together with the pillar cells and Deiters cells, may be unique in the body. Further, as we shall see, there is a relative lack of cytoskeletal structure in the body of the OHC, consistent with a mechanical function rather than a normal epithelial/sensory cell.

On the other hand, some aspects of OHC morphology are not as unusual as is commonly thought. For example, although it is not as well studied as that in OHCs, IHCs also have a cortical structure consisting of actin and spectrin (Jensen-Smith and Hallworth 2006), even though its organization at the electron microscopic level is unknown. IHCs also have subsurface cisternae along their lateral walls, similar to those in IHCs, but not nearly as elaborate.

OHC length varies systematically from cochlear apex to base, with the longest OHCs originating from the most apical (lowest best frequency) locations, which is strikingly different from IHCs, which are uniform in overall dimensions. Indeed, OHC length is well correlated with best frequency *across species* (Dannhof et al. 1991; Pujol et al. 1992). While it is true that the natural frequency of an oscillator also decreases with increasing length, all other things being equal, it is otherwise hard to know what to make of this observation. It is equally possible, for example, that OHCs are larger in the cochlear apex simply because the organ of Corti is also larger. However, OHCs in the apex are also

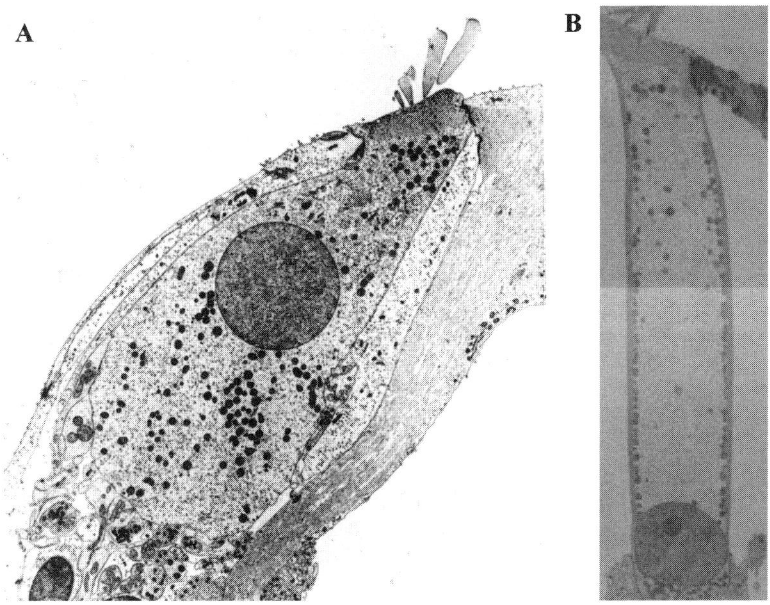

FIGURE 5.1. Anatomical differences between OHCs and IHCs. Electron micrographs of (**A**) IHC in the macaque (courtesy of R. Kimura) and (**B**) OHC in the chinchilla (courtesy of B. Bohne).

more prominently tilted, both radially and longitudinally, than in the base, which suggests than there may be a more fundamental reason for the longer apical OHCs in the apical turn.

2.2 Structural Proteins

2.2.1 Microfilaments

The smallest but most abundant filament class in hair cells is microfilaments, which are paired twisted filaments of the globular protein actin. Microfilament based structures (which contain several actin binding proteins) may be organized as gel-like networks (the cuticular plate is a good example of such a network), or as more or less parallel bundles of filaments, such as filopodia. Networks may be dense and stable, like the cuticular plate, or viscous and dynamic (such as at the leading edge of lamellopodia). The actin binding proteins correspond to, and may determine, the type of network and its stability.

Actin monomers are found in three isoforms: the muscle form α, and the nonmuscle isoforms β and γ. These isoforms are encoded by separate genes but exhibit only slight differences in their amino acid sequences. The functional significance of the sequence differences, if any, is not known. The high level of amino acid sequence conservation in evolution suggests that actin filament structure–function relationships have also been conserved. The β actin isoform is prominent in the cuticular plate and is the predominant form in stereocilia, while the γ isoform is predominant in the lateral wall (Furness et al. 2005). Given the conservation of actin sequences, it is extraordinary that a nonsyndromic deafness associated with γ actin (ATG1) has been discovered. It suggests that γ actin may have a unique role in cochlear hair cells.

Long OHCs in several species, including guinea pig and chinchilla, exhibit a prominent infracuticular network of actin filaments (Carlisle et al. 1988). This tapering network, which extends centrally from the cuticular plate partway down the OHC, does not appear to connect to any other cytoskeletal feature. Its function is completely unknown.

2.2.2 Microtubules

Microtubules are the largest class of cytoskeletal filaments normally found in hair cells. In the electron microscope, individual microtubules are hollow and 24–28 nm in diameter. Each microtubule consists of (usually) 13 parallel longitudinal filaments, which are in turn assembled from dimers of the globular proteins α and β tubulin. Cellular functions of microtubules include transport of cellular materials and organelles via the motor proteins dynein and kinesin. Microtubules are also polarized. They are assembled in one direction, from the minus end where assembly starts to their plus ends. The polarization has physiological significance, since the motor protein kinesin moves only toward the plus end, while the motor protein dynein moves only toward the minus end. Microtubules also provide the scaffold of the axoneme, the internal structure of motile cilia, flagella, and the kinocilium of noncochlear hair cells. The mechanical activity of motile cilia is driven by various isoforms of dynein. Mature cochlear hair cells do not possess motile cilia. However, they do have short primary cilia in postmitotic development (in gerbil up to P20; Jensen-Smith et al. 2003).

OHC microtubules consist of diffuse single filament strands originating under the remnant kinocilium at an apical organelle called the basal body (Steyger et al. 1989; Hallworth et al. 2000; Jensen-Smith et al. 2003 (see particularly Fig. 5.1C of the latter reference)). These microtubules radiate randomly toward the basal pole of the cell and do not anchor conspicuously to any structure. In guinea pig, but not in gerbil, there is also a ring of microtubules associated with the apical tight junctions of OHC (Steyger et al. 1989).

Tubulin α and β isoforms exist in several single gene product variants or isotypes (Ludueña 1998). In mammals there are six α and seven β isotypes. The different functional roles, if any, of these isotypes are unclear. Mature cochlear hair cells show a different distinctive panel of isotypes: OHCs have β_{I} and β_{IV} tubulin in all their microtubules, while IHCs have β_{I} and β_{II} tubulin (Hallworth et al. 2000). The mature isotype distribution is established quite late in development (after P20 in the gerbil) by selective reduction from the three isotypes (β_{I}, β_{II}, and β_{IV} tubulin) common to all organ of Corti cells early in postnatal development (Jensen-Smith et al. 2003).

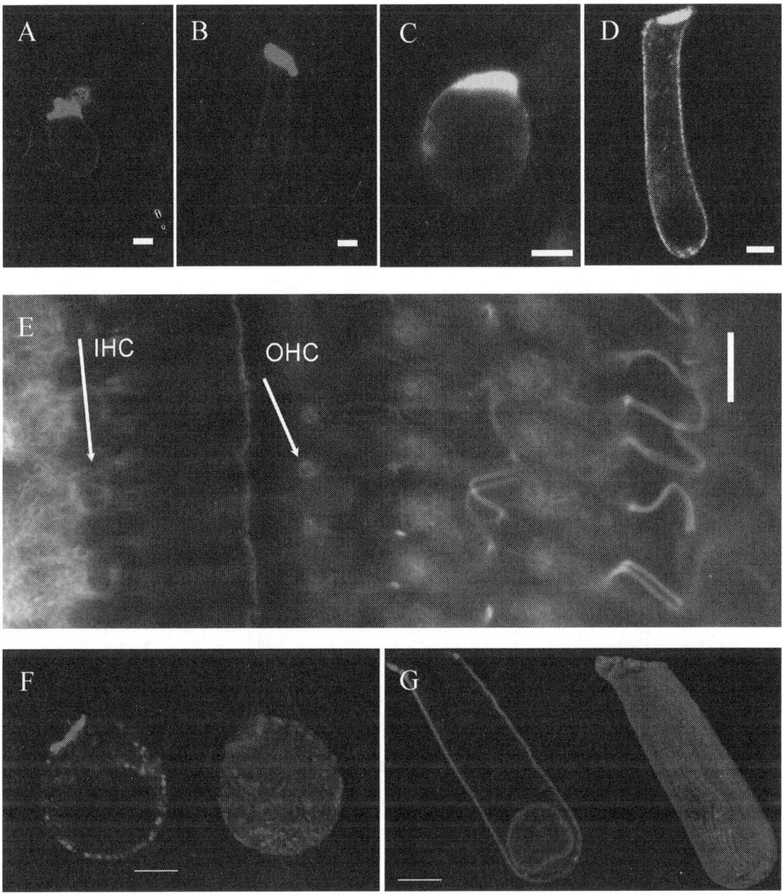

FIGURE 5.2. Molecular composition of the cytoskeleton and PM in OHCs. (**A, B**) F-actin in the lateral wall and cuticular plate of isolated gerbil OHCs at P6 (**A**) and mature (**B**, P18). (**C, D**) Spectrin in the lateral wall and cuticular plate of isolated gerbil OHCs at P6 (**C**) and P18 (**D**). Spectrin label during development extends basally from the cuticular plate during development. Scale bars represent 5 μm for **A–D**. (**E**) Tubulin in the cytoskeleton of organ of Corti cells in a whole mount of gerbil tissue. Whole mount was labeled with an antibody to β_I tubulin. Note how microtubules in the IHC originate at a ring at the level of the cuticular plate, while microtubules in the OHC originate at the basal body. The scale bars represent 10 μm. (**F, G**) Prestin in the lateral wall and under the cuticular plate in isolated OHCs from developing (**F**, P9) and mature (**G**, P21) gerbil. Each panel shows a cross section (left) and a collapsed three-dimensional representation (right) of an isolated OHC labeled with an antibody to prestin (red in the on-line version) and the nuclear stain DAPI. Prestin incorporation in the PM during development is in patches (**F**). Prestin completely occupies the basal PM (**G**), including the nuclear compartment. The scale bars in **F** and **G** represent 5 μm.

Tubulin isotypes, both α and β, undergo a number of carboxy terminal posttranslational modifications after incorporation into microtubules, including detyrosination, acetylation, glutamylation, and glycylation (MacRae 1997). Posttranslational modifications usually accompany changes in the stability of tubulin in microtubules, although they are not causative (Kreis 1987). Microtubules in hair cells are not modified and are therefore thought to be more dynamic, while those in supporting cells are modified and therefore probably stable (Slepecky et al. 1995).

2.2.3 Intermediate Filaments

Intermediate filaments (IFs) are intermediate in size between microfilaments and microtubules. EM evidence for their presence in OHCs is nonexistent. Five classes of proteins exist: neurofilament proteins (NFPs), cytokeratins, vimentin, glial fibrillary acidic protein (GFAP), and desmin. Of these, NFPs are primarily restricted to neurons, although an intermediate molecular weight form has been observed in avian hair cells (Oesterle et al. 1997). Vimentin, GFAP, and cytokeratin are found in supporting cells but not hair cells (Raphael et al. 1987; Oesterle et al. 1993; Mogensen et al. 1998; Rio et al. 2002): indeed, the presence of these proteins has been used as a marker for supporting cells. Incidentally, although absent from the mammalian cochlea, vimentin is found in mammalian vestibular hair cells and bird hair cells of all types (Usami et al. 1993; Oesterle et al. 1997).

2.3 Structural Components of the Lateral Wall

The OHC must possess a significant amount of axial stiffness for the expression of somatic motility (Holley and Ashmore 1988b; Tolomeo et al. 1996; He et al. 2003b). Given that the interior cytoskeletal structures of OHCs are relatively sparse, it is usually assumed that the rigidity and the cylindrical shape of the OHC is conferred by lipid and protein structures associated with the lateral wall. The lateral wall consists of the plasma membrane (PM), including embedded proteins such as prestin and ion channels, some linking structures of unknown molecular identity commonly referred to as pillars, a latticework of actin and spectrin filaments known as the cortical lattice, or cortex, and one or more layers of membrane structures called subsurface cisternae (SSC). The components are discussed below and depicted in Figure 5.3.

2.3.1 Plasma Membrane

As discussed later, the PM of mature OHCs is host to a high concentration of the integral membrane particles, which presumably indicates a large concentration of integral membrane protein. Unfortunately, little is known about the distribution of lipids in the OHC PM. OHC lateral wall membranes are variously said to be cholesterol-containing (Forge 1991) and cholesterol-poor (Nguyen and Brownell 1998). The fact that membrane mechanical properties can be altered by PM

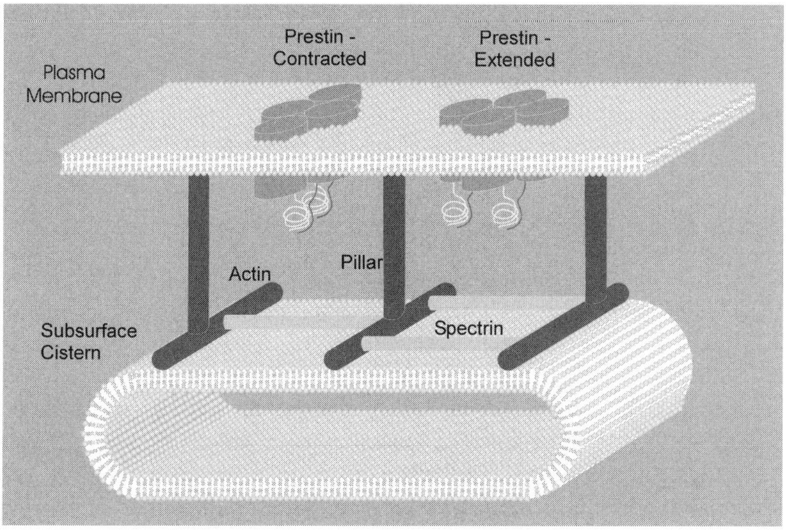

FIGURE 5.3. Diagram of the structure of the OHC lateral wall. The diagram includes tetramers of prestin in possible contracted and extended configurations. The carboxy terminus of prestin is rendered in a light shade, while the amino terminus is in a darker shade. A color version of this figure is available on-line.

lipid and cholesterol composition suggests that a significant contribution of PM lipid content to OHC electromotility is possible (Holley 1996). In fact, the PM is frequently considered to be the primary contributor to OHC axial stiffness (Holley and Ashmore 1988b; Tolomeo et al., 1996; He et al., 2003b), which is critical for transmission of motor force to the organ of Corti (see later). Comparison of the stiffness moduli of the intact lateral wall and isolated cortical lattice indicates that the PM is a major contributor to the total axial stiffness of the intact OHC (Tolomeo et al. 1996). Others have, however, contended that the cortical lattice is the primary contributor to OHC axial stiffness (Adachi and Iwasa 1997; Oghalai et al. 1998).

The density of membrane particles in the OHC lateral wall PM approaches $6000/\mu m^2$ in freeze–fracture observations (Forge 1991; Kalinec et al. 1992), an extraordinary number, comparable to the density of nicotinic acetylcholine receptors at the rat neuromuscular junction ($9000/\mu m^2$; Grohovaz et al. 1982). The membrane particles in the OHC are distributed over a much greater area, of course, than at the neuromuscular junction, thus the overall number of particles is very large, several millions in the longest guinea pig OHCs. The particles have been equated in number with the motor protein prestin (discussed later) (Kalinec et al. 1992). However, the particles, which are 10–11 nm in diameter, are too large to be composed of a single protein moiety. However, while the OHC PM must also contain many common membrane proteins such as ion channels, these cannot account for more than a few thousand particles per cell, thus the motor protein is still the preferred candidate. A specialized glucose transporter,

GLUT5, is also present in the mature OHC PM (Nakazawa et al. 1995b) (see later), but the density of GLUT5 in the membrane is unclear.

It is interesting to note that while this density of large particles in the OHC membrane is remarkable, it is only a factor of two different from the density of equally large particles in IHCs and in bird and lizard OHCs (Köppl et al. 2004). Thus these particles, probably representing a protein complex rather than a single protein, may be a common feature of hair cells rather than a special feature of OHCs.

The OHC maintains its unique cylindrical shape by way of the actin-spectrin cortical lattice underlying its PM. The cortical lattice of mature mammalian OHCs has been well described (Bannister et al. 1988; Holley and Ashmore 1988a, 1990a,b; Arima et al. 1991; Forge 1991; Leonova and Raphael 1999; Le Grimellec 2002). In the cortical lattice, filamentous actin (F-actin) has been identified as 5- to 8-nm-diameter filaments circumferentially arranged in the OHC (Holley and Ashmore 1990b), while cross-linking spectrin fibers have been identified as 2- to 3-nm-diameter filaments extending longitudinally between and orthogonal to actin filaments.

The OHC and erythrocyte cytoskeletons are frequently compared because of the substantial overlap in their components. These comparisons should be made cautiously, as these two cytoskeletal structures are quite different in organization. First, the erythrocyte cytoskeleton is primarily composed of spectrin, whereas the OHC cytoskeleton is proposed to consist of comparable amounts of F-actin and spectrin. Second, the organizations of F-actin and spectrin in the two structures are very different. In the erythrocyte, spectrin tetramers bind to short actin filaments, generating a spoke and hub organization of spectrin and actin, respectively (Lodish et al. 2002). As previously mentioned, F-actin and spectrin in the OHC cortical lattice are organized into circumferential and longitudinal filaments, respectively (Holley and Ashmore 1988a, 1990a,b; Arima et al. 1991; Nishida et al. 1993). As we will see later, these differences in cytoskeletal organization complicate our ability to interpret comparative studies of these superficially similar, but in fact very different, structures.

The isolated OHC cortical lattice is a highly orthotropic structure, which is to say that its mechanical properties differ depending on the axis along which they are measured. The axial and circumferential stiffness moduli of the cortical lattice are respectively 5×10^{-4} N/m and 3×10^{-3} N/m, an order of magnitude difference (Tolomeo et al. 1996). While axial stiffness of the lattice is small relative to that of the intact OHC (3×10^{-3} N/m), the circumferential stiffness is similar, suggesting that the cortical lattice directs prestin mediated length changes to the longitudinal direction. Indeed, several researchers have hypothesized that the OHC cortical lattice influences the mechanical properties of OHCs by providing the directional component to OHC electromotility (Holley and Ashmore 1988a; Huang and Santos-Sacchi 1993, 1994; Tolomeo et al. 1996; Adachi and Iwasa 1999; Zheng et al. 2000).

2.3.2 The Cortical Lattice

Pressurized cylindrical cells such as the OHC usually require reinforcement by circumferentially oriented filaments. The OHC cortical lattice is organized to provide the necessary reinforcement. F-actin and spectrin in the OHC cortical lattice are organized into roughly circumferential and longitudinal filaments, respectively (Holley and Ashmore 1988a, 1990a,b; Arima et al. 1991; Nishida et al. 1993). Note that relatively inextensible actin filaments are poised to restrict circumferential expansion while the more pliable spectrin tetramers are similarly poised to permit longitudinal expansion during prestin-mediated electromotility. On close inspection, cross-linked actin-spectrin arrays can be observed to exist in discrete domains (Holley and Ashmore 1990b; Holley et al. 1992; Kalinec et al. 1992).

2.3.3 Pillars, Protein 4.1, and Other Putative Components of the Lateral Wall

Ultimately, the actin–spectrin cortical lattice can alter OHC mechanical properties only if it is attached to the PM. In the erythrocyte, spectrin, F-actin, ankyrin, protein 4.1 (band 4.1), adducin, tropomodulin, and tropomyosin form a spectrin–actin junctional complex. Ankyrin and protein 4.1 bind specific integral membrane proteins (band 3 and glycophorin, respectively) to connect the PM to the cortical lattice. Protein 4.1's role in stabilizing the erythrocyte PM is well established. Reduced protein 4.1R expression in humans is linked to hereditary elliptocytosis, a condition characterized by hereditary hemolysis and abnormally shaped red blood cells (Tchernia et al. 1981). Protein 4.1 null mice exhibit a similar phenotype (Shi et al. 2001). Protein 4.1's FERM, SABD, and CTD domains mediate membrane–cytoskeleton interactions via integral membrane protein interactions (Sun et al. 2002). Of particular interest is protein 4.1's ability to potentiate spectrin and F-actin interactions through its SABD domain. Protein 4.1 has also been shown to bind tubulin via its SABD domain and may therefore also regulate microtubule architecture (Correas and Avila 1988).

Band 3, protein 4.1, and ankyrin have all been detected in OHCs (Knipper et al. 1995) and whole organ of Corti preparations (Zine and Schweitzer 1997). Although band 3 and protein 4.1 have been observed in the OHC lateral wall using immunofluorescence (Zine and Schweitzer 1997) and immuno–electron microscopy, ankyrin has been identified only in Western blots of intact organ of Corti (Knipper et al. 1995). Although, as in the erythrocyte, band 3 and protein 4.1 have been hypothesized to play a role in stabilizing the cortical lattice, Western blots of intact organ of Corti failed to detect glycophorin (Knipper et al. 1995), suggesting that direct associations between erythrocyte and OHC cytoskeletal structures must be cautiously pursued.

The ubiquitously synthesized membrane protein adducin is another key element of the erythrocyte (Matsuoka et al. 2000). Adducin functions as an assembly factor for the actin-spectrin network and caps the fast growing ends of actin filaments (Gardner and Bennett 1987; Bennett et al. 1988; Fukata et al. 1999). Adducin binds calmodulin and is a substrate for protein kinase C and Rho kinase. Recently, Zhang et al. (2003) identified adducin, RhoA, and its downstream target ROCK in the OHC cortical lattice. This is discussed later, but clearly alterations in OHC PM cytoskeleton associations have the potential to regulate both fast and slow OHC electromotility.

In the OHC, thousands of 25 nm long, electron-dense, rodlike pillar structures have been observed between the PM of the lateral wall and the actin-spectrin cortical lattice (Lim et al. 1989; Arnold and Anniko 1990; Forge 1991; Pujol et al. 1991; Weaver et al. 1993; Zine and Schweitzer 1997). These 7- to 10 nm-diameter pillars have been hypothesized to link the PM to the actin–spectrin cortical lattice (Gulley and Reese 1977; Flock et al. 1986; Bannister et al. 1988; Tolomeo et al. 1996). Zine and Schweitzer (1997) have also suggested that band 3 and/or protein 4.1 may comprise these yet unidentified pillar structures. In gerbil OHCs, pillars are first observed between P8 and P10, just before the onset of hearing and electromotility (Souter et al. 1995; Forge et al. 1997). Although these structures are first observed at P8, they are few in number and widely separated (Forge et al. 1997). The number of pillar structures gradually increases during post natal development, but mature levels are not reached until after P16 (Souter et al. 1995). Thus the appearance of pillars correlates approximately with the appearance of electromotility (He et al. 1994). Souter et al. (1995) reported a decrease in pillar to pillar distance from 53.9 ± 8.2 nm at P10 to 38.6 ± 6.2 nm at P16, which is close to the pillar spacing of 35.4 ± 5.03 nm observed in mature OHCs. Some have suggested that the pillars are responsible for the transmission of force between the motor protein and the cortical lattice, generating a vectorial component to prestin-mediated electromotility (Iwasa 1994; Tolomeo et al. 1996). Flexoelectricity models suggest that the pillars may also play an important role in the transmission and possible generation of forces in the PM (Raphael et al. 2000).

2.3.4 Subsurface Cisternae

Subsurface cisternae (SSC) are closed lamellar intracellular membrane structures first described in neurons (Rosenbluth 1962). OHC SSCs are organized into fenestrated sheets and/or tubule networks that extend from the cuticular plate to the nucleus. These cisternae are hypothesized to be involved in diverse cellular functions including the generation of cochlear microphonics (Engstrom 1955), the establishment of synapse-maintaining transport (Ekstrom von Lubitz 1981), the maintenance of intracellular gradients (Spoendlin 1966; Lutz and Schweitzer 1995), the provision of mechanical support (Saito 1983; Oghalai et al. 1998), and intracellular calcium storage (Zine and Schweitzer 1996). SSCs are frequently associated with ribosomes (Kimura 1975; Saito 1983) and the rough endoplasmic reticulum, suggesting that the cisternae may be derived from the endoplasmic reticulum (Spoendlin 1966; Saito 1983). The Golgi labeling fluorescent probe C6-NBD-Cerimide and endoplasmic reticulum labeling fluorescent probe DiOC6 have been shown to label OHC lateral wall membranes, suggesting that the

SSC exhibits some features of both the Golgi and the endoplasmic reticulum (Pollice and Brownell 1993; Oghalai et al. 1998, 1999). If so, the structural and biochemical maintenance of the OHC may be mediated in part by the SSC.

SSCs exhibit great variability in morphology and layers both within individual OHCs and between OHCs from a given species (Tolomeo et al. 1996). For example, one layer has been described in man, cats, and rats, while multiple cisternal layers have been described in guinea pigs and gerbils (Ekstrom von Lubitz 1981; Saito 1983; Weaver and Schweitzer 1994; Souter et al. 1995). Abundant calcium pumps in the SSC suggest a calcium storage mechanism with the potential to regulate slow motility via alterations in intracellular calcium concentration (Zine and Schweitzer 1996). However, no other evidence for a calcium storage role for the SSC has been presented. Although no functional role has been confirmed, it is noteworthy that the first cisternal layer develops concomitantly with somatic electromotility (Pujol et al. 1991; He et al. 1994). However, removal of the SSC from the inner surface of the OHC cortical lattice does not inhibit electromotility, which casts doubt on this idea (Kalinec et al. 1992).

A neglected area of research is the cytoskeletal anchors that maintain the SSCs in their position and conformation. Lamellar lipid structures such as the SSCs are energetically unfavorable and could not be maintained in that configuration if not for some kind of cytoskeletal framework. We can presume that the cytoskeletal underpinnings of the SSCs are probably actin and/or spectrin, since they are the only cytoskeletal proteins that have been observed in close apposition to the OHC lateral membrane. However, their organization, and how they might be remodeled, has not been given significant attention.

2.4 Glycogen

In contrast to the low levels of cytoskeletal proteins, the OHC cytoplasm contains an unusually large concentration of glycogen granules (Thalmann 1972; Duvall and Hukee 1976; Beneyto et al. 1997). Glycogen granules are readily detectable in both light and electron microscopes. A functional role for glycogen is suggested by the temporal appearance of glycogen granules in the cytoplasm at P10 in the gerbil, which coincides with the detection of electromotility (He et al. 1994; Nakazawa et al. 1995b). The need for glycogen storage is not obvious; prestin requires no ATP, and Na/K-ATPase is almost nonexistent in cochlea hair cell PMs (Nakazawa et al. 1995a). Perhaps the largest metabolic demand is the stereocilia calcium pump PMCA2a (Dumont et al. 2001), which has been stated to account for one half of all ATPase activity in stereocilia. However, the rationale for enhanced glucose storage in OHCs but not IHCs is unclear. It is interesting to note that some infants with a type II glycogen storage disease (Pompe's disease) have a 30- to 90-dB hearing loss with an absence of otoacoustic emissions that is probably attributable to severe OHC dysfunction or damage (Kamphoven et al. 2004).

OHCs, especially basal turn OHCs, are more vulnerable to ischemic insult and to noise exposure (Billett et al. 1989; Dehne et al. 2002). Is there a higher metabolic demand in OHCs than IHCs? Are higher frequencies more metabolically demanding than lower frequencies? No one knows.

3. Electromotility

In this second part of the chapter, what is known about the various forms of somatic motility exhibited by OHCs is discussed, with an emphasis on fast electromotility.

3.1 Fast Electromotility: Where Are We Now?

Since the initial observations of electrically evoked length changes in isolated OHCs (Brownell et al. 1985), fast electromotility has been thoroughly characterized (Ashmore 1987; Santos-Sacchi and Dilger 1988). Electrically evoked length changes are very rapid, so much so that they challenge the upper limits of the instrumentation used to measure them (Dallos and Evans 1995b; Frank et al. 1999). One published report indicates that cyclic length changes at 70 kHz are possible, and operation at greater than 100 kHz, while not yet demonstrated, seems well within reach.

Of the methods used to evoke electrical stimulation, voltage clamp application via a patch pipette gives by far the most reliable measure of the voltage to length change function and dynamic range, but the upper frequency limit is constrained by pipette and cell capacitance (Santos-Sacchi 1992). The voltage-to-length-change function, as measured by step voltage commands, is bidirectional and nonlinear, in that much more contraction can be evoked by depolarization than elongation by hyperpolarization (Fig. 5.4A, B). Further, limits to extension and contraction are seen, which suggests a finite state transition mechanism for the motor protein. The voltage to length change function can generally be fit by a first-order Boltzmann function of the form

$$\partial L = \frac{\partial L_{\text{max}}}{1 + e^{ez(V - V_{\text{m}})/kT}} - \partial L_0, \tag{5.1}$$

where δL is the length change, δL_{max} is the total length change observed from maximum extension to maximum contraction, δL_0 is an offsetting constant, V the membrane potential during the command, V_{m} the holding potential, e is a constant called the slope factor, z is the valence, and k and T have their usual thermodynamic meanings (Fig. 5.4B) (Santos-Sacchi 1991). When the length changes are normalized to δL_{max} (with the base length set to the length at the most negative command potential), the resulting curve looks like Fig. 5.4C. Note that

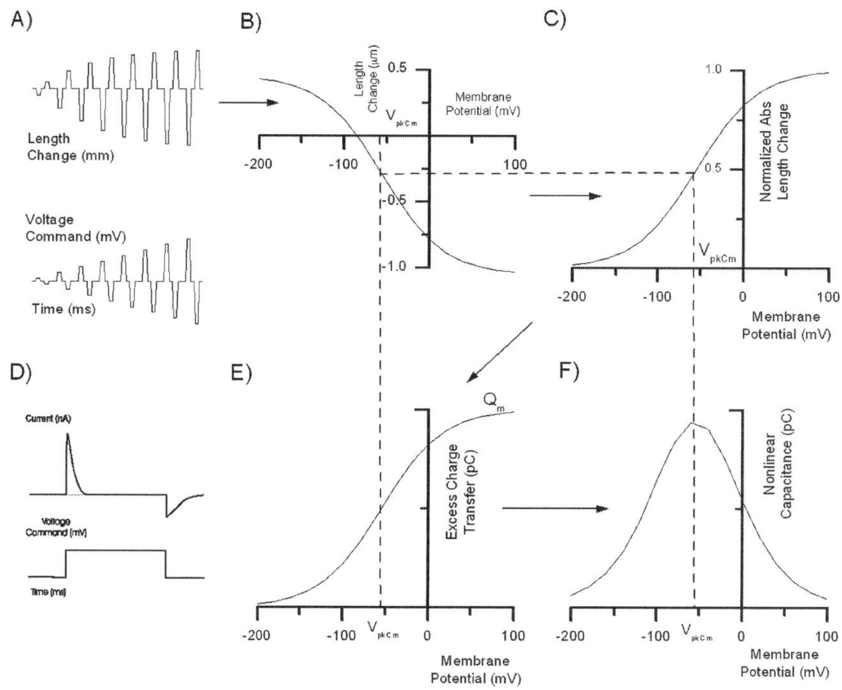

FIGURE 5.4. Simulations of the relationship between OHC electromotility and NLC. (**A**) Diagram of OHC length changes (above) in responses to step voltage commands of increasing amplitude (below). (**B**) Membrane potential ("voltage") to length change function of OHC electromotility plotted from the responses in A. (**C**) Normalized absolute value of the length changes in **B** as a function of membrane potential. (**D**) Example of current waveform (above) in an OHC evoked by a depolarizing step voltage command (below), after subtraction of the linear component. (**E**) Charge transfer as a function of membrane potential calculated by integration of the current transient (shaded area in **D**). Note the correspondence with the normalized absolute value of the length change function in C. (**E**) NLC as a function of membrane potential. V_{pkCm} represents the membrane potential at which the NLC is a maximum. Note that V_{pkCm} also corresponds to the half-maximal value of charge transfer (in **E**) and the maximum slope (and therefore gain) of electromotility (**B, C**), as shown by the dashed lines. V_{pkCm} was set to be −55 mV for these simulations.

the maximum slope gain of the function occurs at membrane potentials somewhat depolarized compared to the normal resting membrane potential of OHCs.

The Boltzmann function relationship of length change to membrane potential implies that the motor protein operates by undergoing a conformation change between two states, extended and contracted, much as an ion channel does in opening or closing. OHC diameter change was also observed to occur in antiphase to length change (Hallworth et al. 1993), even in the microchamber configuration, for which this would not be expected. It was therefore inferred that the motor protein undergoes both contraction along the length axis and extension along the circumferential axis. This in turn implies that the motor proteins have a specific orientation with respect to the long axis of the OHC.

3.2 Nonlinear Capacitance: The Fast Motility Marker

Nonlinear capacitance (NLC) is an electrophysiological signature that is always associated with prestin conformation change. In brief, a depolarizing voltage pulse evokes more outward current in the first few milliseconds than a hyperpolarizing pulse of the same magnitude evokes inward current. This is effectively a sign dependent (or nonlinear) membrane capacitance. The study of NLC has a long history in neurons. For example, a large NLC is intimately associated with the activation of voltage-dependent sodium channels, where it is known as gating charge. Unlike the sodium current, sodium channel associated NLC was not blocked by tetrodotoxin. The phenomenon could be explained as either transmembrane movement of charges associated with the closed to open conformation change of the channel molecule, or as dipole movement. The excess charge transfer Q due to NLC can be fit by a sigmoidal function of the form

$$Q = \frac{Q_0}{1 + e^{ez(V - V_\mathrm{h})/kT}}, \tag{5.2}$$

where Q_0 is the total charge transfer, V the membrane potential, V_h the membrane potential of the half-maximum charge transfer, e is a constant called the slope factor, and z is the valence. The value of ez is about 6 for the sodium channel (Armstrong 1981), which could represent movement of three particles, each of valence 2, across the entire membrane (or six particles of valence 2 halfway across the membrane, or any combination thereof).

OHCs have an unusually large NLC that is intimately associated with fast electromotility. Indeed, separation of NLC from OHC motility has proved impossible: treatments that abolish NLC always abolish electromotility. The nonlinear component of OHC capacitance is very large, about 25 pF in long, apical turn guinea pig OHCs, or about the same as the linear component (Kakehata and Santos-Sacchi 1996). Thus NLC is often used as a proxy for fast motility.

Measurable NLC in OHCs has been demonstrated from −180 mV to 80 mV, which corresponds to the membrane potential range of electromotility. One difference from sodium channel NLC is that ez is about 1 for the OHC (Kakehata and Santos-Sacchi 1995). As a consequence, prestin has a much larger voltage range than sodium channels (240 mV versus 60 mV). A representation of an NLC current transient in response to depolarization is shown in Fig. 5.4D. The charge transfer function of OHC NLC looks very much like the voltage-to-length change function of OHC electromotility, especially when compared to the normalized form of Fig. 5.4C. Differentiating the charge transfer function with respect to V gives the capacitance, which is a Gaussian function with its maximum at $V = V_\mathrm{h}$ (also referred to as V_{pkCm}) (Fig. 5.4F). Note that, at least in theory (since NLC and electromotility are rarely measured simultaneously), the peak nonlinear capacitance occurs at the same membrane potential as the maximum electromotile gain (dashed lines in Fig. 5.4).

NLC in OHCs is subject to manipulation of membrane tension. Inflation of OHCs by up to 1 kPa shifts V_{pkCm} by 25 mV in the depolarizing direction (Kakehata and Santos-Sacchi 1995). Similarly, OHC stretch induces currents in voltage clamped OHCs that are at least partly attributable to changes in motor state (Gale and Ashmore 1994). The latter phenomenon suggests that local OHC activity can influence that activity of other OHCs, and such has been demonstrated (Zhao and Santos-Sacchi 1999).

3.3 Functional Compartmentalization of the OHC

The OHC can be anatomically and functionally divided into three compartments, although there is no physical barrier between them. The apical compartment (above the tight junctions) consists of the cuticular plate and the stereocilia and clearly contributes forward transduction, but also has a high rate of endocytosis from and exocytosis to scala media, perhaps in common with IHCs (Meyer et al. 2001; Griesinger et al. 2004). The somatic compartment consists of the cylindrical cell above the nucleus but below the tight junctions, including the lateral wall plasma membrane, cortical lattice, SSC, and the somatic cytoplasm. The subnuclear compartment consists of the nucleus and the synaptic pole, and is apparently the location of most voltage dependent ion channels (Huang and Santos-Sacchi 1993).

Motor activity is uniformly distributed along cell length, except in the subnuclear and apical compartments, by several measures (Dallos et al. 1991; Hallworth et al. 1993; Huang and Santos-Sacchi 1993). Further, the large membrane particles seen in freeze fracture, and thought to be associated with electromotility, are seen only in the somatic compartment (Kalinec et al. 1992). Thus it has been assumed that the somatic compartment, with its low density of cytoskeletal features that might impede elastic distention, is optimized for length change generation.

3.4 Slow Motility: Is It Real?

Slow motility refers to gross changes in OHC length that occur on a time scale of seconds to minutes and are independent of membrane potential change. Seduced perhaps by the ease with which the experiments could be conducted, many laboratories in the 1980s reported slow motile changes (invariably contractions). However, contradictory results abounded. Slow motility was variously reported in meeting proceedings and elsewhere to be calcium dependent or independent, and induced (or not) by application of the efferent neurotransmitter acetylcholine, e.g., Slepecky et al. 1988. One laboratory tried to have it both ways (Zenner 1986a). Laboratories did at least agree that high external potassium induced contraction, and that the diameter always increased (Zenner et al. 1985; Zenner 1986b; Dulon et al. 1988). The changes were very large, much greater than the

2 % to 3 % associated with fast motility. The reversibility of the induced changes was modest at best.

Gross shape changes in OHCs would, it was proposed, alter the relationship between the OHC stereocilia and the TM, perhaps adjusting its sensitivity range. Some credence was given to this idea when large DC shifts were reported in BM position on overstimulation; however, these results proved impossible to reproduce. The general irreproducibility of the results, and perhaps the casual observation that "slow motility"-like changes occur spontaneously and inevitably in isolated OHCs, led to neglect of the field. However, the idea of physiologically relevant slow changes in OHC mechanical properties has been revived to some extent lately (Zhang et al. 2003; Matsumoto and Kalinec 2005).

3.5 Acoustically Evoked Motility

Extracellularly applied acoustic vibration applied directly to an isolated OHC evoked a slow peristimulus contraction that was also frequency sensitive (Brundin et al. 1989, 1992). Further, the most sensitive frequency of the contraction correlated with OHC length and therefore anatomical best frequency. This result, not often cited today, was at one time the basis for an entirely new theory of cochlea tuning (Brundin et al. 1992), one that bypassed basilar membrane vibration entirely. The experimental findings were later repeated with different methods that permitted higher frequency measurements of length change (Brundin and Russell 1994), and the slow (DC) component was again found to be tuned, but strangely not the AC component. These findings, although they are no longer considered relevant to transduction, quite possibly have something to say about OHC mechanics, and they still await understanding.

4. Prestin and Electromotility

In a tour de force of molecular biology, the Dallos group in 2000 identified the gene underlying the OHC motor protein, which they named prestin (Zheng et al. 2000). Their strategy, which involved successive rounds of subtractive hybridization between IHC and OHC gerbil cDNA libraries, yielded a gene product of 744 amino acids with a strong resemblance to the anion transporter family SLC26, and is now referred to as SLC26A5. When expressed in cell lines such as human embryonic kidney (HEK) cells, the protein demonstrated NLC with similar characteristics to native prestin, although much smaller in overall amplitude. When transfected cells were distended to a quasicylindrical shape, axial voltage-evoked length changes of the appropriate polarity were observed. Thus the identity of prestin with the motor protein has become accepted. Here we describe what is known about prestin and relate it where possible to electromotility.

4.1 Prestin

Prestin sequences from other species (mouse, rat, human) exhibited a high degree of homology (Ludwig et al. 2001; Adler et al. 2003). Rat and mouse prestin each have 744 amino acids; human prestin has 746 (Liu et al. 2003). The structure of prestin was initially proposed to consist of 12 membrane spanning regions, consistent with other members of the SLC26 family by the standard analyses (Kyle-Doolittle, TMPred) (Zheng et al. 2001). The carboxy and amino termini were quickly established to be in the cytoplasm by epitope tagging and antibody labeling (Belyantseva et al. 2000; Zheng et al. 2001), which constrains the number of membrane spanning regions to be even, consistent with the 12 membrane spanning regions originally suggested.

The carboxy terminus is quite large (246 amino acids). Common to the carboxy termini of all SLC26A family members, including prestin, is an approximately 100 amino acid sequence called the STAS domain, named for sulfate transporters and antisigma factor antagonists from its resemblance to certain bacterial antisigma factor antagonists (Aravind 2000). Antisigma factor proteins (protein kinases) regulate bacterial sigma factors (transcription initiation factors). Some antisigma factor antagonists have NTP binding and NTPase activity. It is hard to see a functional role for this motif. Mutations in this domain are associated with disease (in Pendred syndrome and in diastrophic dysplasia/achondrogenesis, which involves SLAC26A2), and have been suggested as indicative of its functional significance (Mount and Romero 2004). However, given the difficulty of membrane insertion associated with any carboxy terminal mutation of prestin, such may be coincidental. Certainly, Pendred syndrome loci are spread throughout the gene (Friedman and Griffith 2003).

More recently, two alternate prestin topologies have been proposed (Fig. 5.5). Based on the intracellular location of cGMP-mediated sites and the presumed extracellular location of two glycosylation sites, the Dallos group proposed a 10 membrane spanning region structure with two consecutive reentrant regions (their previous membrane spanning regions 5 and 6) and a large intracellular loop between them (Deak et al. 2005). Reentrant regions such as these are not unprecedented. The familiar P loop of sodium and calcium channels is just such a structure (Catterall 1993), which enters and emerges from the extracellular side of each six transmembrane region segment between regions 5 and 6. Similar single reentrant figures have been suggested for some transporter molecules (Seal et al. 2000; Grunewald et al. 2002; Ottolia et al. 2005). Although the reentrant regions were numbered along with the membrane spanning regions in Deak et al. (2005), making 12 in all, the membrane spanning regions have been renumbered 1–10 in Figure 5.5 to accord with common practice in the field.

In contrast, the Santos-Sacchi group (Navaratnam et al. 2005), having failed to find glycosylation at the sites suggested by the Dallos group, proposed a different 10 membrane spanning region structure more in accordance with the computer predictions. The extracellular location of the segments between loops 5 and 6 and between loops 7 and 8 was confirmed using epitope-specific antibodies. It

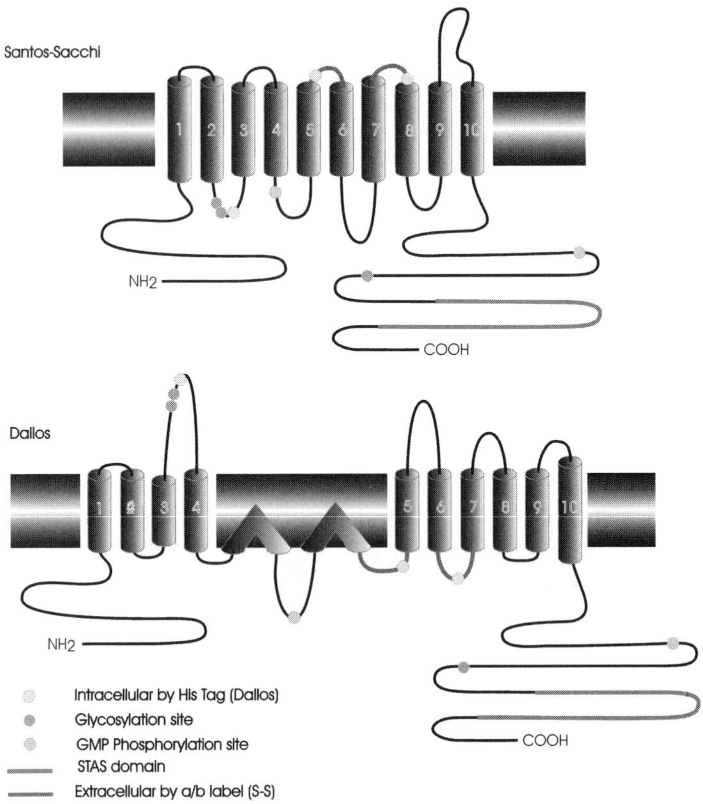

FIGURE 5.5. Proposed prestin 10 membrane spanning region topologies from Santos-Sacchi (above) and Dallos (below). The lengths of membrane spanning regions and intra- and extracellular loops are approximately to scale, based on published information and sequence analysis using TMPred and Kyle–Doolittle. The Santos-Sacchi structure is from Navaratnam et al. (2005); the Dallos structure is inferred from Deak et al. (2005). The numbering of the membrane spanning regions in the Dallos structure has been changed to accord with common practice.

was interesting, however, that the antibodies prepared against those segments only weakly labeled native prestin in OHCs (their Fig. 1). The region between loops 5 and 6 was independently observed to be extracellular by Adler et al. (2003).

The Brownell group has very recently drawn attention to two membrane spanning regions close to the amino terminus, which they inferred from an analysis called evolutionary trace, using a large number of published prestin-like and prestin-unlike sequences from the SLC26 family (Rajagopalan et al. 2006). Site directed mutagenesis in these regions demonstrated the importance of some residues in formation of NLC. Interestingly, the two membrane-spanning regions identified in this study only partly correspond to regions identified by the two other groups.

Thus even the most elementary facts about the structure of prestin remains in considerable doubt at present. The choice of HEK293T cells as the expression system seems to have been particularly unfortunate. Fluorescent protein constructs of prestin appear to accumulate in the cytoplasm of HEK293T cells, while very little is apparent in the membrane. Indeed, so much remains in the synthetic pathways that it is very difficult to distinguish the membrane component. Recall that the prestin NLC in expression systems has been small, around 1 pF, compared to the 25–30 pF routinely observed in apical guinea pig OHCs. A system such as the frog oocyte that routinely incorporates membrane proteins would be a highly desirable alternative.

4.2 Prestin and NLC

Prestin-associated NLC is often used as a proxy for electromotility because they are intimately associated (see earlier). It is the only practical way to study prestin structure–function relationships in a heterologous expression system, because in pseudospherical cells prestin conformation change does not result in a detectable cell shape change. Thus it was of great interest to study the origin of the NLC in a heterologous expression system by site-directed mutagenesis. Deletion of nonconserved charged residues in prestin (i.e., residues not present in other SLC26 family members, which do not exhibit NLC) did nothing to affect NLC in prestin-transfected HEK293T cells (Oliver et al. 2001). This led to the hypothesis that prestin NLC could not be accounted for by the transmembrane movement of charged residues, as for the sodium channel (see earlier), but could instead be explained by the voltage dependent movement of an intracellular anion to and from an interior binding site. Depolarization was posited to expel chloride ions to the cell interior, which would then appear as an outward (positive) current. Indeed, removal of intracellular chloride ions abolished NLC in prestin-transfected cells and motility in isolated OHCs (Oliver et al. 2001; He et al. 2003b; but see Rybalchenko et al. 2003). However, it is not entirely clear how this mechanism would work. If it resembles a voltage dependent block, such as that by magnesium ions on the NMDA channel, then it should not contribute to NLC, as the binding site is always accessible to charged ion species. If, however, a chloride binding site or pocket is revealed by the conformation change induced by hyperpolarization (or obscured by depolarization), then there must be a voltage sensor consisting of charged or polar residues to induce the conformation change that escaped detection in the mutagenesis screen.

Although chloride is the likely prestin-associated anion in OHCs, other monovalent anions can substitute. The relative affinity for the binding site was $I \approx Br > NO_3 > Cl > HCO_3 > F$, which resembles the anion binding affinity of at least some chloride channels, and therefore might suggest structural similarities (Oliver et al. 2001). Larger monovalent anions that can sustain NLC include formate, acetate, propionate, and butyrate. Salicylate ions apparently reversibly compete with chloride and block NLC. Again, it is puzzling that the positive residues that seemingly must be required for the chloride binding site were not detected during the mutagenesis screen. Further, the effect of divalent anions on NLC has received little study. The divalent anion sulfate, for example, would be expected to enhance NLC by increasing z. Instead, intracellular sulfate blocks NLC entirely.

In fact, the two groups working with heterologous prestin have reached differing conclusions about the relationship of Cl ions to prestin. Whereas Oliver et al. (2001) observed complete abolition of NLC after removal of internal Cl ions, Rybalchenko and Santos-Sacchi (2003) found residual effects. They therefore suggested that chloride ions instead allosterically modulate prestin and thereby alter its voltage sensitivity. The origin of NLC would therefore be some internal polar or charged residues still to be determined.

4.3 Structural Features

Gerbil prestin contains nine cystyl residues. Although there has so far been no direct evidence that sulfhydryl linkages contribute to the structure of prestin, at least some cystyls are apparently important in motility. Extracellularly applied *N*-ethylmaleimide had little effect on OHC motility after a 1-hour incubation, but ethacrinic acid reduced it by on average about 40 % and the smaller *p*-chloromercuriphenylsulfonate by about 80 % (Kalinec and Kachar 1993). While it may be presumed that these effects are prestin-specific, they have yet to be confirmed in an expression system. More recently, Zheng et al. (2006a) showed that prestin dimerization, which is resistant to SDS (although tetramerization is not), is not affected by dithiothreitol, which is often used to break sufhydryl linkages, but is abolished by the more hydrophobic reducing agent ethanedithiol (the evidence for prestin multimerization is discussed later). This suggests that there are sufhydryl linkages within relatively inaccessible regions of prestin and that they have a stabilizing effect on dimer formation.

4.4 Multimerization

There are numerous reasons to suspect that prestin exists as a homomultimer or as part of a heteromultimeric motor complex (Geleoc et al. 1999; Belyantseva et al. 2000; Santos-Sacchi et al. 2001; Santos-Sacchi 2002; Dong and Iwasa 2004; Santos-Sacchi and Wu 2004; Navaratnam et al. 2005). The particles observed in the OHC lateral wall are large (about 10 nm) compared to protein subunits, which is consistent with a multimeric structure, and altogether too large to be solely a protein the size of prestin. Further, the large carboxy terminus is suggestive of possible protein–protein interactions. Recently, the Dallos group demonstrated self-association of prestin on gels and in yeast two-hybrid experiments (Zheng et al. 2006a). Prestin dimers, but not tetramers, were resistant to dissociation by SDS. They suggested that tetramers are formed from stable dimers, but this does not explain the prominent appearance of trimers in their Western blots. Further, interactions between fluorescent protein–prestin constructs in HEK cells have been demonstrated by sensitized emission fluorescence resonance energy transfer

(FRET) (Greeson et al. 2006) and by acceptor photobleach FRET (Navaratnam et al. 2005; Hallworth et al. 2006; Wu et al. 2007). It was interesting that the amino terminus rather than the carboxy terminus that was observed to be important for multimerization (Navaratnam et al. 2005).

What are the consequences of multimerization to our understanding of prestin? If prestin tetramerizes as suggested by Zheng et al. (2006a), four times as many copies of prestin are needed as previously supposed. More importantly, prestin state transitions require translation of only one unit charge, in contrast to the sodium channel, for which four are required. Thus either the chloride binding site (or allosteric modulation site) has components from all four subunits, or three of the binding sites are inactivated on tetramerization. Multimerization, however, does open up new possibilities for the conformational change associated with prestin state change. A simple example is shown in Fig. 5.3.

4.5 Prestin Localization

As previously described, the motor protein was expected to be a membrane-spanning protein, based on its direct voltage sensitivity, and several lines of evidence indicated that it was located along the lateral wall but above the nucleus (i.e., about 12 μm from the base and below the cuticular plate by about 5 μm). This would mean that the very shortest basal turn hair cells (overall length about 25 μm in the guinea pig) have less than 10 μm of length devoted to motor protein, i.e., the stiffest OHCs also have the smallest number of motors. Although the identity of prestin and the motor protein are not in question, it has therefore come as a surprise that prestin has been observed by two groups to extend the entire length of the OHC (see Fig. 5.2), in gerbil (Jensen-Smith and Hallworth 2006) and in guinea pig (Yu et al. 2006).

4.6 Control of Prestin Expression

Congenital hypothyroidism is well known to result in a severe high-frequency hearing loss and multiple malformations of the inner ear. Thyroid hormone receptors (TRs), when bound to thyroid hormone, regulate transcription. Of the cochlear TRs, one (TRβ, encoded by *Thrb*) is particularly prominent in the organ of Corti (Knipper et al. 1998). A thyroid hormone response element has been described upstream of the prestin gene that may regulate prestin expression and possibly synthesis and localization (Weber et al. 2002). When pregnant rats were made thyroid deficient by methylmercaptoimidalzole, the offspring showed an immature OHC localization of prestin in the basal pole. Similarly, the offspring of the *hyt/hyt* mouse, which is deficient in TRs, exhibit significantly delayed maturation of auditory brainstem responses. OHC electromotility developed normally in organ of Corti cultured from P0 despite the lack of extrinsic thyroid hormone (He et al. 2003a). However, such a late date of explantation is probably well after TH has performed its normal developmental function.

4.7 Prestin Homologs

Prestin-like gene sequences and expression in mechanosensory organs have recently been reported in *Drosophila* and zebrafish (Weber et al. 2003). The predicted proteins have 30% and 50% identity (49% and 68% similarity), respectively, with gerbil prestin. The sulfate transporter domains are recognizable in the nonmammalian sequences, although the glycosylation and phosphorylation sites are not. The *Drosophila* result is especially interesting, in that active amplification of motion has been observed in the low frequency Johnston's (antennal hearing) organ of the fly (Göpfert and Robert 2003; Gopfert et al. 2005). We eagerly await the results of the *Drosophila* prestin knockout.

4.8 Significance

While the role of prestin in normal hearing is still unclear, its importance to hearing is beyond question. The prestin knockout mouse generated by the Zuo lab (Liberman et al. 2002) exhibited a severe hearing loss over all frequencies, as measured by distortion product emissions or by auditory brainstem responses. However, round window cochlear microphonics in the homozygote were asserted to be essentially normal, which implies intact forward transduction in OHCs. The interpretation of the prestin knockout results at high frequencies was compromised by a progressive degeneration of basal turn OHCs. However, the hearing loss at all frequencies precedes the hair cell loss (Wu et al. 2004). Further support for the significance of prestin in hearing has come from the identification of two human families with nonsyndromic recessive deafness that have splicing defects at the same locus in prestin (exon 3) (Liu et al. 2003).

The results of the heterozygote have been more problematic. Liberman et al. (2002) reported that OHCs from heterozygotic animals of the F1 generation exhibited about half the motility amplitude of the wild type. However, little or no hearing loss was observed in the heterozygote. More recently, Cheatham et al. (2005) found similar hearing losses in the F3/F5 generation homozygote but nearly normal OHC motility and hearing in OHCs.

As previously described, a complication of the prestin knockout is that basal turn OHCs in the homozygote degenerate early in life, compromising hearing loss measurements above about 20 kHz (about a quarter of the cochlea). Whether all OHCs eventually degenerate has not been determined. Wu et al. (2004) have shown that neither OHC loss nor morphological abnormalities are observed prior to P28 in the mouse, whereas ABR thresholds are elevated at all frequencies (up to 32 kHz) as early as P21.

The probable absence of OHCs in the most basal regions of the cochlea in the homozygote also complicates the interpretation of the microphonic results. The microphonics were obtained at 6 to 8 weeks of age, i.e., well after OHC degeneration would have occurred. Microphonic amplitudes in the homozygote were attenuated by only one half to two thirds at low to moderate intensities, compared to the wild type. If the cochlear amplifier were attenuated or abolished by the absence of functional prestin, one would expect that basilar membrane vibration amplitudes (and consequently, cochlear microphonics) would have been attenuated by 30 dB or more. Liberman et al. (2002) assert that the small reduction of the microphonic was due to the loss of OHCs in the basal turn, on the grounds that round window microphonics originate primarily from the most basal OHCs, whose best frequencies are much greater than the 16 kHz test tone used. However, since the most basal OHCs are absent from the homozygote cochlea at this age, the microphonic responses may well have originated from much closer to the 16 kHz region. In that case, the microphonic responses in the homozygote are paradoxically large.

4.9 Prestin-Associated Proteins

The pillar protein or complex may or may not serve as an anchor for prestin to the cortical lattice. As previously described, the identity of the pillar protein or protein complex has not yet yielded to affinity screens. A zinc finger-type transcription factor, promyleocytic leukemia zinc finger protein, was recently isolated by a yeast two-hybrid screen using prestin (Nagy et al. 2005) and was localized by immunolabeling to the PM of OHCs. Possible physiological roles for such a protein remain obscure. The possible relation of GLUT5 to prestin is discussed later in this review.

4.10 Is Prestin a Porter?

The conventional wisdom has been that prestin is a transporter molecule but one in which the transporting function has been lost. However, there remain reports that prestin may transport something, although just what varies. Chambard and Ashmore (2003) have demonstrated transport of sugars into prestin-transfected HEK cells. More recently, Muallem and Ashmore (2006) showed that the behavior of OHC NLC when exposed to internal sulfate ions is inconsistent with either of the standard models. The inference drawn is that prestin may transport sulfate or chloride out of OHCs, i.e., it functions as an antiporter. Direct evidence for sulfate transport, however, is still lacking.

5. OHC Axial Stiffness

OHCs must possess a certain amount of axial stiffness in order to communicate axially directed motor force to the load, which is the organ of Corti (including other OHCs). In this section, we discuss what is known about the passive properties of OHCs and how they may be modulated physiologically.

5.1 Properties and Significance

OHC axial stiffness has usually been measured by compression using glass fibers of known stiffness. Early studies consistently found that axial compliance (the reciprocal of stiffness) was found to be a linearly increasing function of OHC length, implying a uniform specific (per unit length) compliance (Hallworth 1995; Russell and Schauz 1995; Chan and Ulfendahl 1997; Hallworth 1997). However, there was considerable variation in the data presented (Fig. 5.6A). Most potential sources of error, such as misalignment of fiber and cell, would lead to overestimation of compliance. Thus the two best data sets are probably those in Hallworth (1997), using axial compression, and Adachi and Iwasa (1997), using axial stretch (Fig. 5.6B).

When the load on an OHC was increased, the measured compliance remained the same. However, when electrically stimulated, the cell generated less length change but more force (Hallworth 1995). A simple model of the OHC would

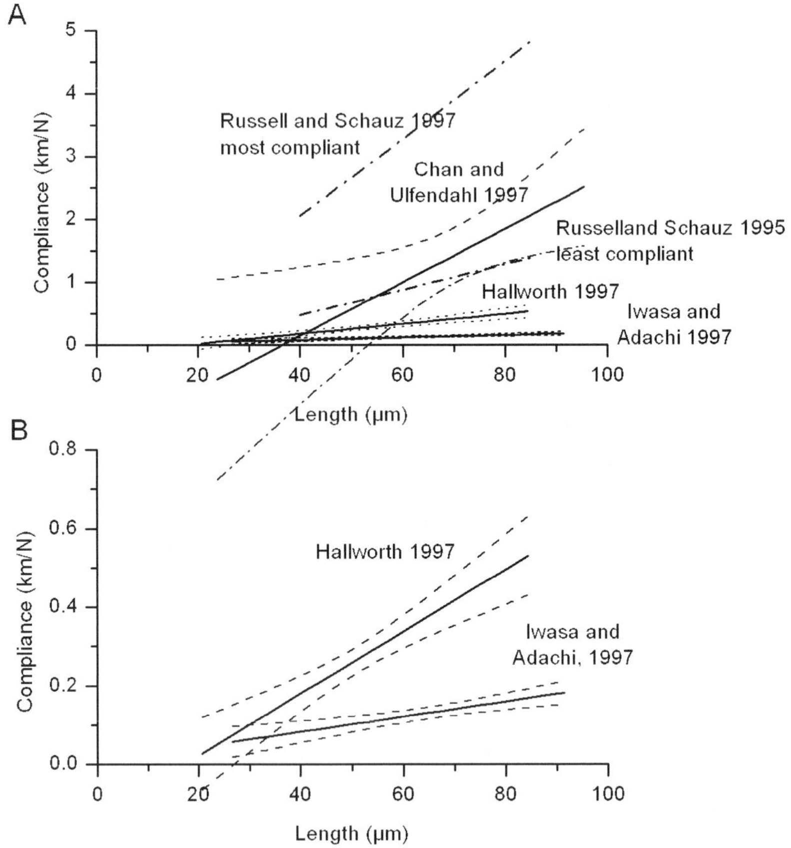

FIGURE 5.6. Axial compliance of guinea pig OHCs as a function of length. Data replotted from the studies indicated, including Chan and Ulfendahl (1997) and Russell and Schauz (1995). The dashed lines represent the 95 % confidence limits of the slope, where such data is available. (**A**) All data sets. (**B**) The two best data sets.

therefore consist of an incompressible length change-generating motor in series with a stiffness that is inversely proportional to cell length. Consistent with the model, the isometric force was found to be independent of OHC length, and therefore of OHC stiffness (Hallworth 1997). Axial stiffness and motor force may be modulated separately: hypoosmotic challenge increased stiffness and force against a constant load, while salicylate ions reduced or abolished force but left axial stiffness unaffected (Hallworth 1997; but see Russell and Schauz 1995).

5.2 Modulation by Acetylcholine

Acetylcholine (ACh) is the major efferent neurotransmitter of OHCs, especially in the basal turn, where the density of efferents is largest (Guinan 1996). OHCs (but not IHCs or cochlear ganglion cells) have an unusual combination of ACh receptor α subunits ($\alpha 9$ and $\alpha 10$) with an unusual pharmacology with both nicotinic and muscarinic characteristics (Elgoyhen et al. 1994, 2001) and strychnine sensitivity (which explained some puzzling early observations about the effect of strychnine on cochlear potentials by Robertson and Johnstone in 1978). Application of ACh by perfusion in near millimolar concentrations was observed to amplify OHC electromotility throughout the voltage range, when applied to the basal pole of short (basal turn) OHCs (Sziklai et al. 1996). The putative apical turn efferent neurotransmitter γ-aminobutyric acid (GABA) had a similar but much smaller effect on long OHCs. Motility amplitude under voltage clamp was also enhanced (Dallos et al. 1997) with a half-maximal concentration to pulse application of 21.3 μM (larger than but comparable to the half-maximal concentration for the outward potassium current (Doi and Ohmori 1993; Kakehata et al. 1993). The compliance measured by fiber compression was decreased by ACh, and the force decreased against a constant load (Dallos et al. 1997). This suggests a model for the action of efferents independent of any electrophysiological action (which is hyperpolarizing but by only a few millivolts (Housley and Ashmore 1991)). Increased compliance would inhibit the transmission of electromotile force to the organ of Corti. Confirmation of the ACh effect on stiffness was obtained by Frolenkov et al. (2000), who also observed calcium dependence.

A proposed mechanism for the compliance increase is G protein-mediated phosphorylation of motor and/or cytoskeletal proteins (Sziklai and Dallos 1993; Frolenkov et al. 1998; Sziklai et al. 2001). Potential phosphorylation sites have also been described on prestin. This implies further that the substantial prestin content of the OHC membrane influences its stiffness. It is possible that such questions could be resolved in the prestin knockout mouse. Although no substrates for actin or spectrin phosphorylation in OHCs have been described in OHCs, phosphorylation of actin and spectrin-associated proteins such as adducin have been proposed to mediate OHC compliance (Zhang et al. 2003; see later).

5.3 Modulation by Diamide

Diamide, a membrane-permeable sulfhydryl reagent, has been reported to decrease OHC axial elongation stiffness (Adachi and Iwasa 1997; Frolenkov et al. 1998), without significantly altering electromotility (Kalinec and Kachar 1993). Diamide specifically oxidizes sulfhydryl groups on accessible cystyl residues resulting in the formation of disulfide bonds (Becker et al. 1986). In the erythrocyte, this bifunctional thiol has been shown to have two main actions. First, diamide induced sulfhydryl oxidation has been shown by several researchers to increase spectrin crosslinking (Haest et al. 1977; Fischer et al. 1978; Maeda et al. 1983). Second, diamide has been shown to decrease actin-spectrin binding mediated by protein 4.1 (Becker et al. 1986). The effects of diamide and other SH-reactive reagents (*p*-chloromercuribenzoate and *N*-ethylmaleimide) on the erythrocyte include induction of spheroid shape, an increase in osmotic fragility, and a decrease in cellular deformability (Jacob and Jandl 1962a,b; Fischer et al. 1978; Haest et al. 1980; Johnson et al. 1980). Thus the reported diamide effect on OHC stiffness, an increase, is paradoxical, to say the least.

5.4 Modulation by Amphipaths

The amphipathic (and antipsychotic) small molecule chlorpromazine (CPZ) readily inserts into plasma membranes. In red blood cells (RBCs), CPZ induces cupping of the plasma membrane (Sheetz and Singer 1974). At much lower concentrations, CPZ makes membrane tethers easier to draw from the lateral wall of isolated OHCs (Murdock et al. 2005; Morimoto et al. 2002), which would suggest an increase in membrane fluidity and therefore a decrease in OHC axial stiffness. Consistent with this view, Oghalai (2004) found that CPZ perfusion of the cochlea increased compound action potential and distortion product otoacoustic emission (DPOAE) thresholds. Zheng et al. (2006) recently reported decreased basilar membrane velocity after infusion of CPZ, However, their proposed model to explain the observations predicted an *increase* in OHC stiffness.

5.5 Modulation by Small GTPases

As previously described, the small GTPases of the Rho family participate in the regulation of actin networks. The Kalinec group recently described parallel RhoA and Rac1 signaling cascades that function as a mechanical homeostatic system. ACh-induced changes in OHC resting length (slow motility) and in amplitude of electromotility (fast motility) were modulated by Rho and their downstream targets Rac and Cdc42 (Kalinec et al. 2000; Zhang et al. 2003). Peculiarly, ACh by itself had no effect on fast or slow motility without addition of the nonhydrolyzable GTP analogue GTPγS to the patch pipette. This result is in stark contrast to those discussed above (Sziklai et al. 1996; Dallos et al. 1997), for which no such manipulation was required. ACh had no effect on fast motility-related NLC in OHCs (Zhang et al. 2003), a result also reported by Frolenkov et al. (2000), which therefore implicates the passive mechanical properties of the OHC. In this model, RhoA activation of ROCK inhibits a constitutively active amplitude-decreasing signaling cascade mediated by Rac1. Taken together with the observation of phosphorylated adducin in the OHC lateral wall (Zhang et al. 2003), the results suggest that the cortical lattice is dynamic and capable of modulation under physiological circumstances. However, the conflicting data from these studies urgently need reconciliation if progress is to be made.

5.6 Voltage-Dependent Stiffness

The Dallos laboratory has repeatedly demonstrated a membrane potential dependent ("voltage dependent") stiffness in isolated OHCs (He and Dallos 1999, 2000; He et al. 2003b). The fraction of axial stiffness that is subject to membrane potential dependent modulation is apparently quite large; in one example (Fig. 5 of He and Dallos 2000), axial stiffness varied from less than 3 mN/m to greater than 7 mN/m over the dynamic range of motility. Voltage-dependent stiffness and motility were intimately associated. Treatments that abolished motility, such as gadolinium treatment (He and Dallos 1999, 2000), or replacement of the high chloride intracellular medium with pentane-sulfonate (He et al. 2003b), also abolished voltage-dependent stiffness (He et al. 2003b), although its absence has not yet been demonstrated in the prestin knockout mouse. Modeling also indicated that the voltage dependence resides in the motor molecule rather than in accessory structures (Dallos and He 2000; but see Iwasa 2000). While the data presented are extensive, and several technical approaches were used, it should be pointed out that no other laboratory has reported the phenomenon.

What contribution does voltage-dependent stiffness make to cochlear mechanics? On the rarefaction phase of an acoustic stimulus (when the BM travels toward the TM), the receptor potential is depolarizing and thus the stiffness should decrease. A stiffness reduction would look like an amplifying force. Models employing a negative impedance have been advanced (Neely 1993), although it is uncertain whether a purely resistive change in impedance is enough. In the absence of data on high-frequency impedance changes in OHCs, it is hard to know whether such a phenomenon could play a role. OHC axial impedance is evidently constant out to acoustic frequencies (Frank et al. 1999). Further, the bandwidth problem again applies. (He and Dallos 1999) argue for a tonic contribution from the DC component of receptor potential. But this has not been modeled, nor does it explain what the cycle by cycle stiffness changes would contribute to transduction.

6. Possible Relation of OHC Motility to Cilia Motion

It has never been clear how the up and down motion of the OHC is translated into amplified lateral motion of the OHC stereocilia and the consequent amplified OHC receptor potential. The suggestion that OHC electromotility may cause the cuticular plate of the OHC to pivot goes back to the early days of isolated OHC electromotility studies (Zenner et al. 1987, 1988). The idea is that the stereocilia would thereby be displaced to a set position closer to the middle of their dynamic range, which would increase their gain. Recently the notion has been partly revived by Jia and He (2005) in observations of OHC stereocilia motion evoked by depolarization from isolated adult gerbil apical organ of Corti fragments. The motion was directly attributable to somatic motility and was not present in sulfate-dialyzed OHCs or in organ fragments from prestin knockout mice. Involvement of transducer currents was ruled out by the use of streptomycin to block transduction. The direction of the motion was toward the kinocilium (i.e., the excitatory direction) and was consistent with the predicted direction of motion of the angled OHCs of the cochlea apex. The preparation did not permit determination of whether cilia motion relative to the cuticular plate was also involved, but this could be crucial. If the motion were purely a secondary consequence of contraction of long, angled OHCs, it would be far less significant in the basal turn where OHCs are shorter and closer to the orthogonal.

In stark contrast, Kennedy et al. (2005) observed a rapid kinopetal motion of stereocilia in response to step depolarization in a similar preparation of immature rat cochlea (7–10 days postnatal). The motion was sensitive to external Ca ions, which is consistent with the transducer channel adaptation process and not consistent with a role for prestin. However, it is not known with any certainty whether OHC electromotility is even present at this relatively early stage of rat cochlea development. In gerbil at 7 days post natal, electromotility is just appearing (He et al. 1994; Oliver and Fakler 1999).

7. Problems of Somatic Motility

Basic models of the action of OHC motility propose that the cycle by cycle component receptor potential drives OHC length change in a feed-forward manner (Dallos 1992). A fundamental problem with this idea is that OHC receptor potentials show no detectable cyclic component above about 3 kHz. Even allowing for the filtering characteristics of intracellular electrodes, it appears unlikely that the OHC PM has a significant high-frequency receptor potential component. The bandwidth of OHCs could in principle be improved by reducing membrane resistance, and indeed basal turn OHCs have an order of magnitude smaller resistance than apical turn OHCs (Axelsson et al. 1990; Housley and Ashmore 1992), but this would come at the expense of receptor potential amplitude. Thus the cycle by cycle driving potential needed for OHC motility is probably lacking in the basal turn.

Feed-forward amplification mechanisms are inherently unstable. In the cochlea, spatial spread of excitation is of special concern. This has not been addressed in models to date, which are invariably of a single cochlear section, as if there were no longitudinal coupling along the BM. Intuitively, if electromotility amplifies the vibration amplitude at a cochlear location, it also amplifies the vibration at nearby points. In a feed-forward system, amplification should therefore cascade simultaneously toward the cochlear apex and base. Of course, this is not what happens. In the hemicochlea, the longitudinal length constant of mechanical perturbation has been stated as being 21 μm (Emadi et al. 2004), which is two OHCs per side, or five total. This is more than enough to start a propagating amplification of adjacent cochlear regions. Further, there is no obvious way in which such a mechanism would turn itself off as stimulus amplitude increases. These problems are fundamental to theories that incorporate OHC electromotility into the cochlear amplifier and are not currently being addressed in a systematic fashion, either experimentally or in models.

Attempts to overcome the severe theoretical problem of bandwidth have consisted of four threads, which are discussed in sections 7.1–7.4.

7.1 The Extracellular Potential as the Drive

Dallos and Evans (1995a,b) proposed that the drive to OHC motility in the basal cochlea turn is the extracellular potential (the cochlear microphonic [CM] as recorded in the organ of Corti) rather than the bandwidth-limited receptor potential. This suggestion has the advantage that the CM is not bandwidth-limited, but it also has a number of problems. If the extracellular potential is dominant in the basal turn, but the receptor potential is dominant in the apical turn, what goes on in the middle turn? Do the two mechanisms compete? Another problem is that the length constant for electrical potential spread along the basal turn is on the order of millimeters (Johnstone et al. 1966), while the cochlear amplifier is, as we have discussed, much more localized. In keeping with this, Fridberger et al. (2004) found that the basal turn extracellular potential requires a subsequent filtering step to account for the basilar membrane motion at best frequency. Further, the action of efferents then becomes paradoxical. Efferent stimulation enhances the microphonic (Desmedt 1962) but attenuates the cochlear amplifier, as measured by IHC receptor potentials (Brown and Nuttall 1984) and basilar membrane vibration (Murugasu and Russell 1996; Russell 1997).

7.2 Piezoelectricity, Flexoelectricity, and Resonance

A second approach to the bandwidth problem exploits the piezoelectric description of OHC mechanics. A piezoelectric solid converts mechanical stress into an electric field, or, reciprocally, demonstrates strain when electrically polarized. Stress and strain and charge movement are linked by piezoelectric coupling coefficients; thus

$$Q = c_{11}E + c_{12}F \tag{5.3}$$

and

$$L = c_{12}E + c_{22}F, \tag{5.4}$$

where Q and L are the charge transfer and mechanical displacement respectively, E and F are any applied electrical field or force, and c_{11},c_{12}, and c_{22} are piezoelectric coupling coefficients.

As we have seen, the OHC exhibits strain when electrically polarized, and there is also some change in transmembrane charge (due to prestin state change) when OHC membrane tension is changed, either by stretch (Gale and Ashmore 1994) or internal pressure change (Kakehata 1995). Thus the OHC has piezoelectric properties, albeit with coupling coefficients that are in turn functions of electrical polarization, and the calculated piezoelectric coefficient (in the voltage to strain direction, at least) is unusually large.

While this is purely descriptive and of no help in understanding the underlying mechanism of electromotility, it does suggest that if the loading is properly matched to the intrinsic mechanical properties of the OHC an amplifying resonance between the OHC and the BM could be induced. An electromechanical resonance to high frequency electrical stimulation has been demonstrated in isolated unloaded OHCs by Rabbitt et al. (2005), and has also been modeled in the microchamber configuration at plausible acoustic frequencies (Liao et al. 2005), although the relevance of this particular unloaded electromechanical configuration to the in vivo situation is questionable. However, the load on the OHC in situ, which consists of the rest of the organ of Corti, including other OHCs, will undoubtedly participate in any resonance. Without reliable knowledge of the load forces and masses, it is unclear whether the proposed amplification is feasible. It is also not clear what would prevent ambient physiological noise from setting the resonant amplification into oscillation.

In a variation on this approach, Brownell and colleagues have advocated an entirely independent mechanism of electromotility, called flexoelectricity (Raphael et al. 2000). Flexoelectricity is an inherent property of liquid crystals, of which lipid bilayer membranes are one form. Local curvature of a lipid bilayer, by differentially distributing their surface charges over the different radii of curvature between inside and out, would generate a transmembrane electric field that could be meaningfully large if the curvature is sufficiently small. Alternatively, imposition of a transmembrane field would modulate curvature, although it would not create curvature where none previously existed. In applying this theory to the OHC, Raphael et al. (2000) proposed that the nanoscale wrinkles in the OHC LW membrane (which are regularly observed in electron microscopy) are sufficiently curved to be susceptible to flexoelectric manipulation. The model was able to demonstrate nonlinear, saturating OHC length changes of the appropriate sign, assuming that the dipole moments formed of the surface charges are sufficiently strong. However, little progress has been made toward establishing values for the appropriate parameters. The model, which does not require or specify a prestin-like molecule, could incorporate prestin if it were assumed to be critical in forming or maintaining the local membrane curvatures. The proposed mechanism by itself does not overcome the bandwidth argument posed earlier, because it is membrane potential driven. Also, it is possible that the membrane wrinkles observed in fixed OHCs are a secondary consequence of fixation-induced shrinkage, rather than a feature of living OHCs. The difficulty of resolving this question recalls the energetic disputes over the continuity, or lack thereof, of SSC in times past (Brownell 1990; Evans 1990).

7.3 Mechanically Activated Chloride Channel

Santos-Sacchi and colleagues have identified a nonspecific stretch-sensitive conductance in the lateral wall of isolated guinea pig OHCs that passes both monovalent cations and the anion chloride (Rybalchenko and Santos-Sacchi 2003). The current becomes evident after replacement of both intracellular and extracellular cations with impermeant substitutes. The authors suggest that chloride ions can accumulate intracellularly in a restricted space close to the PM, and therefore close to prestin. They also suggest that modulation of prestin by chloride is allosteric rather than direct, that is, it increases the rate at which prestin changes conformation from extended to contracted rather than participating directly in the conformation change. This would not be consistent with the observation that neither V_h nor z changes with internal chloride concentration (Oliver et al. 2001), but would be consistent with the observed changes in V_h and z. This is a critical issue, and it is disappointing that it has not yet been resolved. The underlying thesis, then, requires that BM vibration stretches the OHC, which in turn allows chloride entry and thereby modifies motor state. This mechanism would not be subject to the bandwidth limitations due to membrane potential filtering, although its kinetics may be ultimately limited by diffusion, even if the channel and the prestin molecules are especially close. The mechanism also requires a small number of chloride ions to have a dramatic effect on prestin state. While this is feasible in a confined space, direct evidence of such a space is lacking. In particular, the rapid action of chloride substitutes perfused from a patch pipette (Rybalchenko 2003) argues against a substantial barrier to chloride diffusion. If such a barrier existed, would not sodium and potassium ions also be so confined?

7.4 The Gain-Bandwidth Argument

If instead of a feedforward approach OHC somatic electromotility is considered as part of a feedback mechanism, then the product of the loop again and the bandwidth (the gain-bandwidth product) may be in turn considered a fundamental constant of the mechanism. Trading gain for bandwidth can, in principle, increase the frequency response of the system. This approach, first advanced in a qualitative sense by Mountain and Hubbard (1989), has recently been revived in a detailed model (Lu et al. 2006). The TM is thought to move the OHC stereociliary bundle toward the modiolus, i.e., it reduces the shear between organ of Corti and TM on the upward (depolarizing) phase of each vibratory cycle (Fig. 5.7). Presumably, it would enhance the shear on the hyperpolarizing phase, but not nearly as much. While high-frequency responses are indeed enhanced in this model, as would be expected, it is not clear if the proposed negative feedback

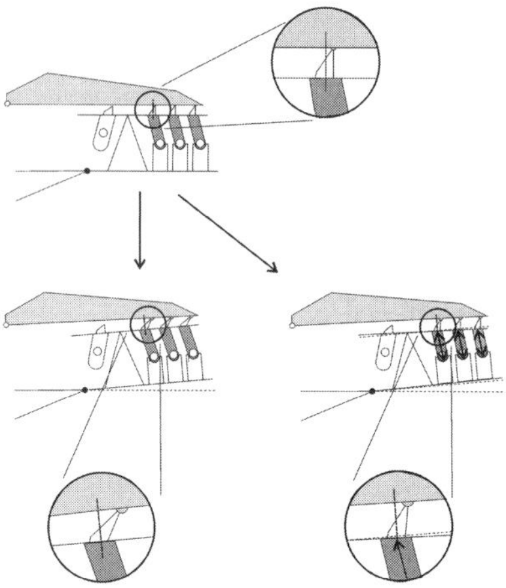

FIGURE 5.7. Relative motion of the basilar membrane and tectorial membrane during the excitatory (rarefaction) phase of basilar membrane motion, showing how negative feedback might arise via reduction of the deflection angle of the OHC stereocilia (adapted from Lu et al. 2006). The insets show the position of the stereociliary bundle with reference to the rest position (depicted by a dashed line orthogonal to the apical surface of the OHC).

actually occurs. The assumption rests on a hypothetical geometrical relationship between the organ of Corti and TM (depicted in Fig. 5.7) and also assumes that OHC contraction substantially deforms the organ of Corti. Optical flow analysis of the relative motion of the organ of Corti and tectorial membrane in the hemicochlea (Cai et al. 2003) suggests that the shear exists, but it did not reveal any modulation by electromotility, possibly because it is beyond the resolution of the method. Although stimulus intensity relationships were not reported, this model does have the advantage that the enhanced feedback at high stimulus levels would help turn it off. The negative feedback would also prevent the amplification of ambient noise. It does not suggest, however, how high amplification would occur at low stimulus intensities, at which the feedback contribution would be minimal. Incidentally, the existence of a feedback component was previously suggested from in vivo measurements (Zinn et al. 2000).

7.5 GLUT5

A cell that accumulates sugars as glycogen must perforce have a facilitated glucose transporter. In OHCs, the transporter appears to be GLUT5 (Nakazawa et al. 1995b). One peculiarity of GLUT5 is that it preferentially transports fructose rather than glucose (Buchs et al. 1998), the only member of the glucose transporter to do so. In a series of experiments, Ashmore and colleagues have shown that extracellular fructose (but not glucose) induces a rapid, reversible shortening of isolated OHCs (Geleoc et al. 1999; Ashmore et al. 2000). However, these studies raise a question—there is glucose in perilymph (Thalmann 1972), but where is the fructose?

While the density of GLUT5 in the OHC PM is hard to estimate from the immunohistochemistry so far published, it seems to be reasonably abundant and possibly colocalized with prestin. Could it be a component of the apparent motor complex? Prestin does not apparently require GLUT5 coexpression for function, nor does heterologous cosynthesis of prestin and GLUT5 significantly modify prestin function (Ludwig et al. 2001). So far, this leaves GLUT5 an orphan, with no apparent role in electromotility or transduction.

7.6 Where Is the Vulnerability?

The cochlear amplifier is extraordinarily vulnerable to overstimulation, metabolic compromise, and drugs that inhibit the generation of the endolymphatic potential, e.g., furosemide. The nonlinear frequency specific amplification of the BM vibration disappears within minutes of such treatment (Ruggero and Rich 1991). What is the source of this vulnerability? The evidence suggests that it resides in OHCs, but where? OHC electromotility itself is very far from vulnerable. Indeed, deterioration of isolated OHC condition initially leads to enhanced electromotility amplitude via depolarization and the consequent increase in the slope gain of the voltage to length change function (Evans et al. 1991) (refer to Fig. 5.4). Accompanying furosemide treatment is a rapid decrease in the endolymphatic potential. However, loss of the endolymphatic potential merely halves the amplitude of the OHC receptor potential (i.e., reduces it by 6 dB); not nearly enough to account for the loss of cochlear amplification (which is more like 30 dB). Similarly, efferent stimulation of OHCs, which can hyperpolarize the OHC by at most a few millivolts to the potassium equilibrium potential, also abolishes most low level amplification at best frequency (but not below best frequency) (Brown and Nuttall 1984; Murugasu and Russell 1996). It is clear that the locus of vulnerability of the cochlear amplifier lies elsewhere than OHC electromotility.

7.7 What Is the Role of OHC Electromotility?

Given the manifest and continuing difficulties in understanding the role of OHC electromotility in auditory transduction, it is tempting to refute a role for electromotility entirely, as many have done (Hudspeth 2005). In this view, the remarkable length changes undergone by OHCs in vitro are presumably an artifact of unphysiological electrical manipulations of a protein (prestin) that has some entirely different function. However, it is hard to escape the fact that electrical stimulation of OHCs in intact organ of Corti does induce deformation of the apical surface of the organ and therefore is poised to influence transduction by stereocilia (Mammano and Ashmore 1993). Further, the prestin knockout mouse, whatever its problems, does argue persuasively for a role for prestin in cochlear transduction, whatever that role might be.

8. Summary

Spoendlin in 1974 showed us the sparse and diffuse innervation of the OHC in the mammalian cochlea, and Dallos long ago drew attention to what he termed the "paradox of the outer hair cell." Our knowledge of the anatomy, physiology, and even molecular biology of the OHC is vastly increased over Spoendlin's day, but even if the wording of the paradox has changed, the enigmatic nature of the OHC remains. No one doubts the importance of OHCs to hearing, but we are still confronted with a molecule (prestin) and a physiological phenomenon (electromotility) that cannot possibly be the solution to the fundamental problem of the cochlear amplifier in the manner originally envisaged. To the disinterested outside observer, the current intense study of prestin must seem like the search for a lost coin in the dark under a street light, knowing full well that the coin was lost elsewhere. Enlightenment almost certainly lies elsewhere, but where?

Acknowledgments. The authors' research is supported by DC02053 (N.I.D.C.D., N.I.H.), RR17417-01 (N.C.R.R, N.I.H), EPS-0346476 (CFD 47.076) (N.S.F. & Nebraska EPSCoR), and a grant to H.C.J.-S. from the Deafness Research Foundation. We thank Richard Ludueña and Jing Zheng for the gift of antibodies, Robert Kimura for the electronmicrograph in Fig. 5.1A and Barbara Bohne for the electronmicrograph in Fig. 5.1B. We also thank members of the Hallworth laboratory for comments on the manuscript.

References

Adachi M, Iwasa KH (1997) Effect of diamide on force generation and axial stiffness of the cochlear outer hair cell. Biophys J 73:2809–2818.

Adachi M, Iwasa KH (1999) Electrically driven motor in the outer hair cell: effect of a mechanical constraint. Proc Natl Acad Sci USA 96:7244–7249.

Adler HJ, Belyantseva IA, Merritt RC, Jr., Frolenkov GI, Dougherty GW, Kachar B (2003) Expression of prestin, a membrane motor protein, in the mammalian auditory and vestibular periphery. Hear Res 184:27–40.

Aravind L, Koonin EV (2000) The STAS domain—a link between anion transporters and antisigma-factor antagonists. Curr Biol 10:R53–55.

Arima T, Kuraoka A, Toriya R, Shibata Y, Uemura T (1991) Quick-freeze, deep-etch visualization of the "cytoskeletal spring" of cochlear outer hair cells. Cell Tissue Res 263:91–97.

Armstrong CM (1981) Sodium channels and gating currents. Physiol Rev 61:644–683.

Arnold W, Anniko M (1990) Structurally based new functional interpretations of the subsurface cisternal network in human outer hair cells. Acta Otolaryngol 109:213–220.

Ashmore JF (1987) A fast motile response in guinea-pig outer hair cells: the cellular basis of the cochlear amplifier. J Physiol 388:323–347.

Ashmore JF, Geleoc GS, Harbott L (2000) Molecular mechanisms of sound amplification in the mammalian cochlea. Proc Natl Acad Sci USA 97:11759–11764.

Axelsson A, Nuttall AL, Miller JM (1990) Observations of cochlear microcirculation using intravital microscopy. Acta Otolaryngol 109:263–270.

Bannister LH, Dodson HC, Astbury AR, Douek EE (1988) The cortical lattice: a highly ordered system of subsurface filaments in guinea pig cochlear outer hair cells. Prog Brain Res 74:213–219.
Becker PS, Cohen CM, Lux SE (1986) The effect of mild diamide oxidation on the structure and function of human erythrocyte spectrin. J Biol Chem 261:4620–4628.
Belyantseva IA, Adler HJ, Curi R, Frolenkov GI, Kachar B (2000) Expression and localization of prestin and the sugar transporter GLUT-5 during development of electromotility in cochlear outer hair cells. J Neurosci 20:RC116.
Beneyto M, Sala JC, Sala ML, Merchan JA, Prieto JJ (1997) Structure of glycogen particles in organ of Corti's outer hair cells in three rodent species. J Hirnforsch 38:341–351.
Bennett V, Gardner K, Steiner JP (1988) Brain adducin: a protein kinase C substrate that may mediate site-directed assembly at the spectrin-actin junction. J Biol Chem 263:5860–5869.
Billett TE, Thorne PR, Gavin JB (1989) The nature and progression of injury in the organ of Corti during ischemia. Hear Res 41:189–197.
Brown MC, Nuttall AL (1984) Efferent control of cochlear inner hair cell responses in the guinea-pig. J Physiol 354:625–646.
Brownell WE (1984) Microscopic observation of cochlear hair cell motility. Scan Electron Microsc1401–1406.
Brownell WE (1990) Outer hair cell electromotility and otoacoustic emissions. Ear Hear 11:82–92.
Brownell WE, Bader CR, Bertrand D, de Ribaupierre Y (1985) Evoked mechanical responses of isolated cochlear outer hair cells. Science 227:194–196.
Brundin L, Russell I (1994) Tuned phasic and tonic motile responses of isolated outer hair cells to direct mechanical stimulation of the cell body. Hear Res 73:35–45.
Brundin L, Flock A, Canlon B (1989) Sound-induced motility of isolated cochlear outer hair cells is frequency-specific. Nature 342:814–816.
Brundin L, Flock A, Khanna SM, Ulfendahl M (1992) The tuned displacement response of the hearing organ is generated by the outer hair cells. Neuroscience 49:607–616.
Buchs AE, Sasson S, Joost HG, Cerasi E (1998) Characterization of GLUT5 domains responsible for fructose transport. Endocrinology 139:827–831.
Cai H, Richter CP, Chadwick RS (2003) Motion analysis in the hemicochlea. Biophys J 85:1929–1937.
Carlisle L, Zajic G, Altschuler RA, Schacht J, Thorne PR (1988) Species differences in the distribution of infracuticular F-actin in outer hair cells of the cochlea. Hear Res 33:201–205.
Catterall WA (1993) Structure and function of voltage-gated ion channels. Trends Neurosci 16:500–506.
Chambard JM, Ashmore JF (2003) Sugar transport by mammalian members of the SLC26 superfamily of anion-bicarbonate exchangers. J Physiol 550:667–677.
Chan E, Ulfendahl M (1997) Relationship between stiffness, internal cell pressure and shape of outer hair cells isolated from the guinea-pig hearing organ. Acta Physiol Scand 161:533–539.
Cheatham MA, Zheng J, Huynh K, Du GG, Gao J, Zuo J, Navarrete E, Dallos P (2005) Cochlear function in mice with only one copy of the prestin gene. J Physiol 560: 821–830.
Correas I, Avila J (1988) Erythrocyte protein 4.1 associates with tubulin. Biochem J 255:217–221.

Dallos P (1992) The active cochlea. J Neurosci 12:4575–4585.

Dallos P, Evans BN (1995a) High-frequency outer hair cell motility: corrections and addendum. Science 268:1420–1421.

Dallos P, Evans BN (1995b) High-frequency motility of outer hair cells and the cochlear amplifier. Science 267:2006–2009.

Dallos P, He DZ (2000) Two models of outer hair cell stiffness and motility. J Assoc Res Otolaryngol 1:283–291.

Dallos P, Evans BN, Hallworth R (1991) Nature of the motor element in electrokinetic shape changes of cochlear outer hair cells. Nature 350:155–157.

Dallos P, He DZ, Lin X, Sziklai I, Mehta S, Evans BN (1997) Acetylcholine, outer hair cell electromotility, and the cochlear amplifier. J Neurosci 17:2212–2226.

Dannhof BJ, Roth B, Bruns V (1991) Length of hair cells as a measure of frequency representation in the mammalian inner ear? Naturwissenschaften 78:570–573.

Deak L, Zheng J, Orem A, Du GG, Aguinaga S, Matsuda K, Dallos P (2005) Effects of cyclic nucleotides on the function of prestin. J Physiol 563:483–496.

Dehne N, Rauen U, de Groot H, Lautermann J (2002) Involvement of the mitochondrial permeability transition in gentamicin ototoxicity. Hear Res 169:47–55.

Desmedt J (1962) Auditory-evoked potentials from cochlea to cortex as influenced by activation of the efferent olivo-cochlear bundle. J Acoust Soc Am 34:1478–1496.

Doi T, Ohmori H (1993) Acetylcholine increases intracellular Ca^{2+} concentration and hyperpolarizes the guinea-pig outer hair cell. Hear Res 67:179–188.

Dong XX, Iwasa KH (2004) Tension sensitivity of prestin: comparison with the membrane motor in outer hair cells. Biophys J 86:1201–1208.

Dulon D, Aran JM, Schacht J (1988) Potassium-depolarization induces motility in isolated outer hair cells by an osmotic mechanism. Hear Res 32:123–129.

Dumont RA, Lins U, Filoteo AG, Penniston JT, Kachar B, Gillespie PG (2001) Plasma membrane Ca^{2+-} ATPase isoform 2a is the PMCA of hair bundles. J Neurosci 21: 5066–5078.

Duvall AJ, 3rd, Hukee MJ (1976) Delineation of cochlear glycogen by electron microscopy. Ann Otol Rhinol Laryngol 85:234–246.

Ekstrom von Lubitz DK (1981) Subsurface tubular system in the outer sensory cells of the rat cochlea. Cell Tissue Res 220:787–795.

Elgoyhen AB, Johnson DS, Boulter J, Vetter DE, Heinemann S (1994) Alpha 9: an acetylcholine receptor with novel pharmacological properties expressed in rat cochlear hair cells. Cell 79:705–715.

Elgoyhen AB, Vetter DE, Katz E, Rothlin CV, Heinemann SF, Boulter J (2001) alpha10: a determinant of nicotinic cholinergic receptor function in mammalian vestibular and cochlear mechanosensory hair cells. Proc Natl Acad Sci USA 98:3501–3506.

Emadi G, Richter CP, Dallos P (2004) Stiffness of the gerbil basilar membrane: radial and longitudinal variations. J Neurophysiol 91:474–488.

Engstrom H (1955) Morphological studies on the possible origin of cochlear microphonics. Rev Laryngol Otol Rhinol (Bord) 76:808–816.

Evans BN (1990) Fatal contractions: ultrastructural and electromechanical changes in outer hair cells following transmembraneous electrical stimulation. Hear Res 45: 265–282.

Evans BN, Hallworth R, Dallos P (1991) Outer hair cell electromotility: the sensitivity and vulnerability of the DC component. Hear Res 52:288–304.

Fischer TM, Stohr-Lissen M, Schmid-Schonbein H (1978) The red cell as a fluid droplet: tank tread-like motion of the human erythrocyte membrane in shear flow. Science 202:894–896.

Flock A, Flock B, Ulfendahl M (1986) Mechanisms of movement in outer hair cells and a possible structural basis. Arch Otorhinolaryngol 243:83–90.

Forge A (1991) Structural features of the lateral walls in mammalian cochlear outer hair cells. Cell Tissue Res 265:473–483.

Forge A, Souter M, Denman-Johnson K (1997) Structural development of sensory cells in the ear. Semin Cell Dev Biol 8:225–237.

Frank G, Hemmert W, Gummer AW (1999) Limiting dynamics of high-frequency electromechanical transduction of outer hair cells. Proc Natl Acad Sci USA 96: 4420–4425.

Fridberger A, de Monvel JB, Zheng J, Hu N, Zou Y, Ren T, Nuttall A (2004) Organ of Corti potentials and the motion of the basilar membrane. J Neurosci 24:10057–10063.

Friedman TB, Griffith AJ (2003) Human nonsyndromic sensorineural deafness. Annu Rev Genom Hum Genet 4:341–402.

Frolenkov GI, Atzori M, Kalinec F, Mammano F, Kachar B (1998) The membrane-based mechanism of cell motility in cochlear outer hair cells. Mol Biol Cell 9:1961–1968.

Frolenkov GI, Mammano F, Belyantseva IA, Coling D, Kachar B (2000) Two distinct $Ca(^{2+})$-dependent signaling pathways regulate the motor output of cochlear outer hair cells. J Neurosci 20:5940–5948.

Fukata Y, Oshiro N, Kaibuchi K (1999) Activation of moesin and adducin by Rho-kinase downstream of Rho. Biophys Chem 82:139–147.

Furness DN, Katori Y, Mahendrasingam S, Hackney CM (2005) Differential distribution of beta- and gamma-actin in guinea-pig cochlear sensory and supporting cells. Hear Res 207:22–34.

Gale JE, Ashmore JF (1994) Charge displacement induced by rapid stretch in the basolateral membrane of the guinea-pig outer hair cell. Proc R Soc Lond B Biol Sci 255:243–249.

Gardner K, Bennett V (1987) Modulation of spectrin-actin assembly by erythrocyte adducin. Nature 328:359–362.

Geleoc GS, Casalotti SO, Forge A, Ashmore JF (1999) A sugar transporter as a candidate for the outer hair cell motor. Nat Neurosci 2:713–719.

Göpfert MC, Robert D (2003) Motion generation by *Drosophila* mechanosensory neurons. Proc Natl Acad Sci USA 100:5514–5519.

Gopfert MC, Humphris AD, Albert JT, Robert D, Hendrich O (2005) Power gain exhibited by motile mechanosensory neurons in *Drosophila* ears. Proc Natl Acad Sci USA 102:325–330.

Greeson JN, Organ LE, Pereira FA, Raphael RM (2006) Assessment of prestin self-association using fluorescence resonance energy transfer. Brain Res 1091:140–150.

Griesinger CB, Richards CD, Ashmore JF (2004) Apical endocytosis in outer hair cells of the mammalian cochlea. Eur J Neurosci 20:41–50.

Grohovaz F, Limbrick AR, Miledi R (1982) Acetylcholine receptors at the rat neuromuscular junction as revealed by deep etching. Proc R Soc Lond B Biol Sci 215: 147–154.

Grunewald M, Menaker D, Kanner BI (2002) Cysteine-scanning mutagenesis reveals a conformationally sensitive reentrant pore-loop in the glutamate transporter GLT-1. J Biol Chem 277:26074–26080.

Guinan J (1996) Physiology of Olivocochlear Efferents. In: Dallos P, Popper A, Fay R (eds) The Cochlea. New York: Springer, pp 435–502.

Gulley RL, Reese TS (1977) Regional specialization of the hair cell plasmalemma in the organ of corti. Anat Rec 189:109–123.

Haest CW, Kamp D, Plasa G, Deuticke B (1977) Intra- and intermolecular cross-linking of membrane proteins in intact erythrocytes and ghosts by SH-oxidizing agents. Biochim Biophys Acta 469:226–230.

Haest CW, Driessen GK, Kamp D, Heidtmann H, Fischer TM, Stohr-Liesen M (1980) Is "deformability" a parameter for the rate of elimination of erythrocytes from the circulation? Pflugers Arch 388:69–73.

Hallworth R (1995) Passive compliance and active force generation in the guinea pig outer hair cell. J Neurophysiol 74:2319–2328.

Hallworth R (1997) Outer hair cell stiffness and force and their modulation by agents known to affect hearing. In: Lewis E, Long G, Lyon R, Narins P, Steele C, Hecht-Poinar E (eds) Diversity in Auditory Mechanics. Singapore: World Scientific, pp 524–530.

Hallworth R, Ludueña RF (2000) Differential expression of beta tubulin isotypes in the adult gerbil cochlea. Hear Res 148:161–172.

Hallworth R, Evans BN, Dallos P (1993) The location and mechanism of electromotility in guinea pig outer hair cells. J Neurophysiol 70:549–558.

Hallworth R, McCoy M, Polan-Curtain J (2000) Tubulin expression in the developing and adult gerbil organ of Corti. Hear Res 139:31–41.

Hallworth R, Currall B, Nichols MG, Wu X, Zuo J (2006) Studying inner ear protein-protein interactions using FRET and FLIM. Brain Res 1091:122–131.

He DZ, Dallos P (1999) Somatic stiffness of cochlear outer hair cells is voltage-dependent. Proc Natl Acad Sci USA 96:8223–8228.

He DZ, Dallos P (2000) Properties of voltage-dependent somatic stiffness of cochlear outer hair cells. J Assoc Res Otolaryngol 1:64–81.

He DZ, Evans BN, Dallos P (1994) First appearance and development of electromotility in neonatal gerbil outer hair cells. Hear Res 78:77–90.

He DZ, Jia S, Feng F (2003a) Thyroid hormone is not necessary for the development of outer hair cell electromotility. Hear Res 175:183–189.

He DZ, Jia S, Dallos P (2003b) Prestin and the dynamic stiffness of cochlear outer hair cells. J Neurosci 23:9089–9096.

Holley M (1996) Outer Hair Cell Motility. In: Dallos P, Popper A, Fay R (eds) The Cochlea. New York: Springer-Verlag, pp 383–434.

Holley MC, Ashmore JF (1988a) On the mechanism of a high-frequency force generator in outer hair cells isolated from the guinea pig cochlea. Proc R Soc Lond B Biol Sci 232:413–429.

Holley MC, Ashmore JF (1988b) A cytoskeletal spring in cochlear outer hair cells. Nature 335:635–637.

Holley MC, Ashmore JF (1990a) A cytoskeletal spring for the control of cell shape in outer hair cells isolated from the guinea pig cochlea. Eur Arch Otorhinolaryngol 247:4–7.

Holley MC, Ashmore JF (1990b) Spectrin, actin and the structure of the cortical lattice in mammalian cochlear outer hair cells. J Cell Sci 96:283–291.

Holley MC, Kalinec F, Kachar B (1992) Structure of the cortical cytoskeleton in mammalian outer hair cells. J Cell Sci 102:569–580.

Housley GD, Ashmore JF (1991) Direct measurement of the action of acetylcholine on isolated outer hair cells of the guinea pig cochlea. Proc R Soc Lond B Biol Sci 244:161–167.

Housley GD, Ashmore JF (1992) Ionic currents of outer hair cells isolated from the guinea-pig cochlea. J Physiol 448:73–98.

Huang G, Santos-Sacchi J (1993) Mapping the distribution of the outer hair cell motility voltage sensor by electrical amputation. Biophys J 65:2228–2236.

Huang G, Santos-Sacchi J (1994) Motility voltage sensor of the outer hair cell resides within the lateral plasma membrane. Proc Natl Acad Sci USA 91:12268–12272.

Hudspeth AJ (2005) How the ear's works work: mechanoelectrical transduction and amplification by hair cells. Curr Rev Biol 328:155–162.

Iwasa KH (1994) A membrane motor model for the fast motility of the outer hair cell. J Acoust Soc Am 96:2216–2224.

Iwasa KH (2000) Effect of membrane motor on the axial stiffness of the cochlear outer hair cell. J Acoust Soc Am 107:2764–2766.

Jacob HS, Jandl JH (1962b) Effects of sulfhydryl inhibition on red blood cells. I. Mechanism of hemolysis. J Clin Invest 41:779–792.

Jacob HS, Jandl JH (1962a) Effects of sulfhydryl inhibition on red blood cells. II. Studies in vivo. J Clin Invest 41:1514–1523.

Jensen-Smith HC, Hallworth R (2006) Outer hair cell mechanics are altered by developmental changes in lateral wall protein content. In: Nuttall A, Gillespie P, Ren T, Boer E, Grosh K (eds) Auditory Mechanisms: Processes and Models. Singapore: World Scientific, pp 187–193.

Jensen-Smith HC, Eley J, Steyger PS, Ludueña RF, Hallworth R (2003) Cell type-specific reduction of beta tubulin isotypes synthesized in the developing gerbil organ of Corti. J Neurocytol 32:185–197.

Jia S, He DZ (2005) Motility-associated hair-bundle motion in mammalian outer hair cells. Nat Neurosci 8:1028–1034.

Johnson RM, Taylor G, Meyer DB (1980) Shape and volume changes in erythrocyte ghosts and spectrin-actin networks. J Cell Biol 86:371–376.

Johnstone BM, Johnstone JR, Pugsley ID (1966) Membrane resistance in endolymphatic walls of the first turn of the guinea-pig cochlea. J Acoust Soc Am 40: 1398–1404.

Kakehata S, Santos-Sacchi J (1995) Membrane tension directly shifts voltage dependence of outer hair cell motility and associated gating charge. Biophys J 68:2190–2197.

Kakehata S, Santos-Sacchi J (1996) Effects of salicylate and lanthanides on outer hair cell motility and associated gating charge. J Neurosci 16:4881–4889.

Kakehata S, Nakagawa T, Takasaka T, Akaike N (1993) Cellular mechanism of acetylcholine-induced response in dissociated outer hair cells of guinea-pig cochlea. J Physiol 463:227–244.

Kalinec F, Kachar B (1993) Inhibition of outer hair cell electromotility by sulfhydryl specific reagents. Neurosci Lett 157:231–234.

Kalinec F, Holley MC, Iwasa KH, Lim DJ, Kachar B (1992) A membrane-based force generation mechanism in auditory sensory cells. Proc Natl Acad Sci USA 89: 8671–8675.

Kalinec F, Zhang M, Urrutia R, Kalinec G (2000) Rho GTPases mediate the regulation of cochlear outer hair cell motility by acetylcholine. J Biol Chem 275:28000–28005.

Kamphoven JH, de Ruiter MM, Winkel LP, Van den Hout HM, Bijman J, De Zeeuw CI, Hoeve HL, Van Zanten BA, Van der Ploeg AT, Reuser AJ (2004) Hearing loss in infantile Pompe's disease and determination of underlying pathology in the knockout mouse. Neurobiol Dis 16:14–20.

Kennedy HJ, Crawford AC, Fettiplace R (2005) Force generation by mammalian hair bundles supports a role in cochlear amplification. Nature 433:880–883.

Kimura RS (1975) The ultrastructure of the organ of Corti. Int Rev Cytol 42:173–222.

Knipper M, Zimmermann U, Kopschall I, Rohbock K, Jungling S, Zenner HP (1995) Immunological identification of candidate proteins involved in regulating active shape changes of outer hair cells. Hear Res 86:100–110.

Knipper M, Bandtlow C, Gestwa L, Kopschall I, Rohbock K, Wiechers B, Zenner HP, Zimmermann U (1998) Thyroid hormone affects Schwann cell and oligodendrocyte gene expression at the glial transition zone of the VIIIth nerve prior to cochlea function. Development 125:3709–3718.

Köppl C, Forge A, Manley GA (2004) Low density of membrane particles in auditory hair cells of lizards and birds suggests an absence of somatic motility. J Comp Neurol 479:149–155.

Kreis TE (1987) Microtubules containing detyrosinated tubulin are less dynamic. EMBO J 6:2597–2606.

Le Grimellec C, Giocondi MC, Lenoir M, Vater M, Sposito G, Pujol R (2002) High-resolution three-dimensional imaging of the lateral plasma membrane of cochlear outer hair cells by atomic force microscopy. J Comp Neurol 451:62–69.

Leonova EV, Raphael Y (1999) Application of a platinum replica method to the study of the cytoskeleton of isolated hair cells, supporting cells and whole mounts of the organ of Corti. Hear Res 130:137–154.

Liao Z, Popel AS, Brownell WE, Spector AA (2005) Modeling high-frequency electromotility of cochlear outer hair cell in microchamber experiment. J Acoust Soc Am 117:2147–2157.

Liberman MC, Gao J, He DZ, Wu X, Jia S, Zuo J (2002) Prestin is required for electromotility of the outer hair cell and for the cochlear amplifier. Nature 419:300–304.

Lim DJ, Hanamure Y, Ohashi Y (1989) Structural organization of the outer hair cell wall. Acta Otolaryngol 107:398–405.

Liu XZ, Ouyang XM, Xia XJ, Zheng J, Pandya A, Li F, Du LL, Welch KO, Petit C, Smith RJ, Webb BT, Yan D, Arnos KS, Corey D, Dallos P, Nance WE, Chen ZY (2003) Prestin, a cochlear motor protein, is defective in non-syndromic hearing loss. Hum Mol Genet 12:1155–1162.

Lodish HH, Berk A, Zipursky SL, Matsudaira P, Baltimore D, Darnell JE (2002) Molecular Cell Biology, 4th ed. New York: W.H. Freeman and Company.

Lu TK, Zhak S, Dallos P, Sarpeshkar R (2006) Fast cochlear amplification with slow outer hair cells. Hear Res 214:45–67.

Ludueña RF (1998) Multiple forms of tubulin: different gene products and covalent modifications. Int Rev Cytol 178:207–275.

Ludwig J, Oliver D, Frank G, Klocker N, Gummer AW, Fakler B (2001) Reciprocal electromechanical properties of rat prestin: the motor molecule from rat outer hair cells. Proc Natl Acad Sci USA 98:4178–4183.

Lutz C, Schweitzer L (1995) Longitudinal and radial differences in the subsurface cisternal system in the gerbil cochlea. Hear Res 84:12–18.

MacRae TH (1997) Tubulin post-translational modifications—enzymes and their mechanisms of action. Eur J Biochem 244:265–278.

Maeda N, Kon K, Imaizumi K, Sekiya M, Shiga T (1983) Alteration of rheological properties of human erythrocytes by crosslinking of membrane proteins. Biochim Biophys Acta 735:104–112.

Mammano F, Ashmore JF (1993) Reverse transduction measured in the isolated cochlea by laser Michelson interferometry. Nature 365:838–841.

Matsumoto N, Kalinec F (2005) Prestin-dependent and prestin-independent motility of guinea pig outer hair cells. Hear Res 208:1–13.

Matsuoka Y, Li X, Bennett V (2000) Adducin: structure, function and regulation. Cell Mol Life Sci 57:884–895.

Meyer J, Mack AF, Gummer AW (2001) Pronounced infracuticular endocytosis in mammalian outer hair cells. Hear Res 161:10–22.

Mogensen MM, Henderson CG, Mackie JB, Lane EB, Garrod DR, Tucker JB (1998) Keratin filament deployment and cytoskeletal networking in a sensory epithelium that vibrates during hearing. Cell Motil Cytoskeleton 41:138–153.

Morimoto N, Raphael RM, Nygren A, Brownell W E (2002) Excess plasma membrane and effects of ionic amphipaths on mechanics of outer hair cell lateral wall. Am J Physiol Cell Physiol 282:C1076–1086.

Mount DB, Romero MF (2004) The SLC26 gene family of multifunctional anion exchangers. Pflugers Arch 447:710–721.

Mountain DC, Hubbard AE (1989) Rapid force production in the cochlea. Hear Res 42:195–202.

Muallem D, Ashmore J (2006) An anion antiporter model of prestin, the outer hair cell motor protein. Biophys J 90:4035–4045.

Murdock DR, Ermilov SA, Spector AA, Popel AS, Brownell WE,Anvari B. (2005) Effects of chlorpromazine on mechanical properties of the outer hair cell plasma membrane. Biophys J 89:4090–4095.

Murugasu E, Russell IJ (1996) The effect of efferent stimulation on basilar membrane displacement in the basal turn of the guinea pig cochlea. J Neurosci 16:325–332.

Nagy I, Bodmer M, Schmid S, Bodmer D (2005) Promyelocytic leukemia zinc finger protein localizes to the cochlear outer hair cells and interacts with prestin, the outer hair cell motor protein. Hear Res 204:216–222.

Nakazawa K, Spicer SS, Schulte BA (1995a) Ultrastructural localization of Na,K-ATPase in the gerbil cochlea. J Histochem Cytochem 43:981–991.

Nakazawa K, Spicer SS, Schulte BA (1995b) Postnatal expression of the facilitated glucose transporter, GLUT 5, in gerbil outer hair cells. Hear Res 82:93–99.

Navaratnam D, Bai JP, Samaranayake H, Santos-Sacchi J (2005) N-terminal mediated homo-multimerization of prestin, the outer hair cell motor protein. Biophys J 89: 3345–3352.

Neely ST (1993) A model of cochlear mechanics with outer hair cell motility. J Acoust Soc Am 94:137–146.

Nguyen TV, Brownell WE (1998) Contribution of membrane cholesterol to outer hair cell lateral wall stiffness. Otolaryngol Head Neck Surg 119:14–20.

Nishida Y, Fujimoto T, Takagi A, Honjo I, Ogawa K (1993) Fodrin is a constituent of the cortical lattice in outer hair cells of the guinea pig cochlea: immunocytochemical evidence. Hear Res 65:274–280.

Oesterle EC, Tsue TT, Reh TA, Rubel EW (1993) Hair-cell regeneration in organ cultures of the postnatal chicken inner ear. Hear Res 70:85–108.

Oesterle EC, Lurie DI, Rubel EW (1997) Neurofilament proteins in avian auditory hair cells. J Comp Neurol 379:603–616.

Oghalai JS (2004) Chlorpromazine inhibits cochlear function in guinea pigs. Hear Res 198:59–68.

Oghalai JS, Patel AA, Nakagawa T, Brownell WE (1998) Fluorescence-imaged microdeformation of the outer hair cell lateral wall. J Neurosci 18:48–58.

Oghalai JS, Tran TD, Raphael RM, Nakagawa T, Brownell WE (1999) Transverse and lateral mobility in outer hair cell lateral wall membranes. Hear Res 135:19–28.

Oliver D, Fakler B (1999) Expression density and functional characteristics of the outer hair cell motor protein are regulated during postnatal development in rat. J Physiol 519 Pt 3:791–800.

Oliver D, He DZ, Klocker N, Ludwig J, Schulte U, Waldegger S, Ruppersberg JP, Dallos P, Fakler B (2001) Intracellular anions as the voltage sensor of prestin, the outer hair cell motor protein. Science 292:2340–2343.

Ottolia M, Nicoll DA, Philipson KD (2005) Mutational analysis of the alpha-1 repeat of the cardiac $Na^{(+)}$-Ca^{2+} exchanger. J Biol Chem 280:1061–1069.

Pollice PA, Brownell WE (1993) Characterization of the outer hair cell's lateral wall membranes. Hear Res 70:187–196.

Pujol R, Zajic G, Dulon D, Raphael Y, Altschuler RA, Schacht J (1991) First appearance and development of motile properties in outer hair cells isolated from guinea-pig cochlea. Hear Res 57:129–141.

Pujol R, Lenoir M, Ladrech S, Tribillac F, Rebillard G (1992) Correlation between the length of outer hair cells and the frequency coding of the cochlea. In: Cazals Y, Demany L, Horner KC (eds) Auditory Physiology and Perception. Oxford: Pergamon, pp 45–51.

Rabbitt RD, Ayliffe HE, Christensen D, Pamarthy K, Durney C, Clifford S, Brownell WE (2005) Evidence of piezoelectric resonance in isolated outer hair cells. Biophys J 88:2257–2265.

Rajagopalan L, Patel N, Madabushi S, Goddard JA, Anjan V, Lin F, Shope C, Farrell B, Lichtarge O, Davidson AL, Brownell WE, Pereira FA (2006) Essential helix interactions in the anion transporter domain of prestin revealed by evolutionary trace analysis. J Neurosci 26:12727–12734.

Raphael RM, Popel AS, Brownell WE (2000) A membrane bending model of outer hair cell electromotility. Biophys J 78:2844–2862.

Raphael Y, Marshak G, Barash A, Geiger B (1987) Modulation of intermediate-filament expression in developing cochlear epithelium. Differentiation 35:151–162.

Rhode WS, Robles L (1974) Evidence from Mossbauer experiments for nonlinear vibration in the cochlea. J Acoust Soc Am 55:588–596.

Rio C, Dikkes P, Liberman MC, Corfas G (2002) Glial fibrillary acidic protein expression and promoter activity in the inner ear of developing and adult mice. J Comp Neurol 442:156–162.

Robertson D, Johnstone BM (1978) Efferent transmitter substance in the mammalian cochlea: single neuron support for acetylcholine. Hear Res 1:31–34.

Robles L, Ruggero MA (2001) Mechanics of the mammalian cochlea. Physiol Rev 81:1305–1352.

Robles L, Ruggero MA, Rich NC (1997) Two-tone distortion on the basilar membrane of the chinchilla cochlea. J Neurophysiol 77:2385–2399.

Rosenbluth J (1962) Subsurface cisterns and their relationship to the neuronal plasma membrane. J Cell Biol 13:405–421.

Ruggero MA, Rich NC (1991) Furosemide alters organ of corti mechanics: evidence for feedback of outer hair cells upon the basilar membrane. J Neurosci 11:1057–1067.

Russell IJ, Murugasu E (1997) Medial efferent inhibition suppresses basilar membrane responses to near characteristic frequency tones of moderate to high intensities. J Acoust Soc Am 102:1734–1738.

Russell IJ, Nilsen KE (1997) The location of the cochlear amplifier: spatial representation of a single tone on the guinea pig basilar membrane. Proc Natl Acad Sci USA 94: 2660–2664.

Russell I, Schauz C (1995) Salicylate ototoxicity: Effects on the stiffness and electromotility of outer hair cells isolated from the guinea pig cochlea. Auditory Neuroscience 1:309–319.

Rybalchenko V, Santos-Sacchi J (2003) Cl^- flux through a non-selective, stretch-sensitive conductance influences the outer hair cell motor of the guinea-pig. J Physiol 547: 873–891.

Saito K (1983) Fine structure of the sensory epithelium of guinea-pig organ of Corti: subsurface cisternae and lamellar bodies in the outer hair cells. Cell Tissue Res 229:467–481.

Santos-Sacchi J (1991) Reversible inhibition of voltage-dependent outer hair cell motility and capacitance. J Neurosci 11:3096–3110.

Santos-Sacchi J (1992) On the frequency limit and phase of outer hair cell motility: effects of the membrane filter. J Neurosci 12:1906–1916.

Santos-Sacchi J (2002) Functional motor microdomains of the outer hair cell lateral membrane. Pflügers Arch 445:331–336.

Santos-Sacchi J, Dilger JP (1988) Whole cell currents and mechanical responses of isolated outer hair cells. Hear Res 35:143–150.

Santos-Sacchi J, Wu M (2004) Protein- and lipid-reactive agents alter outer hair cell lateral membrane motor charge movement. J Membr Biol 200:83–92.

Santos-Sacchi J, Shen W, Zheng J, Dallos P (2001) Effects of membrane potential and tension on prestin, the outer hair cell lateral membrane motor protein. J Physiol 531:661–666.

Seal RP, Leighton BH, Amara SG (2000) A model for the topology of excitatory amino acid transporters determined by the extracellular accessibility of substituted cysteines. Neuron 25:695–706.

Sheetz MP, Singer SJ (1974) Biological membranes as bilayer couples. A molecular mechanism of drug-erythrocyte interactions. Proc Natl Acad Sci USA 71: 4457–4461.

Shi X, Ren T, Nuttall AL (2001) Nitric oxide distribution and production in the guinea pig cochlea. Hear Res 153:23–31.

Slepecky N, Ulfendahl M, Flock A (1988) Shortening and elongation of isolated outer hair cells in response to application of potassium gluconate, acetylcholine and cationized ferritin. Hear Res 34:119–126.

Slepecky NB, Henderson CG, Saha S (1995) Post-translational modifications of tubulin suggest that dynamic microtubules are present in sensory cells and stable microtubules are present in supporting cells of the mammalian cochlea. Hear Res 91:136–147.

Souter M, Nevill G, Forge A (1995) Postnatal development of membrane specialisations of gerbil outer hair cells. Hear Res 91:43–62.

Spoendlin H (1966) The organization of the cochlear receptor. Fortschr Hals Nasen Ohrenheilkd 13:1–227.

Steyger PS, Furness DN, Hackney CM, Richardson GP (1989) Tubulin and microtubules in cochlear hair cells: comparative immunocytochemistry and ultrastructure. Hear Res 42:1–16.

Sun CX, Robb VA, Gutmann DH (2002) Protein 4.1 tumor suppressors: getting a FERM grip on growth regulation. J Cell Sci 115:3991–4000.

Sziklai I, Dallos P (1993) Acetylcholine controls the gain of the voltage-to-movement converter in isolated outer hair cells. Acta Otolaryngol 113:326–329.

Sziklai I, He DZ, Dallos P (1996) Effect of acetylcholine and GABA on the transfer function of electromotility in isolated outer hair cells. Hear Res 95:87–99.

Sziklai I, Szonyi M, Dallos P (2001) Phosphorylation mediates the influence of acetylcholine upon outer hair cell electromotility. Acta Otolaryngol 121:153–156.

Tchernia G, Mohandas N, Shohet SB (1981) Deficiency of skeletal membrane protein band 4.1 in homozygous hereditary elliptocytosis. Implications for erythrocyte membrane stability. J Clin Invest 68:454–460.

Thalmann R (1972) Recent refinements of quantitative microchemical analysis of tissues and cells of the inner ear. Acta Otolaryngol 73:160–174.

Tolomeo JA, Steele CR, Holley MC (1996) Mechanical properties of the lateral cortex of mammalian auditory outer hair cells. Biophys J 71:421–429.

Usami S, Hozawa J, Shinkawa H, Saito S, Matsubara A, Fujita S (1993) Immunocytochemical localization of intermediate filaments in the guinea pig vestibular periphery with special reference to their alteration after ototoxic drug administration. Acta Otolaryngol Suppl 506:7–13.

Weaver SP, Schweitzer L (1994) Development of gerbil outer hair cells after the onset of cochlear function: an ultrastructural study. Hear Res 72:44–52.

Weaver SP, Hoffpauir J, Schweitzer L (1993) Actin distribution along the lateral wall of gerbil outer hair cells. Brain Res Bull 31:225–228.

Weber T, Zimmermann U, Winter H, Mack A, Kopschall I, Rohbock K, Zenner HP, Knipper M (2002) Thyroid hormone is a critical determinant for the regulation of the cochlear motor protein prestin. Proc Natl Acad Sci USA 99:2901–2906.

Weber T, Gopfert MC, Winter H, Zimmermann U, Kohler H, Meier A, Hendrich O, Rohbock K, Robert D, Knipper M (2003) Expression of prestin-homologous solute carrier (SLC26) in auditory organs of nonmammalian vertebrates and insects. Proc Natl Acad Sci USA 100:7690–7695.

Wu X, Gao J, Guo Y, Zuo J (2004) Hearing threshold elevation precedes hair-cell loss in prestin knockout mice. Brain Res Mol Brain Res 126:30–37.

Wu X, Currall B, Yamashita T, Parker LL, Hallworth R, Zuo J (2007) Prestin-prestin and prestin-GLUT5 interactions in HEK293T cells. Dev Neurobiol 67:483–497.

Yu N, Zhu ML, Zhao HB (2006) Prestin is expressed on the whole outer hair cell basolateral surface. Brain Res 1095:51–58.

Zenner HP (1986a) Motile responses in outer hair cells. Hear Res 22:83–90.

Zenner HP (1986b) K+-induced motility and depolarization of cochlear hair cells. Direct evidence for a new pathophysiological mechanism in Ménière's disease. Arch Otorhinolaryngol 243:108–111.

Zenner HP, Zimmermann U, Schmitt U (1985) Reversible contraction of isolated mammalian cochlear hair cells. Hear Res 18:127–133.

Zenner HP, Zimmermann U, Gitter AH (1987) Fast motility of isolated mammalian auditory sensory cells. Biochem Biophys Res Commun 149:304–308.

Zenner HP, Zimmermann R, Gitter AH (1988) Active movements of the cuticular plate induce sensory hair motion in mammalian outer hair cells. Hear Res 34:233–239.

Zhang M, Kalinec GM, Urrutia R, Billadeau DD, Kalinec F (2003) ROCK-dependent and ROCK-independent control of cochlear outer hair cell electromotility. J Biol Chem 278:35644–35650.

Zhao HB, Santos-Sacchi J (1999) Auditory collusion and a coupled couple of outer hair cells. Nature 399:359–362.

Zheng J, Shen W, He DZ, Long KB, Madison LD, Dallos P (2000) Prestin is the motor protein of cochlear outer hair cells. Nature 405:149–155.

Zheng J, Long KB, Shen W, Madison LD, Dallos P (2001) Prestin topology: localization of protein epitopes in relation to the plasma membrane. NeuroReport 12:1929–1935.

Zheng J, Du GG, Anderson CT, Keller JP, Orem A, Dallos P, Cheatham M (2006a) Analysis of the oligomeric structure of the motor protein prestin. J Biol Chem. 281:19916–19924.

Zheng J, Deo N, Zou Y, Grosh K, Nuttall AL (2006b) Chlorpromazine alters cochlear mechanics and amplification: in vivo evidence for a role of stiffness modulation in the organ of Corti. J. Neurophysiol 97:994–1004.
Zine A, Schweitzer L (1996) Development of intracellular Ca-ATPase in the gerbil outer hair cell lateral wall. Brain Res 721:1–10.
Zine A, Schweitzer L (1997) Localization of proteins associated with the outer hair cell plasma membrane in the gerbil cochlea. Neuroscience 80:1247–1254.
Zinn C, Maier H, Zenner H, Gummer AW (2000) Evidence for active, nonlinear, negative feedback in the vibration response of the apical region of the in-vivo guinea-pig cochlea. Hear Res 142:159–183.

6
Active Processes in Insect Hearing

MARTIN C. GÖPFERT AND DANIEL ROBERT

1. Introduction

Active mechanical processes in hearing are not restricted to vertebrates: as shown by recent research, such processes are found in insects as well. The investigation of these amplificatory processes in insect hearing has scarcely begun. Many questions are outstanding, including perhaps the most intriguing one, that is, whether, reflecting the need for sensitive hearing, the mechanisms that effect active amplification in the ears of insects and vertebrates are related or have evolved independently. In either case, the discovery of active auditory processes in insects means that the structurally diverse hearing organs of insects can now be used to investigate the mechanisms and the structural basis of these processes and their physical implementation into ears.

The scope of this chapter is to review our current knowledge of active mechanical processes in insect hearing. Starting with a brief description of the anatomy of insect ears and auditory mechanosensory cells, evidence for the existence of active auditory mechanics in insects is summarized and information about its cellular and molecular basis is discussed. Readers interested in more-detailed accounts on the anatomy, the physiology, the genetics, and the evolutionary history of hearing in insects may want to refer to some recent reviews dedicated to these topics (Hoy and Robert 1996; Eberl 1999; Yager 1999; Caldwell and Eberl 2002; Jarman 2002; Robert and Göpfert 2003, 2004) and to a volume on hearing in insects that has appeared earlier in this series (Hoy et al.1998).

2. Insect Ears and Mechanosensory Cells

2.1 Diversity of Insect Ears: Antennal and Tympanal Types

An almost bewildering diversity of dedicated hearing organs is found in insects (Fig. 6.1). Within insects, hearing organs have evolved many times independently and occur on various body locations (Hoy and Robert 1996; Yager 1999; Yack 2004). Insect ears also present very diverse external anatomies, especially with respect to the structure of their sound receivers, often reflecting specific adaptations to specialized auditory tasks. The fruit fly *Drosophila melanogaster*, for

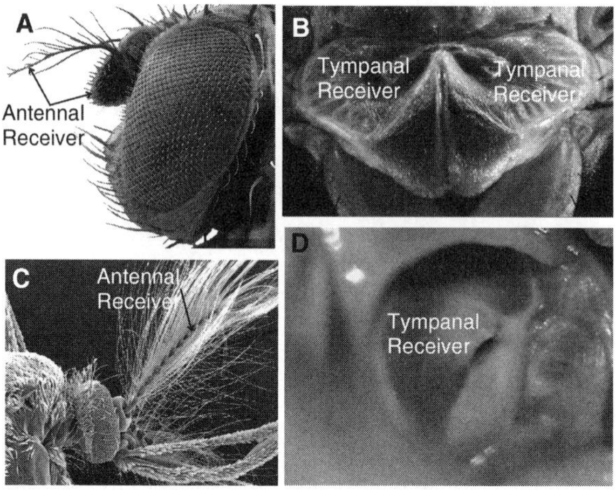

FIGURE 6.1. Scanning electron micrographs of insect ears with antennal and tympanal sound receivers. (**A**) Fruit fly *Drosophila melanogaster*: the club-shaped distal segment of the antenna and its feather-like arista (arrows) sympathetically vibrate in response to sound. (**B**) Parasitoid fly *Ormia ochracea*: a pair of tympanal ears with sound-receiving eardrums is positioned on the neck. (**C**) Mosquito *Anopheles gambiae*: the distal segment of the antenna extends into a long plumose flagellum that rocks about its socket in the presence of sound. (**D**) Migratory locust *Schistocerca gregaria*: ears with sound-receiving tympana are positioned laterally on the thorax. (**A** reproduced with permission from Göpfert and Robert 2003, Fig. 1A.)

example, uses the sense of hearing in the context of intraspecific communication; male flies produce songs by beating one of their wings to court nearby females (e.g., Bennet-Clark 1971; von Schilcher 1976; Hall 1994). Yet another fly, the parasitoid *Ormia ochracea*, uses hearing to acoustically detect and locate its hosts, singing crickets (e.g., Robert and Hoy 1998). The hearing organs of *Drosophila* and *Ormia* are located on different body parts (respectively the head versus the neck) (Fig. 6.1A, B), differ in their anatomy (antennal hearing organs versus tympanal hearing organs), and even respond to different physical components of sound. Like vertebrate ears, the ears of *Ormia* are tympanal ears endowed with an eardrum that vibrates in response to the pressure component of sound (Fig. 6.1B). The antennal hearing organs of *Drosophila*, in turn, are movement receivers (Fig. 6.1A); they respond to the particle velocity component of sound, which is advantageous for communicating acoustically at close range (Bennet-Clark 1971).

2.2 Relative Uniformity at the Cellular Level

Compared to the diverse external anatomies of insect ears, the mechanoreceptor cells comprised by these ears are surprisingly uniform. Hearing in insects seems always mediated by chordotonal organs, internal stretch receptors that are serially arranged throughout the insect body (French 1988; Boyan 1993; Field and Matheson

1998; Yack 2004). Spanning different regions of the cuticular exoskeleton and monitoring relative cuticular movements, these organs primarily serve proprioreceptive functions. Auditory function is obtained if one of the cuticular regions to which the organ connects is made sufficiently compliant to vibrate in response to sound. Within insects, such functional transition from proprioreception to audition has happened multiple times, whereby chordotonal organs in different parts of the body have been modified for hearing (Boyan 1993; Hoy and Report 1996; van Staaden and Römer 1998). According to this evolutionary scenario, it is little surprising that almost any part of the insect body can principally host an ear.

2.3 The Multicellular Mechanoreceptive Units

Anatomically, chordotonal organs are clusters of specialized sensory units, termed chordotonal sensilla or scolopidia (Field and Matheson 1998; Yack 2004) (Fig. 6.2). Each chordotonal sensillum typically consists of one to three primary sensory neurons and three accessory cells: an attachment, a scolopale, and a glial sheath cell. The sensory neurons themselves are bipolar and monodendritic, with a proximal axon projecting to the central nervous system and a single, distal dendrite. The dendrite is subdivided into an inner and an outer segment, the latter representing a modified cilium. The cilium originates in the inner dendritic segment as a series of ciliary rootlets that coalesce to basal bodies,

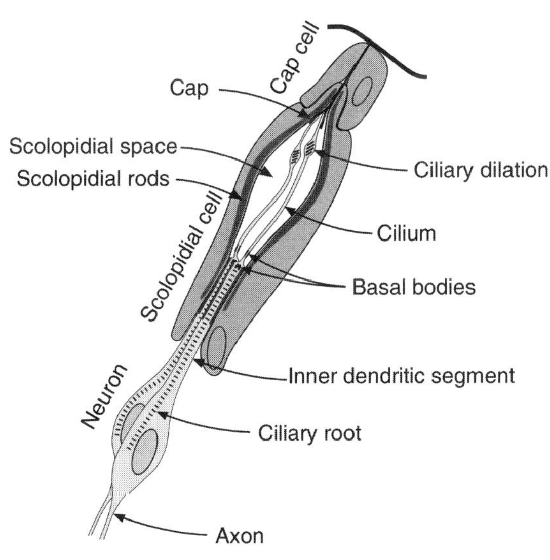

FIGURE 6.2. Structure of insect chordotonal sensilla. Schematic drawing of a single auditory sensillum of *Drosophila melanogaster* with two mechanosensory neurons. Cuticular vibrations are coupled via the cap to the outer dendritic segment, the cilium, which presumably houses the mechanotransduction machinery. Electrical responses are then directly propagated to the central nervous system via axonal projections. (Modified with permission from Sarpal et al. 2003, Fig. 3A.)

centriole-like structures that together form the ciliary base. The cilium itself bears the "$9 \times 2 + 0$" axonemal structure characteristic of primary cilia: its axoneme comprises nine circularly arranged microtubule doublets, but no central one. Adjacent doublets are linked via dynein arms along the length of the cilium, except for a swollen region of the cilium, the ciliary dilation, where dynein arms are missing. Apically, the cilium is surrounded by an extracellular cap produced by the attachment cell. Laterally, in turn, the cilium is surrounded by a barrel-shaped sleeve of actin-based rods produced by the scolopale cell. The apical cap and the lateral rods together form a spindle-shaped cavity that encloses the dendrite. This cavity, the scolopale space, has been reported to contain a K^+-rich receptor lymph, similar to the scala media of the vertebrate cochlea (Thurm and Küppers 1980).

2.4 Mechanoreceptive Function at the Molecular Level

While the anatomy of chordotonal organs has been studied in some detail, their actual workings are little understood. As pointed out by Yack (2004), the mechanical properties of chordotonal sensilla, the way they propagate mechanical forces to the transducer channels, and the identities of the transducer channels themselves remain unclear. Two putative mechanotransducer channels, both formed by members of the transient receptor potential (TRP) family of ion channels, are reportedly expressed in the chordotonal neurons of the *Drosophila* ear: the TRPN channel No mechanoreceptor potential C (NompC, also known as TRPN1) and a heteromultimeric channel formed by the interdependent TRPV subunits Inactive (Iav) and Nanchung (Nan) (Walker et al. 2000; Kim et al. 2003; Gong et al. 2004). Both channels contribute to the afferent response carried by the auditory nerve; the response is reduced by about one-half by the loss of *nompC* function (Eberl et al. 2000) and completely eliminated by mutations in *iav* and *nan* (Kim et al. 2003; Gong et al. 2004). The strength of the effect seems to favor the Iav/Nan channel as mechanotransducer for *Drosophila* hearing (Gong et al. 2004). As long as it remains unclear whether NompC and Nan/Iav are truly mechanically gated channels (Gillespie and Walker 2001; Gong et al. 2004), however, no definite statements about the molecular identities of the auditory mechanotransducers can be made. Because of its molecular structure, a mechanotransducing role of NompC would be particularly interesting: bearing a large number of N-terminal ankyrin repeats that are predicted to form a molecular spring, NompC and its vertebrate relatives (Sidi et al. 2003) are predestined to constitute spring-operated mechanotransducer channels, as have been implicated in vertebrate hearing (Howard et al. 1988; Howard and Bechstedt 2004; Corey et al. 2004).

3. Evidence for Active Processes in Insect Hearing

To determine whether an ear is endowed with active processes, diagnostic tools are required that allow one to probe the mechanical activity inside the ear. Over the past decades, the in-depth investigation of active processes in vertebrate hearing has uncovered four essential characteristics that, in vertebrates, arise from active processes and thus can be principally used as operational criteria to test the existence of equivalent processes in other animals. The four essential characteristics are (1) a compressive nonlinearity, which condenses a wide range of stimulus amplitudes into a narrow range of mechanical responses, (2) a frequency-specific amplification, which selectively improves the sensitivity to sounds of certain frequencies, (3) self-sustained mechanical oscillations or their acoustic analogue, spontaneous otoacoustic emissions, that result from excess in mechanical feedback, and (4) power gain, reflecting the expenditure of biological energy required to promote amplification (e.g., Manley 2001; Robles and Ruggero 2001; Chan and Hudspeth 2005; see also Kemp, Chapter 1; Manley et al., Chapter 2; Jülicher and Duke, Chapter 3; Martin, Chapter 4; Manley and van Dijk, Chapter 5; Russell and Lukashkin, Chapter 9). Because all four characteristics depend on metabolic processes, they are physiologically labile and ultimately disappear with death. Notably, not all characteristics mentioned are equally suited to determine whether active processes exist. Nonlinearity, for example, does not necessarily reflect activity; active processes introduce nonlinearities, yet purely passive systems may be nonlinear as well (e.g., Yates, 1995). A better criterion is large-amplitude self-sustained oscillations or spontaneous otoacoustic emissions, the mere occurrence of which provides compelling evidence for mechanical activity inside the ear (Kemp, Chapter 1). In a strict sense, establishing the existence of active processes finally requires the demonstration of power gain, that is, that more energy is coming out of the system than is initially fed in. By testing for violations of fundamental principles in thermodynamics, the fluctuation – dissipation or the equipartition theorems, power gain can be assessed (Bialek, 1987; Martin et al. 2001; Göpfert et al. 2005; see also Jülicher and Duke, Chapter 3; Martin, Chapter 4).

Equipped with operational criteria for the identification of active processes in hearing, we now will examine how these criteria are met by insect ears. We will start this investigation with a brief historical account of the study of active auditory processes in insect hearing before discussing the recent findings that have lead to the conclusion that such processes exist in insects.

3.1 Historical Overview

Until recently, insect ears were generally considered to be purely passive mechanical systems, yet the idea that the insect chordotonal neurons are mechanically active is not new. Chordotonal neurons lack actin-based stereociliar hair bundles, yet they are endowed with a dendritic cilium reminiscent of the kinocilium of vertebrate hair cells (Thurm et al. 1983). Inspired by structural similarities between these sensory cilia and motile cilia, Atema (1973, 1975) surmised that the former could be motile as well. This idea was subsequently specified by Moran et al. (1977), who proposed that the dendritic cilia of chordotonal neurons would actively bend during mechanotransduction, thereby focusing mechanical forces to the sites where transduction takes place. This hypothesis, though supported by histological evidence (Moran et al. 1977), remained problematic because the sensory cilia of chordotonal neurons display a "$9 \times 2+0$" axomemal structure: lacking central microtubules, such primary cilia were generally considered to be immotile (French 1988; Field and Matheson 1998). Even the general involvement of sensory cilia in mechanosensory transduction was called into question when studies on other types of arthropod sensilla revealed that depletion of ciliary microtubules does not necessarily affect mechanotransduction (Erler 1983; Kuster et al. 1983; Höger and Seyfarth 2001).

Notwithstanding these objections, the idea that the dendritic cilia of insect chordotonal neurons are motile has been revived by four interesting observations. First, although displaying a "9+0" axonemal structure, oviductal cilia of rabbits (Odor and Blandau 1985) and nodal cilia of mice (Nonaka et al. 1998) and zebrafish (Kramer-Zucker et al. 2005) were found to be motile, demonstrating that primary cilia can be motile as well. Second, immuno-histochemical studies revealed that the chordotonal neurons express centrin. Centrins are a highly conserved EF-hand subfamily of calcium-binding basal body proteins (Salisbury et al. 1984; Schiebel and Bornens 1995) that, participating in calcium-dependent flagellar rootlet contractions in green algae, have been surmised to promote mechanical activity in chordotonal neurons (Wolfrum 1991, 1992). Third, mutations in the *Drosophila* genes *beethoven* (*btv*) and *touch-insensitve-larva-B* (*tilB*) were found to cause ciliary defects and, in addition, to affect hearing (Eberl et al. 2000). Mutations in *btv* affect the structural integrity of the ciliary dilation (Eberl et al. 2000), while *tilB* mutants, in turn, besides being deaf, turned out to lack ciliary dynein arms and to produce immotile sperm (Kernan and Zuker 1995; Eberl et al. 2000). As documented by these findings, the dendritic cilia of chordotonal neurons are essential for mechanosensory transduction and possibly serve an active role. Fourth, recordings of distortion-product otoacoustic emissions from the ears of locusts and moths provided the first experimental evidence for active mechanical processes in insect hearing (see Section 3.2.3), the existence of which was subsequently confirmed by the investigation of the auditory mechanics in mosquitoes and flies (see Sections 3.2.1 and 3.2.2).

3.2 Experimental Evidence

Evidence for the existence of active mechanical processes in insect hearing has been reported for the tympanal ears of locusts and moths and the antennal hearing organs of mosquitoes and flies. The investigation of auditory mechanics in mosquitoes has provided a first comprehensive set of mechanical evidence that active processes support hearing in insects (see Section 3.2.1). Information about the cellular and molecular basis of these processes was obtained by linking the examination of the auditory mechanics with genetic manipulations in *Drosophila* (Section 3.2.2). The examination of auditory nonlinearities in locusts and moths finally has provided evidence that active processes assist tympanal hearing at ultrasonic frequencies (Section 3.2.3).

3.2.1 Mosquitoes: Existence of Active Processes in an Insect Ear

The antennal hearing organs of mosquitoes (Fig. 6.1C) are triumphs of complexity, sensitivity, and miniaturization. The second segment of the antenna houses the auditory chordotonal organ, Johnston's organ. Comprising some 7,500 neurons in females and about twice that number in males (Boo and Richards 1975), this organ contains about as many sense cells as the human cochlea packed into a volume just about the size of the head of a pin. The neurons directly connect to, and transduce the vibrations of, the antenna's third, distal, segment, which extends into a long flagellum that forms the sound receiver proper (Göpfert et al. 1999) (Fig. 6.1C). As shown by simultaneous mechanical measurements and electrophysiological recordings, sound-induced deflections of this receiver by as little as ± 0.1 millidegree are sufficient to elicit an electrical response (Göpfert and Robert 2000; Robert and Göpfert 2002). This auditory sensitivity allows male mosquitoes to detect their females by the faint buzzing sounds produced by the wing-stroke during flight (e.g., Tischner 1953).

Laser Doppler vibrometry measurements of sound-induced antennal vibrations in the species *Toxorhynchites brevipalpis* revealed that the exquisite sensitivity of mosquito hearing is achieved by means of active auditory mechanics (Göpfert and Robert 2001a). In dead animals, the mechanical response of the antennal receiver was found to be linear, displaying a moderately damped (Q_{3dB} values around 1.6) resonance at ca. 430 Hz in males and about half that frequency in females. A similar tuning was observed in live animals when the stimulus intensity was high. As the intensity declined, however, the response nonlinearly sharpened at frequencies around resonance (Q_{3dB} values increasing to about 2.6), displaying a compressive nonlinearity that boosts the sensitivity to faint sounds (Fig. 6.3A). At resonance, the nonlinear gain in sensitivity was highest, reaching figures between approximately 1.5 and 2. This gain seems surprisingly small if compared to the ca. 5- to 10-fold gain reported for isolated vertebrate hair cells (Martin et al. 2001) and the ca. 100-fold gain reported for the mammalian cochlea (e.g., Robles and Ruggero 2001). At least some of this discrepancy may reflect the use of different stimuli: in vertebrates the nonlinear sensitivity gain is conventionally measured in response to sinusoidal stimulation at best frequency, so active energy contributions will be concentrated into a single frequency band. In mosquitoes, however, the use of chirp sounds (FM-sweeps) may have spread these contributions across a wider frequency range, leading to lower estimates of the gain. This methodological effect is supported by measurements in *Drosophila*, where the gain is ca. 1.5 for chirp stimuli and ca. 10 for pure tones, (Göpfert et al. 2006), the latter figure being close to that reported for mechanically active hair-bundle movements of vertebrate hair cells (Martin and Hudspeth 2001).

The compressive nonlinearity displayed by the receiver's mechanics, its frequency-specificity, and its physiological vulnerability together suggest that some metabolic processes boosts the receiver's mechanical response in an intensity- and frequency-specific way. Observations of self-sustained receiver oscillations (Göpfert and Robert 2001a) document that this metabolic process is able to mechanically drive the receiver and accordingly must be active: in some

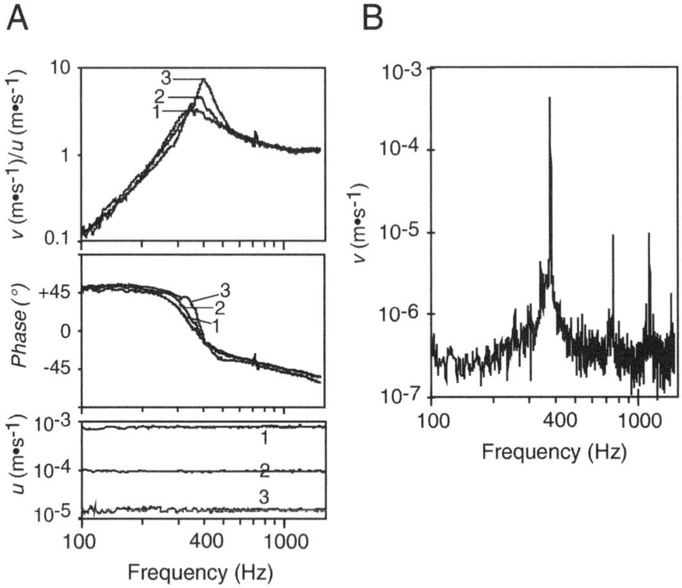

FIGURE 6.3. Nonlinear response (**A**) and self-sustained oscillations (**B**) of the antennal receiver of the mosquito *Toxorhynchites brevipalpis*. (**A**) Response amplitude (upper panel) and phase (middle panel) for three different stimulus intensities (1–3, lower panel). As intensity declines (1–3), the tuning sharpens and the resonance slightly shifts up in frequency. (**B**) Self-sustained oscillations measured in the absence of sound. The frequency spectrum of the receiver's vibration velocity displays a sharp peak at some 360 Hz and higher harmonics. Response amplitudes in (**A**) and (**B**) are presented as vibration velocities (v, m/s) normalized to the stimulus particle velocity (u, m/s), the phase of the response is given in degrees.

animals, the antennal receiver displayed large-amplitude, self-sustained oscillations in the absence of sound. Such self-sustained oscillations occasionally occurred spontaneously (Fig. 6.3B) and were reliably induced by thoracic injection of dimethyl sulfoxide (DMSO). DMSO is a local analgesic known to affect mechanotransduction by chordotonal sensilla. Its detailed effects on these sensilla remain unclear, yet because of the close resemblance of spontaneous and DMSO-induced oscillations, DMSO has provided a valuable tool for systematically evaluating the properties of self-sustained oscillations in the mosquito ear (Göpfert and Robert 2001a). This evaluation revealed that the oscillations are sharply tuned (Q_{3dB} up to 30) to frequencies around the receiver's resonance (Fig. 6.3A,B). Peak displacement amplitudes (up to ca. 400 nm) considerably exceed the background noise (air-particle displacements <1 nm), making it unlikely that the oscillations solely represent thermal fluctuations. Moreover, much like SPOAEs in vertebrates, the spontaneous oscillations of the mosquito receiver are transiently suppressed by external tones at nearby frequencies and vanish with death, documenting their nonlinearity and their physiological origin.

Accordingly, at least three of the four essential characteristics of active processes in vertebrate hearing—nonlinear compression, frequency-specific amplification, and self-sustained oscillations—are shared by the hearing organs of mosquitoes, which thus are deemed to be endowed with active processes as well.

Where do the active processes reside? Is the receiver actively driven by muscles or by chordotonal mechanosensory cells? Systematic vibration measurements along the length of the antenna revealed that during self-sustained oscillations, the receiver rocks about its proximal suspension in the second segment of the antenna. Because this segment lacks muscles, this pattern of antennal movement suggests that the receiver is driven by mechanical activity of the auditory chordotonal sensilla that connect to its base (Göpfert and Robert 2001a). As outlined in the following section, this interesting possibility has been tested by exploiting experimental advantages provided by the *Drosophila* ear.

3.2.2 *Drosophila*: Tracing the Cellular and Molecular Basis of Active Amplification

The question whether active processes in insect hearing are linked to mechanical activity of chordotonal sensilla was tested in *Drosophila* by linking the examination of the auditory mechanics with genetic manipulations. As in mosquitoes, hearing in *Drosophila* is based on Johnston's organ transducing sound-induced vibrations of the antenna's third segment, which forms the sound receiver. Compared to mosquitoes, the number of chordotonal neurons comprised by the fly's Johnston's organ is lower (between 400 and 500, corresponding to ca. 150–200 chordotonal sensilla with 2 to 3 neurons each [Caldwell and Eberl 2002]), and the receiver displays a different shape: instead of extending into a long flagellum as is the case in mosquitoes, the club-shaped third segment of the fly's antenna laterally bears a stiffly coupled, feather-like arista (Fig. 6.1A) that twists the entire third segment about its longitudinal axis in the presence of sound (Göpfert and Robert 2001b, 2002).

The fly's antennal sound receiver, like that of mosquitoes, displays signatures of mechanical activity, including nonlinearity, frequency-specific amplification, and self-sustained oscillations. In dead wild-type flies, the mechanical response of the receiver is linear and passive, closely resembling that of a moderately damped harmonic oscillator tuned to frequencies around 800 Hz (Fig. 6.4). Intense stimuli evoke an equivalent response in live flies, yet as the stimulus intensity is decreased, the resonance nonlinearly shifts down in frequency toward approximately 200 Hz (the dominant frequency component of the fly's songs), and the sensitivity at resonance goes up (Fig. 6.4). This nonlinear effect, which is transiently suppressed during exposure to CO_2, associates with spontaneous oscillations: in the absence of external stimuli, the receiver displays noisy twitches that may build up to large-amplitude limit cycles, both spontaneously and after thoracic injection of DMSO. The origin of the receiver's nonlinearity and oscillation activity was traced by analyzing the mechanical effects caused by mutations in *no mechanoreceptor potential A* (*nompA*), which encodes an extracellular linker protein that is expressed in the extracellular cap that connects the

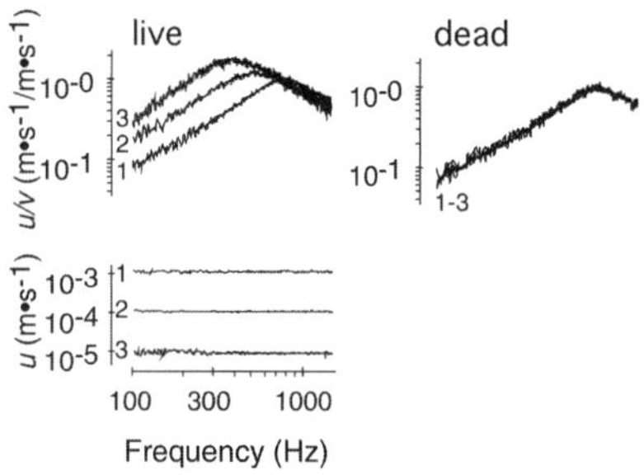

FIGURE 6.4. Physiologically labile nonlinearity of the *Drosophila* receiver. Responses (*upper panels*) to stimuli of three different intensities (1–3, *lower panel*) were measured in vivo and postmortem. The responses nonlinearly shift down in frequency in vivo, but superimpose after death. Units as in Fig. 6.3. (Modified with permission from Göpfert and Robert 2003, Fig. 1B.)

chordotonal sensory neurons to the antennal receiver. Disrupting this connection, mutations in *nompA* isolate the neurons from the antennal receiver, leading to conductive hearing loss. This genetic dissection eliminated the receiver's nonlinearity (Fig. 6.5) and oscillation activity, which are thus deemed to be introduced by the neurons. Similar mechanical defects were also caused by mutations in *nompC*, *tilB*, and *btv* (Fig. 6.5), supporting the neural origin of active processes in the auditory system of the fly. Especially, the close resemblance between the

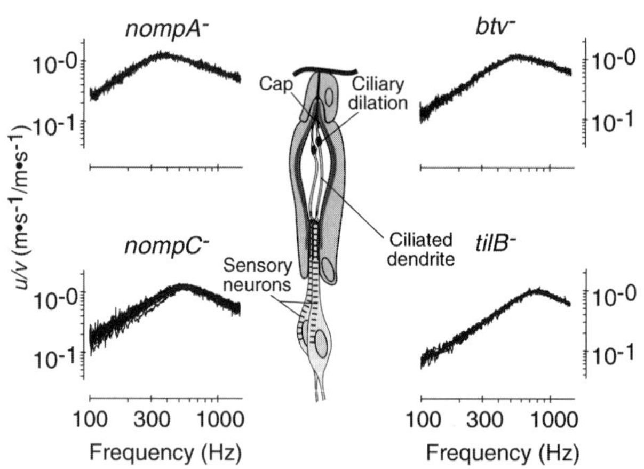

FIGURE 6.5. Linearization of the response of the *Drosophila* receiver by mechanosensory mutations. Responses to acoustic stimuli of different intensities superimpose (intensity range as in Fig 6.4). Mutations in *nompA* disconnect the sensory neurons from the receiver; mutations in *btv* disrupt the structural integrity of the neurons' ciliary dilations; mutations in *tilB* eliminate ciliary dynein arms; mutations in *nompC* abolish a putative mechanotransducer channel. (Reproduced with permission from Göpfert and Robert 2003, Fig. 4, and Sarpal et al. 2003, Fig. 3A, inset.)

receiver's mechanics in live *tilB* mutants and dead wild-type flies nicely illustrated that the integrity of the neurons is both necessary and sufficient to explain the signatures of active processes observed in the mechanics of the antennal receiver.

The possibility to genetically disrupt the active processes in *Drosophila* hearing has provided an opportunity to assess the power gain exhibited by chordotonal neurons by analyzing the mechanical fluctuations of the receiver to which they connect. Power gain was assessed by testing for violations of the equipartition theorem, which posits that the fluctuations of a passive system such as the receiver of dead flies will be at thermal equilibrium with its surroundings. Equilibrium conditions were defined by analyzing the receiver's fluctuations in dead WT flies, and were then used as a background to test for violations of the equipartition theorem in live wild-type flies, mechanosensory mutants, and controls. Quite intriguingly, the fluctuations of the antennal receiver's of live *nompA*, *tilB*, and *btv* mutants were found to obey the equipartition theorem, demonstrating that in flies with defective chordotonal neurons, the receiver solely fluctuates in response to thermal noise. In live wild-type and control flies with intact chordotonal neurons, however, the receiver fluctuations were found to violate the equipartition theorem, demonstrating that the neurons actively lift the energy to the receiver's fluctuations above thermal noise. In wild-type flies, the mean energy contributed by the neurons was 19 zJ (19×10^{-21} joules), which is about 1/20th the energy of a single green photon. Larger energies (200 zJ) are associated with self-sustained receiver oscillations, indicating that the flies control the mechanical performance of their antennal receivers by adjusting the neural energy contribution. Hence, much as motile hair cells improve hearing in vertebrates, hearing in at least some insects is improved by the motility of chordotonal neurons.

3.2.3 Tympanal Ears: Active Amplification at Ultrasonic Frequencies

The work on antennal hearing organs of mosquitoes and flies has shown that the ears of insects employ active mechanical feedback to improve the detection of low-pitched sounds at subkilohertz frequencies. Evidence for high-frequency amplification has been reported for the ears of locusts and noctuid moths. These insects are endowed with tympanal ears that, at least partly, serve the detection of bat echolocation signals and operate at ultrasonic frequencies of up to ca. 30 kHz (locusts) and ca. 100 kHz (noctuid moths) (Robert 1989; Miller and Surlykke 2001). Two-tone stimulation of the tympanal membranes in locusts and moths has unveiled prominent distortion product otoacoustic emissions (DPOAEs) (Fig. 6.6), demonstrating nonlinear auditory mechanics (Kössl and Boyan 1998a,b; Coro and Kössl 1998, 2001). The DPOAEs recorded from these insects were found to display characteristics similar to those known from vertebrate ears, including maximum DPOAE levels (and minimum DPOAE thresholds) at biologically relevant sound frequencies (Kössl and Boyan 1998a,b), monotonic DPOAE growth functions that are indicative of nonlinear compression (Coro and Kössl 2001) (Fig. 6.7), and, most strikingly, physiological lability;

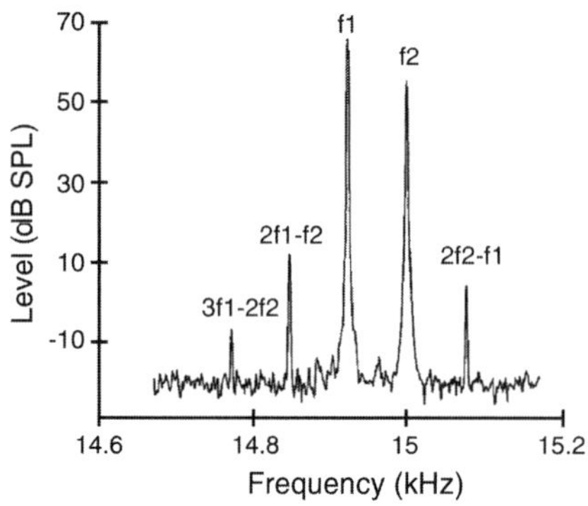

FIGURE 6.6. DPOAEs recorded from the tympanal ear of the grasshopper *Locusta migratoria*. Two-tone stimulation of the tympanum at approximately 14.95 kHz (f_1) and 15 kHz (f_2) elicits prominent distortion products, the most prominent one being found at $2f_1 - f_2$, as is the case in vertebrates. (Reproduced with permission from Kössl and Boyan, 1998b, Fig. 1B.)

in both locusts and moths, the DPOAEs transiently vanished during anesthetization with CO_2 or ethylether and disappeared with death (Coro and Kössl 1998, 2001). The DPOAE growth functions determined in moths were further found to display shapes similar to those reported for the cochlea of gerbils. Also, the nonlinear gain in sensitivity was similar (some 17 to 21 dB vs. 25 dB in gerbils; Fig 6.7) (Coro and Kössl 2001; Mills and Rubel 1996). Collectively, these findings have demonstrated that the mechanics of insect eardrums is nonlinearly improved by a physiological process that—as in *Drosophila*—may reside in the auditory mechanosensory cells. Such a possibility has interesting implications. First, provided the nonlinearity is introduced by an active

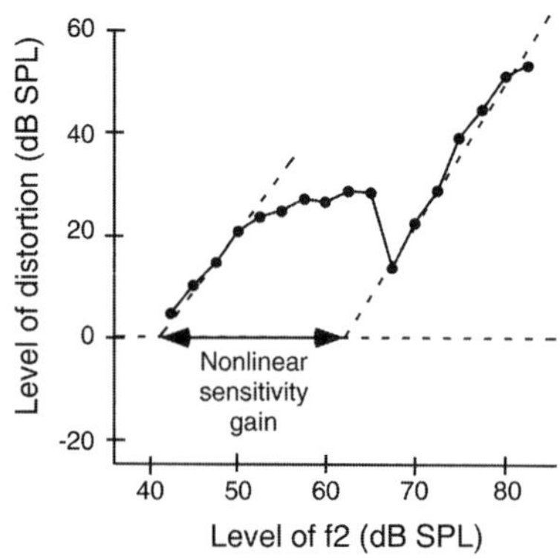

FIGURE 6.7. Growth function of $2f_1 - f_2$ DPOAEs from the tympanal ear of a moth. The function reveals a ca. 11-fold nonlinear sensitivity gain ($\approx$21 dB), indicating active amplification. (Redrawn with permission from Coro and Kössl 2001, Fig. 1.)

process, this process will operate at ultrasonic frequencies, much like the cochlear amplifier of bats. Second, because the ear of noctuid moths houses only one to two auditory neurons (Roeder and Treat 1957; Surlykke 1984), this system would allow one to investigate the active modulation of the ear's mechanics by single mechanosensory cells. Third, the tympanal ears of locusts (Fig. 6.1D) are more complex than those of moths, comprising distinct populations of neurons that respond to different frequencies of sound (Michelsen 1968). Because this frequency discrimination involves traveling waves propagated by the eardrum (Windmill et al. 2005), this auditory system promises an opportunity to study the mechanisms of traveling-wave amplification in an insect.

4. Candidate Mechanisms Effecting Active Processes in Insects

Work on vertebrate hearing has uncovered candidate mechanisms that are likely to account for the motility of vertebrate hair cells and active amplification in vertebrate ears. These motor mechanisms involve the interplay between mechanotransducer channels and myosin-1c adaptation motors, which brings about active hair-bundle motions (e.g., Howard et al. 1988; Fettiplace et al. 2001; Gillespie and Cyr 2003; Martin et al. 2003; Kennedy et al. 2005; see also Jülicher and Duke, Chapter 3, and Martin, Chapter 4), and the prestin-mediated somatic motility of mammalian outer hair cells (Zheng et al. 2000, see also Forge and Ashmore, Chapter 7). Both motor mechanisms may principally also promote the motility of insect chordotonal neurons. First, mechanotransduction by insect chordotonal sensilla and vertebrate hair cells seems to involve TRP channels with long ankyrin springs (NompC, TRPA1 (Howard and Bechstedt 2004; Corey et al. 2004; Nicolson 2005), raising the possibility that mechanotransduction in the ears of both insects and vertebrate is mediated by spring-operated channels that work in equivalent ways. Second, adaptation characteristics strikingly similar to those of vertebrate hair cells have been documented for sensory bristles in *Drosophila*, suggesting that adaptation motors are found in insect mechanosensors as well. A close relative of the hair-cell adaptation motor myosin1c is comprised by the fly's genome sequence (Myo31-DF). Whether My031-DF is expressed in the fly's auditory neurons and uquired for amplification, however, has not been established yet. Third, homologues of mammalian prestin are expressed in the auditory organs of mosquitoes and flies (Weber et al. 2003), raising the possibility that prestin-like molecules contribute to active processes in insect hearing as well (Weber et al. 2003). Whether these insect homologues constitute prestin-like motors, however, remains to be seen.

Apart from candidate hair-cell motors, it seems plausible to surmise that microtubule-dependent motors contribute to the motility of the ciliated chordotonal neurons of insects. Such microtubule-associated motors may include dynein and, possibly, centrin. The role of dynein in active amplification is supported by the anatomical and functional defects observed in *Drosophila tilB* mutants, which lead to the loss of ciliary dynein arms and abolish the motilities of the auditory neurons and of sperm (Kernan and Zuker 1995; Eberl et al.2000; Göpfert and Robert2003; Todi et al.2004). Correlated deafness and sperm immotility have recently been demonstrated also to arise from mutations in *Drosophila uncoordinated* (*unc*), which encodes a basal body protein required for ciliogenesis (Baker et al.2004). Whether mutations in *unc* abolish active auditory amplification remains to be seen. Another basal body protein that has been implicated in force generation by chordotonal neurons is centrin (Wolfrum 1991, 1992). The *Drosophila* genome sequence comprises one centrin gene, the disruption of which promises insights into the function of centrins in the context of hearing.

5. Evolutionary Considerations

Despite previous indications (see Section 3.1), the discovery of active processes in insect hearing is rather unexpected. Among terrestrial animals, hearing is widespread only in vertebrates and insects. Evolutionarily, both groups are distantly related; their last common ancestor, the protostome-deuterostome ancestor, lived some 600 to 1200 Mya years ago (Erwin and Davidson 2002). As expected, due to this long-lasting evolutionary separation, the ears of both groups are not homologous: insect and vertebrate hearing organs differ with respect to their anatomical position and internal structure, and even the mechanosensory cells that mediate hearing seem vastly different. Notwithstanding these anatomical differences, accumulating evidence suggests that the vertebrate hair cells and insect chordotonal neurons are derived from a common ancestor: both types of cells are developmentally specified by homologous genes (Hassan and Bellen 2000) and share conserved molecular networks involved in mechanosensory cell formation and function (Fritzsch and Beisel 2004; Boekhoff-Falk 2005). Accordingly, it seems in principle possible that at least some of the molecular components and mechanisms that support the mechanical activity of vertebrate hair cells and insect chordotonal neurons are evolutionarily conserved. Even if this should turn out not to be the case, the possibility remains that insects and vertebrates have invented their active processes independently and in parallel by employing conserved molecular components and modules. At least as long as active processes appear to be linked only to hearing, the question of the single or multiple origins is unlikely to be resolved. In effect, only if active processes constitute a widespread and general capacity shared by a multitude of mechanosensory systems would it seem conceivable that they evolved only once.

6. Summary

Work over the past years has made a compelling case that active processes in insect hearing exist. The manifestations of these processes are remarkably similar to those known from vertebrates. The origin of mechanical activity has been traced to auditory sensory cells, and some of the molecules involved have been revealed. Quite intriguingly, several of these molecules have vertebrate homologues that are expressed in hair cells, raising the possibility that some of the components and mechanisms that promote active processes in insect and vertebrate hearing are evolutionarily conserved. These unexpected findings provide the stage for using the accessible ears of insects to study active processes in hearing. Many questions are outstanding: What mechanism promotes active mechanical amplification (what molecules are involved, what are their respective contributions, and how do they interact)? To what extent is this mechanism evolutionarily conserved: is this a case of convergence, or are distinct mechanisms employed by different animals' ears? Are these mechanisms particularly adapted to perform specific tasks for the genuine detection of certain input signals? Are active processes linked to hearing only or can they be found in other mechanosensory organs as well? Which specific benefits may have promoted the evolution of active processes in hearing? Were evolutionary constraints at work on dynamic range compression, increase in sensitivity and/or frequency selectivity, or a species-specific combination thereof? The diverse ears of insects provide a rich pool of model systems that may help us sensibly approach these questions, with the prospect of gaining fascinating insights into the refined workings auditory sensors, active auditory mechanics, and the evolutionary history of hearing.

References

Atema J (1973) Microtubule theory of sensory transduction. J Theor Biol 38:181–190.

Atema J (1975) Stimulus transmission along microtubules in sensory cells: an hypothesis. In: Borgers M, deBrabander M (eds) Microtubules and Microtubule Inhibitors. Amsterdam: North-Holland, pp. 247–257.

Baker JD, Adhikarakunnathu S, Kernan MJ (2004) Mechanosensory-defective, male-sterile *unc* mutants identify a novel basal body protein required for ciliogenesis in *Drosophila*. Development 131:3411–3422.

Bennet-Clark HC (1971) Acoustics of insect song. Nature 234:255–259.

Bialek, W (1987) Physical limits to sensation and perception. Annu Rev Biophys Biophys Chem 16:455–468

Boekhoff-Falk G (2005) Hearing in *Drosophila*: development of Johnston's organ and emerging parallels to vertebrate ear development. Dev Dyn 232:550–558.

Boo KS, Richards AG (1975) Fine structure of the scolopidia in Johnston's organ of female *Aedes aegypti* compared with that of male. J Insect Physiol 21:1129–1139.

Boyan GS (1993) Another look at insect audition: the tympanic receptors as an evolutionary specialization of the chordotonal system. J Insect Physiol 39:187–200.

Caldwell, J, Eberl DF (2002) Towards a molecular understanding of *Drosophila* hearing. J Neurobiol 53:172–189.

Chan DK, Hudspeth AJ (2005) Ca^{2+} current-driven nonlinear amplification in vitro. Nat Neurosci 8:149–155.

Corey DP et al. (2004) TRPA1 is a candidate for the mechanosensitive transduction channel of vertebrate hair cells. Nature 432:723–730.

Coro F, Kössl M (1998) Distortion-product otoacoustic emissions from the tympanic organ in two noctuid moths. J Comp Physiol A 183:525–531.

Coro F, Kössl M (2001) Components of the 2f1-f2 distortion-product otoacoustic emission in a moth. Hear Res 162:126–133.

Eberl DF (1999) Feeling the vibes: chordotonal mechanisms in insect hearing. Curr Opin Neurobiol 9:389–393.

Eberl DF, Hardy RW, Kernan MJ (2000) Genetically similar transduction mechanisms for touch and hearing in *Drosophila*. J Neurosci 20:5981–5988.

Erler G (1983) Sensitivity of an insect mechanoreceptor after destruction of dendritic microtubules by means of vinblastine. Cell Tissue Res 229:673–684.

Erwin DH, Davidson EH (2002) The last common bilaterian ancestor. Development 129:3021–3032.

Fettiplace R, Ricci AJ, Hackney CM (2001) Clues to the cochlear amplifier from the turtle ear. Trends Neurosci 24:169–175.

Field LH, Matheson T (1998) Chordotonal organs in insects. Adv Insect Physiol 27:1–28.

French AS (1988) Transduction mechanisms of mechanosensilla. Annu Rev Entomol 33:39–58.

Fritzsch B, Beisel KW (2004) Keeping sensory cells and evolving neurons to connect them to the brain: molecular conservation and novelties in vertebrate ear development. Brain Behav Evol 64:182–197.

Gillespie PG, Cyr JL (2003) Myosin-1c, the hair cell's adaptation motor. Annu Rev Physiol 66:521–545.

Gillespie PG, Walker RG (2001) Molecular basis of mechanosensory transduction. Nature 413:194–202.

Gong Z, Son W, Chung YD, Kim J, Shin DW, McClung CA, Lee Y, Lee H, Chang D-J, Kaang B-K, Cho H, Oh U, Hirsh J, Kernan MJ, Kim C (2004) Two interdependent TRPV channel subunits, Inactive and Nanchung, mediate hearing in *Drosophila*. J Neurosci 41:9059–9066.

Göpfert MC, Albert JT, Nadrowksi B & Kamikouchi A (2006) Specification of auditory sensitivity by Drosophila TRP channels. Nat Neurosci 9:999–1000.

Göpfert MC, Robert D (2000) Nanometre-range acoustic sensitivity in male and female mosquitoes. Proc R Soc Lond B 267:453–457.

Göpfert MC, Robert D (2001a) Active auditory mechanics in mosquitoes. Proc R Soc Lond B 268:333–339.

Göpfert MC, Robert D (2001b) Turning the key on *Drosophila* audition. Nature 411:908.

Göpfert MC, Robert D (2002) The mechanical basis of *Drosophila* audition. J Exp Biol 205:1199–1208.

Göpfert MC, Robert D (2003) Motion generation by *Drosophila* mechanosensory neurons. Proc Natl Acad Sci USA 100:5514–5519.

Göpfert MC, Briegel H, Robert D (1999) Mosquito hearing: sound-induced antennal vibrations in male and female Aedes aegypti. J Exp Biol 202:2727–2738.

Göpfert MC, Humphris ADL, Albert JT, Robert D, Hendrich O (2005) Power gain exhibited by motile mechanosensory neurons in *Drosophila* ears. Proc Natl Acad Sci USA 102:325–330.

Hall JC (1994) The mating of a fly. Science 264:1702–1714.

Hassan BA, Bellen HJ (2000) Doing the MATH: is the mouse a good model for fly development? Genes Dev 14:1852–1865.

Höger U, Seyfarth E-A (2001) Structural correlates of mechanosensory transduction and adaptation in identified neurons of spider split sensilla. J Comp Phys A 187:727–736.

Howard J, Bechstedt S (2004) Hypothesis: a helix of ankyrin repeats of the NOMPC-TRP ion channel is the gating spring of mechanoreceptors. Curr Biol 14:R224–R226.

Howard J, Roberts WM and Hudspeth AJ (1988) Mechanoelectrical transduction by hair cells. Annu Rev Biophys Biophys Chem 17:99–124.

Hoy RR, Robert D (1996) Tympanal hearing in insects. Annu Rev Entomol 41:433–450.

Hoy RR, Popper AN, Fay RR (eds) (1998). Comparative Hearing: Insects. New York: Springer-Verlag.

Jarman AP (2002) Studies of mechanosensation using the fly. Hum Mol Genet 11:1215–1218.

Kennedy HU, Crawford AC, Fettiplace R (2005) Force generation by mammalian hair bundles supports a role in cochlear amplification. Nature 433:880–833.

Kernan M, Zuker C (1995) Genetic approaches to mechanosensory transduction. Curr Opin Neurobiol 5:443–448.

Kim J, Chung YD, Park DY, Choi S, Shin DW, Soh H, Lee HW, Son W, Yim J, Park CS, Kernan MJ, Kim C (2003) A TRPV family ion channel required for hearing in *Drosophila*. Nature 424:81–84.

Kössl M, Boyan GS (1998a) Acoustic distortion products from the ear of a grasshopper. J Acoust Soc Am 104:326–335.

Kössl M, Boyan GS (1998b) Otoacoustic emissions from a nonvertebrate ear. Naturwissenschaften 85:124–127.

Kramer-Zucker AG, Olale F, Haycraft CJ, Yoder BK, Schier AF, Drummond IA (2005) Cilia-driven fluid flow in the zebrafish pronephros, brain and Kupffer's vesicle is required for normal organogenesis. Development132:1907–1921.

Kuster JE, French AS, Sanders EJ (1983) The effect of microtubule dissociating agents on the physiology and cytology of the sensory neuron in the femoral tactile spine of the cockroach, *Periplaneta americana* L. Proc R Soc Lond B 219:397–412.

Manley GA (2001) Evidence for an active process and a cochlear amplifier in nonmammals. J Neurophysiol 86:541–549.

Martin P, Hudspeth AJ (2001) Compressive nonlinearity in the hair bundle's active response to mechanical stimulation. Proc Natl Acad Sci USA 98:14386–14391.

Martin P, Hudspeth AJ, Jülicher F (2001) Comparison of a hair bundle's spontaneous oscillations with its response to mechanical stimulation reveals the underlying active process. Proc Natl Acad Sci USA 98:14380–14385.

Martin P, Bozovic D, Choe Y, Hudspeth AJ (2003) Spontaenous oscillation by hair bundles of the bullfrog's sacculus. J Neurosci 23:4533–4548.

Michelsen A (1968) Frequency discrimination in the locust ear by means of four groups of receptor cells. Nature 220:585–586.

Miller LA, Surlykke A (2001) How some insects detect and avoid being eaten by bats: Tactics and countertactics of prey and predator. Bioscience 51:570–581.

Mills DM, Rubel EW (1996) Development of the cochlear amplifier. J Acoust Soc Am 100:428 441.

Moran DT, Varela FJ, Rowley III JC (1977) Evidence for active role of cilia in sensory transduction. Proc Natl Acad Sci USA 74:793–797.

Nicolson T (2005) Fishing for key players in mechanotransduction. Trends Neurosci 28:140–144.

Nonaka S, Tanaka Y, Okada Y, Takeda S, Harada A, Yoshimitsu K, Kido M, Hirokawada N (1998) Randomization of left-right asymmetry due to loss of nodal cilia generating leftward flow of extraembryonic fluid in mice lacking KIF3B motor protein. Cell 95:829–837.

Odor DL, Blandau RJ (1985) Observations of the solitary cilium of rabbit oviductal epithelium: its motility and ultrastructure. Am J Anat 174:437–453.

Robert D (1989) The auditory behaviour of flying locusts. J Exp Biol 147:279–301.

Robert D, Göpfert MC (2002) Novel schemes for hearing and acoustic orientation in insects. Curr Opin Neurobiol 12:715–720.

Robert D, Göpfert MC (2004) (eds) The Biology of Insect Audition. Micr Res Tech 63:311–412.

Robert D, Hoy RR (1998) The evolutionary innovation of tympanal hearing in Diptera. In: Hoy RR, Popper AN, Fay RR (eds), Comparative Hearing: Insects. New York: Springer-Verlag, pp. 197–227.

Robles L, Ruggero MA (2001) Mechanics of the mammalian cochlea. Physiol Rev 81:1305–1352.

Roeder KD, Treat AE (1957) Ultrasonic reception by the tympanic organ of noctuid moths. J Exp Zool 134:127–157.

Salisbury JL, Baron A, and Surek B (1984) Striated flagellar roots: isolation and characterization of a calcium-modulated contractile organelle. J Cell Biol 99:962–970.

Sarpal R, Todi SV, Sivan-Loukianova E, Shirolikar S, Subramanian N, Raff EC, Erickson JW, Ray K, Eberl DF (2003) *Drosophila* KAP interacts with the kinesin II motor subunit KLP64D to assemble chordotonal sensory cilia, but not sperm tails. Curr Biol 13:1687–1696.

Schiebel E, Bornens M (1995) In search of a function for centrins. Trends Cell Biol 5:197–201.

Sidi S, Friedrich RW, Nicolson T (2003) NompC TRP channel required for vertebrate sensory hair cell mechanotransduction. Science 301:96–99.

Surlykke A (1984) Hearing in notodontid moths—a tympanic organ with a single auditory neurons. J Exp Biol 113:323–335.

Thurm U, Küppers J (1980) Epithelial physiology of insect sensilla. In: Locke M, Smith DS (eds) Insect Biology in Future. New York: Academic Press, pp. 735–763.

Thurm U, Erler G, Goedde J, Kastrup H, Keil T, Voelker W and Vohwinkel B (1983). Cilia specialized for mechanoreception. J Submicrosc Cytol 15:151–155.

Tischner H (1953) Über den Gehörsinn von Stechmücken. Acustica 3:335–343.

Todi SV, Sharma Y, Eberl DF (2004) Anatomical and molecular design of the *Drosophila* antenna as a flagellar auditory organ. Microsc Res Tech 63:388–399.

van Staaden MJ, Römer H (1998) Evolutionary transition from stretch to hearing organs in ancient grasshoppers. Nature 394:773–776.

von Schilcher F (1976) The role of auditory stimuli in the courtship of *Drosophila melanogaster*. Anim Behav 24:18–26.

Walker RG, Willingham AT and Zuker CS (2000) A *Drosophila* mechanosensory transduction channel. Science 287:2229–2234.

Weber T, Göpfert MC, Winter H, Zimmermann U, Kohler H, Meier A, Hendrich O, Rohbock K, Robert D, Knipper M. (2003) Expression of prestin-homologous solute carrier (SLC26) in auditory organs of insects and lower vertebrates. Proc Natl Acad Sci USA 100:7690–7695.

Windmill J, Göpfert MC, Robert D (2005) Tympanal travelling wave in migratory locusts. J Exp Biol 208:157–168.

Wolfrum U (1991) Centrin- and α-actinin-like immunoreactivity in the ciliary rootlets of insect sensilla. Cell Tissue Res 266:321–238.

Wolfrum U (1992) Cytoskeletal elements in arthropod sensilla and mammalian photoreceptors. Biol Cell 76:373–381.

Yack JE (2004) The structure and function of auditory chordotonal organs in insects. Microsc Res Tech 63:315–337.

Yager DD (1999) Structure, development and evolution of insect auditory systems. Microsc Res Tech 47:380–400.

Yates GK (1995) Cochlear structure and function. In: Moore BCJ (ed), Hearing. San Diego, CA: Academic Press, pp. 41–74.

Zheng J, Shen W, He DZ, Long KB, Madison LD, Dallos P (2000) Prestin is the motor protein of cochlear outer hair cells. Nature 405:149–155.

7 Otoacoustic Emissions in Amphibians, Lepidosaurs, and Archosaurs

GEOFFREY A. MANLEY AND PIM VAN DIJK

1. General Introduction

Some of the earliest work on otoacoustic emissions (OAEs) was carried out on amphibians (Palmer and Wilson 1982), establishing right at the beginning of studies of this phenomenon that it is not exclusively the province of mammals. During the 1980s and 1990s, a good deal of work on frogs, lizards, and birds established the almost ubiquitous presence of the various kinds of emissions in these different nonmammalian groups and provided an excellent basis for discussions of the nature and origin of emissions. These data indicate clearly that the mechanisms generating typical emission patterns are not restricted to the hair-cell constellation of the mammalian or avian cochlea, but result from more fundamental hair-cell properties. Thus to understand the dependence of the patterns on the structure of the hearing epithelium, it is desirable to be able to compare detailed data from a number of species with differing morphology.

Can OAEs in nonmammals tell us anything about the origin and mechanisms of active processes in vertebrate hearing organs? A thorough comparison of OAE characteristics of mammals and nonmammals would be very useful in establishing those features that are basic to hair-cell function and those that are derived from unique morphologies or physiologies. Although not every category of emissions has been investigated in each particular species, the wide distribution of otoacoustic emissions among vertebrates favors the idea that the capacity to produce active movements and the generation of otoacoustic emissions are very basic properties of auditory hair-cell epithelia (Köppl 1995). This chapter will provide an outline of this work; detailed information is available in the various publications that are cited.

All the information discussed in this chapter needs to be understood in the light of the source of the energy that manifests itself in OAEs in nonmammals. The few studies available on bird hair cells strongly suggest that a membrane-based motor system as suggested for mammals does not exist (Brix and Manley 1994; He et al. 2003; Köppl et al. 2003, 2004). Instead, for birds (Brix and Manley 1994; Hudspeth et al. 2000), turtles (Crawford and Fettiplace 1985), and lizards (Manley 2000a,b; Ricci et al. 2000; Manley et al. 2001), all evidence points to the existence of a motor system in the hair-cell bundle that is closely integrated into the function of the transduction channels (Hudspeth et al. 2000). Some of this evidence will be discussed below. Interestingly, new studies of mammalian inner (Chan and Hudspeth 2005) and outer hair cells (Kennedy et al. 2003) indicate the presence of bundle motors in these hair cells as well. This may, of course, help to explain why so many of the manifestations of the phenomenon of active processes are so similar between mammals and nonmammals (Manley 2001).

Since modern classifications do not recognize the existence of a group called "reptiles," the present chapter has different sections devoted to those groups that are now established (Fig. 7.1). These are the extant archosaurs (Crocodilia and their closest relatives, the birds), lepidosaurs (snakes and lizards and the Tuatara *Sphenodon*), and Chelonia (tortoises and turtles). As yet, there have been no OAE studies of snakes or of the Tuatara, and our own attempts to measure spontaneous otoacoustic emissions (SOAEs) and distortion-product otoacoustic emissions (DPOAEs) from turtles (*Pseudemys scripta*) were unsuccessful (Gallo and Manley, unpublished data). On the basis of previous lizard work, SOAEs would not be expected in the turtles' hearing range (below 1 kHz). However, the lack of DPOAEs must be attributed to the poor transmission of the turtle middle ear (Wever 1978), since DPOAEs are a universal phenomenon. In the species

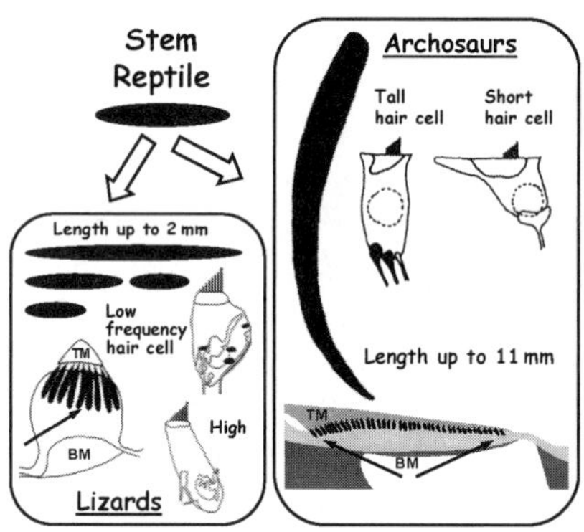

FIGURE 7.1. A schematic diagram of the evolution of different auditory papilla morphologies in (*left*) the lizards and (*right*) the archosaurs from the primitive stem-reptile papilla. In both groups, papilla elongation and hair-cell specialization occurred. In each box, the black area depicts typical papilla shapes as seen from above. The transverse sections show the hair-cell (*arrows*) shapes in the papilla. BM = Basilar membrane; TM = tectorial membrane. The lizard papilla shows a variety of shapes. In each box also, two of the hair-cell types are depicted, together with their innervating afferent (*filled shapes*) and efferent (*open shapes*) nerve fibers. Neither lizard high-frequency hair cells nor bird short hair cells have efferent innervation. Whereas lizard hair-cell types are found in different frequency areas, archosaur hair-cell shape changes across the papilla (*arrows*), whereby this change is more abrupt in Crocodilia than in birds (shown here is the avian situation). (After Manley and Clack 2004, with permission.)

Pseudemys scripta, the best thresholds of auditory-nerve fibers are near 40 dB SPL, and it is likely that a substantial part of this relatively poor sensitivity is due to the middle ear, through which emissions must also pass in the reverse direction. Similar problems would likely be encountered in any future attempt to measure emissions in snakes and the tuatara. Because the fundamental structures of archosaur and lepidosaur ears differ substantially, this chapter describes what is known about emissions of these two groups separately.

As noted above, the first SOAEs were measured from the amphibian ear (Wilson et al. 1990a,b). Rosowski et al. (1984) had already shown that alligator lizards show DPOAEs, a finding that was later confirmed in the bobtail skink. During that latter study it was discovered that lizards also have SOAEs (Manley and Köppl 1992, 1994; Köppl and Manley 1993a, 1994). Since then, various types of OAE have been studied in a variety of amphibians, lizards, and birds. These data, together with a discussion of the anatomical substrate, will be briefly described below.

2. The Different Types of Emissions

2.1 Spontaneous Otoacoustic Emissions

SOAEs are low-level sounds emitted by the ear through the middle ear that were not induced by acoustical, electrical, or any other stimuli.

2.2 Stimulus–frequency Otoacoustic Emissions

Stimulus–frequency otoacoustic emissions (SFOAEs) are emissions that are evoked by pure-tone stimuli, and the emission is evoked at the frequency of the stimulus. The most commonly applied stimulus to record SFOAEs is a long external tone that is swept in frequency. SFOAEs are generally detected by their phase interference with the external tone that leads to peaks and troughs in spectral energy. SFOAEs are also referred to as simultaneous-evoked or synchronously evoked otoacoustic emissions (SEOAEs).

2.3 Distortion-Product Otoacoustic Emissions

When two tones are played into the ear, distortion products that originate from those areas of the hearing organ that process the stimuli can be measured with a microphone connected to the ear. Humans listening to such pairs of pure tones perceive new frequencies that are not present in the eliciting stimulus. The frequencies of the DPOAEs are determined by the frequencies of the so-called primary tones, f_1 and f_2 ($f_2 > f_1$). Difference tones $f_2 - f_1$, summation tones $f_2 + f_1$, and cubic distortion products ("combination tones") such as $2f_1 - f_2$ and $2f_2 - f_1$ can be measured. Intermodulation distortion products at other frequencies have also been described, for example, $3f_1 - 2f_2$, $4f_1 - 3f_2$,

$3f_2 - 2f_1$. Most data are available on the first lower-sideband cubic distortion product $2f_1 - f_2$, which is the most prominent in the ear-canal sound pressure.

2.4 Electrically Evoked Otoacoustic Emissions

The active process in the hair cells of both mammals and nonmammals responds to electrical signals and can thus be driven by injecting current into scala media. The frequency of the resulting electrically evoked otoacoustic emissions (EEOAEs) is determined by the frequency of the electrical current. This phenomenon has been demonstrated in mammals (Hubbard and Mountain 1983; Sun et al. 2000) and lizards (Manley et al. 2001; Manley and Kirk 2002). In mammals, low-level AC electrical currents (5–10 μA) injected into scala media elicit hair-cell motility manifest as EEOAEs in the external ear that have sound pressures up to 30 dB SPL.

3. Otoacoustic Emissions in Amphibians

3.1 The Anatomical Substrate and Inner-Ear Physiology

Most data on otoacoustic emissions in frogs is available from the genera *Rana* and *Hyla*, both of which are "advanced" frogs, with fully developed middle and inner ear structures. Some data is available in more "primitive" frogs, where some of the structures are less developed. The following briefly reviews the structural and function properties for the frog species for which emission data are available.

The fully developed frog outer and middle ear consists of a tympanic membrane just behind the eye, a single middle-ear bone called the columella, and the operculum. The columella and the operculum together close the hole on the lateral wall of the inner ear. Both these structures have a middle-ear muscle attached to them. In ranid and hylid frogs, the tympanic membrane, the columella, and the operculum are present. In the more primitive species discussed below, some structures are absent (Wever 1985). In the oriental fire-bellied toad, *Bombina orientalis*, the tympanic membrane is absent, but the columella is present (Hetherington and Lindquist 1999). The intermediate African clawed frog, *Xenopus laevis,* has a cartilaginous tympanic disk instead of a tympanic membrane. The disk is not visible from the outside. Finally, the intermediate spadefood toad, *Scaphiopus couchii*, has a tympanic middle ear, like hylids and ranids.

The frog inner ear includes two papillae with high sensitivity to airborne sound: the amphibian papilla and the basilar papilla.

The amphibian papilla resides in a recess that has an opening on its rostral side into the saccular space (Lewis 1984; Wever 1985). At its caudal end, a thin contact membrane separates the endolymph in the amphibian papilla recess from perilymph. The amphibian papilla does not contain a basilar membrane. The sensory hair cells are located on the roof of the recess. The hair cell epithelium has an elaborated shape. It consists of an approximately triangular patch of hair cells near the opening to the sacculus, and an extension that points caudally toward the contact membrane. The length of the S-shaped extension varies across frog species (Lewis 1981). In the "primitive" species, it is absent and the amphibian papilla consists of the triangular patch only (size 80–100 μm). In the most "advanced" species, the S-shaped caudal extension is fully developed, and the total length of the papilla is about 500 μm. The species discussed in this chapter all have a fully developed amphibian papilla, except *Bombina orientalis*, *Scaphiopus couchii*, and *Xenopus laevis* (Lewis 1984). In these three species, the papilla has an intermediate form with a length of about 250 μm (*Bombina*) or 400 μm (*Scaphiopus*, *Xenopus*).

The number of hair cells in the amphibian papilla ranges across species from about 170 to about 1200 (reviewed by Fox 1995). The hair cells are covered by a tectorium. The portion of the tectorium that covers the triangular hair cell patch near the saccular space is rather bulky and almost entirely fills the rostral portion of the papillar recess. Toward the caudal end of the papilla, the tectorial membrane is much thinner (Lewis 1984). A tectorial curtain, which in advanced frogs is found approximately halfway along the hair-cell epithelium, blocks the papillar recess. The curtain may play a role in coupling sound vibration in the inner-ear fluid to the tectorial membrane (Wever 1985).

The basilar papilla has a simpler structure. Like the amphibian papilla, the recess of the basilar papilla connects rostrally to the sacculus space and is caudally separated from perilymph by a thin contact membrane (Frishkopf and Flock 1974). The recess can be envisioned as a short tube (about 200 μm) extending from the saccular space to the contact membrane. About 30 to 100 hair cells are embedded in the relatively stiff wall in approximately 50 % of the perimeter of the tube (Fox 1995). A tectorium covers a large portion of the hair cells and spans approximately half the cross section of the tube. Some hair cells at the edge of the epithelium have freestanding stereovillar bundles (Lombard et al. 1999). Since the tectorial membrane does not close the perimeter of the recess, low frequencies will be shunted away from the hair cells of the basilar papilla.

The details of the mechanics of the tectorial membranes of the amphibian and the basilar papilla are unknown. However, sound presumably travels from the middle ear to the amphibian and basilar papillae through the saccular space (Purgue and Narins 2000b). It may then excite the tectorial membranes of the papillae, and thus stimulate the hair cells. The fluid flow will also displace the contact membranes and the round window that lies just caudal to the membranes. In *Rana catesbeiana*, the movement of the amphibian and basilar papilla contact membranes is maximum for stimulation frequencies less the 500 Hz and higher than 1100 Hz, respectively. This indicates a frequency segregation of the sound paths to the papillae (Purgue 2000a).

The auditory neurons that connect to the amphibian and basilar papillae have characteristic frequencies (CFs) in complementary frequency intervals. Neurons of the amphibian papilla are tuned to CFs in the range from about 100 Hz to between 600 and 1600 Hz, where the upper boundary is species specific (e.g., Capranica and Moffat 1975; Ehret and Capranica 1980; Lewis et al. 1982; Ronken 1991). The amphibian papilla exhibits a tonotopic organization, with low-frequency neurons connecting to rostral hair cells near the sacculus, and high-frequency neurons connecting to the caudal extension near the contact membrane (Lewis et al. 1982). Hair cells in the rostral portion of the papilla are also electrically tuned and their resonance frequency parallels the tonotopic organization (Pitchford and Ashmore 1987; Smotherman and Narins 1999).

The range of CFs of neurons from the basilar papilla is species-specific and varies between about 1000 Hz and 3800 Hz (reviewed in Ronken 1991; see also Fig. 7.2). In individual animals, the majority of neurons connected to the basilar papilla have virtually identical tuning properties (Ronken 1991): the basilar papilla seems to function as a single auditory filter. However, some neurons

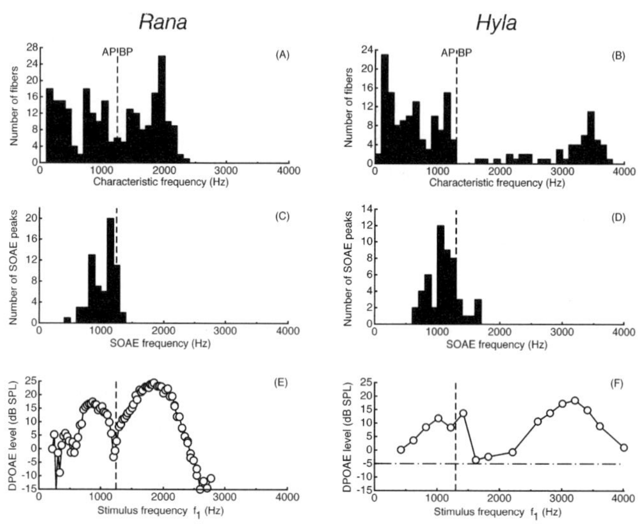

FIGURE 7.2. Comparison between primary-neural characteristic frequencies and emissions frequencies for ranid (*left*) and hylid (*right*) frogs. (**A**) Distribution of characteristic frequencies in *Rana pipiens* (Ronken 1991). (**B**) Distribution of characteristic frequencies in *Hyla cinerea* (Capranica and Moffat 1983). In **A** and **B**, the vertical dashed line indicates the boundary between amphibian papilla and basilar papilla neurons and is also shown in the other panels. (**C**) Distribution of spontaneous otoacoustic emission (SOAE) frequencies in ranid frogs. *Rana esculenta* (van Dijk et al. 1989) and *Rana pipiens* (van Dijk 1996) are shown together. (**D**) Distribution of SOAE frequencies in hylid frogs. *Hyla cinerea*, *Hyla chrysoscelis*, and *Hyla versicolor* are shown together (van Dijk 1996). (**E**) Example of a DP-gram measured in *Rana pipiens* (Meenderink and van Dijk 2004). The noise floor in these measurements was at about −25 dB SPL. (**F**) Example of a DP-gram measured in *Hyla cinerea* (van Dijk and Manley 2001). The horizontal dash-dot line indicates the approximate noise floor of these measurements at −5 dB SPL. SOAE = spontaneous otoacoustic emission, AP = amphibian papilla, BP = basilar papilla.

are tuned to a higher frequency, above the CF of the majority of nerve fibers in an individual animal, and it has been suggested that these may connect to freestanding hair cells (van Dijk et al. 1997a).

The filter quality factor Q_{10dB} of primary auditory neurons in amphibians is typically between 0.5 and 2.5 for both the amphibian and the basilar papillae. For the frequency range of the amphibian papilla, the sharpness of tuning is similar to that in lizards, birds, and mammals (Evans 1975; Manley et al. 2001). In contrast, the Q_{10dB} values of auditory neurons from the basilar papilla is smaller that those in the corresponding frequency range in these other groups (with Q_{10dB} ranging up to 12 in bird; Manley et al. 1997). Thus, the amphibian papilla has a frequency selectivity that is similar to that in other vertebrates, while the frequency selectivity of the basilar papilla is poorer.

3.2 Spontaneous Otoacoustic Emissions in Amphibians

3.2.1 General Characteristics

SOAEs have been recorded from the ears of ranid and hylid frog species (Whitehead et al. 1986; van Dijk et al. 1989; Wilson et al. 1990b; van Dijk et al. 1996a). SOAEs were detected in 50% to 72% of ranid ears and 83% to 100% of hylid ears. The average number of peaks in SOAE spectra of emitting ears is 1.5 for ranids and 1.8 for hylids (see Table 7.1). The highest number of peaks recorded simultaneously from a frog ear is three.

From recorded SOAE signals, an individual SOAE peak can be isolated by bandpass filtering and the statistical distribution of the isolated SOAE pressure

TABLE 7.1. Characteristics of SOAE in two frog and eight lizard genera.

Species (no. of specimens)	Papillar type	Frequency range of SOAE, kHz	Average nr. SOAE per emitting ear	SOAE-amplitude [dB SPL]		Frequency shift, oct/°C at 20–21°C (frogs) or 28–30°C (lizards)
				min	max	
Rana (29)	CTM, no BM	0.45–1.35	1.5	−13	13	0.010–0.069
Hyla (18)	CTM, no BM	0.65–1.68	1.8	−17	8	0.021–0.090
Anolis (6)	No TM	1.3–7.7	7	−7.2	4.7	0.02
Gerrhonotus (2)	No TM	0.99–4.1	6	−7	3.1	0.035
Gekko (7)	STM	0.95–4.55	12	−6	10.3	0.035
Eublepharis (9)	STM	1.53–4.02	4	−6.6	4.9	0.03
Tiliqua (18)	STM	0.93–4.61	10	−10	9.3	0.03
Varanus (10)	CTM	1.08–2.91	3	−2.8	25.8	0.055
Tupinambis (6)	CTM	1.4–4.6	7	−7	15.5	0.08
Callopistes (8)	CTM	2.09–3.73	2.2	−4.5	23.3	0.1

Column 2 describes the structure of the tectorial membrane over the high-frequency hair-cell areas that presumably emit the SOAEs: "CTM, no BM" = continuous tectorial membrane and no basilar membrane; "No TM" = No tectorial membrane; "STM" = tectorial sallets; "CTM" = continuous tectorial membrane.

obtained (Bialek and Wit 1984). The statistical distribution of the instantaneous sound pressure of isolated SOAEs from the frog ear resembles that of a sinusoid (van Dijk et al. 1989). The combination of the narrow spectral width of SOAE peaks and a sinusoidal amplitude distribution shows that the SOAEs cannot be filtered noise (Rice 1954) and must be generated by a (self-) sustained oscillator in the inner ear.

SOAEs are rarely observed below about 500 Hz. Thus, the low-frequency, electrically tuned part of the amphibian papilla does not seem to be involved in SOAE generation. Rather, the high-frequency extension of the papilla caudal to the tectorial curtain presumably generates the SOAEs. However, an alternative interpretation is possible. Low-frequency sound may reach the amphibian papilla through the operculum and the muscle associated with it rather than through the tympanic membrane and the columella (Lombard and Straughan, 1974). In the reverse direction, low-frequency sound generated in the amphibian papilla may not couple efficiently to the tympanic membrane and may thus be missed by the recording microphone.

The upper frequency at which SOAEs have been detected exceeds the upper neural characteristic frequency of the amphibian papilla. Neural characteristic frequencies were usually determined in experiments where the animals' temperature was held below room temperature (e.g., Lewis et al. 1982; Ronken 1991). In contrast, in most SOAE experiments, the emission frequencies were recorded at room temperature. Since both neural tuning in the amphibian and emission frequencies (see later) are temperature sensitive, these experimental differences may account for the observed differences between frequency ranges. The upper SOAE frequency does not, however, correlate with the highest characteristic frequencies from the basilar papilla (see Fig. 7.2). Taken together, these observations strongly suggest that spontaneous emissions from the frog inner ear are generated in the caudal extension of the amphibian papilla.

No emissions were detectable in the *Xenopus laevis* and *Bombina orientalis* (van Dijk et al. 1996a). It is likely that the reduced middle ear structures in these species blocked the transmission of possible spontaneous oscillations from the inner ear. Alternatively or additionally, the absence of SOAE may be related to the reduced length of the caudal extension in these species.

SOAE levels have been found up to 13 dB SPL (van Dijk et al. 1989) and are thus comparable in level to those in lizards, bird, and mammals. The spectral width (FWHM = full width half maximum) of amphibian SOAEs ranges from 1 to 200 Hz, with an average of 38 Hz. Spectral width is negatively correlated with peak height (Long et al. 1996; van Dijk et al. 1996a). However, the relationship between these two parameters is rather complicated, presumably as a result of the complex interaction between multiple SOAE peaks.

An electrical microphonic signal can be recorded with an electrode in the perilymph space (Wit et al. 1989b). The spectral characteristics of this microphonic signal closely correspond to those of an acoustic SOAE that is measured simultaneously. When the tympanic membrane is mechanically obstructed, the acoustic signal is reduced but the electrical signal can be consistently measured. This shows that the emission signal originates within the inner ear. Obstructing the middle ear also results in a frequency increase of the otoacoustic emission frequency of about 3 %. This is similar to the slight frequency changes observed in mammals (Schloth and Zwicker 1983) and lizards (Manley 2005) as a result of a change in ear-canal pressure.

If a DC current is applied through the perilymph electrode, both the amplitude and frequency of SOAE are affected (Wit et al. 1989a). An inward current results in a decrease of emission amplitude and an increase of emission frequency. Outward current has the inverse effect. On average, the frequency shift was 2 Hz/μA. A similar effect has also been shown in mammals and lizards (Sun et al. 2000; Manley and Kirk 2002).

3.2.2 Interactions with External Tones: Suppression Tuning Curves

An external tone can shift SOAE frequency and suppress its amplitude. Isosuppression tuning curves can be measured; they resemble neural threshold tuning curves (Baker et al. 1989). The frequency of a SOAE peak is typically pushed away from the suppressor frequency, except for suppressor tones less than about 100–200 Hz below the SOAE frequency (Baker et al. 1989); for such suppressors, the SOAE frequency is pulled toward the suppressor frequency. An external tone may also cause peak splitting, where one SOAE peak in the spectrum is replaced by two separate peaks (Long et al. 1996). At present, these complex interaction patterns have not been studied comprehensively.

3.2.3 Temperature Dependence

The frequency and the amplitude of amphibian SOAE depend on body temperature (van Dijk and Wit 1987; van Dijk et al. 1989; Wilson et al. 1990b; Long et al. 1996). The frequency of SOAE increases with increasing temperature (Wilson et al. 1986; van Dijk and Wit 1987; van Dijk et al. 1989; 1996a; see Fig. 7.3), with the rate of increase being larger for higher-frequency emissions than for low-frequency emissions (both in Hz/°C and octave/°C [oct/°C]).

The temperature dependence of SOAE amplitude is more complex (Long et al. 1996; van Dijk et al. 1996a). At a high temperature (say 25°C) only two peaks may be present, one near 600 Hz and one near 1000 Hz. Upon cooling the frog, one of these peaks may reversibly disappear and may be replaced by a peak that is 100–200 Hz lower in frequency (e.g., see Fig. 7.4). This pattern of temperature dependence is variable across individuals. The observation that either the 600-Hz SOAE or the 1000-Hz SOAE may be replaced by another peak has been interpreted to indicate that the frog ear may potentially generate four SOAE frequencies (Long et al. 1996).

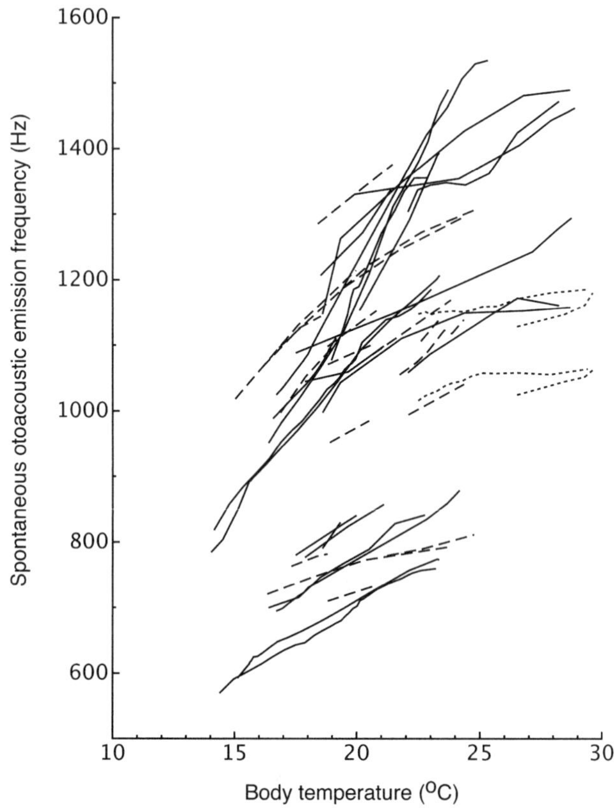

FIGURE 7.3. The dependence of SOAE frequency on body temperature in *Rana temporaria* (Wilson et al. 1986), *Rana esculenta* (van Dijk and Wit 1987; van Dijk et al. 1989), *Hyla cinerea*, *Hyla versicolor*, *Hyla chrysoscelis*, and *Leptodactylus albilabris* (van Dijk 1996). *Coarse dashed lines*: ranid species. *Solid lines*: hylids. *Fine dashed line*: *Leptodactylus*. Smoothed curves are shown.

3.3 Evoked Emissions in Amphibians

SOAEs seem to originate only from the amphibian papilla (Fig. 7.2). In contrast, evoked emissions in the frog are generated at frequencies corresponding to those of both the amphibian and the basilar papillae.

3.3.1 SFOAEs of Amphibians

Stimulus frequency emissions in *Rana esculenta* were observed between 800 and 2000 Hz, and possibly up to 2800 Hz (Palmer and Wilson 1982; Wilson et al. 1990b). This indicates that both the amphibian and the basilar papillae are involved in their generation. In a study in the common frog, *Rana temporaria,* emission frequencies were limited to the interval from 500 to 1000 Hz. At a stimulus level of 5 dB SPL, the emission level is typically about 6 dB below

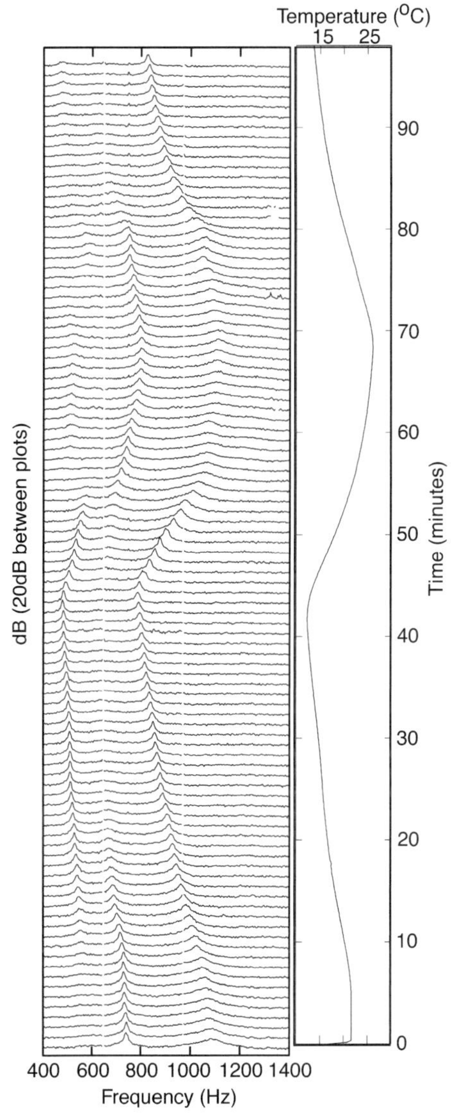

FIGURE 7.4. Typical example of the temperature dependence of a SOAE in the frog *Rana esculenta* (Long et al. 1996 1996). At the beginning of the experiment, two SOAE frequencies were identified (centered at 740 Hz and 1090 Hz). Cooling the animal resulted in a reversible decrease of SOAE frequencies. In addition, the 740-Hz SOAE disappears, and an additional emission peak appears at an even lower frequency. In this individual, the 1090-Hz emission remained present at all temperatures. (Reprinted from Long et al. 1996. Used with permission.)

that of the stimulus. The emission can be as strong as 10 dB SPL, which is slightly weaker than in human (Wilson et al. 1990b). Whitehead et al. (1986) and Wilson et al. (1990b) describe the temperature dependence of SPL minima observed during the frequency sweep of a single tone. These minima correspond to the situation in which the SFOAE is out of phase with the stimulus tone. Similar to the behavior of spontaneous otoacoustic emissions, the frequencies of these minima change with temperature. The rate of change is between 0.01 and 0.03 oct/°C, which is approximately half the rate these authors observed for SOAEs.

3.3.2 DPOAEs of Amphibians

DPOAEs were present in all hylid and ranid species that were investigated in recent studies (e.g., van Dijk and Manley 2001; van Dijk et al. 2002). The range of stimulus frequencies for which DPOAEs can be detected is species dependent and is correlated with the range of hearing sensitivity of the amphibian and basilar papillae (see Fig. 7.2). For ranids, DPOAEs are absent below 400 Hz (Baker et al. 1989), which contrasts with the presence of nonlinear interaction predominantly in low-frequency auditory neurons from the amphibian papilla (Capranica and Moffat 1980; Benedix et al. 1994; van Dijk et al. 1997a). Rather, DPOAE are present for stimuli in the range from 400 to 2500 Hz (Meenderink and van dijk 2005), which matches the frequency ranges of the basilar papilla and the caudal extension of the amphibian papilla. In *Hyla cinerea*, DPOAE are present for stimuli in the frequency ranges between 500–1500 Hz and 2500–3700 Hz (van Dijk and Manley 2001). The gap between these two ranges is consistent with the separation of the sensitivity ranges of the amphibian and basilar papillae in this species (Capranica and Moffat 1983). These results suggest that the upper 1–1.5 octaves of the amphibian papilla (400–1200 Hz in most species) and the entire basilar papilla contribute to DPOAE generation.

In *Xenopus laevis*, DPOAEs are detectable between 600 and 2300 Hz. The relationship to the sensitivity ranges of the amphibian and basilar papillae in *Xenopus* is, however, unknown (van Dijk et al. 2002). In *Xenopus*, the caudal extension of the amphibian papilla is less developed than in ranid and hylid species (Lewis 1984). Thus the presence of DPOAE from *Xenopus* shows that a fully developed caudal extension is not required for the generation of DPOAEs.

No DPOAEs are found in *Bombina orientalis* and *Scaphiopus couchii* (van Dijk et al. 2002). In *Bombina*, the limited development of middle-ear structures is an obvious explanation for the absence of DPOAEs. In *Scaphiopus couchii*, which has a tympanic middle ear and an amphibian papilla similar to that of *Xenopus laevis*, the absence of DPOAEs is hard to understand. It shows that a tympanic middle ear and a developed caudal extension are not sufficient for DPOAE generation.

For strong stimulus tones at 90 dB SPL, DPOAE amplitudes range up to 55 dB SPL in *Rana pipiens*, and up to 18 dB SPL in *Xenopus laevis* (van Dijk et al. 2002). At stimulus levels of 60–70 dB, DPOAE amplitudes range from −10 to 10 dB SPL for *Hyla cinerea*, *Rana pipiens*, and *Rana catesbeiana* (van Dijk and Manley

2001; Meenderink and van Dijk 2004; Vassilakis et al. 2004). DPOAEs are slightly stronger in female animals as compared to males (Vassilakis et al. 2004).

In the following, the DPOAEs in response to low-level stimuli (i.e., below 65 dB SPL) are referred to as low-level DPOAEs; those in response to high-level stimuli are referred to as high-level DPOAEs. For high-level DPOAEs, the rate of increase of level with increasing stimulus level is similar for the amphibian and the basilar papilla (on average 1.8 and 2.0 dB/dB, respectively). In contrast, for low-level DPOAEs, emission amplitudes grow gradually for the amphibian papilla (0.5 dB/dB on average) but remarkably steeply for the basilar papilla (2.0 dB/dB). This suggests that the nonlinear mechanism underlying low-level DPOAE generation is different between these papillae (Meenderink and van Dijk 2004; Vassilakis et al. 2004).

The transition from low- to high-level DPOAEs is accompanied by a change in DPOAE phase, and sometimes by a notch in the DPOAE amplitude (Meenderink and van Dijk 2004, 2005a). For both the amphibian and the basilar papillae, the phase change observed for $2f_1 - f_2$ and $2f_1 - f_2$ is near 0° at the low-frequency end of the papillar range and near 180° at the high-frequency end.

Low- and high-level DPOAE have differing vulnerabilities to physiological insult (van Dijk et al. 2003). When the oxygen supply to the ear is dramatically reduced by cardioectomy, high-level DPOAE remain stable for about 2 hours, whereas low-level DPOAE disappear within six minutes in the amphibian papilla, and within 34 minutes in the basilar papilla. With destruction of the central nervous system, which presumably maintains oxygen supply to the ear, similar results are obtained, but the decay of low-level DPOAEs is slower: the time to full decay of low-level DPOAEs is on average 13 minutes for the amphibian papilla and 58 minutes for the basilar papilla (van Dijk et al. 2003).

In the amphibian papilla, low-level DPOAE levels increase with increasing temperature, but high-level DPOAEs are nearly temperature insensitive. In contrast, both low- and high-level DPOAEs from the basilar papilla do not consistently change with temperature. Apparently, only low-level DPOAEs from the amphibian papilla depend on the metabolic rate, whereas DPOAE generation in the basilar papilla does not involve an active biological mechanism. It must be concluded that DPOAEs from the basilar papilla are generated by passive reflectance in the inner ear (Meenderink and Van Dijk 2006).

DPOAE amplitude and phase also depend on the frequency ratio of both stimulus tones (Meenderink et al. 2005). The relationship between stimulus frequency, stimulus frequency ratio, and DPOAE level and phase is similar to that of a single oscillator (see Fig. 7.5)., This is not surprising for the basilar papilla, since, as was confirmed by neural recordings, it functions as a single auditory filter (Ronken 1991). In agreement with the single-filter hypothesis, neural and DPOAE characteristic frequencies in the basilar papilla closely correspond (Meenderink et al. 2005). For the amphibian papilla, it is unclear how this behavior is consistent with its tonotopic organization. These results from the amphibian ear contrast with those in humans (Knight and Kemp 2000), where the relationship between stimulus frequency and ratio versus DPOAE amplitude is consistent with a traveling-wave, tonotopic model of the cochlea.

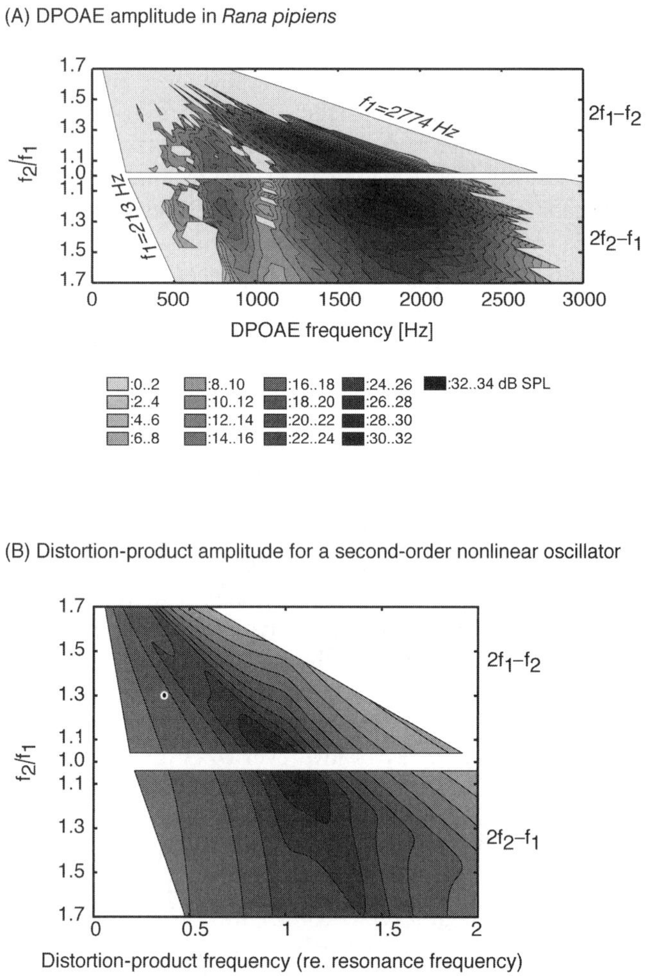

FIGURE 7.5. The relationship between DPOAE frequency, stimulus ratio f_2/f_1, and DPOAE amplitude in frog and human. The amplitude of the distortion product at $2f_1 - f_2$ and $2f_2 - f_1$ is indicated by a gray-scale code. The patterns observed in *Rana pipiens* (**A**) are considerably different from those observed in human (Knight and Kemp 2000). In the frog, the patterns are similar to distortion-product responses obtained in a simple second-order nonlinear oscillator model (**B**). DPOAE = distortion-product otoacoustic emission. (Adapted from Meenderink et al. 2005a, with permission.)

3.3.3 Click-Evoked and Noise-Evoked Emissions in Frogs

Click- and noise-evoked otoacoustic emissions in amphibians have been sparsely documented. Palmer and Wilson (1982) describe a click-evoked emission in a specimen of *Rana esculenta*. The peak response was at a delay of about 4.5 ms. van Dijk et al. (1997b) evaluated noise-evoked otoacoustic emissions in one

Rana esculenta. The corresponding first-order correlation function between the noise stimulus and the ear's response displayed a response component at about 800 Hz that peaked at 5 ms. These delays are about half those observed in man for click-evoked (reviewed in Probst et al. 1991) and noise-evoked (Maat et al. 2000) otoacoustic emissions.

4. Otoacoustic Emissions in Lepidosaurs

4.1 The Anatomical Substrate and Inner-Ear Physiology of Lizards

Most of what we know about the anatomy of the auditory papilla of lizards comes from the work of Wever (1978) and Miller (1980, 1985, 1992). Supplementary information and physiological correlations have been summarized by Köppl and Manley (1992, 2000) and Manley et al. (1990, 1997, 2000a,b,c,d, 2002, 2004a).

The morphology of the auditory papillae of modern lizard families is best understood in the light of its probable evolution, which suggests that the earliest lepidosaurs possessed a papilla not unlike that of modern Tuataras (*Sphenodon*). This is an elongated cellular epithelium located over a thickened basilar membrane and consisting of supporting cells and the sensory hair cells. The ancestral papilla was probably less than 1 mm long and contained at least several hundred hair cells that were covered by a continuous, noncellular tectorial membrane. The hair cell's responses to sounds were at least partly determined by the electrical properties of the cell membrane, and they were arranged tonotopically, probably with the lowest response frequencies apically and the highest (maximally 1 kHz) basally. All hair cells were connected via afferent and efferent nerve fibers to the cochlear nuclei in the medulla oblongata of the brain.

The first step in the evolution of a lizard-like papilla was probably the development of two groups of hair cells at the ends of the papilla that responded to frequencies higher than 1 kHz and whose response properties were mainly determined by the micromechanics of the hair-cell bundle and the tectorial membrane (Manley 2002) rather than by electrical tuning. This development produced hair cells whose upper frequency-response limit was not restricted by the intrinsic limits of electrical tuning (Wu et al. 1995) and is likely to have been the result of the independent development of a tympanic middle ear that transmitted higher frequencies better (Manley and Clack 2004).

Miller (1980, 1985, 1992) noted that in general, low-frequency areas of lizard papillae contain hair cells of a particular morphological type that have their stereovillar bundles oriented in the same direction. He called this type UHC (unidirectional hair-cell type). In cases in which the hair-cell orientation was not exclusively unidirectional but the cell morphology and response characteristics were the same, he referred to UTHC, or unidirectional-type hair-cell areas. Similarly, the high-frequency areas contained hair cells with a different morphology that was consistent across species and whose bundles had either of two orientations, BTHC (bidirectional-type hair-cell area).

The most basal lizard families, the Iguanidae and the Agamidae (Lee 1998) have an auditory papilla with three hair-cell areas, a central (ancestral) area bounded at both ends by hair-cell areas that are mirror images of each other (Fig. 7.6). These areas are also tonotopically organized (see Manley et al. 1990 for a review), with their lowest frequency (about 1 kHz) adjacent to the central area and the highest frequency (4 kHz or higher) at each end of the papilla. Interestingly, these two high-frequency areas have lost their tectorial membrane. The most likely reason for this is that for hair cells that are micromechanically tuned, a coupling to neighboring hair cells binds the hair cells into groups with a common response. When so few hair cells are present (in some species, only something like 30 hair cells in total), the loss of the tectorial membrane allows the hair cells to respond more independently of each other and have their own best response frequencies that range from 1 kHz to 4 kHz or higher. The loss of the tectorial membrane does, however, bring the disadvantages of poorer sensitivity and lower frequency selectivity (Manley 2002, 2004a).

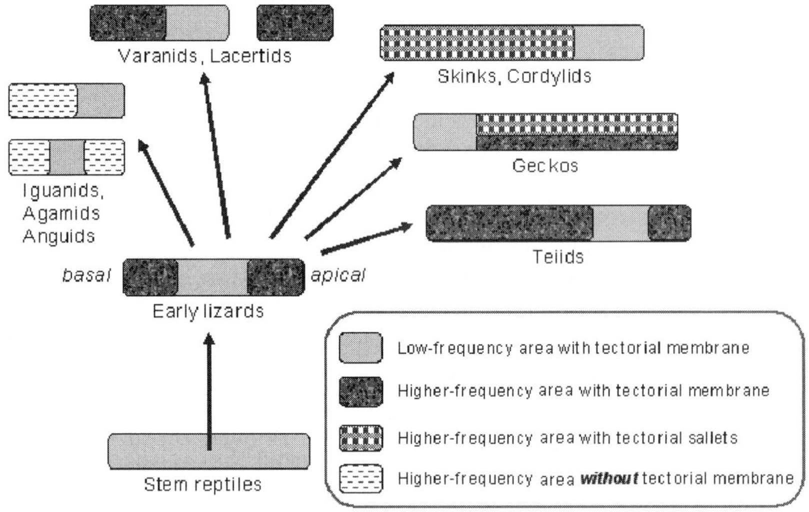

FIGURE 7.6. A schematic diagram showing some of the evolutionary trends in the papillae of lizards. Whereas the stem reptile papilla was probably all low-frequency, the earliest lizards developed higher-frequency responding hair-cell areas at both ends (basal and apical). In different evolutionary lines of lizards, this early lizard papilla was modified in different ways. Whereas iguanids and agamids and the anguids (which are not closely related) kept this arrangement but lost the tectorial covering over the higher-frequency areas, most other groups elongated the papilla, especially the higher-frequency area, and many groups lost the apical high-frequency area. Geckos lost the basal high-frequency area instead. This figure is not intended to show true evolutionary relationships between lizard families. (Modified after Manley 2004a. Used with permission.)

The lizards show a strong adaptive radiation during the late Mesozoic era, and the evolution of the inner ear happened largely independently in the different groups. The main evolutionary trends were as follows (see Fig. 7.6):

1. A trend to papillar elongation, so that in some families (Gekkonidae Varanidae, Teeidae, Scincidae) the papilla is up to 2 mm long and contains up to 2000 hair cells.
2. A trend to the loss of redundancy (mirror symmetry) in the two hair-cell areas. Many lizard families abandoned one or the other high-frequency area. Mostly, the apical high-frequency area is lost, resulting in a tonotopic organization of the papilla that incidentally matches that of birds and mammals, with the high-frequency responses at the base. This loss is found even in some members of primitive families, such as the Anolids among the family Iguanidae. The geckos and the related pygopod lizards have instead abandoned the basal high-frequency area, which results in a reversed tonotopic organization, unique among amniotes (Manley et al. 1999).
3. A different way to deal with the problem of redundant hair cell areas is to change the properties of the hair cells in the different areas such that they no longer overlap in their frequency responses. This effectively increases the space constant for higher frequencies. Interestingly, this effect occurs only if the two areas are physically divided. Constrictions developed in the papilla of some families (Lacertidae, Varanidae) or in one subfamily (the basciliscine Iguanids), which enabled the frequency responses of the two higher-frequency hair-cell areas to drift apart, with the result that one covers the lower and the other the upper half of the frequency range above 1 kHz (Manley 2002, 2004a).
4. A differentiation took place in the shape of the tectorial membrane. As noted above, tectorial membrane loss in small areas of high-frequency hair cells occurred early in the families Iguanidae and Agamidae. The same configuration—independently developed—is found in a more recent family, the Anguidae. Many long papillae retain a continuous tectorial membrane, and there are apparently sufficient hair cells that the advantages of coupling outweigh the disadvantages. Some families with long papillae have, however, developed a different solution. In the Gekkonidae and Pygopodidae, but also independently in the Skinkidae (and skink-like families such as Cordylidae and Gerrhosauridae), the tectorial membrane is shaped like a string of flat pearls. Each of these units is called a sallet, and there can be up to 170 sallets along one high-frequency hair-cell area (as in the Tokay gecko *Gekko gecko*, Köppl and Authier 1995). The sallet optimally balances the advantages and disadvantages of possessing a tectorial membrane. The sallets couple a sufficient number of hair cells and achieve high sensitivity and frequency selectivity, but the small size of the sallets reduces the strength of the coupling to neighboring hair-cell groups outside of the sallet.

The above trends that developed in parallel over time and in several cases independently in different families determine that the morphology of the auditory papillae of lizards is much more varied than in any other group of amniote vertebrates. The lizard auditory papilla was a playground of neutral evolution and provides modern auditory science with a spectacular array of natural experiments (Manley 2002, 2004a).

4.2 Spontaneous Otoacoustic Emissions in Lizards

4.2.1 General Characteristics of Lizard OAEs

Details of the characteristics of SOAEs are available for nine species of lizards from eight different families, and information in less detail is available for a further 20 species. The systematic variation of structure between and even within families as described above provides a unique opportunity to study correlations between anatomy and the characteristics of SOAEs. Some lizard SOAE data are summarized in Table 7.1.

Compared to most other classes of vertebrates, an unusually high percentage of lizard ears, mostly 100%, produces SOAEs (Fig. 7.7). There is no reason to doubt that these SOAEs are a product of the normal function of the hearing epithelium. Peaks in ear-canal sound spectra are identified as SOAEs when their frequencies are temperature-sensitive and their peak level is suppressible by external tones. SOAEs are found even in very tiny papillae that have only about 50 hair cells (Manley et al., 2007b). Across all species, SOAE center frequencies are found in the range between 0.75 and 7.7 kHz (Fig. 7.7). The number of emissions per ear varies from 2 to 15, and the number of peaks is roughly correlated with the structure of the hearing organ. In those species having an unspecialized tectorial membrane covering all hair cells (Table 7.1), such as the northern Teju *Tupinambis teguixin*, the Chile-tegu *Callopistes maculatus* and the cape monitor *Varanus exanthematicus*, only between two and seven SOAE peaks are found in each ear. These peaks are often very large, up to 27 dB SPL. The other species tend to have a larger number of SOAE peaks that are of smaller amplitude, maximally up to 10 dB SPL. This latter pattern correlates with a more specialized hearing-organ structure, showing either the complete absence of a tectorial membrane over the relevant hair-cell area or the presence of specialized tectorial structures such as sallets (STM in Table 7.1).

In those species for which frequency maps of the hearing organ are available (the bobtail skink *Tiliqua rugosa,* the texas alligator lizard *Gerrhonotus leiocephalus*, and *Gekko gecko*; Manley et al. 1990; Manley et al. 1999), the range of the center frequencies of SOAEs lies within the "high-frequency" hair-cell area of the basilar papilla. SOAEs are thus only generated by micromechanically tuned hair-cell groups. It is probably significant in this regard that the hair cells in these areas are generally arranged in one of two bundle orientations that are 180° opposed to each other. If the active process is in the hair-cell bundle, as noted below, twitches of the bundles of one orientation would activate the oppositely oriented bundles just as would a sound stimulus. The oppositely oriented hair-cell groups thus probably activate each other alternately and set up steady oscillations of their hair-cell group that manifest themselves as spectral peaks.

SOAE peaks appear in the averaged spectra to be like very narrow-band noise (Fig. 7.7). The SOAE bandwidths differ across lizard species and overlap

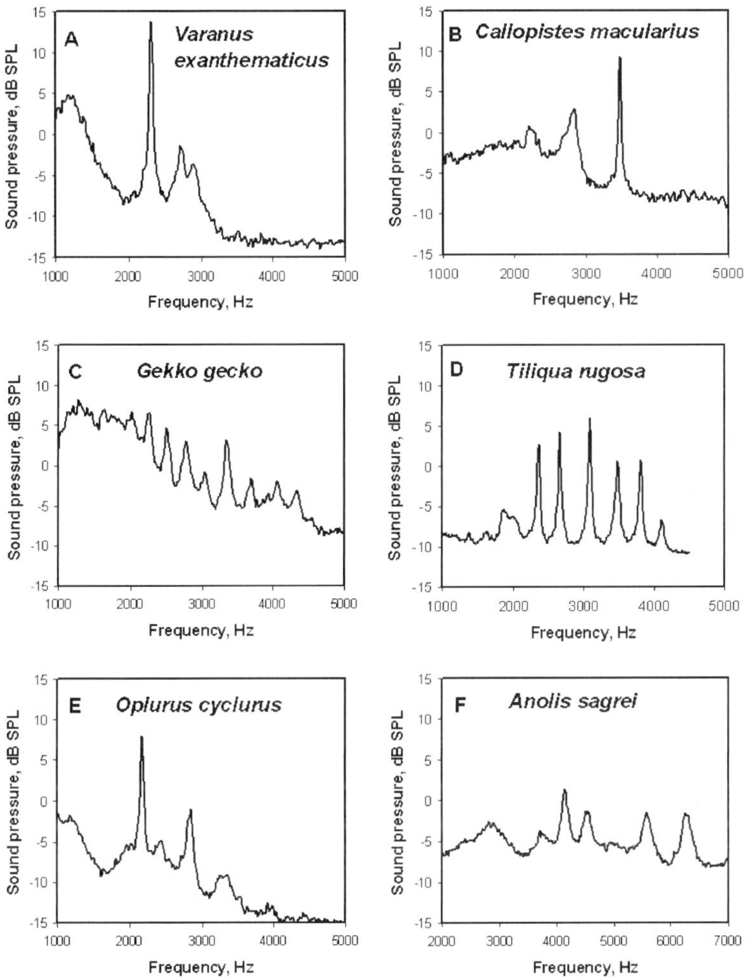

FIGURE 7.7. Examples of SOAE spectra from six species of lizards. In **A** and **B**, the spectra are derived from species that have a continuous tectorial structure over the papilla's high-frequency area. In **C** and **D**, the species' papillae have tectorial sallets, and in **E** and **F**, there is no tectorial structure over the higher-frequency hair cells. Note that in **F**, the frequency scale is different.

with bandwidths measured for other species, including humans, although lizard bandwidths tend to be much wider than most (which in mammals can be as narrow as 1 Hz). In *Gekko gecko*, 3-dB bandwidths were between 44 and 74 Hz, and in the leopard gecko *Eublepharis macularius*, they were from 74 to 170 Hz (Manley et al. 1996). In *Varanus exanthematicus*, the bandwidth was clearly dependent on the relative amplitude of the SOAE peak, and the largest emissions, with peak amplitudes of more than 15 dB above the noise, had bandwidths of only 20 Hz. Similarly, in *Tiliqua rugosa*, the largest SOAE peaks had 3-dB bandwidths of about 10 Hz (Köppl and Manley 1993a). However, most lizard SOAEs have small amplitudes.

Is there evidence that SOAEs represent the output of actively oscillating cell groups? This question was examined in an analysis of the statistical properties of SOAEs from different lizard species and from one bird species, the barn owl *Tyto alba* (van Dijk et al. 1996b). The evidence clearly suggested that the SOAE of these nonmammals have their origin in some kind of active oscillations and that they are not filtered noise. Because of this fact, it is reasonable to expect studies of SOAE to provide useful and even detailed information about the characteristics of the hearing organ.

Since in lizards there is no frequency-selective basilar-membrane traveling wave (e.g., Manley et al. 1990; Köppl and Manley 1992), the origin of SOAEs has to be sought at the local level in terms of motility of particular groups of hair cells and, if they are attached to a tectorial structure, in the driven motility of this structure also. Since studies of isolated lizard papillae indicate that acoustic stimuli activate hair cells through a side-to-side movement of the auditory papilla and its tectorial structure, the origin of SOAEs is presumably in such movements generated spontaneously by the hair cells. A possible source of mechanical energy is the hair-cell bundle itself, which has been shown to undergo active movements, including rapid, biphasic twitches during deflection and even spontaneous twitching (see Manley and Gallo 1997 for references).

Detailed observations of the variation of SOAE frequency and amplitude over time in various species of lizards (e.g., Manley et al. 1996; Manley 2004b) showed that the wide bandwidth in averaged spectra results from generators whose frequency is unstable and varies very rapidly. The differences in SOAE bandwidth in the various classes of vertebrates reflect the frequency stability of the SOAE generators; in lizards, the SOAE generators show poor frequency stability. Thus lizard SOAEs may shift their center frequency a great deal (up to several hundred hertz) under the influence of external tones. Since there is evidence that standing waves in the lizard cochlea play a role in maintaining SOAEs (Manley 2005), the instability may be the result of the presumably very fast feedback times in tiny cochleae. Interactions between SOAE peaks occur with extremely short latencies (van Dijk et al. 1998).

In papillae that lack a tectorial membrane, the spectral patterns of SOAE are not stable over longer time periods. As a consequence, the spectral patterns can look quite different on different days (Manley 2003). This variation is presumably linked to the tectorial material being coupled to specific groups of hair cells, whereas without a tectorial membrane, hair-cell groups can enter into temporary "alliances" that can change over time.

4.2.2 Interactions Between SOAEs and External Tones; Suppression Tuning Curves

Several effects on SOAEs have been reported in response to external tones, all of which are reversible upon silence: amplitude reduction (suppression), amplitude increase (facilitation), and frequency shifting. The relatively high sensitivity of SOAE amplitude and frequency to the presence of external tones is a manifestation of the frequency instability of the generators of lizard SOAEs. External tones can cause large shifts in the center frequency of SOAEs (Köppl and Manley 1994; Manley et al. 1996; Manley 2004b). The magnitude of all effects is dependent on both the frequency distance between the external tone and SOAEs and the tone's level.

4.2.2.1 Level Suppression by External Tones

Externally applied tones influence lizard SOAEs in a strongly frequency-selective way (Fig. 7.8A). In all cases, the most sensitive frequency for SOAE level suppression lies very close to the center frequencies of the SOAEs, in contrast to findings in mammals, where the center frequencies of suppression tuning curves (STCs) lie at a frequency slightly higher than that of the SOAEs (e.g., Schloth and Zwicker 1983). This may reflect the presence in mammals of a highly nonlinear region slightly basal to the center-frequency position. There is no evidence in nonmammals for such a phenomenon. The tone level necessary to reach criterion suppression (generally −2 dB) rises as the tone frequency is moved away in either direction. Thus STCs are V-shaped (e.g., Köppl and Manley 1994; Manley et al. 1996; Manley 2004b) and bear a remarkable similarity to excitatory tuning curves of primary auditory nerve fibers of the same species (for *Tiliqua rugosa*, see Table 7.2). The similarities extend to details of the frequency selectivity, measured as Q_{10dB}, and the slopes of the flanks (Manley and Köppl 1992; Köppl and Manley 1994). A similar correspondence has also been reported for *Gekko gecko* (Manley et al. 1996). As this has also been shown for mammals and birds (e.g., Taschenberger and Manley 1997), it indicates that STCs of SOAE give very good information concerning frequency-selectivity characteristics of the sensory elements in the auditory papilla. It shows that SOAE generation is intrinsically related to the mechanisms underlying frequency selectivity of the inner ear.

4.2.2.2 Facilitation by External Tones

For particular combinations of frequency and level of external tones, facilitation of SOAE level is seen (e.g., Köppl and Manley 1994; Manley et al. 1996; Manley 2004b), in extreme cases up to 10 dB. It is consistently seen only at frequencies and levels that lie below the high-frequency flank of STCs. In some species,

TABLE 7.2. The average values and the ranges of the parameters measured from the STC of SOAE and for primary afferent tuning in *Tiliqua*.

Parameter	Tuning of primary afferents	SOAE suppression tuning
Range of center frequencies	0.2–4.5 kHz	0.91 to 4.18 kHz
Best thresholds	6–78 dB SPL	9–33 dB SPL
Slopes of low-frequency flanks	−10 to −115 dB/octave	−20 to −95 dB/octave
Slope of high-frequency flanks	10–325 dB/octave	55–215 dB/octave
Q_{10dB} values	0.2 to 9	1.7 to 21

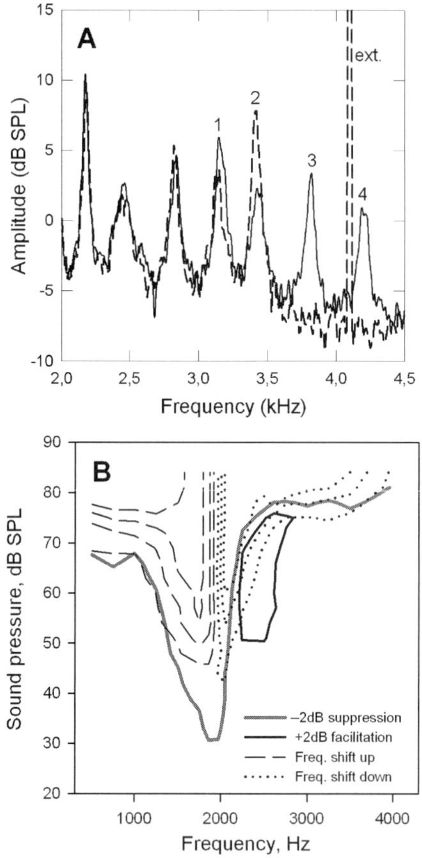

FIGURE 7.8. (**A**) An example of the possible effects of external tonal stimuli on lizard SOAEs. In this case, taken from measurements on the bobtail skink *Tiliqua rugosa*, seven SOAE peaks are visible in an undisturbed spectrum (*continuous line*). When a tone was added at 4.1 kHz ("ext.", the peak of the tone, is off scale), four of the SOAE peaks show an effect. The peaks numbered 3 and 4, which are closest to the tone, are completely suppressed. Peak 2 is facilitated by about 5 dB and its frequency shifted slightly down. Peak 2 is suppressed by about 2 dB and its frequency is shifted down, away from the tone (frequency pushing). (After Köppl and Manley. Used with permission.) (**B**) A diagram illustrating the results of an analysis of suppression, facilitation, and frequency shifting in one single SOAE with a frequency near 1.9 kHz, in this case from *Varanus exanthematicus*. The *thick gray curve* shows the 2-dB suppression contour. Above this line, all tones suppress the SOAE amplitude by at least 2 dB. The *continuous black line* encloses an area of frequency–amplitude space within which tones will facilitate—increase—the emission amplitude by at least 2 dB. The *thin dashed lines* reveal areas within which tones will cause the frequency of the emission to rise by 15 Hz (lowest curve), 45 Hz, 75 Hz, and 105 Hz. Similarly, tones within the *dotted* contours will cause the SOAE frequency to shift down by 15 Hz (lowest curve), 45 Hz, and 75 Hz. In both cases, this represents frequency pushing. (After Manley 2004b. Used with permission.)

a single area of facilitation was observed, with a center frequency about 0.25 octave above the STC center frequency (CF; *Gekko, Eublepharis*, with up to 8 dB and 4 dB facilitation, respectively; Manley et al. 1996) or covering the range from 0.2 to 0.6 octave above the STC's CF (*Tiliqua*, 10-dB facilitation; Köppl and Manley 1994). In *Varanus exanthematicus*, the two areas are centered near 0.4 and 0.9 octave above the CF, with maximally 10 dB of facilitation (Manley 2004b).

Facilitation is especially large in lizards and often occurs at frequencies and levels of tones that match notches or breaks in the slope on the flanks of the STC (Fig. 7.8B). At least in *Tiliqua rugosa*, such notches are also seen in frequency-tuning curves for primary auditory nerve fibers (Manley et al. 1990; Köppl and Manley 1994), suggesting that they reflect some important phenomenon that also influences tuning selectivity at the hair-cell level.

4.2.2.3 Shifts in SOAE Frequency Due to External Tones

External tones can shift the frequency of an SOAE peak (Fig. 7.8B) away from that of the external tone (frequency pushing) or, for frequencies very close to the SOAEs, toward the external tone (frequency pulling). This effect is similar to that in amphibians (see Section 3.2.2). For frequency pushing, the SOAE frequency can shift a great deal, up to 400 Hz in Varanus (Manley Manley), 250 Hz in geckos (Manley et al. 1996) and 330 Hz in *Tiliqua rugosa* (Köppl and Manley 1994). In two species whose high-frequency papillar areas lack a tectorial structure, the maximal frequency shifts were even larger: in the brown anole *Anolis sagrei*, up to −547 Hz and in *Gerrhonotus leiocephalus* up to −469 Hz for tones above the SOAE frequency. Overall, SOAE from nonmammals behave qualitatively similarly to those of mammals in response to an external tone, but with large quantitative differences.

4.2.3 Temperature Dependence of Lizard SOAEs

The amplitudes of SOAEs are not strongly influenced by the temperature, except that at very low and high temperatures, all peaks tend to disappear. The center frequency of lizard SOAEs is, however, systematically influenced by the head temperature, with quantitative differences between species. The center frequency shifts to higher values when the animals are warmed and to lower frequencies when cooled, with the rate of frequency change varying between groups (Table 7.1, Fig. 7.9). The difference between species in this regard depends on anatomical features, much as the spectral features of the SOAEs do. Near their optimal temperatures, species with continuous tectorial membranes exhibit a larger change in SOAE frequency with temperature than species without a continuous tectorium. In *Callopistes macularia* and *Tupinambis nigropunctatus*, but also in *Varanus exanthematicus*, the rate of change is up to 0.071 oct/°C (Manley et al. 1997). In contrast, in species with papillae that have sallets or no tectorial structures, the shift is only 0.015 to 0.06 oct/°C.

The shift of center frequency is often dependent both on the frequency itself and on the temperature range in question. In general, the shift is greater for

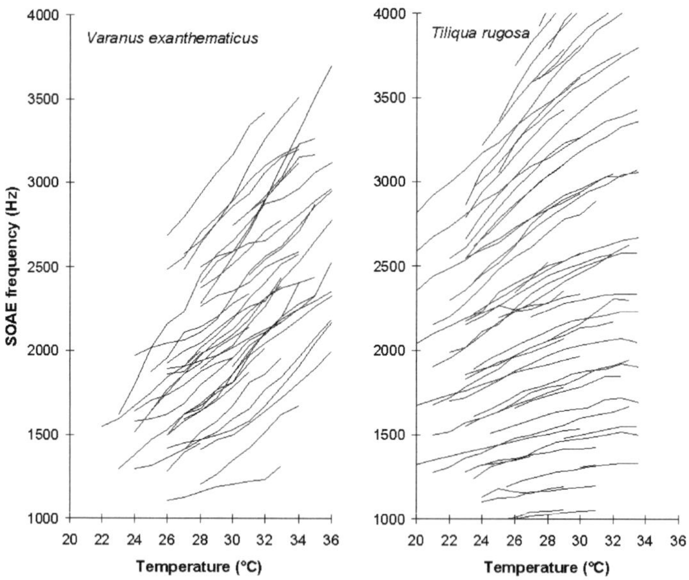

FIGURE 7.9. Examples of the effect of changing the head temperature on the frequencies of SOAE peaks in two species of lizards. On the *left*, each line represents the frequency shift in one single SOAE peak as a function of the head temperature in *Varanus exanthematicus*. In the figure on the right, which has the same scaling, many SOAE peak traces are shown for the same range of temperatures in the bobtail skink *Tiliqua*. The traces for *Varanus* are substantially steeper (the center frequency of each SOAE changes more rapidly for each temperature step), the temperature dependence is thus higher than in *Tiliqua*. The data in each species are pooled across many ears, and measurements were taken in 1°C steps.

SOAEs of higher frequency, both absolutely (i.e., in Hz/°C) and proportionally (in oct/°C). The smallest shifts and highest levels are observed in a temperature range that corresponds—where this parameter is known—to the ecological optimal range. This of course raises interesting questions as to how this range is matched at the cellular level, about which nothing is known.

4.3 Evoked Otoacoustic Emissions

4.3.1 DPOAEs in Lizards

In contrast to SOAEs, which are found at particular frequencies that differ for each ear, DPOAE in lizards can be induced at any frequency, including below 1 kHz, where SOAEs never occur. For DPOAEs, the hearing organ emits at various frequencies that are mathematically related to combinations of the two stimulus frequencies presented. Thus these DPOAE frequencies cannot be affected by external tones or temperature changes.

The first study of DPOAEs in lizards (Rosowski et al. 1984) was carried out before fast and sensitive measurement systems became available and therefore provided mainly data on responses to higher-level stimuli. Rosowski et al. (1984) described three level-dependent components of $2f_1 - f_2$. One component was present at relatively low sound pressure, but it only slightly exceeded the noise level of their system. Rosowski et al. attributed the components appearing at higher SPL to nonlinearities in the *macromechanical* stage of stimulus processing in the inner ear.

The most detailed DPOAE data are from *Tiliqua rugosa* (Köppl and Manley 1993b; Manley et al. 1993). In many ways, DPOAEs behave similarly to SOAEs. They can be suppressed and facilitated by a third tone, and STCs have been measured (Köppl and Manley 1993b, Gallo 1997). The minimal levels of primary tones for the production of DPOAEs in *Tiliqua* correspond closely to the best auditory-nerve-fiber thresholds for this lizard species. DPOAE originating from the low- and high-frequency segments of the basilar papilla differ in a number of respects. DPOAEs from the higher-frequency segment bear a remarkable resemblance to the acoustic DPOAEs of mammals in many details, in spite of substantial structural and micromechanical differences between these hearing organs. In some cases the DPOAE level was clearly affected by the presence of SOAE.

The frequency selectivity of the DPOAE $2f_1 - f_2$ and $2f_2 - f_1$ in *Tiliqua rugosa* was also studied by measuring thresholds of DPOAE detection (the primary-tone level at which the DPOAE level is 2 standard deviations [SD] above the noise) while varying the frequency ratio of the primary tones. The characteristics of tuning in high-frequency DPOAEs showed some remarkable similarities to the tuning curves of single nerve fibers, suggesting that the characteristics of acoustic DPOAEs produced by low-level primary tones reveal the tuning characteristics of the hearing organ at the hair-cell level. Thresholds of DPOAE detection varied both with f_1 and with the primary-tone frequency ratio. Considering only thresholds for optimal primary-tone frequency ratios, the area of greatest sensitivity fell between 1 and 2 kHz for both $2f_1 - f_2$ and $2f_2 - f_1$. This frequency dependence resembles that of the neural audiogram of this species as derived from the thresholds of single primary auditory-nerve fibers (Manley et al. 1990). In addition, the absolute thresholds and the range of thresholds of DPOAE detection closely resemble the values and the range of the neural thresholds (Fig. 7.10). If the DPOAE detection threshold reflects the start of overlap of the primary-tone excitation patterns on the sensory epithelium, then the threshold rise with increasing primary-tone frequency ratio should give an indication of the slopes of these excitation patterns. The slopes of the detection-threshold functions for higher-frequency DPOAEs for both $2f_1 - f_2$ and $2f_2 - f_1$ lie between 60 and 100 dB/oct and are thus equivalent to the steepest slopes of the low-frequency flanks of tuning curves of auditory-nerve fibers of equivalent CF (Manley et al. 1990).

To directly compare thresholds of DPOAEs and auditory-nerve fibers as a function of frequency, DPOAE threshold data for both $2f_1 - f_2$ and $2f_2 - f_1$

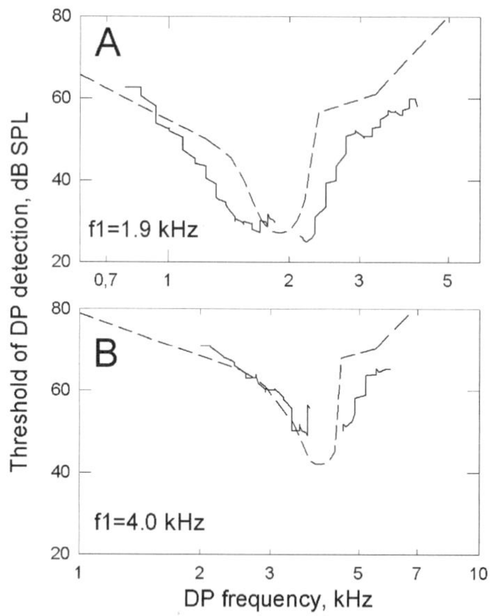

FIGURE 7.10. The threshold for the detection of a DPOAE as a function of the frequency of the distortion product for two fixed lower primary frequencies in all ears of the bobtail lizard *Tiliqua*. For **A** and **B**, the frequency of f_1 is 1.9 and 4.0 kHz, respectively. The different DP frequencies were generated by changing the frequency of f_2. The data for the DPOAE $2f_1 - f_2$ make up the curve on the *left side* in each panel, and data from $2f_2 - f_1$ make up the *right-hand curve* in each case. The solid lines are 11-point weighted moving averages for the two data sets in each panel. These thresholds for the detection of DP as a function of the frequency of the distortion product (*continuous lines*) are compared to idealized rate-threshold tuning curves of single auditory-nerve fibers (*dashed lines*) having characteristic frequencies at the frequency of the respective f_1. Note that while each frequency axis covers exactly one decade, the starting frequency differs between panels. (After Manley et al. 1993. Used with permission.)

as a function of the appropriate DPOAE frequency were plotted together with idealized neural tuning curves (Manley et al. 1990) having a CF the same as the f_1 frequencies used (Fig. 7.10). The absolute threshold values at the center frequency are remarkably similar for both DPOAE and neural data. The similarity of the DPOAEs and single-neural threshold data is greater for primary tones that lie in the high-frequency area, which is micromechanically tuned (Manley et al. 1988, 1989, 1990). Although the DPOAE threshold curves are less well tuned than the single-fiber tuning curves, the DPOAE tuning curves show similar slopes of their flanks. The high-frequency flank also shows a clear break in the slope that is characteristic of the high-frequency neural tuning curves of this species (Manley et al. 1990). Thus DPOAE threshold measurements reflect to a remarkable degree the features of micromechanical tuning in this hearing organ, suggesting that the interaction of the tones producing the DPOAEs at low levels is governed by the same rules that shape the tip region of neural tuning curves

(Köppl and Manley 1993b). The thresholds of DPOAEs–STCs are, however, higher because a higher degree of activation of the hearing organ is necessary to generate DPOAEs than for SOAEs, which of course arise spontaneously. In addition, the two frequencies generating DPOAEs must have a finite distance from each other, and together, they activate a much broader region of the hearing organ than produces a SOAE. Thus it is not unexpected that the STC of DPOAE suppression is less sharply tuned than the STC of SOAE. With these caveats in mind, however, it is clear that the features of data derived from DPOAE generation, suppression, and facilitation reflect the frequency selectivity of the hearing organ.

The best suppressive frequencies for the DPOAE $2f_1 - f_2$ at low primary-tone levels were near the first primary tone (f_1), whereas those for $2f_2 - f_1$ lay near or above f_2. Facilitation of both DPOAEs was also seen, most often over a restricted frequency range above f_2 and at levels below those eliciting suppression. The two DPOAEs showed a number of differences in their behavior in the presence of a third tone that suggest that they do not originate from the same location or source.

4.3.2 EEOAEs in a Lizard

Previous studies of rapid movement of hair-cell bundles come mainly from in vitro studies of nonmammalian vestibular systems and turtle auditory organs and indicate that the molecular mechanisms responsible are directly associated with the transduction channels (review in Hudspeth et al. 2000). Because these channels are directly gated by the stimulus, coupling a source of mechanical power to the channel gating permitted the development of a fast amplification system closely associated with the transduction process itself. In all of these preparations, the response frequencies were well below 1 kHz, often below 100 Hz, and thus very low in the auditory range of vertebrates. The origin of higher-frequency hair-cell motor activity was examined in vivo in *Tiliqua rugosa* and provided the first evidence that in lizards, the hair-cell bundles produce mechanical forces at audio frequencies.

Uniquely, the high-frequency regions of lizard auditory papillae consist of populations of hair cells whose stereovillar bundles are oppositely oriented, that is, the directional sensitivities of the bundles have opposing polarities (Miller 1980, 1992; Manley et al. 1990). This distinctive organization permits examination of the different putative active mechanisms in a way that is impossible in mammals and birds. If EEOAEs could be evoked in lizards, this hair-cell arrangement should make it possible to determine whether the cochlear motors driven by the electrical current are in the lateral hair-cell membrane or are associated with the bundles (Manley et al. 2001). A simple model indicated that if EEOAE could be modulated by adding low-frequency sound, the site of the hair-cell motor could be delineated in vivo. Using AC currents below 5 μA that were injected into scala media above the auditory papilla, small EEOAE signals were recorded that generally had amplitudes below 0 dB SPL. The amplitude of EEOAEs increased in size and was modulated by adding low-frequency sound, and this modulation was abolished after the animal's death but also by introducing the ototoxic agent streptomycin into scala media. The EEOAE spectra differed for each ear, presumably related to the individuality of hair-cell population sizes at different frequencies.

Modulation of the EEOAE amplitude by low-frequency sound was of central importance to testing the hypotheses regarding the site of hair-cell motility assumed to underlie the generation of EEOAEs. In *Tiliqua rugosa,* the high-frequency papillar region has opposing hair-cell populations of roughly equal size that are systematically arranged on opposite sides of the basilar papilla. If electrical current elicited motility from the lateral membranes of these cells as demonstrated in mammalian preparations, EEOAE from all cells would have the same phase (because the voltage polarity and the movement elicited would be independent of the bundle polarity, Fig. 7.11a, c, e, g). However, if the electrical stimulus drove a motor integral to the hair-cell bundle, EEOAEs from oppositely oriented cells would be in phase opposition (Fig. 7.11b, d, f, h) and much of the signal would be lost through cancellation. In the presence of a modulating sound, the EEOAE should increase in level in the second case, and the waveforms should differ clearly and characteristically for the two possible sites of active hair-cell motility. This difference was interpreted as an unambiguous signature for the bundle motor.

Low-frequency tones modulated EEOAE levels, and modulated signals were larger in maximum amplitude than in the unmodulated case, as would be expected if the unmodulated signal is the resultant of phase cancellation between two roughly equal-size and oppositely oriented hair-cell populations. Modulation of 100 % was achieved at many of the combinations of AC frequency and sound modulator frequency that were tried ($f_{\text{electrical}}$ from 1464 to 2928 Hz, $f_{\text{acoustical}}$ from 86 to 689 Hz). The fact that strong modulation was seen is incompatible with an origin of the motor activity in the lateral cell membrane. Remarkably, in a few cases, higher levels of modulating sound induced an overmodulated pattern, with a 180° phase change in the EEOAE signal at every half-cycle of the modulator. These results clearly show the signature expected when the mechanism driving EEOAE in lizards is that of an active process in the stereovillar bundles.

If we assume that SOAEs are driven by the same active process as that involved in EEOAE generation, then SOAEs should be influenced by AC current injection. This proposition was tested by averaging the ear-canal sound field during AC injection, with sampling locked to the phase of the current. In contrast to a spectrum analyzer, which detects spectral peaks regardless of their phase, the computer recognized only those components that were phase-locked to the injected current. Thus, when current strength was set to zero, unsynchronized SOAEs of random phase (whose existence was established using the spectrum analyzer) were not reliably picked up by the computer. With sufficient amplitude of the swept AC current, however, phase-locked peaks at SOAE frequencies showed up in the computer spectra, indicating that the AC current had entrained the SOAE-generator mechanism. The center frequency and bandwidth of these peaks did not necessarily correspond precisely to those measured by the spectrum analyzer.

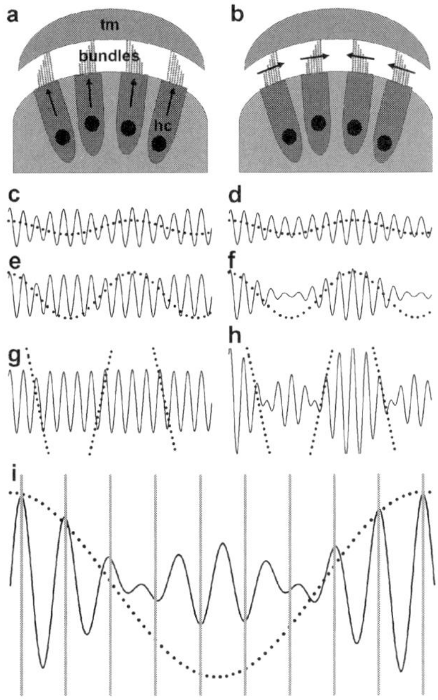

FIGURE 7.11. A model comparing the predicted EEOAE output of a lizard auditory papilla under acoustical modulation. **a, b:** Schematic basal cross sections through the papilla of *Tiliqua rugosa* to illustrate how hair-cell orientation determines EEOAE output for two possible locations of the active process. Note that there are two hair-cell populations, [whose bundles, tallest stereovilli] face the middle of the papilla. Under electrical stimulation, a membrane motor would drive all cells to elongate and shorten in phase (**a**, arrows) and EEOAE from both populations adds constructively. A bundle-based motor, however, would move oppositely oriented bundles 180° out of phase (**b**, arrows) and EEOAE adds destructively. (tm - tectorial membrane; hc - hair cell). Modulation by low-frequency sound decreases the output from hair cells of one orientation while that of the other increases, the force from each type being of the same phase in a and of opposite phase in **b.** A 5% excess of cells of one polarity is assumed, allowing a small EEOAE response at the resting point in **b.** The model uses a frequency ratio between electrical and acoustical stimuli of 9.25: 1, and the time axes cover just over 1.5 cycles of the low-frequency modulator. EEOAE output (*continuous lines*) of the model simulating membrane motors (**c, e, g**) and bundle motors (**d, f, h**) stimulated by AC current and modulated by a low-frequency sound (*dotted line*; the amplitude is arbitrary relative to the EEOAEs). Compared to **c** and **d**, the relative sound pressure of the modulator is increased by 10 and 20 dB in **e, f** and **g, h**, respectively. (**i**) A data recording under the conditions of a loud, low-frequency modulator. In this panel, the vertical lines emphasize the roughly 180° phase shift in the EEOAE in i at each half-cycle of the modulator. These data clearly match the model using a bundle-based active process. (After Manley et al. 2001, copyright 2001, National Academy of Sciences, USA. Used with permission.)

To verify that the AC current was entraining SOAEs, SOAE and EEOAE patterns were examined in the presence of a tone of a frequency close to certain SOAE peaks. As discussed above, tones can suppress, facilitate, and/or shift the center frequency of SOAE peaks in lizards (Manley 2000b). Under suppression, entrained SOAE peaks show the same effects as unsynchronized SOAE measured using the spectrum analyzer under the same conditions (Fig. 7.12). The entrained peak in the EEOAE data in this figure (center at 1.55 kHz) is broader, but shows the same shift to lower frequencies as the SOAE. It even shows, in the presence of the tone at 69 dB SPL, the small level facilitation that occurred in the unsynchronized SOAEs.

Both SOAE and EEOAE peaks were completely suppressed when the tone level reached 79 dB SPL. The higher-frequency peak shifted up in frequency in both cases, but it was noisier in both sets of data. Interestingly, the injection of current at frequencies at which the SOAE peaks were already present did not readily lead to an increase in the emission amplitude. This suggested that SOAEs represent oscillations that are already near saturation for the motor system of the active hair cells, which would be expected from the behavior of a limit-cycle oscillator working near its peak output.

At frequencies between SOAE peaks, there was a clear dependence of amplitude on current strength. This is an important point, since it suggests that the narrow-band AC current mainly affects the hair cells tuned to that frequency

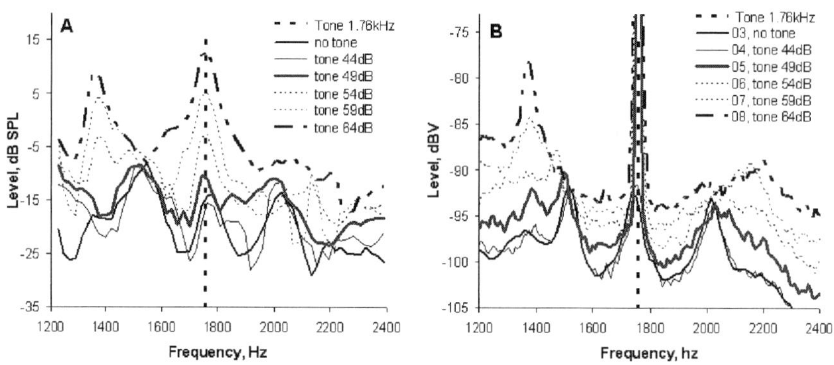

FIGURE 7.12. A comparison of the effect of a tone near 1.7 kHz and played at different attenuation levels on (**A**) EEOAE and (**B**) SOAE in the same ear and over the same frequency range. The frequency of the tone at 1.76 kHz (which is not synchronized with the computer signal and therefore not visible in (**A**) is marked by a *vertical dashed line*. Although the EEOAE data in the left panel are noisier, the same frequency-shifting effects of the tone with increasing tonal level are observed. Note especially the EEOAE that in the absence of the tone is at about 1.5 kHz (*lower arrow* in **A**), which moves away from the tone to be below 1.4 kHz when the tone reaches 64 dB SPL (*upper arrow* in **A**). The same effect is seen for the SOAE in panel B. These data strengthen the idea that these EEOAE are SOAE that were entrained by the current. In **B**, a small SOAE at 2.02 kHz is also pushed away from the tone, rising in frequency. This signal is not clearly visible in **A**. (After Manley and Kirk 2002. Used with permission.)

band. At frequencies where hair cells are not oscillating as groups, the current possibly leads to entrainment of closely spaced—perhaps previously independently vibrating—hair cells, whose coordinated oscillations then appear as an EEOAE.

5. Otoacoustic Emissions in Archosaurs

5.1 The Anatomical Substrate and Inner-Ear Physiology

A great deal of morphological and physiological data on the avian basilar papilla is available (for details and reviews see Manley 1990, 2000b; Manley and Gleich 1992 Fischer 1994; Manley and Köppl 1998; Gleich and Manley 2000; Gleich et al. 2004). The mechanism of hearing in birds is, however, still far less understood than in mammals. Phylogenetically, ratite birds (e.g., the emu, *Dromaius novaehollandiae*) are considered to be more primitive than water birds (e.g., ducks and seagulls), while the most advanced are the land birds (Carroll 1988). Traces of this evolutionary history can be seen in the anatomy of the "cochlea" of birds.

Unlike the coiled mammalian organ of Corti, the avian hearing epithelium (the basilar papilla, BP) is a relatively short, slightly curved band. The length ranges from 2.1 mm in small songbirds up to 10 mm in *Tyto alba*. Its width decreases from the apex to the base (Fig. 7.1). A thick tectorial membrane (TM) covers the entire papilla, and the hair-cell bundles are firmly attached to it. Although the BP size roughly correlates with the animals' sizes, specializations do occur. Thus in *Tyto alba*, the base of the BP is highly elongated, and in the pigeon, the apex is exceptionally broad.

Between 3000 and 17000 hair cells cover the avian papilla, and they are arranged in a mosaic rather than in rows. The number of hair cells across the width varies from as few as four hair cells basally to up to 50 apically, depending on the species. Each hair cell is surrounded by supporting cells, and there are no fluid spaces within the papilla. There is a continuous transition between two basic hair-cell forms across and along the papilla. Neurally and apically are relatively tall and columnar hair cells (THCs). Shorter, cup-shaped hair cells are found abneurally and basally (short hair cells, SHCs). Most THC are situated over the cartilage-like neural limbus, whereas SHC lie over the free basilar membrane (Fig. 7.1). In contrast to THCs, SHCs have no afferent nerve-fiber contacts but only one large, cuplike efferent synapse (Fischer 1994).

The tonotopic organization of the avian BP correlates with species-specific gradients in hair cell morphology over the width and along the length of the papilla, for example, in the hair-cell shape, in bundle morphology and orientation, and in the innervation pattern. In general, apical hair cells are taller than basal ones, and neural hair cells are taller than abneural ones. Hair cells of the neural region give rise to the most sensitive afferent nerve-fiber responses. In *Tyto alba*, the entire basal half of the BP covers only the frequencies between 5 and 10 kHz and is regarded as being an auditory "fovea" (Köppl et al. 1993). In this region, the bundle characteristics are rather constant, and the frequency response changes very slowly.

An additional, unique feature in avian papillae is the systematic change in HC bundle orientation, both across and along the BP. The sensitive axes of stereovillar bundles of neural and abneural HC are oriented perpendicularly to the edge of the BP (the tallest stereovilli face away from the neural edge). HC bundles in the center of the apical half of the BP, however, are rotated toward the apex (up to 90°). The location of maximal rotation and the degree of bundle rotation toward the apex are species-specific (an example is given in Fig. 7.13 and 7.14). Since a bundle shows the highest sensitivity to a deflection toward the kinocilium, the orientation presumably correlates with the direction of maximal sensitivity.

It is likely that both electrical and micromechanical tuning exist in the avian auditory papilla (Manley 1986, 1990; Manley and Gleich 1992). Avian afferent fibers are spontaneously active, and many fibers have an unexpectedly large proportion of interspike intervals whose period is related to their most sensitive frequency. This results in peaks (preferred intervals) in interspike interval histograms, as first described in the starling *Sturnus vulgaris* (Manley 1979; Manley and Gleich 1984; Manley et al. 1985) and subsequently observed in *Gallus* (Manley et al. 1991), *Tyto* (Köppl et al. 1997), and *Dromaius* (Manley et al. 1997). The presence of preferred intervals in the spontaneous activity of primary auditory-nerve fibers and a strong temperature sensitivity of frequency tuning suggest that birds have retained electrical tuning, at least in their low-frequency hair cells. Wu et al. (1995) demonstrated that at avian body temperatures, the upper frequency limit of electrical tuning can extend beyond 4 kHz. The highest measured upper limit for preferred intervals in the spontaneous activity of avian primary auditory-nerve fibers is 4.7 kHz in *Tyto alba* (Köppl et al. 1997).

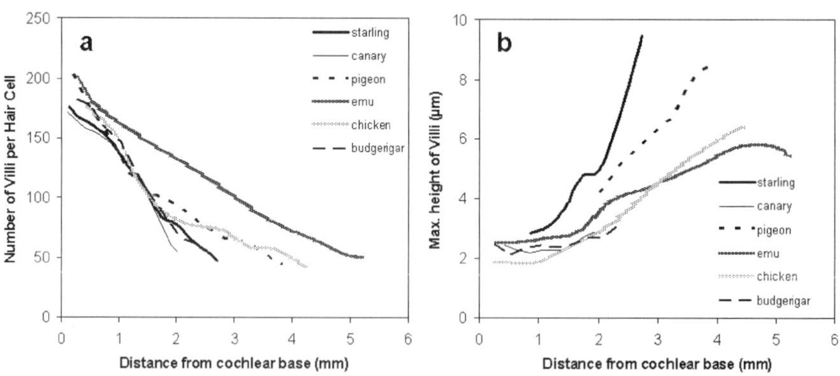

FIGURE 7.13. A graphical representation of the typical changes in two parameters determining the micromechanical properties (and thus also part of the frequency response) of avian hair-cell bundles as a function of the distance of the hair cell from the (high-frequency) base of the papilla basilaris. Thus the stereovillar bundles of basal hair cells contain many stereovilli (**a**), and these stereovilli are shorter (**b**) than in the apex. (After Gleich et al. 2004. Used with permission.)

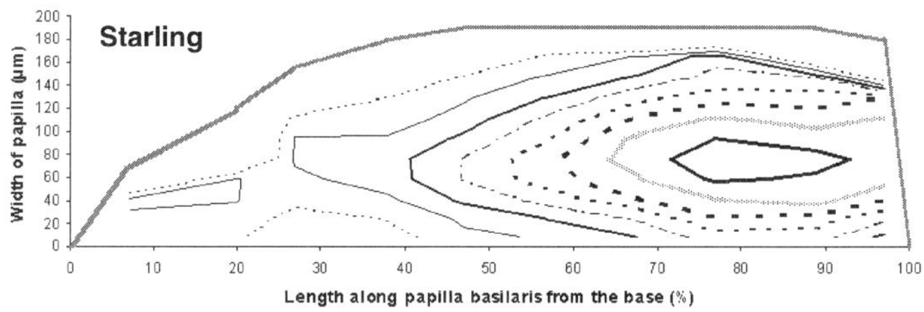

FIGURE 7.14. An example of the isoorientation contours found in an avian basilar papilla, in this case, the European starling. In the schematic plot, the positions of contours are given with reference to the length of the papilla from the basal end and the width of the papilla, which gets wider toward the apex (*right*). The angles given indicate whether the hair-cell bundles at the particular locations are oriented parallel to the neural axis (in this case, the *x*-axis of the graph) or are turned at various angles, up to 70°, toward the apex. There is thus a patch of hair cells in the apical third of the papilla that are rotated at least 70°. (Modified after Gleich and Manley 2000. Used with permission.)

Regarding the role of micromechanical tuning in frequency selectivity, a number of structures may interact in complex ways. The passive mechanical characteristics of the hair-cell stereovillar bundles (as determined, for example, by the number and height of stereovilli) and of the tectorial membrane (stereovillar and tectorial dimensions vary along the papilla) and/or active motile processes in hair cells may all play a role. As shown above, there are several strong morphological gradients in all avian papillae, and these are certainly responsible for much of the gradient of frequency responses along the epithelium.

Rate-intensity responses of primary afferents in *Dromaius novaehollandiae* (Yates et al. 2000) and spontaneous otoacoustic emissions in *Tyto alba* (Taschenberger and Manley 1997) indicated that birds have a cochlear amplifier. Also, spontaneous oscillations of avian hair cell bundles have been observed (Hudspeth et al. 2000). Preferred intervals in spontaneous neural firing may reflect spontaneous oscillations of the hair-cell bundles. The morphological and physiological data suggest that there are specializations of hair cells for active motion on the one hand (short hair cells that lack an afferent innervation) and of tall hair cells as receptor elements on the other (Manley 1995). In our study of hair cells of the domestic chicken *Gallus gallus*, isolated hair cells displayed bundle movements or shape changes upon current injection (Brix and Manley 1994). However, due to the small size of the hair cells, it was not possible to investigate whether *Gallus* hair cells are capable of fast movements at auditory frequencies.

The most common response pattern of avian primary afferent fibers to sound is an increase of the discharge rate. Maximum discharge rates in birds are on average higher than those of mammals for fibers at the same characteristic frequency. In higher-CF afferent fibers of *Dromaius novaehollandiae* (i.e., out of the phase-locking range), rate-level functions at CF were almost invariably of the sloping-saturation type (Köppl et al. 1997), which can be interpreted as indicating that a compressive nonlinearity sets in above a certain sound-pressure level, the "break point." The break points of a large sample of fibers correlated very tightly with fiber sensitivity (Yates et al. 2000). This, and data from *Tyto alba* (Köppl and Yates 1999), suggests that the nonlinearity associated with the cochlear amplifier in birds is local in nature and not global as in mammals (Köppl et al. 1997). The mechanical output of the cochlear amplifier in birds is thus not fed into the motion of the basilar membrane on a large scale but is retained within the local papillar–tectorial complex.

Avian auditory-nerve fibers generally respond to only a restricted frequency range with a modulation of their discharge rate, that is, they are frequency selective. Avian excitatory tuning curves are V-shaped and—on a logarithmic frequency scale—roughly symmetrical around the CF (Manley et al. 1985). The average frequency selectivity increases with the CF, and the frequency selectivity of avian auditory fibers tends to be higher at the equivalent CF than in mammals (Figure 7.15).

5.2 SOAEs in Birds

There is one reported case of a single SOAE in *Gallus gallus* (Burkard et al. 1996). Measured under urethane anesthesia, this emission had a frequency of 3.15 kHz and an amplitude of 3 to 4 dB SPL. The suppression tuning curve was V-shaped and sharply tuned. The SOAE disappeared within 1 min of hypoxia, but did not return.

The only detailed study of avian SOAEs was reported from *Tyto alba* (Taschenberger and Manley 1997). SOAEs were found in 79% of the ears, each emitting ear generating on average 1.9 emissions with center frequencies

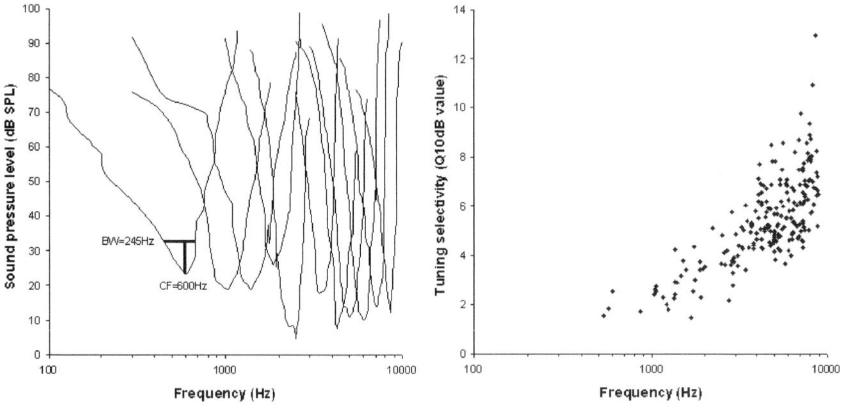

FIGURE 7.15. The frequency selectivity of auditory-nerve fibers increases with the characteristic frequency of the fiber. (**a**) Frequency-threshold tuning curves for 11 auditory-nerve fibers in the barn owl *Tyto alba*. For the lowest-frequency tuning curve, the calculation of the Q_{10dB} value is shown. Here, the CF of 600 Hz is divided by the frequency bandwidth 10 dB above the best threshold (bandwidth 245 Hz), producing a Q_{10dB} value of 2.45. (**b**) Q_{10dB} values for a larger sample of auditory nerve fibers in the barn owl. (From Köppl 1997. Used with permission.)

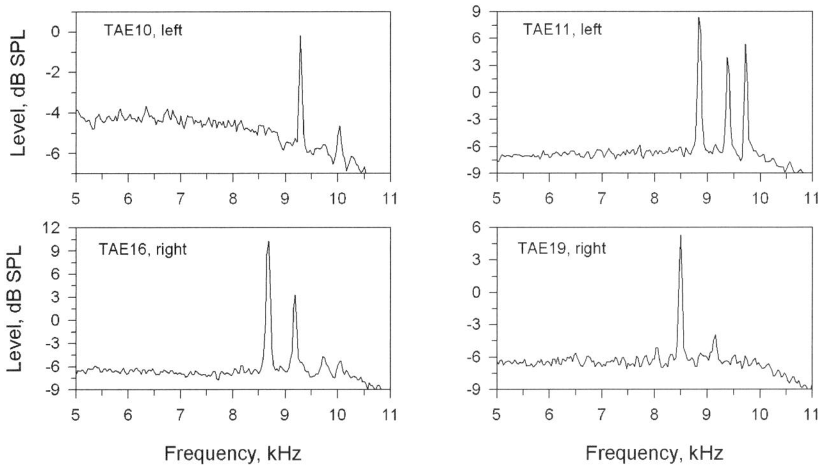

FIGURE 7.16. Four example spectra from the external ear canals of four barn owls, *Tyto alba,* at normal body temperature. The scales in all cases are the same, and it can be seen that some SOAE peaks are small (10, left ear) whereas others can be quite large (16, right ear). It can also be seen that all peaks in these spectra lie within the frequency range of the auditory fovea of this species (5–10 kHz). (After Taschenberger and Manley 1997. Used with permission.)

between 2.3 and 10.5 kHz and peak levels between 5.8 and 10.3 dB (examples in Fig. 7.16). Interestingly, almost all SOAEs had center frequencies above 7.5 kHz and thus within the frequency range of the specialized area of the auditory papilla previously described as being an auditory fovea (Köppl et al. 1993). The median of the frequency distances between neighboring SOAEs was 406 Hz or 0.058 oct. For SOAEs whose level exceeded 10 dB above the noise floor, the 3-dB bandwidths ranged between 4.5 and 11.4 Hz.

SOAE frequencies were also temperature sensitive. Raising the temperature shifted the emissions to higher frequencies, and vice versa (at on average 0.039 oct/°C). External tones could suppress the level of SOAEs, an effect that was highly frequency selective (Taschenberger and Manley 1997). For SOAEs with frequencies between 2.5 and 10.5 kHz, the Q_{10dB} values of 2-dB isosuppression tuning curves (STC) varied from 1.07 to 10.4. The most sensitive tips of the 2-dB STC were generally below 15 dB SPL. Thus the STC strongly resemble the characteristics of the most sensitive afferent nerve fibers (Fig. 7.17, compare Fig. 7.15; see also Köppl et al. 1997).

5.3 *Evoked Emissions in Archosaurs*

Three types of evoked OAEs have been studied in archosaurs. SEOAE were reported from the caiman, *Caiman crocodilus* (Klinke and Smolders 1984) and from a bird, *Sturnus vulgaris* (Manley et al. 1987). DPOAE were studied in

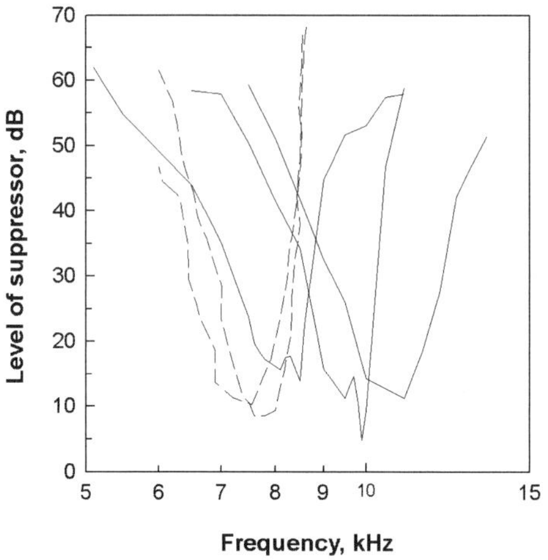

FIGURE 7.17. Suppression-tuning curves with criterion level −2 dB for five different SOAE peaks from different ears of the barn owl *Tyto alba*. The thresholds and tuning-curve selectivity strongly resemble those of primary neural tuning curves (compare Fig. 7.15). (After Taschenberger and Manley 1997. Used with permission.)

three species of birds: *Gallus gallus*, *Sturnus vulgaris* (Norton and Rubel 1990, Kettembeil et al. 1995), and *Tyto alba* (Taschenberger and Manley 1998).

5.3.1 Simultaneous-Evoked OAEs from the Caiman and the Starling

Klinke and Smolders (1984) recorded SFOAEs from *Caiman crocodilus* ear using swept low-level tones (10 dB SPL). They found irregularities in the sound pressure (pressure maxima with bandwidths of about 200 Hz followed by minima when measured in the frequency domain) that were not seen at higher sweep levels or in dead animals. These irregularities could be suppressed by a second, stationary tone. Tones most effectively suppressed irregularities that were observed for nearby frequencies. Similar SFOAE patterns were found in *Sturnus vulgaris* (Manley et al. 1987) using swept frequencies between 0.5 and 9 kHz and tone levels of 12 to 45 dB SPL. Irregularities in the low-level spectra were detected in 34 of 56 ears, and up to 9 were observed in a single ear. The center frequencies of the peaks lay between 1 kHz and 5.5 kHz, but mainly in the upper two-thirds of this range. Their (calculated) sound-pressure levels lay between −43 and +2 dB SPL. With increased sound pressure of the swept tone, the peak pressures of the SFOAE showed a highly nonlinear behavior, increasing over a dynamic range of 16 to 36 dB and then saturating. Interestingly, relatively light anesthesia was able to suppress SFOAEs in *Sturnus vulgaris*, an effect that was also studied using DPOAE in other birds (see later). In four cases, −6-dB isosuppression tuning curves were measured for SFOAE in *Sturnus*. Tones near the SFOAE frequency suppressed at a rate of about 0.75 dB/dB increase in tone level, whereas tones further away suppressed at lower rates, down to 0.2 dB/dB on the high-frequency side. STCs had CF between 2.30 and 5.45 kHz, and their Q10dB sharpness coefficients (from 2.6 to 4.9) compared favorably with those of auditory nerve fibers of *Sturnus* (Manley et al. 1985).

5.3.2 DPOAEs in the Chicken, Starling, and Barn Owl

Norton and Rubel (1990) compared distortion products of a mammal, the gerbil *Meriones unguiculates*, and of *Gallus gallus* at different developmental ages, and described similar basic response patterns (e.g., the shapes of I/O functions) in the two species. Differences between these species were found in the detection thresholds and maximal amplitude of the DPOAEs for the same levels of the primary tones. The authors attributed these differences either to structural differences in the middle ear of the two different species or functional and structural differences between the respective inner ears.

Kettembeil et al. (1995) reported DPOAE measurements from adult *Sturnus vulgaris* and 11-to-24-day-old *Gallus gallus*. Basic characteristics of $2f_1 - f_2$ and $2f_2 - f_1$ were measured in the ear canal of both awake and anaesthetized animals, including the effect of a third suppressive tone and the behavior of DPOAE under anesthesia. The levels of unsuppressed DPOAE varied from 0 to 22 dB SPL ($2f_1 - f_2$, *Sturnus*), 1 to 18 dB SPL ($2f_1 - f_2$, *Gallus*), and −4 to 13 dB SPL ($2f_2 - f_1$, *Sturnus*). DPOAEs have also been reported from adult *Gallus* by Burkard et al. (1996). In general, the behavior of the DPOAEs of both bird species resembled that seen in lizards and mammals, but, as noted by Norton and Rubel (1990), first appeared at somewhat higher primary-tone levels than in mammals.

The suppressive effect of a third tone was strongly frequency selective, the STCs for both $2f_1 - f_2$ and $2f_2 - f_1$ being V-shaped, and STCs for low-frequency primary tones were broader than those for high-frequency tones. The frequency-tuning selectivity of the STCs, even though measurable only over a narrow range of levels, was comparable to that known from *Sturnus* and *Gallus* auditory nerve fibers (Manley et al. 1985, 1991). In general, the CF of STCs for $2f_1 - f_2$ in both *Sturnus* and *Gallus* lay close to the frequency of f_1. By contrast, for the STCs of the DPOAE $2f_2 - f_1$ (measured only in *Sturnus*), the CF was either close to the frequency of f_2, or to that of the DP. Facilitation via a third tone was seen for both DPOAEs, often at levels below those eliciting suppression. For $2f_1 - f_2$, facilitation occurred at frequencies both below and above the center frequency (CF), whereas for $2f_2 - f_1$, facilitation appeared only in a frequency region above the CF. Where facilitation occurred, the flanks of the STC showed breaks and irregularities.

In *Sturnus* and *Gallus*, the output level of the DPOAE showed a remarkable degree of variability that correlated with the level of anesthesia (Kettembeil et al. 1995). As noted for SEOAE in *Sturnus*, deep anesthesia led to a complete loss of DPOAEs in both species. The peak at $2f_1 - f_2$ disappeared shortly after the onset of deep anesthesia and recovered to its original magnitude when the anesthesia was lightened, sometimes, however, with a considerable delay. In simultaneous measurements of DPOAEs and compound action potential (CAP) under halothane in *Sturnus*, the CAP was found to be more sensitive to anesthesia than the DPOAEs. It disappeared just before the DPOAEs and also reappeared later. In some cases, medium levels of anesthesia maintained quasistable states of below-normal DPOAEs and/or CAP levels. The effects observed were not the result of hypoxia and cannot all be explained as the result of pressure changes in the middle ear due to anesthesia.

The DPOAE $2f_1 - f_2$ has also been measured in *Tyto alba* (Taschenberger and Manley 1998). They were elicited by primary tones in 11 frequency regions from 1 to 9 kHz. The highest DPOAE levels and best thresholds were found using an f_1 of 1 kHz and for f_1 from 4 to 7 kHz. In some cases, the DPOAE sound pressures were only 37 dB below the primary-tone levels. At DPOAE threshold (−10 dB SPL), the highest values for the optimal f_2/f_1 frequency ratio (1.47) were measured for $f_1 = 5$ kHz. The regions of high DPOAE output correspond to areas of known anatomical specializations, especially the frequency region of *Tyto*'s auditory fovea from about 5 to 10 kHz (Fischer 1994; Köppl et al. 1993). At lower and higher f_1, the value of the most effective ratio gradually decreased, and the smallest optimal ratios were measured at the lowest (1 kHz) and highest f_1 (9 kHz).

STCs for the DPOAE $2f_1 - f_2$ in *Tyto* were generally V-shaped with a single minimum—the center frequency (CF). STCs from six different frequency regions are illustrated in Fig. 7.18. As in other nonmammals, the CFs were near the f_1 frequency, suggesting DPOAE generation near the characteristic place of this primary tone. Tuning sharpness of STCs becomes more sharply tuned with increasing suppression criterion. On average, the highest Q_{10dB} values were obtained for the suppression criterion of 12 dB, where it increased with the frequency of the f_1 primary tone and ranged from 1.16 ($f_1 = 2$ kHz) to 15.83 ($f_1 = 9$ kHz). However, the frequency tuning of DPOAE-STC was generally below the tuning sharpness of nerve fibers (Taschenberger and Manley 1998).

DPOAEs are thought to be generated in that region of the cochlea where both primary tones overlap. The optimal frequency separation between primary tones could therefore provide information about the frequency tuning of *Tyto*'s cochlea, which is an interesting question, given the fact that *Tyto*'s cochlea shows an auditory fovea (Köppl et al. 1998). The optimal f_2/f_1 ranged up to 1.45,

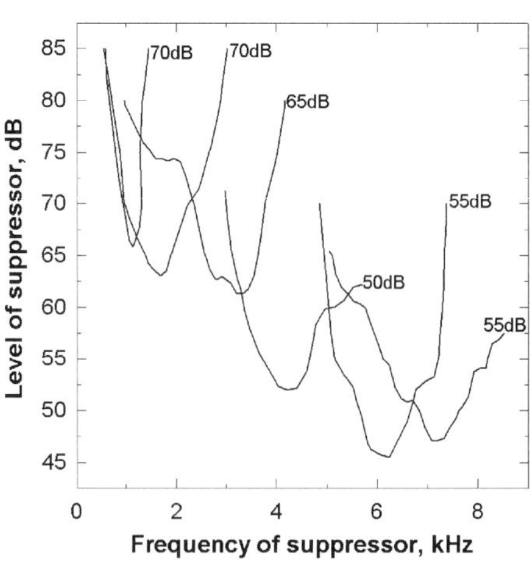

FIGURE 7.18. Examples of 2-dB suppression tuning curves for the DPOAE $2f_1 - f_2$ based on six different combinations of f_1 and f_2. In each case, the most sensitive point on the curve is near the f_1 frequency, and the primary-frequency ratio lay between 1.2 and 1.25 in all six cases. The level of the primary tones is given at the top right of each curve. (After Taschenberger and Manley 1998. Used with permission.)

indicating that the frequency tuning is not exceptional. These data confirm what has previously been seen in auditory nerve-fiber recordings from *Tyto* (Köppl et al. 1997).

6. Mechanisms of OAE in Nonmammalian Species

6.1 The Nature of the Active Process in Nonmammals: Evidence from OAE

Previous attempts to induce somatic motility in bird hair cells (it has not yet been attempted in lizards) were unsuccessful (Brix and Manley 1994; He et al. 2003). In some hair-cell types, such as the tiny SHC of the base of the cochlea of *Tyto alba* (height 3 to 4 μm), a somatic motility is hardly conceivable. In addition, somatic motility in mammals is observed only in the specialized outer hair cells, and hair cells with such specializations do not occur in nonmammals. Hair-cell membranes of geckos and barn owls, both of which are known to produce SOAEs, do not show the high densities of membrane particles that are typical of mammalian outer hair cells (Köppl et al. 2003, 2004). On the other hand, an active motility of the stereovillar bundle of *Gallus* hair cells has been reported (Hudspeth et al. 2000).

In addition to these indicators, the characteristics of nonmammalian OAEs provide some evidence on the nature of the active process in these hair cells. The acoustic modulation of lizard EEOAEs (Manley et al. 2001), for example, can be easily explained only if the active process is in the hair-cell bundle (see preceding text). It should perhaps be emphasized that in contrast to EEOAE generation, under normal acoustic stimulation, the activation of bundle motors in the two oppositely oriented hair-cell populations of lizards would not be simultaneous. Thus the presumed contributions to amplification of the two hair-cell groups would not cancel, but alternate in a push–pull fashion and thus add constructively. An active process in the bundle would thus be very suitable for amplifying low-level sound in lizards, where during sound stimulation the auditory papilla rocks from side to side (review in Manley et al. 1990). Since the lizard basilar membrane shows no frequency selectivity, the high selectivity of the hair cells and afferent nerve fibers must result from the local micromechanics of the papilla and tectorial membrane (Manley 1988, 1989). The ease with which active bundle movements could modify such papillar oscillations may indeed explain the unique presence of oppositely oriented hair-cell groups in all lizards (Miller 1980) and in the amphibian basilar papilla (Lewis et al. 1982).

6.2 Comparison to Humans and Other Mammals

One motivation for a comparative approach to hearing research is that it enhances understanding of the relationship between structure and function of the ear. If certain functional characteristics across species correlate with the presence of particular structures, it may be concluded that these structures are important for that particular function. For example, the presence of a few (two to four) peaks in emission spectra of amphibians and certain lizards correlates with the presence of a continuous tectorial membrane that covers their hair cells. In contrast, lizards with a tectorial membrane that consists of subunits ("sallets") have a larger number of emission peaks over the same frequency range. Apparently, the structure, and thus presumably the mechanical properties, of the tectorial membrane are important for the properties of otoacoustic emissions in these ears.

This example also illustrates that differences in otoacoustic emission properties do not necessarily mean that the active mechanism that generates the emission is different. It is highly unlikely that the hair-cell active mechanism that underlies otoacoustic emission generation is different for different lizards. Thus, although the emission patterns differ significantly, the cellular hair-cell mechanisms are presumably the same.

In spite of the differences that exist between species, however, the similarities are numerous and striking. These include:

1. All SOAEs are physiologically vulnerable.
2. Each ear shows a specific pattern of SOAE peak frequencies that is—with the exception of lizards lacking a tectorial structure—maintained over long periods of time.
3. SOAEs, DPOAEs, and SEOAEs are suppressible in a frequency-selective fashion, whereby the form of the STC strongly resembles that of threshold tuning of primary auditory nerve fibers.
4. The SOAE center frequency is temperature-dependent.
5. The statistical characteristics of the SOAE amplitude distributions indicate an origin in a sinusoidal oscillator (Bialek and Wit 1984; van Dijk et al. 1989; 1996b).
6. Emission–emission interactions occur as mutual suppression or as mutual facilitation in approximately the same number of emission pairs in both lizards (van Dijk et al. 1998) and humans (van Dijk et al. 1998). In spite of the fact that lizard papillae rarely exceed 2 mm in length, the octave distance between SOAE peaks that show correlated amplitude fluctuations is essentially the same in lizards and humans (van Dijk et al. 1998).
7. DPOAE amplitude grows with a rate less than 1 dB/dB for low-level stimulus tones. The only known exception are DPOAEs from the frog basilar papilla, which originate from passive nonlinear responses (Meenderink and Van Dijk 2006).
8. DPOAE amplitude for low-level stimulus levels depends on temperature. Here again, the frog basilar papilla is an exception.

In addition to the above points, SOAE amplitudes and frequencies are sensitive to the state of the middle ear (see Manley 2006a for references). Changes to the characteristics of the middle ear due to, for example, differential pressure across the tympanum can change both amplitude and level of SOAEs. For amplitude this is hardly surprising, as amplitude is obviously affected by the sound transmission through the middle ear. The effect on the frequency of SOAEs indicates that the oscillating components of the inner ear in some way interact with the middle ear. This is consistent with two views on the SOAE generation mechanism, which can be viewed as two extremes of a spectrum of possibilities: (1) If SOAEs set up standing waves between the middle ear window (round window in mammals) and the place of the characteristic frequency, the frequency effect may reflect the effect the middle ear has on the impedence at the middle ear window, and thus on the frequency of the SOAEs (e.g., Shera 2003). (2) Alternatively, if an SOAE frequency is primarily determined by a local resonance mechanism of an oscillating hair-cell group, the SOAE characteristics, including oscillation frequency, will depend on the acoustic load on the oscillator, and the middle ear is part of that load.

It is possible that the two mechanisms account for the difference between the high frequency stability of SOAEs in humans on the one hand, and the much weaker stability in nonmammalian species on the other hand. For standing-wave SOAEs, which have specifically been suggested for humans (e.g., Shera 2003), the strict standing-wave conditions determine the SOAE frequency. This leads to very stable inner ear oscillation (Bialek and Wit 1984; van Dijk et al. 1998). In contrast, when a local feedback mechanism controls SOAE generation, the frequency stability could be less. This situation apparently occurs in nonmammalian vertebrates (van Dijk et al. 1989), nonhuman mammals (Ohyama et al. 1991, 1992), and some human cases (Whitehead et al. 1993).

One of the few differences between the SOAE patterns in nonmammals and mammals is in the STC. In mammals, the best frequency of the STC is displaced to a frequency above the center frequency of the SOAEs. In nonmammals, by contrast, the emission center frequency and the best frequency of the STC match well. This difference is almost certainly the result of the involvement of the basilar membrane in the movement of the mammalian organ of Corti and the strongly asymmetrical shape of its traveling wave. A similar difference across species is seen in the DPOAE frequency that is most affected by suppression. In mammals, the characteristics of STCs for the DPOAEs $2f_1 - f_2$ suggest generation of this distortion product either near the geometric mean of both primary-tone frequencies or near f_2 (e.g., Kummer et al. 1995). In nonmammals, the best sensitivity of STC for the DPOAE $2f_1 - f_2$ lies near the frequency of the f_1 primary tone (Köppl and Manley 1993b; Kettembeil et al. 1995).

In addition to the above, the relationship between DPOAE amplitude and the ratio and frequency of the stimulus tones is markedly different between mammals (Knight and Kemp 2000) and frogs (Meenderink et al. Meenderink and van dijk 2005; see Fig. 7.5). Again, the traveling wave on the basilar membrane presumably accounts for the pattern observed in humans, while there is no evidence for such a wave in frogs and lizards.

Across species, there is a correspondence between STCs of otoacoustic emissions and neural threshold tuning curves. For example, in *Tyto alba*, a transition from more shallow to steeper STCs was observed with increasing frequency (Taschenberger and Manley 1998). This tendency has also been described for other species and resembles the increasing frequency selectivity of auditory nerve fibers toward higher frequencies (e.g., Köppl et al. 1997; Köppl and Manley 1993b). The use of significantly smaller f_2/f_1 yielded notably higher Q_{10dB} values of STC in *Tyto*. The highest tuning of STC at the smallest primary-tone frequency ratios was also observed in humans by Brown and Kemp (1984). Taken together, one can conclude that at low primary-tone levels and small f_2/f_1 frequency ratios, and with the use of a sensitive suppression criterion, the tuning of DPOAE-STCs will be similar to neuronal tuning curves in both mammals and nonmammals. Some of the differences seen between neural and DPOAE-STC tuning selectivity undoubtedly results from the wider excitation pattern generated by the primary tones than by the single, threshold-level tones used to study neural tuning.

Data reported to date fail to show an anesthesia effect on OAEs in mammals equivalent to that in birds. The loss of DPOAEs in birds occurs after neural transmission in the auditory nerve has either been so desynchronized that a CAP (Compound action potential) measurement is no longer possible or, more likely, has ceased completely. Thus anesthesia affects the hair-cell synapses before the hair cells cease to transduce sound and lose the ability to generate DPOAEs. A similar effect of anesthesia was discussed by Köppl et al. (1990) for primary auditory nerve fibers in *Tiliqua* in which at deep levels of anesthesia, no responses can be recorded from nerve fibers. Avian and lizard hair-cell function is more physiologically vulnerable than that of mammals. Indeed, in mammals, evoked OAE become *larger* under anesthesia, an effect that is attributed to the suppression of efferent input to the cochlea by anesthesia (Harel et al. 1997). While the different anesthesia sensitivities of these animal groups may seem remarkable, this difference is only one of several that have been reported between nonmammalian and mammalian hearing systems (e.g., Manley 1990; Klinke and Smolders 1993). The differential effect of anesthesia on otoacoustic emissions supports the notion that despite widespread similarities in OAEs of vertebrates across species, some fundamental metabolic processes of hair cells do differ between mammals and nonmammals.

Finally, a difference across species exists in the distribution of SOAE frequencies across the hearing range of the animal. In humans, SOAE are fairly randomly distributed across the auditory frequency range, but a minimum spacing between peaks exists (Schloth and Zwicker 1983; Talmadge et al. 1993). In frogs and lizards with a continuous tectorial membrane, only a couple of emissions peaks are present (see earlier). In lizards with a tectorial membrane that consists of sallets, many SOAE peaks occur with a relatively regular frequency spacing (Köppl and Manley 1993a).

7. Summary

OAE can be measured in all nonmammalian vertebrates and especially SOAE patterns reveal interesting and group-specific differences that can be readily compared to the patterns observed in humans and other mammals. The main differences in OAE patterns across taxonomic boundaries presumably can be accounted for by the different structural and macromechanical properties of the respective hearing epithelia. Based on the available evidence, there is no reason to believe that the active hair-cell mechanism is different across species. Current evidence points to a single mechanism in nonmammals (active stereovillar bundles), whereas both this mechanism and somatic motility may be involved in mammals. Whatever our future view on active mechanisms will be, it will have to account for the broad similarities across species of the characteristics of its primary indicator, otoacoustic emissions.

Acknowledgments. We thank Christine Köppl, Ulrike Sienknecht, and Rick Schoffelen for comments on an earlier version of the manuscript.

References

Baker RJ, Wilson JP, Whitehead ML (1989) Otoacoustic evidence for nonlinear behaviour in frog's hearing: suppression but no distortion products. In: Wilson P, Kemp DT (eds), Cochlear Mechanisms: Structure, Function and Models. New York: Plenum Press, pp. 349–356.

Benedix Jr JH, Pedemonte M, Velluti R, and Narins PM (1994) Temperature dependence of two-tone rate suppression in the northern leopard frog, *Rana pipiens pipiens*. J Acoust Soc Am 96:2738–2745.

Bialek W, Wit HP (1984) Quantum limits to oscillator stability: theory and experiments on acoustic emissions from the human ear. Phys Let 104A:173–178.

Brix J, Manley GA (1994) Mechanical and electromechanical properties of the hair-cell bundles of isolated and cultured hair cells of the chicken. Hear Res 76:147–157.

Brown AM, Kemp DT (1984) Suppressibility of the $2f_1$ – fs stimulated acoustic emissions in gerbil and man. Hear Res 13:29–37.

Burkard R, Salvi R, Chen L (1996) 2f1–f2 distortion product otoacoustic emissions in White Leghorn chickens (*Gallus domesticus*): effects of frequency ratio and relative level. Audiol Neurootol 1996 1:197–213.

Capranica RR, Moffat AJM (1975) Selectivity of the peripheral auditory system of spadefoot toads (*Scaphiopus couchi*) for sounds of biological significance. J Comp Physiol 100:231–249.

Capranica RR, Moffat AJM (1980) Nonlinear properties of the peripheral auditory system of anurans. In: A. N. Popper and R. R. Fay (eds), Comparative Studies of Hearing in Vertebrates. Berlin: Springer-Verlag, pp. 139–165.

Capranica RR, Moffat, AJM (1983) Neurobehavioural correlates of sound communication in anurans. In: Ewert JP, Capranica RR, Ingle DJ (eds), Advances in Vertebrate Neuroethology. New York: Plenum Press, pp. 701–730.

Carroll RL (1988) Vertebrate Paleontology and Evolution. New York: Freeman.

Chan DK, Hudspeth AJ (2005) Ca^{2+} current-driven nonlinear amplification by the mammalian cochlea in vitro. Nat Neurosci 8:149–155.

Crawford AC, Fettiplace R (1985) The mechanical properties of ciliary bundles of turtle cochlear hair cells. J Physiol 364:359–379.

Ehret G, Capranica RR (1980) Masking patterns and filter characteristics of auditory nerve fibers in the green treefrog (*Hyla cinerea*). J Comp Physiol 141:1–12.

Evans EF (1975) The sharpening of cochlear frequency selectivity in the normal and abnormal cochlea. Audiology 14:419–442.

Fischer FP (1994) Quantitative TEM analysis of the barn owl basilar papilla. Hear Res 73:1–15.

Fox JH (1995) Morphological correlates of auditory sensitivity in anuran amphibians. Brain Behav Evol 45:327–338.

Frishkopf LS, Flock Å (1974) Ultrastructure of the basilar papilla, an auditory organ in the bullfrog. Acta Otolaryngol 77:176–184.

Gallo L (1997) Otoakustische Emissionen bei Reptilien. Doctoral dissertation, Technische Universität München.

Gleich O, Manley GA (2000) The hearing organ of birds and *Crocodilia*. In: Comparative Hearing: Birds and Reptiles, Dooling R, Popper AN, Fay RR (eds) Evolution of the Vertebrate Auditory System. New York: Springer-Verlag, pp. 70–138.

Gleich O, Fischer FP, Köppl C, Manley GA (2004) Hearing organ evolution and specialization: Archosaurs. In: Manley GA, Popper A, Fay RR (eds) Evolution of the Vertebrate Auditory System. New York: Springer-Verlag, pp. 224–255.

Harel N, Kakigi A, Hirakawa H, Mount RJ, Harrsion RV (1997) The effects of anesthesia on otoacoustic emissions. Hear Res 110:25–33.

He DZZ, Beisel KW, Chen L, Ding DL, Jia S, Fritzsch B, Salvi R (2003) Chick hair cells do not exhibit voltage-dependent somatic motility. J Physiol 546:511–520.

Hetherington TE, Lindquist ED (1999) Lung-based hearing in an "earless" anuran amphibian. J Comp Physiol A 184:395–401.

Hubbard AE, Mountain DC (1983) Alternating current delivered into the scala media alters sound pressure at the eardrum. Science 222:510–512.

Hudspeth AJ, Choe Y, Mehta AD, Martin P (2000) Putting ion channels to work: mechanoelectrical transduction, adaptation, and amplification by hair cells. Proc Natl Acad Sci USA 97:11765–11772.

Kennedy HJ, Evans MG, Crawford AC, Fettiplace R (2003) Fast adaptation of mechanoelectrical transducer channels in mammalian cochlear hair cells. Nat Neurosci 6: 832–836.

Kettembeil S, Manley GA, Siegl E (1995) Distortion-product otoacoustic emissions and their anaesthesia sensitivity in the European starling and the chicken. Hear Res 86: 47–62.

Klinke R, Smolders J (1984) Hearing mecanisms in caiman and pigeon. In: Bolis L, Keynes R, Maddrell S (eds) Comparative Physiology of Sensory Systems. Cambridge, UK: Cambridge University Press, pp. 195–211.

Klinke R, Smolders JWTh (1993) Performance of the avian inner ear, In: Allum JHJ, Allum-Mecklenburg DJ, Harris FP, Probst R. (eds) Progress in Brain Research, Vol. 97, pp. 31–43.

Knight RD, Kemp DT (2000) Indications of different distortion product otoacoustic emission mechanisms from a detailed f_1, f_2 area study. J Acoust Soc Am 107:457–473.

Köppl C (1995) Otoacoustic emissions as indicators of active cochlear mechanics: a primitive property of vertebrate auditory organs. In: Manley GA, Klump GM, Köppl C, Fastl H, Oeckinghaus H (eds), Advances in Hearing Research. Hackensack, NJ: World Scientific, pp. 200–209.

Köppl C (1997) Frequency tuning and spontaneous activity in the auditory nerve and cochlear nucleus magnocellularis of the barn owl, *Tyto alba*. J Neurophysiol 77: 364–377.

Köppl C, Authier S (1995) Quantitative anatomical basis for a model of micromechanical frequency tuning in the Tokay gecko, *Gekko gecko*. Hear Res 82:14–25.

Köppl C, Manley GA (1992) Functional consequences of morphological trends in the evolution of lizard hearing organs. In: Fay, RR, Popper AN, Webster DB (eds) The Evolutionary Biology of Hearing. New York: Springer-Verlag, pp. 489–509.

Köppl C, Manley GA (1993a) Spontaneous otoacoustic emissions in the bobtail lizard. I: General characteristics. Hear Res 71:157–169.

Köppl C, Manley GA (1993b) Distortion-product otoacoustic emissions in the bobtail lizard. II: Suppression tuning characteristics. J Acoust Soc Am 93:2834–2844.

Köppl C, Manley GA (1994) Spontaneous otoacoustic emissions in the bobtail lizard. II: Interactions with external tones. Hear Res 72:159–170.

Köppl C, Manley GA (2000) Anatomy of the cochlea and physiology of auditory afferents in lizards. In: Manley GA, Fastl H, Kössl M, Oeckinghaus H, Klump GM (eds), Auditory Worlds: Sensory Analysis and Perception in Animals and Man. Weinheim: Wiley-VCH, pp. 52–59.

Köppl C, Yates GK (1999) Coding of sound pressure level in the barn owl's auditory nerve. J Neurosci 19:9674–9686.

Köppl C, Manley GA, Johnstone BM (1990) Peripheral auditory processing in the bobtail lizard *Tiliqua rugosa*: V. Seasonal effects of anesthesia. J Comp Physiol A 167:139–144.

Köppl C, Gleich O, Manley GA (1993) An auditory fovea in the barn owl cochlea. J Comp Physiol A 171:695–704.

Köppl C, Yates GK, Manley GA (1997) The mechanics of the avian cochlea: Rate-intensity functions of auditory-nerve fibres in the emu. In: Lewis ER, Long GR, Lyon RF, Narins PM, Steele CR (eds) Diversity in Auditory Mechanics. Singapore: World Scientific, pp. 76–82.

Köppl C, Klump GM, Taschenberger G, Dyson M, Manley GA (1998) The auditory fovea of the barn owl—no correlation with enhanced frequency resolution. In: Palmer AR, Rees A, Summerfield AQ, Meddis R, (eds) Psychophysical and Physiological Advances in Hearing. London: Whurr, pp. 153–159.

Köppl C, Forge A, Manley GA (2003) No correlates for somatic motility in freeze-fractured hair-cell membranes of lizards and birds. In: Gummer AW, Dalhoff E, Nowotny M, Scherer MP (eds) Biophysics of the Cochlea: From Molecules to Models. Singapore: World Scientific, pp. 185–186.

Köppl C, Forge A, Manley GA (2004) Low density of membrane particles in auditory hair cells of lizards and birds suggests an absence of somatic motility. J Comp Neurol 479:149–155.

Kummer P, Janssen T, Arnold W (1995) Suppression tuning characteristics of the $2f_1 - f_2$ distortion-product otoacoustic emission in humans. J Acoust Soc Am 98:197–210.

Lee, M.S.Y. (1998) Convergent evolution and character correlation in burrowing reptiles: towards a resolution of squamate relationships. Biol J Linn Soc 65:369–453.

Lewis ER (1981) Suggested evolution of tonotopic organization in the frog amphibian papilla. Neurosci Let 21:131–136.

Lewis ER (1984) On the frog amphibian papilla. Scan Electr Microsc IV:1899–1913.

Lewis ER, Leverenz EL, Koyama H (1982) The tonotopic organization of the bullfrog amphibian papilla, an auditory organ lacking a basilar membrane. J Comp Physiol 145:437–445.

Lombard RE, Straughan, IR (1974) Functional aspects of anuran middle ear structures. J Exp Biol 61:71–93.

Lombard RE, Lewis ER, Narins PM (1999) The acoustic periphery of amphibians. In: Fay RR, Narins PM (eds) Comparative Hearing: Fish and Amphibians. Berlin: Springer-Verlag, pp. 101–154.
Long GR, Van Dijk P, Wit HP (1996) Temperature dependence of spontaneous otoacoustic emissions in the edible frog (*Rana esculenta*). Hear Res 98:22–28.
Maat B, Wit HP, Van Dijk P (2000) Noise-evoked otoacoustic emissions in humans. J Acoust Soc Am 108:2272–2280.
Manley GA (1979) Preferred intervals in the spontaneous activity of primary auditory neurones. Die Naturwiss 66:582–583.
Manley GA (1986) The evolution of the mechanisms of frequency selectivity in vertebrates. In: Moore B, Patterson R (eds) Auditory Frequency Selectivity. New York: Plenum Press, pp. 63–72.
Manley GA (1990) Peripheral Hearing Mechanisms in Reptiles and Birds. Heidelberg: Springer-Verlag.
Manley GA (1995) The avian hearing organ: a status report. In: Manley GA, Klump GM, Köppl C, Fastl H, Oeckinghaus H (eds) Advances in Hearing Research. Singapore: World Scientific, pp. 219–229.
Manley GA (1997) Diversity in hearing-organ structure and the characteristics of spontaneous otoacoustic emissions in lizards. In: Lewis ER, Long GR, Lyon RF, Narins PM, Steele CR (eds) Diversity in Auditory Mechanics. Singapore: World Scientific, pp. 32–38.
Manley GA (2000a) Do non-mammals have a cochlear amplifier? In: Wada H, Takasaka T, Ohyama K, Ikeda K, Koike T (eds) Recent Developments in Auditory Mechanics. Singapore: World Scientific, pp. 499–505.
Manley GA (2000b) Cochlear mechanisms from a phylogenetic viewpoint. Proc Natl Acad Sci USA 97:11736–11743.
Manley GA (2000c) Otoacoustic emissions in lizards In: Manley, GA, Fastl H, Kössl M, Oeckinghaus H, Klump GM (eds) Auditory Worlds: Sensory Analysis and Perception in Animals and Man. Weinheim: Wiley-VCH, pp. 93–102.
Manley GA (2000d) The hearing organs of lizards. In: Dooling R, Popper AN, Fay RR (eds) Comparative Hearing: Birds and Reptiles. New York: Springer, pp. 139–196.
Manley GA (2001) Evidence for an active process and a cochlear amplifier in nonmammals. J Neurophysiol 86:541–549.
Manley GA (2002) Evolution of structure and function of the hearing organ of lizards. J Neurobiol 53:202–211.
Manley GA (2003) The tectorial membrane stabilizes spontaneous otoacoustic emissions. In: Gummer AW, Dalhoff E, Nowotny M, Scherer MP (eds) Biophysics of the Cochlea: From Molecules to Models. Singapore: World Scientific, pp. 480–487.
Manley GA (2004a) The lizard basilar papilla and its evolution. In: Manley GA, Popper A, Fay RR (eds), Evolution of the Vertebrate Auditory System. New York: Springer-Verlag, pp. 200–223.
Manley GA (2004b) Spontaneous otoacoustic emissions in monitor lizards. Hear Res 189:41–57.
Manley G. A (2006a) Spontaneous otoacoustic emissions in lizards, air pressure effects on them and the question of point sources and global standing waves. In: Nuttall, A, Ren T, Gillespie, P, Grosh, K, de Boer, E (eds) Auditory Mechanisms: Processes and Models. Singapore: World Scientific. pp. 369–376.
Manley, G. A. (2006b) Spontaneous otoacoustic emissions from free-standing stereovillar bundles of ten species of lizard with papillae. Hearing Research 212, 33–47.

Manley GA, Clack JA (2004) An outline of the evolution of vertebrate hearing organs. In: Manley GA, Popper A, Fay RR (eds) Evolution of the Vertebrate Auditory System. New York: Springer-Verlag, pp. 1–26.

Manley GA, Gallo L (1997) Otoacoustic emissions, hair cells and myosin motors. J Acoust Soc Am 102:1049–1055.

Manley GA, Gleich O (1984) Avian primary auditory neurons: the relationship between characteristic frequency and preferred intervals. Die Naturwiss 71:592–594.

Manley GA, Gleich O (1992) Evolution and specialization of function in the avian auditory periphery. In: Fay RR Popper AN, Webster DB (eds) The Evolutionary Biology of Hearing. New York: Springer-Verlag, pp. 561–580.

Manley GA, Kirk D (2002) The influence of injected ac and dc current on spontaneous otoacoustic emissions in the bobtail lizard. J Assoc Res Otolaryngol 3:200–208.

Manley GA, Köppl C (1992) A quantitative comparison of peripheral tuning measures: Primary afferent tuning curves versus suppression tuning curves of spontaneous and distortion-product otoacoustic emissions. In: Cazals Y, Demany L, Horner K (eds) Auditory Physiology and Perception. Oxford: Pergamon, pp. 151–157.

Manley GA, Köppl C (1994) Spontaneous otoacoustic emissions in the bobtail lizard. III: Temperature effects. Hear Res 72:171–180.

Manley GA, Köppl C (1998) Phylogenetic development of the cochlea and its innervation. Curr Opin Neurobiol 8:468–474.

Manley GA, Gleich O, Leppelsack HJ, Oeckinghaus H (1985) Activity patterns of cochlear ganglion neurons. J Comp Physiol A 157:161–181.

Manley GA, Schulze M, Oeckinghaus H (1987) Otoacoustic emissions in a song bird. Hear Res 26:257–266.

Manley GA, Yates G, Köppl C (1988) Auditory peripheral tuning: evidence for a simple resonance phenomenon in the lizard *Tiliqua*. Hear Res 33:181–190.

Manley GA, Köppl C, Yates GK (1989) Micromechanical basis of high-frequency tuning in the bobtail lizard. In: Wilson JP, Kemp D (eds) Mechanics of Hearing. New York: Plenum Press, pp. 143–150.

Manley GA, Köppl C, Johnstone BM (1990) Peripheral auditory processing in the bobtail lizard *Tiliqua rugosa*: I. Frequency tuning of auditory-nerve fibres. J Comp Physiol A 167:89–99.

Manley GA, Kaiser A, Brix J, Gleich O (1991) Activity patterns of primary auditory-nerve fibres in chickens: development of fundamental properties. Hear Res 57:1–15.

Manley GA, Köppl C, Johnstone BM (1993) Distortion-product otoacoustic emissions in the bobtail lizard: I: General characteristics. J Acoust Soc Am 93:2820–2833.

Manley GA, Gallo L, Köppl C (1996) Spontaneous otoacoustic emissions in two gecko species, *Gekko gecko* and *Eublepharis macularius*. J Acoust Soc Am 99: 1588–1603.

Manley GA, Köppl C, Yates GK (1997) Activity of primary auditory neurones in the cochlear ganglion of the Emu *Dromaius novaehollandiae*: spontaneous discharge, frequency tuning and phase locking. J Acoust Soc Am 101:1560–1573.

Manley GA, Köppl C, Sneary M (1999) Reversed tonotopic map of the basilar papilla in *Gekko gecko*. Hear Res 131:107–116.

Manley GA, Kirk D, Köppl C, Yates GK (2001) In-vivo evidence for a cochlear amplifier in the hair-cell bundle of lizards. Proc Natl Acad Sci USA 98:2826–2831.

Meenderink SWF, van Dijk P (2004) Level dependence of distortion product otoacoustic emissions in the leopard frog, *Rana pipiens pipiens*. Hear Res 192:107–118.

Meenderink SWF, van Dijk P (2005) Characteristics of distortion product otoacoustic emissions in the frog from L_1, L_2-maps. J Acoust Soc Am 118, 279–286.

Meenderink SW, van Dijk P (2006) Temperature dependence of anuran distortion product otoacoustic emissions J. Assoc Res otolaryngol. 2006 Sep; 7(3): 246–52. on distortion product otoacoustic emissions in the frog. Hear Res In press

Meenderink SWF, Narins PM, Van Dijk P (2005a) Detailed f_1, f_2 area study of distortion product otoacoustic emissions in the frog. JARO 6:37–47.

Meenderink SWF, Van Dijk P, Narins PM (2005b) Comparison between distortion product otoacoustic emissions and nerve fiber responses from the basilar papilla of the frog. J Acoust Soc Am 117:3165–3173.

Miller MR (1980) The reptilian cochlear duct. In: Popper AN, Fay RR (eds) Comparative Studies of Hearing in Vertebrates. New York: Springer-Verlag, pp. 169–204.

Miller MR (1985) Quantitative studies of auditory hair cells and nerves in lizards. J Comp Neurol 232:1–24.

Miller MR (1992) The evolutionary implications of the structural variations in the auditory papilla of lizards. In: Webster DB, Fay RR, Popper AN (eds), The Evolutionary Biology of Hearing. New York: Springer-Verlag, pp. 463–487.

Norton SJ, Rubel EW (1990) Active and passive components in mammalian and avian ears, In: Dallos P, Geisler CD, Matthews JW, Ruggero MA, Steele CR (eds), The Mechanics and Biophysics of Hearing. New York: Springer-Verlag, pp. 219–226.

Ohyama K, Wada H, Kobayashi T, Takasaka T (1991) Spontaneous otoacoustic emissions in the guinea pig. Hear Res 56:111–121.

Ohyama K, Sato T, Wada H, Takasaka T (1992) Frequency instability of the spontaneous otoacoustic emissions in the guinea pig. Abstracts of the 15th Midwinter Meeting, Association for Research in Otolaryngology, p. 150.

Palmer AR, Wilson JP (1982) Spontaneous and evoked acoustic emissions in the frog *Rana esculenta*. J Physiol 324:66P.

Pitchford S, Ashmore JF (1987) An electrical resonance in hair cells of the amphibian papilla of the frog *Rana temporaria*. Hear Res 27:75–83.

Probst R, Lonsbury-Martin BL, Martin GK (1991) A review of otoacoustic emissions. J Acoust Soc Am 89:2027–2067.

Purgue AP, Narins PM (2000a) Mechanics of the inner ear of the bullfrog (*Rana catesbeiana*): the contact membranes and the periotic canal. J Comp Physol A 186:481–488.

Purgue AP, Narins PM (2000b) A model for energy flow in the inner ear of the bullfrog (*Rana catesbeiana*). J Comp Physiol A 186:489–495.

Ricci AJ, Crawford AC, Fettiplace R (2000) Active hair bundle motion linked to fast transducer adaptation in auditory hair cells. J Neurosci 20:7131–7142.

Rice SO (1954) Mathematical analysis of random noise. In: Wax N (ed), Selected Papers on Noise and Stochastic Processes. New York: Dover, pp. 133–294.

Ronken DA (1991) Spike discharge properties that are related to the characteristic frequency of single units in the frog auditory nerve. J Acoust Soc Am 90:2428–2440.

Rosowski JJ, Peake WT, White JR (1984) Cochlear nonlinearities inferred from two-tone distortion products in the ear canal of the alligator lizard. Hear Res 13:141–158.

Schloth E, Zwicker E (1983) Mechanical and acoustical influences on spontaneous otoacoustic emissions. Hear Res 11:285–293.

Shera CA (2003) Mammalian spontaneous otoacoustic emissions are amplitude-stabilized cochlear standing waves. J Acoust Soc Am 114:244–262.

Smotherman MS, Narins PM (1999) The electrical properties of auditory hair cells in the frog amphibian papilla. J Neurosci 19:5275–5292.

Sun W, Ding D, Reyes S, Salvi RJ. (2000) Effects of AC and DC stimulation on chinchilla SOAE amplitude and frequency. Hear Res 150:137–148.

Talmadge C, Long G, Murphy W, Tubis A (1993) New off-line method for detecting spontaneous otoacoustic emissions in human subjects. Hear Res 71:170–182.

Taschenberger G, Manley GA (1997) Spontaneous otoacoustic emissions in the barn owl. Hear Res 110:61–76.

Taschenberger G, Manley GA (1998) General characteristics and suppression tuning properties of the distortion-product otoacoustic emission $2f_1 - f_2$ in the barn owl. Hear Res 123:183–200.

van Dijk P, Manley GA (2001) Distortion product otoacoustic emissions in the tree frog *Hyla cinerea*. Hear Res 153:14–22.

van Dijk P, Wit HP (1987) Temperature dependence of frog spontaneous otoacoustic emissions. J Acoust Soc Am 82:2147–2150.

van Dijk P, Wit HP (1998) Correlated amplitude fluctuations of spontaneous otoacoustic emissions. J Acoust Soc Am 104:336–343.

van Dijk P, Wit HP, Segenhout JM (1989) Spontaneous otoacoustic emissions in the European edible frog (*Rana esculenta*): Spectral details and temperature dependence. Hear Res 42:273–282.

van Dijk P, Narins PM, Wang J (1996a) Spontaneous otoacoustic emissions in seven frog species. Hear Res 101:102–112.

van Dijk P, Manley GA, Gallo L, Pavusa A, Taschenberger G (1996b) Statistical properties of spontaneous otoacoustic emissions in one bird and three lizard species. J Acoust Soc Am 100:2220–2227.

van Dijk P, Wit HP, Segenhout JH (1997a) Dissecting the frog inner ear with Gaussian noise: I. Application of high-order Wiener-kernel analysis. Hear Res 114:229–242.

van Dijk P, Maat A, Wit HP (1997b) Wiener kernel analysis of a noise-evoked otoacoustic emission. Br J Aud 31:473–477.

van Dijk P, Manley GA, Gallo L (1998) Correlated amplitude fluctuations of spontaneous otoacoustic emissions in five lizard species. J Acoust Soc Am 104:1559–1564.

van Dijk P, Mason MJ, Narins PM (2002) Distortion product otoacoustic emission in frogs: correlation with middle and inner ear properties. Hear Res 173:100–108.

van Dijk P, Narins PM, Mason MJ (2003) Physiological vulnerability of distortion product otoacoustic emissions from the amphibian ear. J Acoust Soc Am 114:2044–2048.

Vassilakis PN, Meenderink SWF, Narins PM (2004) Distortion product otoacoustic emissions provide clues to hearing mechanisms in the frog. J Acoust Soc Am 116: 3713–3726.

Wever EG (1978) The Reptile Ear. Princeton: Princeton University Press.

Wever EG (1985) The Amphibian Ear. Princeton: Princeton University Press.

Whitehead ML, Wilson JP, Baker RJ (1986) The effect of temperature on otoacoustic emission tuning properties. In: Moore BCJ, Patterson RD (eds) Auditory Frequency Selectivity. London: Plenum Press, pp. 39–46.

Whitehead ML, Kamal N, Lonsbury-Martin BL, Martin GK (1993) Spontaneous otoacoustic emissions in different racial groups. Scand Audiol 22:3–10.

Wilson JP, Baker RJ, Whitehead ML (1990a) Otoacoustic emissions in frogs. Adv Audiol 7:47–56.

Wilson JP, Baker RJ, Whitehead ML (1990b) Otoacoustic emission in frogs. In: Grandori F, Cianfrone G, Kemp DT (eds) Cochlear Mechanisms and Otoacoustic Emissions. Basel: Karger, pp. 47–56.

Wit HP, Van Dijk P, Segenhout JM (1989a) DC injection alters spontaneous otoacoustic emission frequency in the frog. Hear Res 41:199–204.

Wit HP, Van Dijk P, Segenhout JM (1989b) An electrical correlate of spontaneous otoacoustic emissions in a frog, a preliminary report. In: Wilson JP, Kemp DT (eds) Cochlear Mechanics: Structure, Function, and Models. New York: Plenum Press, pp. 341–347.

Wu Y-C, Art JJ, Goodman MB, Fettiplace R (1995) A kinetic description of the calcium-activated potassium channel and its application to electrical tuning of hair cells. Prog Biophys Mol Biol 63:131–158.

Yates GK, Manley GA C Köppl (2000) Rate-intensity functions in the emu auditory nerve. J Acoust Soc Am 107:2143–2154.

8 Otoacoustic Emissions: Basic Studies in Mammalian Models

Brenda L. Lonsbury-Martin and Glen K. Martin

1. Introduction

The following discussion about research on otoacoustic emissions (OAEs) in laboratory mammal models begins with a general introduction to the present thinking regarding the basis of emission generation. This overview is a fitting introduction to the discussion that follows, since most of what is known about the generation of OAEs was uncovered using animal models. The overall summary is followed by a more detailed discussion of the specific OAE "types" and their properties, which has been organized within a framework based on current considerations that describes the sources of emissions that comprise two fundamentally distinct generation mechanisms. Within this framework, further information is provided about the similarities and differences between the various classes of OAEs in humans and the laboratory mammals commonly used in hearing research. Finally, a review of the use of OAEs in mammalian models is presented in terms of how this work has contributed to increasing the knowledge base that describes many of the common problems associated with pathological hearing.

Pragmatically, to effectively focus on relating the above information, an important topic in the animal emissions field that is not addressed here is the area of electrically evoked OAEs. The reader who has a particular interest in this subject should consult the comprehensive review by Yates and Kirk (1997). Moreover, the appreciable research that uses OAEs to better understand the cochlear mechanics underlying the unique behavioral hearing abilities of specialized mammalian groups such as the bat (e.g., Kössl and Vater 2000) is not addressed for similar reasons. Likewise, only a limited amount of information about using OAEs to examine the role of the descending cochlear efferent system in normal hearing is presented. For a more complete discussion of this particular topic, the reader is referred to a recent review by Guinan (2006).

1.1 Background

A major discovery within the past several decades is that the normal cochleas of humans and other animals produce audio-frequency sounds, called OAEs, in response to acoustic stimuli, and, in some ears, spontaneously as well (Kemp 1978, 1979; see also, Kemp, Chapter 1, and Cooper et al., Chapter 2). OAEs are diminished, or abolished, by manipulations that reduce or eliminate the metabolic energy supply to the cochlea. Thus, metabolic energy is required for the generation of evoked OAEs. The existence of spontaneous OAEs (SOAEs), i.e., sound energy produced by the ear in the absence of deliberate acoustic stimulation, also suggests that metabolic energy can be converted into sound energy within the cochlea.

Another significant observation of the past few decades comes from direct measurements of sound-evoked basilar membrane (BM) motion, in vivo, in mammalian models (e.g., Robles and Ruggero 2001; see also, Cooper et al., Chapter 2). These observations showed that in healthy ears, i.e., ears with normal thresholds, sound-elicited BM vibration at the best or characteristic frequency (CF) is sharply tuned and compressively nonlinear. However, following cochlear damage, sharp tuning is lost and BM responses at the CF become linear. At the time of these initial observations, the unknown process(es) responsible for such sensitivity and frequency selectivity was called the "cochlear amplifier" (CA) (Davis 1983). The properties of this metabolically dependent component of BM motion suggest that the action of the CA consists of a cycle-by-cycle, i.e., audio-frequency, application of force to the cochlear partition. This inference is consistent with the existence of OAEs at audio frequencies.

A third major breakthrough that contributed to understanding the underlying basis of the CA was the observation that outer hair cells (OHCs) isolated from the mammalian cochlea exhibited electromotility, that is, in vitro, they change shape in response to electrical stimuli, on a cycle-by-cycle basis at audio frequencies (Brownell et al. 1985; see also Martin, Chapter 4, and Hallworth and Jensen-Smith, Chapter 5. It is the changes in the receptor potential of the OHC, that is, the voltage drop across the cell membrane, that causes it to change its shape. Based on these in vitro findings, the consensus is that in the intact ear, sound stimulation results in voltage changes across the OHC membrane, so that these receptors generate a mechanical force on a cycle-by-cycle basis in response to sound stimuli. It is currently known that this motility depends on a unique motor protein called prestin that resides in the basolateral wall of the OHC (Zheng et al. 2000).

Consequently, in the present view of cochlear function, OHCs act primarily as motor rather than sensory elements. Such a motor role for OHCs is consistent with the observation that, whereas there are approximately three times as many OHCs as IHCs, OHCs have little afferent innervation along with a substantial efferent innervation. Moreover, direct electrical stimulation of the efferent-nerve supply to the OHCs reduces sound-evoked BM motion (Dolan et al. 1997) and alters OAEs (Mountain 1980). It is the IHCs that receive the great majority of the afferent innervation of the cochlea, and convert cochlear-partition vibration to auditory nerve excitation, which represents the final output of the ear. The IHCs, then, are thought to passively transduce their input into neural activity. To summarize, the actions of the OHC-based CA are responsible for the normal hearing sensitivity, frequency selectivity, temporal resolution, and dynamic range of the cochlea.

1.2 Classification of OAEs

Traditionally, OAEs were classified based primarily on the characteristics of the acoustic stimulation required to evoke them (e.g., Lonsbury-Martin et al. 1991). In this schema, OAEs were typically divided into two general categories: spontaneous and evoked emissions. SOAEs, which occur in the absence of any deliberate stimulation of the ear, are customarily measured in normal hearing subjects by sealing a sensitive miniature microphone into the external ear canal. In contrast, evoked OAEs occur in response to the presentation of acoustic stimuli to the ear. Consequently, a sound source must also be sealed into the outer ear canal to present the sounds necessary for eliciting evoked emissions. On the basis of the stimuli used to elicit them, evoked emissions are customarily categorized into three different subtypes. That is, transient-evoked OAEs (TEOAEs) are elicited by an acoustic transient such as a click or tone burst; stimulus-frequency-evoked OAEs (SFOAEs) are elicited by a single, continuous pure tone; and distortion-product OAEs (DPOAEs) are generated in response to two continuous pure tones, separated in frequency by a prescribed difference (in Hz). These three types of evoked OAEs are found in essentially all ears exhibiting normal hearing.

Fairly recently it has become more useful to understand OAEs in terms of the mechanisms that are responsible for their generation rather than classifying them according to the type of evoking stimuli (Shera and Guinan 1999; see also Shera and Guinan, Chapter 8). That is, Shera and Guinan (1999) proposed that there are two fundamental mechanisms involved in the generation of OAE components based largely on their phase characteristics. Specifically, the emission taxonomy of Shera and Guinan (Shera) is categorized in terms of linear coherent reflection and nonlinear distortion sources. The reflection-emission components (SFOAEs, TEOAEs) arise from linear reflection from structural irregularities in the cochlea. Thus, the phase of these emissions changes rapidly as a function of frequency as different "reflectors", such as the numbers and orientations of the OHCs, are encountered. And within this schema, SOAEs can be viewed as a special case of self-exciting reflection emissions.

On the other hand, DPOAEs arise from nonlinear distortion sources induced by the traveling wave itself. Thus, the DPOAE generators move as the generation stimuli are swept in frequency. This model predicts relatively constant phases for DPOAEs, because a constant f_2/f_1 ratio sweep maintains the relative phases of the f_1, f_2, and DPOAE due to cochlear scale invariance. Thus, DPOAE phase remains relatively constant for a particular f_2/f_1 ratio sweep.

It is now known that DPOAEs, especially in humans, are often composed of two components that behave as if they are generated by the above two mechanisms. The notion that DPOAEs consist of two emission-generation components, that is, one originating at the f_2 place and the other arising from the DPOAE place (fdp), was first proposed by Kim et al. (1980). The now commonly accepted two-source model for apical DPOAE generation (fdp $< f_1$, e.g., the $2f_1 - f_2$ DPOAE) proposes that within the cochlea there are two separate contributors to the DPOAE level and phase measured in the ear canal when primary tones at moderate levels are used.

Knight and Kemp (2001) established the terms "wave-fixed" emission to encompass the OAE source that is physically associated with the traveling wave peak within the cochlea, and "place-fixed" OAE emission to designate the source that is physically associated with fixed structures within the cochlea. As described above, Shera and Guinan (1999) extended the two-source model further by suggesting that not only do DPOAEs arise from two distinct source locations in the cochlea, but they also arise from two distinct mechanisms. In their taxonomy, the initial source near the f_2 place arises from the active nonlinearity of OHCs, and a second source originates from a coherent linear reflection from the frequency place of the DPOAE. It is now common to reserve the use of the terms DPOAE "source" to refer to the cochlear location or region of the BM that contributes energy to the emission, and to distinguish "mechanism" as the underlying physiological basis for the emitted energy (Shera et al. 2004). The Shera and Guinan (1999) model predicts unique phase characteristics of the DPOAE, depending on the site of emission generation. For example, for the wave-fixed component, a constant f_2/f_1 ratio sweep maintains the relative phases of the f_1 and f_2 primaries, and the DPOAE. Thus, because of cochlear scale invariance, DPOAE phase remains relatively constant for a particular f_2/f_1 ratio sweep.

Because the DPOAE place component comes from the fdp location along the BM, for a particular f_2/f_1 sweep, the phase of this component changes rapidly and behaves like the phases of SFOAEs. Thus, when DPOAE phase is plotted as a function of DPOAE frequency, as shown in Figure 8.1B, a constant vertical phase band for a given DPOAE frequency is indicative of a place-fixed source (lower section of plot). In contrast, a constant horizontal phase band for an f_2/f_1 ratio sweep, as a function of DPOAE frequency, is indicative of a wave-fixed source (upper section of plot). For convenience, Shera and Guinan (1999) "linear reflection" versus "nonlinear distortion" terminology is used throughout the remainder of this chapter. In this context, the reader may consider the "wave-fixed" emission sources to be consistent with "distortion emissions" and "place-fixed" to be consistent with "linear reflection emissions."

These relationships can be clearly seen in the recent experiments by Knight and Kemp (2001). Specifically, using the above sweep techniques, they showed, as illustrated in Figure 8.1B, that for the $2f_1 - f_2$ DPOAE, place-fixed generation dominated at small f_2/f_1 ratios, whereas at the more traditional f_2/f_1 ratios, the ear-canal signal consisted predominately of wave-fixed DPOAEs. Their findings

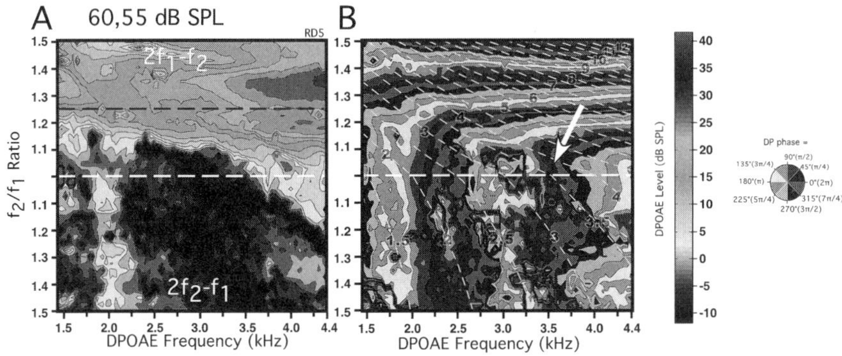

FIGURE 8.1. Example of DPOAE level (**A**) and phase (**B**) plotted as a function of f_2/f_1 ratio and DPOAE frequency obtained from a normal rabbit ear. DPOAE level and phase were obtained for constant f_2/f_1 ratio sweeps with DPOAE frequency increasing in 44-Hz steps and f_2/f_1 ratios in 0.05 steps. Figure **A** shows levels for the $2f_1 - f_2$ (*top*) and $2f_2 - f_1$ (*bottom*) DPOAEs. At the primary-tone levels of $L_1 = 60$ and $L_2 = 55$ dB SPL, robust $2f_1 - f_2$ DPOAEs can be measured, but $2f_2 - f_1$ DPOAEs are largely absent (see magnitude legend at right of Figure B). In the phase plots of **B**, vertical phase bands indicate "wave-fixed" emissions and are easily appreciated for f_2/f_1 ratios greater than 1.1. In the rabbit, although some "place-fixed" emissions can be observed (*arrow*), they are not well organized. The $2f_2 - f_1$ when present at higher primary-tone levels shows strong "place-fixed" behavior in rabbits. In this figure, the small region of $2f_2 - f_1$ DPOAEs at the left side of the lower portion of the plot in **A** displays the vertical banding indicative of the "place-fixed" emissions in the lower portion of **B**. *White dashed lines* in **B** track f_2 for the DPOAEs indicated along the abscissa.

provide a unique means of observing not only the DPOAE level space (Fig 8.1A), but also the phase characteristics (Fig 8.1B) associated with different DPOAE sources. Knight and Kemp (2001), using both inverse fast Fourier transform (IFFT) and time-windowing methods, further showed that the $2f_2 - f_1$ DPOAE is predominately a place-fixed DPOAE, a finding that agrees with an extensive study of this emission using suppression, noise-exposure, magnitude correlation, and latency techniques (Martin et al. 1998a). It seems reasonable to suggest that as OHCs degenerate, more reflection points would be introduced and place-fixed emissions would become more salient. This notion is supported by the appearance of place-fixed DPOAEs in the noise damaged rabbit ear (Martin et al. 2005), a representative laboratory species that normally exhibits small, if any, place-fixed $2f_1 - f_2$ DPOAEs, presumably because of the orderly OHC arrays observed for the rabbit under normal "hearing" conditions (Martin et al. 1983).

Ultimately, the DPOAE recorded in the ear canal is an acoustic signal comprising both the "generator" and "reflector" sources and is contingent upon the stimulus conditions and individual subject characteristics for the relative phases and magnitudes recorded. The fine structure recorded in DPOAEs (He and Schmiedt 1993) is then attributed to the interference between the two different sources (e.g., Dhar et al. 2002).

Although the linear reflection and the nonlinear distortion mechanisms proposed by Shera and Guinan (1999) can account for the phase behavior of the various emission types, their hypothesis assumes that OAEs are propagated to the stapes by a reverse traveling-wave mechanism. This notion implies that OAE latencies reflect roundtrip travel times equal to around twice the forward travel time to the place of emission generation. It is important to note that at the present time, this formulation of OAE propagation is controversial in that there are a growing number of observations that suggest that emissions may reach the stapes almost instantaneously as pressure waves in the cochlear fluids (e.g., Siegel et al. 2005). However, exactly how OAEs are propagated to the ear canal is still being actively debated (Shera et al. 2004). But regardless of the eventual outcome of this controversy, the phase behavior of OAEs remains unchanged, and for now, the Shera and Guinan (1999) taxonomy remains a useful framework within which to view various OAE observations.

2. Measures of OAEs

Measurements of OAEs are typically displayed in terms of their frequency and level properties. They are initially detected within the ear-canal signal using narrow-band filtering, which conventionally involves spectral analysis. Whereas SOAEs and SFOAEs are based on spectral averaging in which a number of fast Fourier transforms (FFTs) of the waveforms are summed and then averaged, the TEOAEs and DPOAEs are detected conventionally using standard stimulus-locked averaging. Consequently, in the latter cases, the spectral analysis is performed on the average time waveform. Whereas most studies that measure conventional OAEs in humans use commercially available equipment, for reasons that will become apparent in the discussion below, emissions in experimental mammals are typically detected using customized instrumentation that usually extends the stimulus frequency range and, at times, the number of stimuli that can be presented simultaneously.

2.1 General Properties of OAEs in Mammalian Models

In humans, as noted above, DPOAEs are typically measured using one of the many commercial devices currently available in the marketplace. However, at times, perhaps inappropriately, the same commercially available equipment for OAE measurement in humans is used for testing emissions in laboratory mammals, even though this approach may not be adequate, because of the differences between the hearing ranges of humans and most experimental mammals. For example, the range of best hearing for most of the common laboratory species used in auditory research, for example, guinea pigs, rabbits, rats, and mice, extends to frequencies that are much higher than in humans. The plot of Figure 8.2 illustrates this problem by comparing average DP-grams, for equilevel primaries at $L_1 = L_2 = 75$ dB SPL, for humans (open circles) and mice (filled

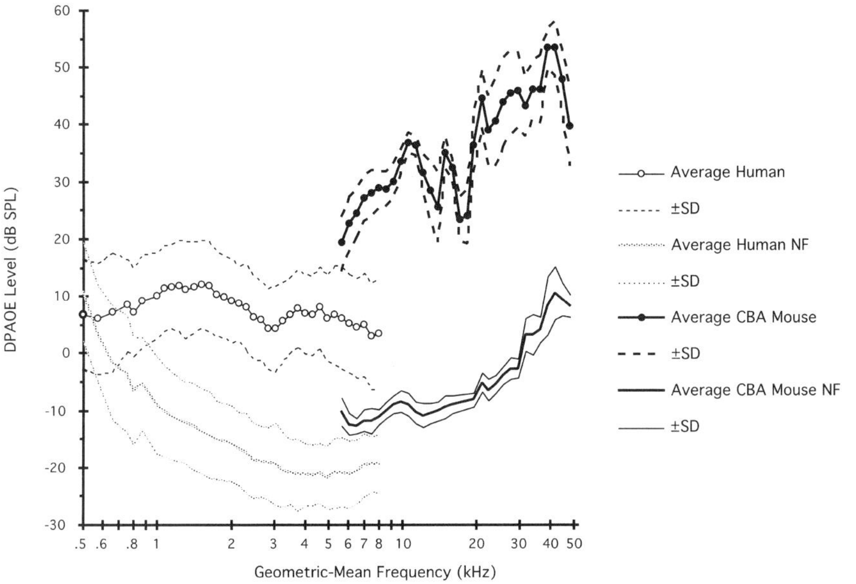

FIGURE 8.2. Average DP-grams for 10 mouse (*solid circles*) and 36 human (*open circles*) ears in response to $L_1 = L_2 = 75$ dB SPL primary tones. Note the minimal overlap in the frequency extent between the low-frequency human and high-frequency mouse ears. Also, note the lower noise floors [*shaded lines* (*left*) or *lines without symbols* (*right*) in lower portion of the plot depict mean ±1 standard deviation (SD) for human and mouse, respectively] for the human compared to the mouse, but the smaller variability for the mouse than for the human ears. Clearly, mice have much larger DPOAEs on average than humans, some part of which may be accounted for by the smaller ear canals of mice.

circles). Here, as is the case for other small laboratory mammal models, it is clear that mice do not generate DPOAEs very well in the lower-frequency region, i.e., below about 4 kHz, which represents the human range of best hearing. In fact, there is only a slight overlap for the two species between 6 and 8 kHz. In contrast to humans, to adequately test OAEs in mice, the high-frequency extent must be extended to at least 50–60 kHz (Martin et al. 2006). Thus, as Brown (1987) established earlier for several types of rodents, OAE levels are, in general, largest at the frequencies of best hearing in each species, which likely reflects the influence of factors related to both the middle and inner ears. In the past, many commercially available devices were designed to operate primarily at about 8 kHz and below, and thus they were relatively restricted in terms of being capable of testing over the frequency range of hearing for most common laboratory animals. Consequently, such instruments are not ideal for measuring OAEs in experimental mammals. More recently, several manufacturers have extended the upper frequency range of their instruments, making them more suitable for testing OAEs in some laboratory mammal species (e.g., chinchilla).

Another major difference between humans and laboratory mammal models is that the animals are frequently anesthetized during OAE measurements. Thus, often, Eustachian tube function is impaired during anesthesia, resulting in a buildup of negative pressure in the middle ear. Middle-ear pressure can greatly reduce or abolish OAEs in the ear canal, even when cochlear function is normal (Naeve et al. 1992). This factor can be a serious problem in guinea pigs and gerbils, in particular, but does not seem to be as problematic in rabbits, rats, and mice. Interestingly, reductions of OAE levels in humans have been reported during general anesthesia, particularly for gas anesthetics such as nitrous oxide. But such influences appear to be primarily due to middle-ear factors (Hauser et al. 1992). In addition, especially for small mammals such as the mouse, body temperature can substantially decrease during anesthesia (Martin et al. 2006), and hypothermia is also associated with reduced cochlear function (Brown and Nuttall 1984). Thus, it is important that the anesthetized animal's body temperature be monitored and maintained as close to the normal range as possible when measuring OAEs.

The normative properties of the SOAEs, TEOAEs, SFOAEs, and DPOAEs in the common laboratory mammals are reviewed below and compared to those of humans.

2.2 Spontaneous OAEs

Spontaneous OAEs (SOAEs) are narrow-band, often almost tonal, acoustic signals generated within the cochlea in the absence of deliberate sound stimulation. These OAEs occur at one or more discrete frequencies in more than 80% of normal adult human ears (e.g., Penner and Zhang 1997), predominantly in the 1–3 kHz region, where TEOAEs and SFOAEs are largest (Kemp 1981).

SOAEs are presumed to be produced by the same process as TEOAEs and SFOAEs, apparently, as noted above, as a result of feedback of the output of the TEOAE/SFOAE generator, due to reflection off the stapes, into its input (Kemp 1981). At frequencies where this feedback is positive, self-sustaining oscillation results if the loop gain is sufficient. The resulting self-sustaining oscillation is observed in the ear canal as an SOAE. Thus, SOAEs can be thought of as continuously self-eliciting evoked OAEs. Consistent with this view, SOAEs are found in frequency regions of strong evoked response, that is, where the gain of the emission generator is presumably high (Kemp 1981), and they rarely exceed 20 dB SPL, due to the compressive nonlinearity of the OAE generator.

In addition, in humans, there are rare, "atypical" SOAEs that differ in several respects from the SOAEs typically observed in normal human ears (e.g., Mathis et al. 1991). Specifically, these rare SOAEs are much larger (up to 60 dB SPL) than typical SOAEs, occur over a higher frequency range than the large majority of representative SOAEs, and tend to be associated with regions of audiometric abnormality. These atypical SOAEs also differ from usual SOAEs in that they can demonstrate greater short-term instability of frequency and level, their phase is not synchronized by low-level click stimuli (e.g., Mathis et al. 1991), and they may have broad, multilobed rather than sharp, single-lobed isosuppression contours (e.g., Wilson and Sutton 1983).

In monkeys, SOAEs are relatively common. Although they demonstrate a lower prevalence than in humans (Lonsbury-Martin et al. 1988), their properties are essentially identical to the SOAEs normally found in humans (Martin et al. 1988). Figure 8.3A (bottom trace E) illustrates for the left ear of an 8-year-old male monkey (*Macaca mulatta*) several strong SOAEs at around 1400 and 1550 Hz that were stable over a 1-year monitoring period (Martin et al. 1988). Although not as robust, this monkey also exhibited at times smaller SOAEs around 1100, 1700, and 2200 Hz. In rodents, although in an early study Zurek and Clark (1981) found no SOAEs in normal hearing chinchillas, Murphy and Johnson (1999) later observed these emissions in a subset of normal chinchillas originating from a breeding colony at the National Institute of Occupational Safety and Health. Further, Ohyama et al. (1991) reported SOAEs in about 20% of the ears of a Hartley strain of albino guinea pigs that displayed characteristics similar to those of SOAEs in humans. Thus, it seems that SOAEs in rodents have a substantially lower prevalence than in humans or monkeys, but are otherwise similar to those typically present in humans.

In addition, there have been descriptions of individual SOAEs from guinea pigs (Brown et al. 1990), chinchillas (e.g., Clark et al. 1984), and dogs (e.g., Ruggero et al. 1984). Whereas some of these SOAEs displayed properties similar to the typical SOAEs in human ears (Brown et al. 1990), most appeared to resemble the rare, high-level, atypical SOAEs reported for some pathological human ears. In particular, these animal SOAEs tended to be very large (up to 59 dB SPL), and were typically unstable in frequency and level, and/or demonstrated relatively broad or multilobed isosuppression contours. In addition, some of these SOAEs were associated with cochlear lesions (Clark et al. 1984), or apparent audiometric abnormalities (Ruggero et al. 1984). Thus, whereas humans and nonprimate mammals demonstrate both the typical low-level and atypical high-level SOAEs, it appears that in the nonprimates, the representative type is less prevalent than in humans or monkeys, whereas the atypical type, although rare, may be more common in nonprimates than in humans or monkeys. Because SOAEs likely arise from a linear reflection mechanism, their prevalence in humans and other primates may in some way relate to the cochlear irregularities that are typically inherent in their cochleas (e.g., Manley 1983). In contrast, such inhomogeneities with respect to OHC arrays are less common in most other laboratory mammals such as rodents and rabbits, in particular (e.g., Martin et al. 1989).

2.3 Transient Evoked OAEs

Essentially, all normal hearing humans exhibit transient evoked OAEs (TEOAEs) as measured with the commonly available commercial instruments (e.g., Gorga et al. 1993). In general, TEOAE levels for humans are largest around 1–1.5 kHz, and decrease at both the lower and higher frequencies, such that they are typically very small or immeasurable below 0.5 kHz and above about 4 kHz,

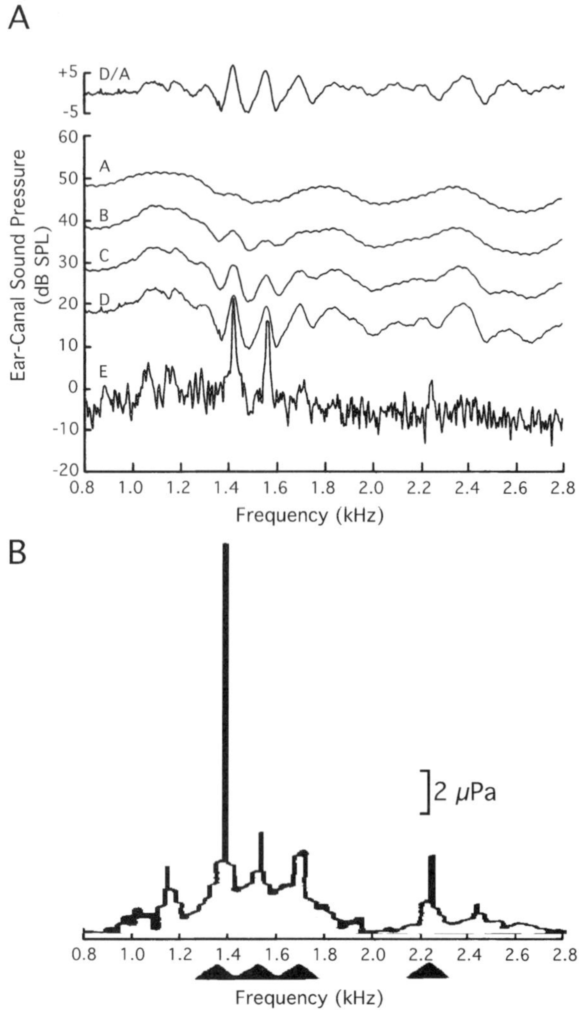

FIGURE 8.3. SFOAEs, SOAEs, and TEOAEs measured for the left ear of an 8-year-old male rhesus monkey (*Macaca mulatta*) that was lightly anesthetized with a dissociative anesthetic regimen. (**A**) SFOAEs elicited with moderate (traces *A, B*) and low-level (traces *C, D*) continuous tones swept in frequency from 0.8 to 2.8 kHz. SFOAEs, which were detected with a dynamic signal analyzer that provided a real-time Fourier analysis of the ear-canal signal, appeared as added sound energy, represented here as peaks. As the nonlinear emission-generating process(es) saturated with increasing stimulus levels, SFOAEs disappeared, as observed in trace A at the 50 dB SPL level. *Top trace* (D/A) represents an SFOAE record that resulted when the passive influence of the sound system was eliminated by dividing the response to the lowest-level stimulus (trace *D*) by that to the highest-level stimulus (trace *A*). Trace *E* at the bottom of the plot represents SOAEs recorded in the absence of sound stimulation. Note the larger SOAEs at around 1.4 and 1.55 kHz, with smaller ones present around 1.1, 1.2, 1.7, and 2.25 kHz. (**B**) FFT of average click-evoked TEOAE for this monkey using a customized laboratory system that

particularly in adult ears. The decline below 1 kHz is likely due to middle-ear effects and contamination by subject-generated noises such as breathing and body movements. The decline at high frequencies may also be partially due to the middle ear, but it likely reflects other factors as well, including the effects of latency, that is, the time between presentation of the stimulus and detection of the response, as well as restrictions inherent in the measuring equipment itself (e.g., the frequency response of the earspeakers). For example, the latency of TEOAEs in humans varies from <3 to 20 ms and, on average, decreases with increasing frequency (Kemp 1978). Thus, at high frequencies, it is difficult to separate TEOAE energy from the stimulus-induced ringing of the speakers and middle ear (see also Withnell et al. 2000). For example, it is not uncommon in protocols controlled by commercial equipment that the tail of the click stimulus is eliminated by zeroing out the first 2.5 ms of the response. In this manner, the onset of the response window is shaped so that there is a loss of some high-frequency TEOAE energy. The dependence of latency on frequency is thought to arise, at least partially, from a progressively more apical origin of OAEs within the cochlea as frequency decreases.

Except perhaps at the lowest stimulus levels, growth of TEOAE level with stimulus intensity is compressively nonlinear, that is, TEOAE levels increase at less than 1 dB per dB increase in stimulus level, and saturate at moderate stimulus levels such that click-evoked TEOAEs rarely exceed 20–25 dB SPL. Typically, in adult ears, TEOAEs are around 10 dB SPL, and are about 10 dB greater in the ears of infants and young children. The saturation of TEOAEs at moderate stimulation levels is consistent with the view that they are generated by the action of the CA, because the workings of this process are also thought to saturate at moderate stimulus levels.

Within the 0.5–4 kHz frequency region in which TEOAEs are typically detected in normal human ears, emission levels vary across frequency. That is, TEOAEs are strong, weak, or absent at different frequencies, which are constant in any one ear but vary greatly between individual ears. This idiosyncratic distribution pattern of TEOAE magnitude is particularly clear at the lower stimulus levels. The cause of these unique TEOAE patterns probably results to some extent from unique distributions of cochlear irregularities and the consequential multiple internal reflections that result in enhancements at some frequencies and cancellations at others.

In macaque and patas monkeys, TEOAEs are very similar to those in humans, although with slightly shorter latencies, which may be expected from the shorter cochleas of monkeys. Additionally, TEOAEs in monkeys demonstrate shorter

FIGURE 8.3. (Continued) eliminated linear stimulus-ringing components, while retaining the nonlinear evoked emission. Note that the prominent reemitted frequencies in the TEOAE spectrum in response to a series of clicks were associated with the majority of the SOAE frequencies (*solid triangles*) indicated along the abscissa. Also, note that the major extent of the TEOAEs for this monkey was observed between 1 and 2 kHz.

durations than in humans, perhaps indicating greater damping of the emission generator (e.g., Martin et al. 1988). The TEOAE displayed in Figure 8.3B was recorded with customized laboratory equipment from the macaque monkey of Figure 8.3A. Although this plot is illustrated using linear scaling, it is clear that the majority of the TEOAEs in primates are elicited from approximately the 1- to 2-kHz region of the cochlear partition. In addition, there is a clear association of the sharp peaks in the amplitude spectrum with both the large SOAEs around 1400 and 1550 Hz, and the smaller SOAEs around 1700 and 2200 Hz.

In nonprimate mammals, TEOAEs have been measured in cats (Wilson 1980), guinea pigs (e.g., Avan et al. 1990), rats (Hatzopoulos et al. 2002), and rabbits (personal observations). In guinea pigs, TEOAEs have much shorter latencies and durations than in humans and monkeys. Thus, when commercial equipment such as the ILO88 (Otodynamics, Ltd.) is used, its response window is typically set from 1 to 7 ms postclick (with a 2.5-ms raised-cosine onset and offset ramp), rather than the 2.5–20 ms window used with humans, in order to improve the signal-to-noise ratio. The TEOAE time waveform (left) and amplitude spectra (right) plots of Figure 8.4 compare these emissions recorded with the ILO88 commercial system for a macaque monkey (*Macaca nemestrina*) (Fig. 8.4A) and a guinea pig (*Cavia porcellus*) (Fig. 8.4B). As noted above, it is clear that the guinea pig TEOAEs are much shorter and smaller than those measured from the monkey.

The TEOAE spectrum displayed in Figure 8.5 was obtained from a rabbit (*Oryctolagus cuniculus*) of the New Zealand red breed using commercially available instrumentation (DPOAE System, Mimosa Acoustics, Inc.) that was developed relatively recently. It is clear from these data that rabbit TEOAEs represent very small emissions that likely are measurable only with low-noise equipment. Such TEOAEs, although much smaller than their human counterparts, typically disappear following deliberate exposures to moderately intense noises, and usually recover within a few days following such adverse events, thereby attesting to their authenticity (personal observations). Interestingly, in the past, TEOAEs were immeasurable in the rabbit using older commercial devices (personal observations). The better frequency response of newer transducers compared to those used in older equipment likely explains the recent success of measuring TEOAEs in rabbits.

TEOAEs in guinea pigs and other nonprimate mammals show compressively nonlinear growth, with saturation at moderate stimulus levels. From the literature, it is difficult to compare absolute TEOAE levels between species, because of variations in the measurement techniques used in different studies. However, it appears that if only a short period of time is considered in which a particular frequency component is largest, guinea pig TEOAEs often have approximately similar levels to those of humans (Ueda et al. 1992). It is important to note, though, that presumably the smaller ear-canal volume of guinea pigs than of humans results in a relative enhancement of TEOAE sound pressures for the guinea pig. This consideration suggests that at its source, the guinea-pig TEOAE may actually be somewhat smaller than the human TEOAE. The short latency

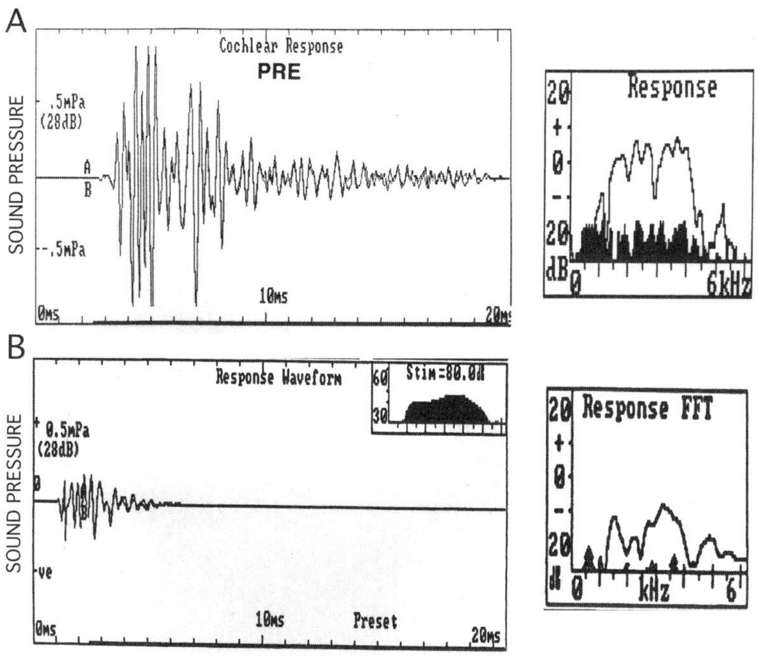

FIGURE 8.4. TEOAEs measured with a commercially available device (ILO88, Otodynamics Ltd) for a 4-year-old male pigtail monkey (*Macaca nemestrina*) (**A**) and a guinea pig (*Cavia porcellus*) (**B**). For the click-evoked TEOAEs, the average-time waveforms are shown on the left, with the related amplitude spectra displayed on the right. The conventional human settings including the 2.5-ms delay time were used for the monkey record, while only a 1-ms delay time was utilized for the guinea pig measures. Note the much larger and longer lasting TEOAEs measured for the monkey than for the guinea pig.

and duration of TEOAEs in rodents may account for the detection of tone-burst-evoked TEOAEs in only some guinea-pig ears (e.g., Avan et al. 1990). These characteristics also likely account for the failure of other studies to find click-evoked TEOAEs in guinea pigs (Wit and Ritsma 1980), gerbils (Schmiedt and Adams 1981), or rabbits (personal observations) in contrast to the almost 100% prevalence of TEOAEs in primate ears. In addition, it is important to recognize that TEOAEs are almost certainly a mixture of components generated by both structural irregularities along the organ of Corti and processes related to nonlinear distortion (Withnell and McKinley 2005).

2.4 Stimulus-Frequency OAEs

A stimulus-frequency OAE (SFOAE) is a constant low-level tone at the frequency of a continuous pure-tone stimulus that is assumed to be generated by the linear reflection mechanism that produces TEOAEs in response to transient

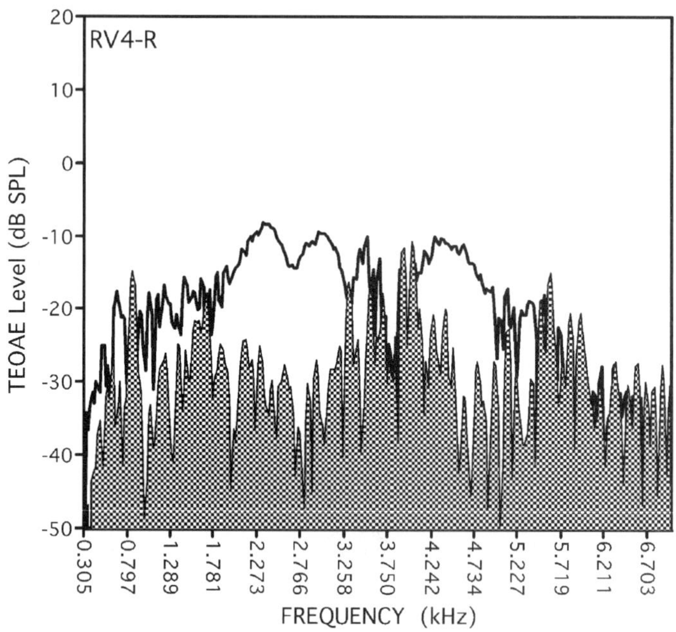

FIGURE 8.5. TEOAEs elicited from a rabbit (RV4-R) with clicks using a recently developed commercially available system (see text). Note the low-level emission, which is similar to the one exhibited for the guinea pig above in Figure 8.4B. Newer systems such as this instrument have much lower noise floors on average, which enhances the measurability of small TEOAEs from rodent and rabbit ears.

stimuli (Kemp and Chum 1980). The properties of SFOAEs are similar to those of TEOAEs in that SFOAEs are present in nearly all normally hearing human ears, are robust at those frequencies where TEOAEs are strong, and demonstrate latencies and a compressive nonlinearity similar to those of TEOAEs (Kemp and Chum 1980). Thus, it appears that the SFOAE is the steady-state response and the TEOAE the transient response arising from the presumed coherent linear reflection mechanism (Shera and Guinan 1999).

In monkeys, as illustrated in Figure 8.3A, SFOAEs are very similar to those in humans, although the frequency spacing of the ripples caused by the interference of the SFOAE and the stimulus tone is greater in the monkey (Martin et al. 1988). This greater frequency spacing in monkeys is to be expected, as frequency spacing is inversely proportional to the latency of the emission, and monkey TEOAEs demonstrate shorter latencies than those of humans (e.g., Martin et al. 1988).

SFOAEs have also been measured in cats (Guinan 1986), guinea pigs (e.g., Avan et al. 1990), and gerbils (Kemp and Brown 1983). In guinea pigs, the frequency spacing of the interference ripples is much greater than that in humans or monkeys, consistent with the shorter TEOAE latencies in this species. In gerbils, SFOAEs with properties similar to those observed in humans have been reported in response to stimulus levels below 40 dB SPL (Kemp and Brown 1983). These SFOAEs demonstrated compressive nonlinearity, with maximum levels of about −5 dB SPL, approximately 30 dB below maximum levels in humans and monkeys. Moreover, their latencies were comparable to those of TEOAEs in guinea pigs (e.g., Zwicker and Manley 1981), and much shorter than those of SFOAEs and TEOAEs in human and nonhuman primates.

2.5 Distortion-Product OAEs

Distortion-product OAEs (DPOAEs) are typically elicited by two long-lasting tonebursts that are related in frequency, such that the higher-frequency f_2 tone is about 1.2–1.3 times the lower-frequency f_1 tone (e.g., Johnson et al., Chapter 12, and Janssen and Muller, Chapter 13). The largest DPOAEs in all mammalian species occur at $2f_1 - f_2$, although DPOAEs are often present at other frequencies including other cubic (e.g., $3f_1 - 2f_2$, $2f_2 - f_1$) and quadratic distortion products (e.g., $f_2 - f_1$) as well. The levels of the $2f_1 - f_2$ DPOAEs vary systematically with the parameters of the two simultaneously applied stimulus tones, including the absolute frequency of f_1 and f_2, the frequency separation of f_1 and f_2 (i.e., the f_2/f_1 ratio), the absolute levels of f_1 (L_1) and f_2 (L_2), and the level difference of f_1 and f_2 (L_1-L_2). For example, in humans, the level of the $2f_1 - f_2$ DPOAE is greatest when the f_2/f_1 ratio is approximately 1.22 (Harris et al. 1989), and when L_1-$L_2 = 0$ dB at high stimulus levels. However, at low stimulus levels and at the same f_2/f_1 ratio, DPOAEs are larger in humans when L_1-$L_2 = 30$ dB or more (Whitehead et al. 1995). Thus, off-setting the primary tone levels as stimulus intensity decreases is required to maintain the BM excitation pattern at the f_2 place approximately constant. That is, maintaining equal level input to the nonlinearity produces the largest $2f_1 - f_2$ DPOAEs.

In humans, DPOAEs are most commonly measured as DP-grams (see Fig. 8.2) that depict DPOAE levels as a function of the frequency of the primary tones, i.e., typically at f_2, with L_1 and L_2 constant at $L_1 = 65$ and $L_2 = 55$ dB SPL. On average, the DP-gram is somewhat flat at about 10 dB SPL, while the signal-to-noise level is frequency-dependent in that it is smallest for the noisier lower frequencies and greatest at the highest frequencies. DPOAEs are also measured as response growth or input/output (I/O) plots that provide details about response magnitude as well as detection threshold, i.e., the primary tone levels at which the DPOAE increases to a criterion amount (e.g., 6 dB) above the related noise floor, and the rate of emission growth. In fact, with increasing $L_1 = L_2$ in regular steps of 2 or 5 dB, the $2f_1 - f_2$ DPOAE increases in level at a rate of <1 dB/dB, and saturates above $L_1 = L_2 =$ 65–75 dB SPL in human ears (Lonsbury-Martin et al. 1990).

It is important to note that although the DP-gram is the most common form of a DPOAE measure, along with the I/O function, several other response parameters are also used, particularly for experimental study. These more uncommon DPOAE responses include phase, which is extracted from the FFT analysis along with level (e.g., Telischi et al. 1999); microstructure, noted above, which is known as the fine-frequency DP-gram (He and Schmiedt 1993); latency in both onset (Whitehead et al. 1996) and group-delay forms (O'Mahoney and Kemp 1995); and suppression tuning curves (STCs) (Abdala et al. 1996) and suppression response areas (SRAs) (Martin et al. 1999; see also Section 2.6 below). Figure 8.6 illustrates the onset latency for the $2f_1 - f_2$ DPOAE (Fig. 8.6A) at an f_2 frequency of 2.239 kHz adapted from Whitehead et al. (1996). In this example from a rabbit ear, the onset of the emission (bold line in Fig. 8.6A) is determined by its synchrony with a sinusoidal template (dotted line in background) at the DPOAE frequency. Here, onset latency is established either by identifying by "eyeball" when the two traces first "line up" (arrow), or, as illustrated in Figure 8.6B, by computing a correlation between the two functions, so that the latency is defined as the latest time at which the correlation equals 0.75.

DPOAEs have also been measured in a wide variety of mammalian species including monkeys (e.g., Martin et al. 1988), cats (e.g., Kim 1980), rabbits (e.g., Lonsbury-Martin et al. 1987), mice (e.g., Jimenez et al. 1999), and several

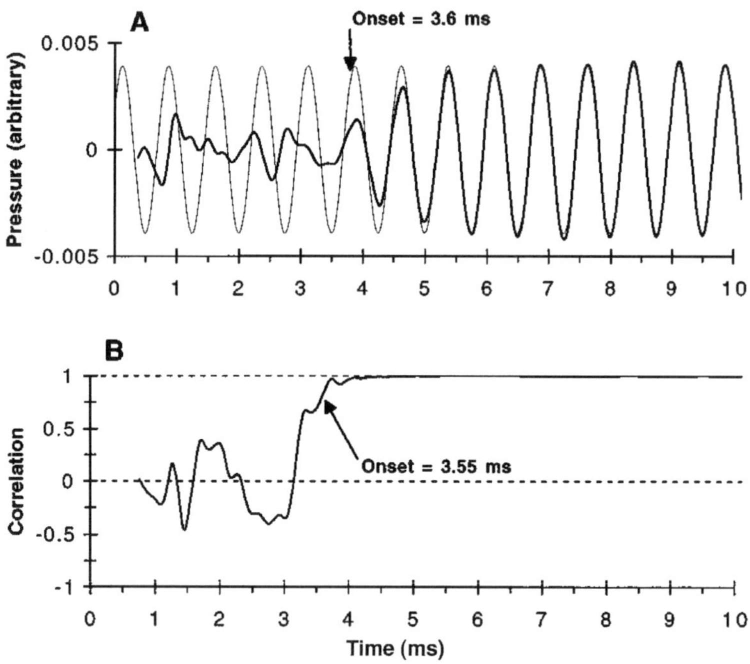

FIGURE 8.6. Visualization of the onset latency of the $2f_1 - f_2$ DPOAE from a rabbit ear. (**A**) Primary tones were $f_1 = 1.787$ kHz and $f_2 = 2.239$ kHz with $L_1 = L_2 = 60$ dB SPL. *Heavy line* is the DPOAE waveform after low- and high-pass filtering. *Faint dotted line* is a synthesized sinusoid at the same level and phase as the DPOAE at steady state. Latency is determined by "eyeball" when the two waveforms align at 3.6 ms (*arrow*). (**B**) Time plot indicating when the correlation between the two waveforms reaches 0.75 at 3.65 ms (*arrow*).

other rodents such as the chinchilla (e.g., Harding et al. 2002), guinea pig (e.g., Brown et al. 1990), rat (e.g., Henley et al. 1989), and gerbil (e.g., Kemp and Brown 1984). As noted above, for humans, the ideal f_2/f_1 ratio is around 1.22 (Harris et al. 1989), whereas for common laboratory mammals such as the rat, guinea pig, and gerbil (Brown 1987), rabbit (Martin et al. 1987), or mouse (Jimenez et al. 1999), it is around 1.25.

In addition, there are clear species differences in the dependence of DPOAE level on frequency, in that, as mentioned earlier, DPOAEs, like all emissions, tend to be largest in the regions of best hearing sensitivity in each species, and these regions vary between species (Brown 1987). In general, systematic variations in DPOAE parameters such as $L_1 = L_2$ and L_1-L_2, and f_2/f_1 produce qualitatively similar changes in emission levels in humans, monkeys, cats, rabbits, and rodents (Whitehead et al. 1992a). These similarities occur despite the quantitative differences in particularly the f_2/f_1 ratio that elicits the largest DPOAEs, which is greater in rabbits and rodents ($f_2/f_1 = 1.25$) than in humans ($f_2/f_1 = 1.22$).

However, several significant species differences of the parametric properties of DPOAEs are particularly noteworthy. First and most remarkably, the absolute magnitudes of DPOAEs are much larger in rabbits and rodents than in humans or monkeys. Whereas DPOAEs can be detected in essentially all normal human ears (e.g., Lonsbury-Martin et al. 1990), they are typically small, at about 5–10 dB SPL, i.e., usually about 55–65 dB below stimulus tones. In contrast, in rabbits and rodents, $2f_1 - f_2$ DPOAEs are considerably larger than in humans and monkeys, with the difference being about 25–30 dB for moderate-level stimuli at frequencies within the best hearing-sensitivity range of each species. The species differences of absolute levels of DPOAEs are in contrast to the situation for TEOAEs and SFOAEs, which tend to be similar, or somewhat smaller, in rodents than in primates (note the approximately -10 dB SPL TEOAEs for the guinea pig in Fig. 8.4B and rabbit in Fig. 8.5). In humans, the relationship of DPOAE levels to TEOAE or SFOAE levels is approximately 30 dB smaller than the corresponding association in guinea pigs.

The second exceptional difference between DPOAEs in primates, compared to rabbits and rodents, is the shape of certain DPOAE I/O functions. In individual ears of all species, DPOAE I/O functions for some primary-tone frequencies, but not others, are nonmonotonic, in that they demonstrate a minimum, or notch, at certain stimulus levels. However, the properties of these notches differ between humans and monkeys, and nonprimate mammals. For example, in rabbits, only specific primary-tone frequency/level pairs yield growth-function minima, and these stimulus–tone pairs yield minima in all rabbit ears, i.e., the minima are highly repeatable across ears (Whitehead et al. 1992a) for given frequency/level combinations. In contrast, in primates, the occurrence of notches is highly idiosyncratic in that the primary-tone frequency pairs that yield nonmonotonic I/O functions vary greatly across ears. That is, in rabbits and guinea pigs, I/O function notches are restricted to stimulus levels of around 55 to 70 dB SPL. Similar behaviors of response growth functions appear to be present in guinea pigs and gerbils (Mills et al. 1993). In contrast, in humans (Lonsbury-Martin et al. 1990) and monkeys (Martin et al. 1988), growth-function minima occur over a wide range of stimulus levels, rather than only in a specific range of stimulus levels.

Detailed studies in rabbits and rodents strongly suggest that the I/O function minima arise from a cancellation of out-of-phase components. This observation indicates that the $2f_1 - f_2$ DPOAE may actually consist of two discrete components, each of which has a differential dependence on stimulus parameters, and each of which dominates the total ear-canal $2f_1 - f_2$ at different stimulus levels (e.g., Whitehead et al. 1992a,1992b). These earlier studies equated the two sources as representing "active" and "passive" DPOAEs. More recent evidence suggests that this strict formulation is probably incorrect, because these sources may be generated by harmonics of the primaries as stimulus level is increased (see Section 2.6 for a detailed discussion of this notion). However, because of the different properties of the I/O function notches in humans, it is not clear whether the $2f_1 - f_2$ DPOAE in primates consists of two discrete components, as may be the case in rabbits and rodents. It is possible that some of the nonmonotonic irregularities observed in DPOAE I/O functions in primates reflect interactions of the DPOAEs with the mechanism that produces SFOAEs (Shera and Guinan 1999), because in humans and monkeys, SFOAEs are much larger relative to DPOAEs than in rodents and rabbits.

An issue noted above that must be discussed regarding DPOAEs is the notion that there are "active" versus "passive" DPOAEs. This conceptualization originated from the results of earlier studies reported by Norton and Rubel (1990) and Whitehead et al. (1992a,1992b). In these investigations, administration of loop diuretics, such as ethacrynic acid or fursosemide, eliminated low-level DPOAEs, while DPOAEs evoked by high-level tones remained relatively unaffected [see Fig. 3 in Whitehead et al. 1992b]. Observations like these led to the notion that DPOAEs evoked by high-level tones were not relevant to cochlear function, and as a consequence, many clinical studies in patients have focused on low-level primaries in the 55–65 dB SPL range. However, early studies in humans (e.g., Martin et al. 1990) clearly indicated that equilevel $L_1 = L_2 = 75$ dB SPL primaries can accurately track the pattern of hearing loss in patients with impaired hearing. More recently, experimental findings in mice with age-related hearing loss (AHL) indicated that both moderate and high-level primaries accurately follow the progressive degeneration of high-frequency OHCs observed in these animals (Jimenez et al. 1999).

Similarly, a brief exposure to damaging levels of noise will affect not only low-level DPOAEs, but high-level DPOAEs as well (Howard et al. 2002). Comparable findings were utilized by Avan et al. (2001) to develop a model that permits both high" and low-level DPOAEs to arise from a single source. Thus, more current thinking assumes that there are not two sources of DPOAEs that consist of a low-level "active" one, along with another high-level "passive" source. Rather, the present assumption is that all DPOAEs are generated in the nonlinear aspects of the mechanoelectrical transduction process whereby stereocilia deflection opens the transduction channels to allow potassium ions into the OHC. There is strong evidence that this process is nonlinear and that it is the source of all distortion products. In the traditional view, distortion products generated during this process modulate the OHC receptor potential to be fed back into BM motion, and they appear in the ear canal signal as DPOAEs (Liberman et al. 2004). These notions can explain many of the paradoxical findings such as the fact that high-level DPOAEs can be measured for some time after death (e.g., Whitehead et al. 1992b). Thus, as long as there is an electrochemical gradient across the OHC, even if it is greatly reduced, high-level stimuli can still modulate this potential to produce DPOAEs. As previously mentioned, noise overexposure can also affect DPOAEs evoked by both high- and low-level stimuli. Because it is well known that intense sounds can directly damage OHC stereocilia, if the stereocilia are the generation source of DPOAEs, then it is not difficult to appreciate how DPOAEs evoked by all levels of stimulation would be affected by excessive sounds.

Recently, Liberman et al. (2004) recorded DPOAEs to high-level stimuli in genetically produced knockout mice in which the motor protein of the OHCs, i.e., prestin, was eliminated. This finding somewhat complicates the traditional view of DPOAE generation described above in that apparently the stereocilia are coupled to the BM in such a way that they can produce measurable DPOAEs in the ear canal without the action of the OHC "gain" process, that is, without the CA, that is attributed to OHC electromotility. Apparently, in some situations, high-level stimulation is sufficient to vibrate the BM and deflect the stereocilia in the absence of an inherent amplifier. That is, such stimulation is capable of operating the nonlinear stimulus transduction apparatus to evoke DPOAEs. The importance of these recent findings is that the presence or absence of DPOAEs, regardless of the level of stimulation, assuming equipment artifacts are not involved, can significantly contribute to understanding cochlear pathology. If the OHC stereocilia are severely damaged by, for example, exposure to an intense noise, DPOAEs will also be absent. Thus, DPOAEs must be viewed as a window into the complicated OHC transduction process, and disruption anywhere along this chain of transduction-related events could potentially result in the reduction or absence of DPOAEs.

2.6 Other DPOAEs: High-Frequency DPOAE Sources

Mills (1998) and Martin et al. (1999) observed, respectively, in gerbils and rabbits that when a suppressor tone (f_3) was above f_2, substantial suppression/enhancement could be obtained for the measured DPOAE. Similar phenomena can also be observed in humans (Martin et al. 1998b), although the suppression/enhancement effects at high frequencies, in this latter case, are not as prominent as those observed for laboratory mammals. Figure 8.7 illustrates for a rabbit measures of SRAs, i.e., plots where each 3-dB contour is equivalent to an STC obtained with a different suppression criterion. Such SRAs reveal suppression/enhancement phenomena for the $2f_1 - f_2$ DPOAE for three f_2/f_1 ratios, and clearly indicate that enhancement (stippling) predominated at

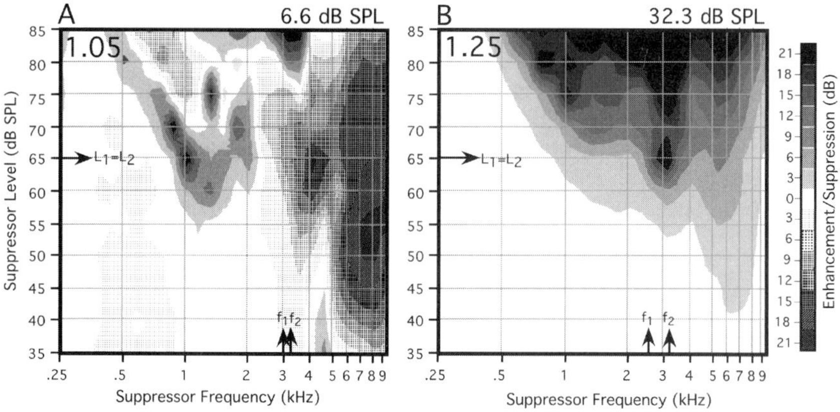

FIGURE 8.7. Rabbit suppression response areas (SRAs) shown as contour plots and obtained by keeping the f_1 and f_2 primary tones fixed in frequency and level and plotting the effects on DPOAE level as a suppressor tone (f_3) is swept in level and frequency. At narrow f_2/f_1 ratios of 1.05 (**A**), regions of strong enhancement (*stippled areas*) are often observed in rabbit ears above f_2. At the standard ratio of 1.25 (**B**), this region of enhancement often becomes one of suppression above f_2. A possible explanation of these phenomena is that DPOAE sources generated basal to f_2 (see text) were adding either in phase or out of phase with the measured DPOAE. When the suppressor "removes" one of these sources, either suppression or enhancement then results.

low ratios around 1.05 (Fig. 8.7A), while at more normal f_2/f_1 ratios of 1.25 (Fig. 8.7B), suppression above f_2 was more often observed. It is not clear from our present understanding of cochlear mechanics how a tone more than an octave above f_2 can affect a DPOAE presumably generated at the lower f_2 place. To explain these unexpected effects, it was initially proposed that suppression/enhancement might be produced by the interaction of two DPOAE sources (Martin et al. 1999).

Relevant to the above observations, Fahey et al. (2000) hypothesized that a second DPOAE source could be based on even-order nonlinearities. One possibility, for example, is that a harmonic of one of the primary tones interacts with the other primary tone to produce an emission. That is, the second harmonic of f_1 (i.e., $2f_1 = f_{1h}$) might interact with f_2 to produce a $2f_1 - f_2$ DPOAE (eg, $f_{1h} - f_2 = 2f_1 - f_2$, etc). In another possible scenario, the f_3 tone might act in a more complex manner as a catalyst [e.g., $(f_3 - f_2) - (f_3 - f_1) + f_1$] to produce a $2f_1 - f_2$ DPOAE that is based on two difference-tone emissions that interact with one of the primary tones. Concerning the harmonic mechanism, it can be hypothesized that a DPOAE is either adding to, or subtracting from, the emission being measured. Thus, the second source would be present even in the absence of f_3. In contrast, with respect to the catalyst mechanism, the presumed second source is present only when the f_3 is presented. It is important to emphasize that these effects occur when primary tone levels exceed 65–70 dB SPL, which is consistent with the generation of harmonics. The fact that noise exposures affecting the region above f_2 eliminate the suppression or enhancement strongly suggests that under certain circumstances, DPOAEs are generated basal to the f_2 place. The issue of other DPOAE sources basal to the f_2 place is still under investigation in a number of laboratories. However, it is expected that basal sources will be uncovered that contribute to the ear signal, especially at higher primary-tone levels.

2.7 Other DPOAEs: Lower and Upper Sideband DPOAEs

Presentation of two tones to the ear produces a family of DPOAEs referred to as lower sideband (LSB) and upper sideband (USB) DPOAEs. Most DPOAE literature has focused on the parameters necessary to elicit the largest and most prominent LSB DPOAE at the $2f_1 - f_2$ frequency. Several laboratories have also studied the corresponding USB $2f_2 - f_1$ DPOAE. Collectively, these studies convincingly demonstrated that the $2f_2 - f_1$ DPOAE originates from the DPOAE frequency place (e.g., Moulin and Kemp 1996). This finding does not agree with the physical and theoretical models that predict that all DPOAEs should originate at the f_2 place, where the maximal interaction of the f_1 and f_2 primary tones occurs. However, this perplexing result has been supported by studies that make use of STCs, along with onset latency measures, and temporary (TTS) and permanent threshold shift (PTS) noise exposures to the higher-frequency region above f_2. In fact, DP-grams of the $2f_1 - f_2$ versus $2f_2 - f_1$ DPOAEs, swept over a region of basal cochlear damage, never align with one another with respect to the damage pattern, even though both emissions are plotted for the same primary tone pair. Moreover, in damaged ears, the $2f_2 - f_1$ DPOAE can be enhanced at the same time that the $2f_1 - f_2$ DPOAE is reduced. Figure 8.8 illustrates a rabbit-ear-canal amplitude spectrum (Fig. 8.8B), where a number of both LSB and USB emissions can be observed. These emissions originate from both a normal control ear and its contralateral counterpart following a noise exposure. The DP-gram plot in Figure 8.8A shows the changes observed for the $2f_1 - f_2$ DPOAE, with the expected loss above the noise exposure band. The DP-grams in Figure 8.8C show similar plots for the $2f_2 - f_1$ DPOAE. In contrast to the LSB emission, this USB DPOAE is enhanced following noise exposure. These findings illustrate both the diverse behaviors of the two DPOAEs and the fact that by ignoring DPOAEs other than the $2f_1 - f_2$, a considerable amount of information regarding the cochlear nonlinearity is discarded. For example, as more energy is applied to the nonlinearity, or as f_2/f_1 ratios are decreased, more sideband DPOAEs are produced (e.g., Moulin 2000). In an early study by Brown et al. (1989), it was discovered that, following gentamicin treatment in guinea pigs, LSB emissions produced at small f_2/f_1 ratios were dramatically reduced, or their growth patterns substantially altered. Comparison of the pre-versus postgrowth functions for the various DPOAEs measured in Brown et al. (1989) clearly reveals abnormalities that would have been much less apparent if only a single DPOAE had been monitored. It is likely that in the future, as the $2f_1 - f_2$ DPOAE becomes more fully understood, attention will be shifted to the

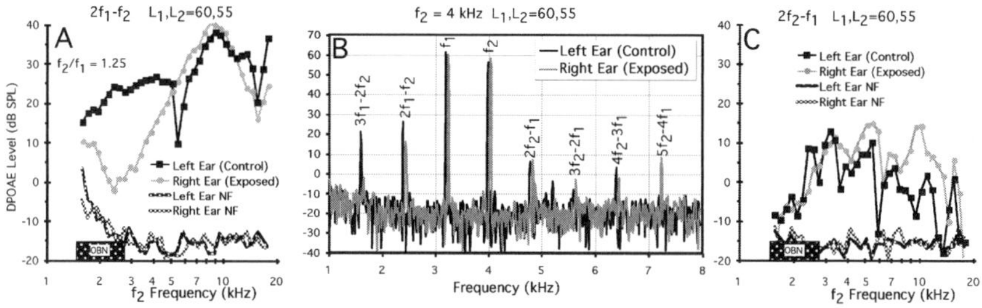

FIGURE 8.8. Example of the effects of an OBN exposure (*stippled bar* centered at 2 kHz along abscissa in **A**) on lower sideband (LSB) and upper sideband (USB) DPOAEs measured in a rabbit ear. (**A**) $2f_1-f_2$ DP-grams at L_1, $L_2=60$, 55 dB SPL for the exposed right ear (*gray circles*) and the unexposed or control left ear (*solid squares*). As expected, this DPOAE showed significant losses at and above the exposure band. (**B**) An example of the ear-canal amplitude spectrum obtained with an f_2 of 4 kHz in both the control unexposed left ear (*black trace*) and exposed right ear (*gray trace*). Two LSB DPOAEs ($2f_1-f_2$, $3f_1-2f_2$) and four USB DPOAEs ($2f_2-f_1$, $3f_2-2f_1$, $4f_2-3f_1$, $5f_2-4f_1$) can easily be measured. (**C**) Same data for the simultaneously measured $2f_2-f_1$ DPOAE. This emission showed little change at low frequencies, but was enhanced following noise exposure between 7 and 15 kHz. When the f_2 producing the $2f_1-f_2$ DPOAE was in the region of maximal loss for this emission, that is, around 3 kHz, the $2f_2-f_1$ DPOAE was unaffected. This outcome is consistent with the notion that these two DPOAEs are generated at unique BM places.

study of other LSB and USB emissions to derive a more complete description of cochlear nonlinearities in both normal and pathological ears.

2.8 Other Measures of DPOAEs: DPOAE Suppression

An important aspect of DPOAE generation is related to a well-recognized aspect of compressive nonlinearities in which one tone has the ability to interfere with or "suppress" the response produced by another. For DPOAEs, the suppression process can be used to produce STCs (Brown and Nuttall 1984) based on sweeping a third f_3 tone in frequency and level around the f_1 and f_2 primary tones used to elicit the DPOAE of interest. Investigation of this phenomenon as a means of measuring cochlear frequency selectivity has received considerable attention in both human (e.g., Abdala 2004) and laboratory mammal studies (e.g., Martin et al. 1998b). In some circumstances such as during human development, small changes in frequency selectivity have been observed (Abdala 2004). However, in animal studies under very controlled conditions, STCs remain relatively unchanged following cochlear insults including both administration of loop diuretics (Martin et al. 1998b) and sound overexposure episodes that produce either temporary (Howard et al. 2002) or permanent cochlear dysfunction (Howard et al. 2003). Figure 8.9 shows contour plots of rabbit SRAs, which were measured before (Fig. 8.9A) and after (Fig. 8.9B) administration of the loop diuretic, ethacrynic acid. It is clear that even though

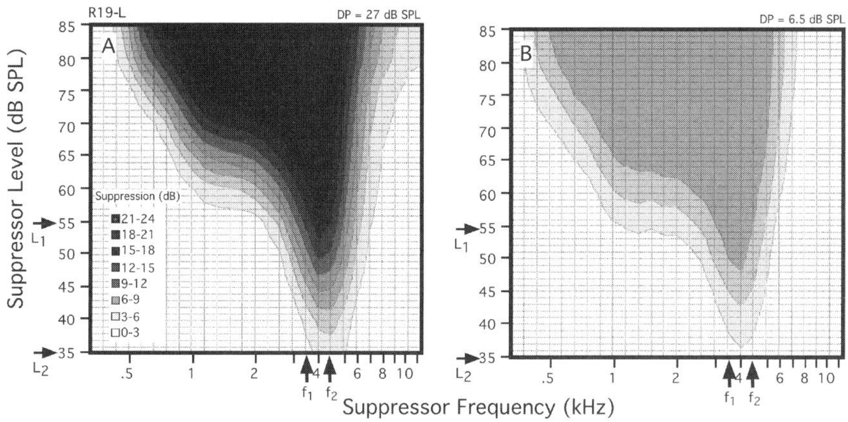

FIGURE 8.9. An example of the aftereffects of administering the loop diuretic ethacrynic acid on rabbit $2f_1 - f_2$ SRAs. (**A**) Predrug SRA obtained prior to diuretic injection at a time when the DPOAE level was 27 dB SPL. The successive 3-dB contours indicating the amount of suppression distinguish the STCs that would be obtained for different suppression criteria (e.g., 3 dB, 6 dB). (**B**) Postdrug SRA following ethacrynic acid injection showing a dramatic drop in DPOAE level to 6.5 dB SPL. Although the number of contours that can be obtained is clearly reduced, the overall shapes of the STCs were essentially unchanged. Examples like this demonstrate that DPOAE STCs are relatively insensitive to agents known to alter cochlear frequency tuning.

DPOAE levels were reduced by approximately 20 dB, Figure 8.9B reveals no noticeable changes in the overall shape of the SRA, or STCs that could have been extracted for the various 3-dB contours. This finding is representative of the effects of various cochlear insults on measures of DPOAE STCs in that STCs acquired by keeping primary-tone levels constant before and after cochlear induced damage do not sensitively detect the presumed changes in cochlear frequency selectivity. Moreover, the growth of suppression can also be measured for various suppressors as vertical slices through the SRAs. Again, under controlled conditions in rabbits, this measure was also relatively insensitive to cochlear damage induced by noise overexposure (Porter et al. 2006). Overall, measures of DPOAE suppression, although interesting, do not appear to hold significant promise for the development of sensitive tests of cochlear abnormality, at least as they are presently implemented.

3. Otoacoustic Emissions in Hearing Pathology

Studies conducted in various human patient groups have shown that OAEs can be used clinically to detect sensorineural hearing losses resulting from a number of factors (e.g., Janssen and Muller, Chapter 13). Such hearing losses include those resulting from the administration of ototoxic drugs (e.g., Katbamna et al. 1999), acoustic trauma or overexposure (e.g., Engdahl and Kemp 1996), Meniere's disease (e.g., Harris and Probst 1992), sudden idiopathic sensorineural hearing loss (e.g., Sakashita et al. 1991), acoustic neuroma (e.g., Telischi et al. 1995), presbycusis (e.g., Stover and Norton 1993), and hereditary hearing disorders (e.g., Cohn et al. 1999). OAE testing has also proven useful in dealing with various aspects of other more general disease states such as multiple sclerosis (e.g., Cevette et al. 1995), Alzheimer's disease (e.g., Gates et al. 1995), autism (e.g., Grewe et al. 1994), and cerebral hypertension (e.g., Avan et al. 1998). In addition, OAE testing has contributed to a greater understanding of the role of prenatal hormones in influencing sexual orientation (McFadden 2002). Finally, special tests of OAEs evaluating the cochlear efferent system have also been useful in determining the site of lesion in disorders involving a central auditory neuropathy (e.g., Berlin et al. 1994).

Studies in laboratory mammal models have similarly investigated the influence on OAEs of genetics and aging, and induced cochlear traumas including hypoxia, ischemia, ototoxicity, acoustic overexposure, and endolymphatic hydrops. The unique advantage of animal over human studies is that they allow direct comparison of the modifications in OAEs to changes in electrophysiological measures of cochlear function, as well as to the underlying histopathology (e.g., Martin et al. 1988). In the discussion below, the results of studies of a variety of traumas on OAEs in animal models are reviewed in order to provide insight into the relationship of reductions in OAEs to impairments of cochlear structure and function.

3.1 Middle-Ear Disorders

Because the eliciting stimuli and the OAEs must both pass through the middle ear, evoked OAEs are reduced in cases of impaired sound conduction through the middle ear. This dependence of OAE detection on robust middle-ear function has been demonstrated in a variety of human patient groups exhibiting a number of middle-ear disorders (e.g., Lonsbury-Martin et al. 1994), and by experimental treatments of the middle ear by applying positive or negative air pressure to the outer ear canal in both humans (e.g., Kemp 1981) and laboratory mammal models (e.g., Whitehead et al. 1991), or manipulating postural position, particularly in humans (e.g., de Kleine 2001). Thus, if OAEs are used in clinical or experimental settings to assess cochlear status, care must be taken in both humans and animal models to ensure normal middle-ear function.

3.2 Retrocochlear Hearing Loss

Visually confirmed surgical sections of the auditory portion of the eighth cranial nerve resulted in no change in DPOAEs in rabbits for periods of up to 10 weeks, by which time significant retrograde neural degeneration would be expected to have occurred (Martin et al. 1999). Laboratory mammal studies like these have demonstrated that DPOAEs are independent of primary afferent activity, thus clearly indicating that DPOAEs are generated in the cochlea, with no direct neural, or retrocochlear, involvement. The finding that a strain of mutant mice with central deafness, but apparently normal cochlear function, demonstrated DPOAEs similar to those observed in normal mice (Horner et al. 1985) supports this conclusion.

3.3 Hypoxia, Anoxia, and Ischemia

In experimental studies, limiting an animal's oxygen intake effectively limits the metabolic energy supply to the ear. Brief hypoxia results in rapid, reversible reductions of both the endolymphatic potential (EP) and the sound-evoked cochlear potentials including the compound action potential (CAP). Longer-term hypoxia or anoxia results in irreversible functional losses and structural trauma to multiple elements in the cochlea. In guinea pigs, temporary hypoxia produced by limiting the animal's oxygen intake for a few minutes caused a rapid, reversible reduction of TEOAEs (Zwicker and Manley 1981), SFOAEs (Avan et al. 1990), and SOAEs (Ohyama et al. 1991). Moreover, in cat (Evans et al. 1981) and guinea pig (Ueda et al. 1992), lethal anoxia rapidly abolished these same OAEs.

Whereas lethal anoxia in gerbils (e.g., Schmiedt and Adams 1981) and rabbits (Lonsbury-Martin et al. 1987) caused a rapid reduction in DPOAE levels evoked by stimulus tones below 60–70 dB SPL, DPOAEs elicited by stimulus tones above these intensities declined much more slowly postmortem. At low primary-tone levels (e.g., 45 dB SPL), the DPOAEs were abolished within about 3 minutes, whereas at high primary-tone levels (e.g., 75 dB SPL), there was no significant change in DPOAE level over this same time interval (Whitehead et al. 1992b). These observations provide evidence as discussed previously that as long as a sufficient electrochemical gradient can be maintained across the OHCs, high-level stimuli can still modulate this potential to produce DPOAEs.

3.4 Ototoxic Drugs

3.4.1 Loop Diuretics

Acute administrations of the loop diuretics furosemide and ethacrynic acid, used in the treatment of kidney failure in human patients, produce a temporary hearing loss. These drugs are thought to act primarily on the stria vascularis in such a way as to produce a rapid and reversible reduction of the EP and corresponding decreases in the sound-evoked responses of the cochlea (Rybak 1986). The major consequent morphological changes occur in the stria vascularis, although some hair-cell damage has been reported for high dosages of loop diuretics.

There have been no published studies of the effects of loop diuretics on OAEs in humans. However, acute administrations of furosemide or ethacrynic acid produced temporary reductions of TEOAEs in monkeys (Anderson and Kemp 1979). Further, acute administrations of furosemide temporarily reduced TEOAEs in cats (Wilson and Sutton 1983) and guinea pigs (Ueda et al. 1992), SOAEs in guinea pigs (Ohyama et al. 1991), and DPOAEs in gerbils (e.g., Kemp and Brown 1984) and rabbits (Whitehead et al. 1992b). In rabbits (Whitehead et al. 1992b) and rodents (Mills et al. 1993), as observed in hypoxia, DPOAEs evoked by low-level stimuli ($L_1 = L_2 \leq 60$ dB SPL) were greatly reduced, whereas those elicited by high-level stimuli ($L_1 = L_2 = 75$ dB SPL) were less affected. These findings imply that the low-level DPOAE component is more dependent on the EP, whereas the high-level DPOAE element is relatively independent of this cochlear potential.

3.4.2 Salicylate

Oral aspirin (acetylsalicylic acid) administration reduces TEOAEs, SFOAEs, and SOAEs in humans (e.g., Long and Tubis 1988). Wier et al. (1988) discovered that aspirin had less effect on $2f_1 - f_2$ DPOAEs evoked by low-level stimulus tones than on SOAEs at nearby frequencies in the same ears. Similarly, in monkeys, acute injections of sodium salicylate that greatly reduced or abolished TEOAEs, SFOAEs, and SOAEs had little effect on $2f_1 - f_2$ DPOAEs over a wide range of stimulus levels (Martin et al. 1988). Thus, in humans and monkeys, it appears that salicylate reduces TEOAEs, SFOAEs, and SOAEs, but has relatively little effect on DPOAEs, even at low stimulus levels. Again, this finding suggests some dissociation of the mechanism(s) underlying the generation of DPOAEs versus TEOAEs, SFOAEs, and SOAEs, which clearly supports the proposition of Shera and Guinan (1999) discussed above.

In nonprimate mammals, salicylate has been shown to greatly reduce the low-level component of DPOAEs. In guinea pigs, cochlear perfusion of salicylate decreased DPOAEs, with a much larger reduction of DPOAEs evoked by low-level stimuli than of those evoked by high-level stimuli (Kujawa et al. 1992). This finding is consistent with the finding that salicylate suppresses the activity of single auditory neurons (Evans and Borerwe 1982), and the CAP of the auditory nerve, at low, but not high, intensities of stimulation (Puel et al. 1989). In cats, DPOAEs evoked by 70 dB SPL stimulus tones were reduced by $\approx$5 dB by injections of sodium salicylate (Stypulkowski 1990). The finding that salicylate had little or no effect on DPOAEs in monkeys, but reduced DPOAEs in other species, could also be due to different dosages administered and, possibly, to species differences in the metabolism of salicylate and its transport into and out of the fluids of the inner ear.

Whereas the mechanism of salicylate ototoxicity has not yet been established, salicylate is known to influence OHC ultrastructure both in vivo (Douek et al. 1983) and in vitro (Dieler et al. 1991), and to reduce OHC motility in response to electrical stimuli in vitro (Shehata et al. 1991). The finding that OAEs can be reduced by salicylate, therefore, is consistent with the hypothesis that the generation of OAEs is dependent on the normal function of OHCs.

3.4.3 Aminoglycoside Antibiotics

In guinea pigs chronically treated with the aminoglycoside gentamicin, Brown et al. (1989) demonstrated that in frequency regions where there was substantial

OHC pathology, DPOAEs elicited by stimulus tones below 60–70 dB SPL were greatly reduced. However, those produced by higher-level stimuli were relatively unaffected. Henley et al. (1996) observed similar findings in rats treated chronically with kanamycin, another aminoglycoside.

Aminoglycoside antibiotics and loop diuretics administered in combination are known to have a powerful ototoxic effect in humans, which can result in profound, permanent hearing loss (West et al. 1973). Laboratory mammal studies have demonstrated that acute administrations of these drugs in combination have a very disruptive effect on the function and anatomy of the organ of Corti, resulting in severe decrements of sound-evoked cochlear potentials, and in widespread hair-cell loss (West et al. 1973). In rabbits, ethacrynic acid administered intravenously 2 hours after an injection of the aminoglycoside antibiotic gentamicin caused a reduction of DPOAEs at all stimulus levels measured up to $L_1 = L_2 = 75$ dB (Whitehead et al. 1992b). Indeed, DPOAEs were essentially eliminated by a few hours after the ethacrynic acid injection. The time course of this effect was similar to that seen for the decline of cochlear electrophysiological responses in guinea pigs after administration of an aminoglycoside antibiotic and a loop diuretic (e.g., Hayashida et al. 1989).

3.4.4 Other Drugs and Ototoxins Reported to Reduce OAE Levels

Quinine causes a reversible, mild to moderate hearing loss in humans through an unknown mechanism. In studies using small numbers of normal volunteer subjects, quinine reduced TEOAEs, SFOAEs, and DPOAEs (e.g., McFadden and Pasanen 1994). However, to date, there have been no reported results of the effects of quinine on OAEs in intact animal models.

Cisplatin is a known ototoxin utilized in antitumor pharmacological therapy, which causes a permanent high-frequency hearing loss in some human patients (e.g., Pasic and Dobie 1991). In fact, a patient study by Ress et al. (1999) that compared the effects of cisplatin on hearing thresholds versus DPOAE levels discovered that for every 3-dB increase in the ototoxic-induced hearing loss, emissions were reduced by 1 dB. Interestingly, this metric was also noted for a population of patients with hearing losses due to a variety of cochlear-based disorders (Whitehead et al. 1993) as well as for the C57BL/6J (C57) mouse strain that exhibits AHL (Martin et al. 2002). These clinical data on the effects of cisplatin on emissions are consistent with the results of McAlpine and Johnstone (1990) that described decreased DPOAEs and elevated I/O detection thresholds in guinea pigs administered cisplatin. In addition, high-frequency DPOAEs were also reduced by cisplatin in rats (e.g., Hatzopoulos et al. 2001).

The solvent toluene also caused reductions in DPOAEs in guinea pigs (McWilliams et al. 2000). Moreover, several days of inhalation of toluene by rats resulted in elevated auditory brainstem response (ABR) thresholds and reduced DPOAE levels (Johnson and Canlon 1994). The frequencies of these DPOAE level reductions coincided with the frequencies of ABR threshold elevation and OHC loss. However, the hearing loss caused by toluene inhalation in rats is unusual in that it appears to be most pronounced at mid-frequencies, rather than at high frequencies. In other studies, the organic solvent ethyl benzene was ototoxic in rats in that it produced elevated detection thresholds for both DPOAE I/O functions and other cochlear potentials (Cappaert et al. 2000). Furthermore, Laskey et al. (2002) showed that DPOAEs measured in rats were sensitive to the ototoxicity produced by exposure to Aroclor 1254, a common polychlorinated biphenyl (PCB). And, finally, Pouyatos et al. (2002) demonstrated in a rat model that DPOAEs detected the ototoxicity induced by inhaling the synthetic chemical styrene, which is commonly used in the production of plastic products.

3.5 Summary of the Effects of Ototoxic Drugs on OAEs

Experiments involving ototoxic drugs and certain toxic industrial compounds demonstrate that OAEs can provide information regarding the time course of ototoxicity in that emission level tends to decrease approximately in parallel with electrophysiological measures of cochlear function. However, recovery of the DPOAE sometimes differs from recovery of electrophysiological measures (e.g., Rebillard and Lavigne-Rebillard 1992). Moreover, these experiments demonstrated that OAEs provide information regarding the frequency extent of ototoxicity. Thus, drugs producing predominantly high-frequency hearing loss, e.g., cisplatin, also cause predominantly high-frequency reduction of OAEs (e.g., Ress et al. 1999), whereas drugs causing hearing loss over the whole frequency range, e.g., loop diuretics, cause OAE reductions over the whole frequency range (e.g., Whitehead et al. 1992b).

4. After-Effects of Different Acoustic Overexposures

4.1 Temporary Threshold Shifts

Sound exposures in human ears producing a temporary behavioral threshold elevation also reduced the levels of SOAEs (e.g., Kemp 1981), TEOAEs (e.g., Kemp 1981), and $2f_1 - f_2$ DPOAEs (Sutton et al. 1994). These OAE reductions were frequency specific, and demonstrated a correlate of the psychophysical half-octave shift effect, i.e., the maximum functional reduction tended to occur at a frequency approximately 1/2-octave above the exposure frequency. Additionally, the recovery of the postexposure OAE levels followed, in general, a time course similar to that of the corresponding behavioral threshold.

In animal models, temporary reductions of OAEs by noise overexposure were reported for both TEOAEs in monkeys (Anderson and Kemp 1979) and cats (Wilson and Sutton 1983), and SFOAEs in guinea pigs (Avan et al. 1990). Further, recovery of $2f_1 - f_2$ DPOAEs from overstimulation effects was observed in gerbils (Schmiedt 1986), cats (e.g., Kim et al. 1980), and rabbits (e.g., Mensh et al. 1993).

4.2 Permanent Threshold Shifts

In human patient groups, Probst et al. (1987) determined that high-frequency-noise-induced hearing loss (NIHL) was associated with a reduction in the number of prominent peaks in the spectra of TEOAEs. In addition, a number of investigators (e.g., Martin et al. 1990) have demonstrated frequency-specific reductions of DPOAEs in frequency regions of behaviorally measured NIHL in humans.

Irreversible reductions of DPOAEs resulting from chronic noise overexposure have been described in chinchilla (e.g., Harding et al. 2002), rabbit (e.g., Franklin et al. 1991), and several mouse strains (e.g., Jimenez et al. 2001). In general, $2f_1 - f_2$ DPOAEs were reduced at the frequencies of corresponding behaviorally measured threshold increases (e.g., Franklin et al. 1991), or in frequency regions associated with areas of histologically observed damage to the organ of Corti (e.g., Vazquez et al. 2004). In rabbits, the time course of the DPOAE reduction during chronic-noise overexposure and the recovery thereafter reflected the temporal behavior of the behavioral threshold elevations observed in the same ears, over periods of days and weeks (Franklin et al. 1991).

The results of the noise overstimulation experiments therefore indicate that DPOAEs have the ability to track, within both the frequency and time domains, the temporary and permanent impairments of cochlear function resulting from exposure to intense acoustic stimuli. The results of several of these studies also demonstrated a tendency for noise overexposures to have less effect on DPOAEs evoked by higher-level stimuli (e.g., Schmiedt 1986). Figure 8.10 shows this effect for an intense overstimulation in the mouse consisting of an exposure to a 1-hour, 105 dB SPL octave band of noise (OBN) centered at 10 kHz. In this figure, plots of the absolute DP-grams are shown in the left panels, while the pre/post difference plots (i.e., postexposure DPOAEs minus preexposure DPOAEs) are displayed in the right panels. Here, it is clear that the noise exposure produced greater effects on DPOAEs elicited by low-level primaries of 55 dB SPL (bottom) than those evoked by the higher-level primary tones at 65 (middle) and 75 dB SPL (top). This outcome is especially apparent for DPOAEs obtained at 1 d postexposure (open triangles). Although this stimulus level dependent sensitivity to acoustic overexposure was not as clear-cut as that seen upon anoxia or administration of loop diuretics, this observation again underscores the need to exercise care in the choice of stimuli to be used in monitoring cochlear condition.

Subramaniam et al. (1994) discovered in chinchillas that DPOAEs over a wide range of stimulus levels were normal, or near normal, despite quite substantial OHC damage 4 week after intermittent exposure (3 hours on/9 hours off for 15 days) to a 95 dB SPL, 0.5 kHz OBN. In these animals, after each exposure session, DPOAEs were reduced, and evoked potentials measured from the inferior colliculus demonstrated threshold elevation. At low, but not high, frequencies, the threshold elevation decreased, despite continued overexposure. Similarly, DPOAEs showed reduced levels, which recovered at low, but not high, frequencies. Thus, the DPOAE deficits tracked the loss and then recovery of evoked potential thresholds. Five days postexposure, both evoked potential

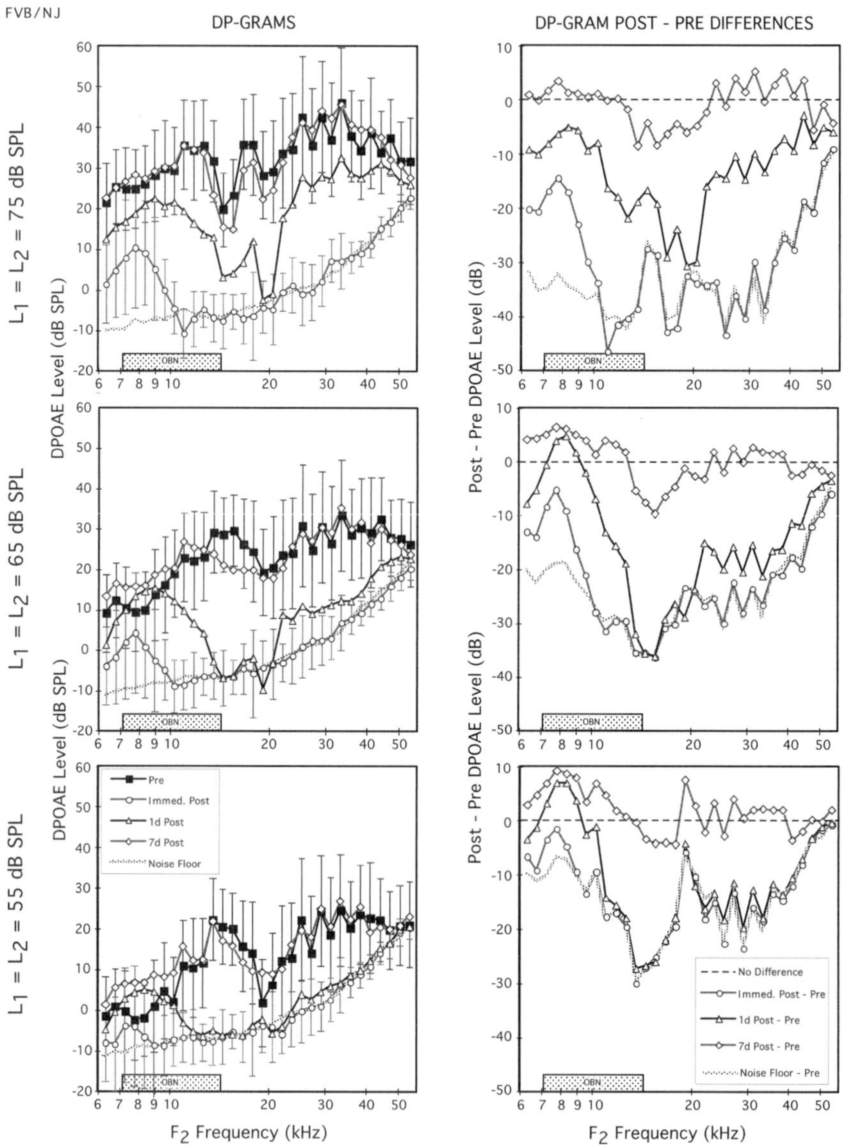

FIGURE 8.10. Effects of a 105-dB SPL OBN exposure centered at 10 kHz (*stippled bar* on abscissa) and delivered for 1 hour on DP-grams obtained at three primary-tone levels [$L_1 = L_2 = 75$ (*top*), 65 (*middle*), and 55 (*bottom*) dB SPL] in the mouse. Left DP-gram panels show absolute DPOAE levels preexposure (*solid squares*) and immediately postexposure (*open circles*) as well as at 1 day (*open triangles*) and 7 days (*open diamonds*) postexposure. These same data are presented in the *right panels* as corresponding "difference" DP-grams, that is, postexposure DP-grams minus preexposure DP-grams. It is clear that DPOAEs evoked by lower primary-tone levels of 55 dB SPL (*bottom*) were affected more than those elicited by higher-level primaries, particularly for the 75-dB SPL primary tones (*top*). *Dotted lines* at "0 dB" in the "difference" plots at the right indicate "no change" from preexposure DPOAE levels.

threshold and DPOAE levels were normal. However, there was substantial OHC disruption observed upon electron microscopy. Thus, DPOAEs can be normal in the presence of substantial OHC disruption, similar to the normal CAP thresholds Robertson et al. (1980) measured in the presence of substantial trauma to OHCs in guinea pigs. In fact, previous reports (e.g., Clark and Bohne 1978) indicated that behavioral hearing thresholds can be normal in the presence of a moderate (<30 %) loss of OHCs in the apical turns of the chinchilla cochlea. These observations suggest that DPOAEs are quite robust to relatively small amounts of OHC damage that also might include disruption of their stereocilia. However, in contrast, other studies reported clear correlations between missing OHCs and reduced DPOAE levels (e.g., Clark et al. 1984).

Canlon et al. (1988) first purposely showed that lengthy exposures to moderate-level sound alter the ear's response to subsequent acoustic overstimulation. This "conditioning" protection against noise-induced cochlear damage has since been shown in a variety of laboratory mammal models using emission measures (see the review in Yoshida et al. 2000). The current view is that conditioning likely changes the strength of the medial olivocochlear reflex to protect the ear from acoustic overexposure (Maison and Liberman 2000).

5. Other Pathological Hearing Problems

5.1 Ménière's Disease and Endolymphatic Hydrops

Measurements of evoked emissions in human patient groups with sensorineural hearing losses of various etiologies have demonstrated that OAEs are typically absent in frequency regions in which the hearing level is greater than 25–30 dB (e.g., Kemp et al. 1986). This observation is also true for OAEs in some Ménière's ears. However, TEOAEs have been measured in other Ménière's ears with corresponding hearing thresholds, at the time of measurement, exceeding 40 dB HL (Harris and Probst 1992). Similarly, in many Ménière's ears, DPOAE levels were reduced below population averages at frequencies corresponding to those of increased behavioral thresholds, consistent with data obtained from ears with sensorineural hearing loss of other etiologies. However, again, in some patients, DPOAE levels are normal, or near normal, in the presence of equally large increases in behavioral thresholds (e.g., Perez et al. 1997). One interpretation of these contrasting findings is that only in some cases of Ménière's disease, i.e., those demonstrating reduced OAEs, is OHC activity compromised. In contrast, in Ménière's ears with normal emissions, the hearing loss probably does not result from OHC pathology, but from dysfunction of other sensory or neural components of the cochlea.

Experimental endolymphatic hydrops produced by surgical destruction of the endolymphatic sac and duct is commonly used as an animal model of Ménière's. For example, in hydropic rabbits, behavioral threshold measures showed an appreciable low-frequency hearing loss, which often oscillated over several days postsurgery (Martin et al. 1989). In general, measurements of

DPOAEs in these hydropic rabbits demonstrated changes qualitatively parallel to those observed behaviorally and electrophysiologically, again suggesting that DPOAEs are capable of providing useful information concerning cochlear function.

5.2 Genetically Determined Hearing Loss

In patients with both autosomal dominant- and recessive-based hereditary hearing loss, DPOAEs have been shown to be reduced below population means, or absent (e.g., Martin et al. 1990). More recently, some patients with familial-related hearing loss, including Waardenburg's syndrome (Liu and Newton 1997) and the nonsyndromic condition caused by mutations in the connexin 26 gene (Cohn et al. 1999), showed reduced emissions in the presence of normal hearing. These results support the view that OAEs can detect subtle dysfunction in OHCs before they become clinically apparent.

In experiments using the mouse model, Horner et al. (1985) reported that whereas normal mice demonstrated strong $2f_1 - f_2$ DPOAEs, no DPOAEs were detected in strains of mice with congenital dysplasia of the stria vascularis, or of the whole cochlea. In contrast, mutants with central deafness, but anatomically normal cochleas, demonstrated DPOAEs similar to those in normal mice. Further, Schrott et al. (1991) studied DPOAEs in the Bronx waltzer mutant mouse, which has about 70% missing IHCs, but a normal number of OHCs that show some anatomical and morphological disruption. In this mutant, $2f_1 - f_2$ DPOAEs were detectable, although reduced by 10–20 dB, and the detection thresholds for DPOAE I/Os were elevated relative to normal control mice (Schrott et al. 1991). It is probable that such abnormal DPOAE attributes resulted from the disruption of OHCs rather than from the absence of a majority of the IHCs.

Schrott et al. (1991) also discovered that in homozygous W^v/W^v mutant mice with selective OHC loss, but IHCs that appeared normal by light microscopy, DPOAEs could not be detected, except in one ear of a single animal, in which both the DPOAEs and the OHCs appeared normal. This result indicates that OHCs are critical for DPOAE generation, and is consistent with the effects on DPOAEs of chronic gentamicin treatment (Brown et al. 1989), in which it was shown that DPOAEs were reduced when the gentamicin selectively damaged the OHCs.

DPOAEs have been useful, in particular, in determining cochlear function in a number of inbred mutant mice including the deafness *dn/dn* (Huang et al. 1995), CD1 (Le Calvez et al. 1998), hypothyroid hyt/hyt (Li et al. 1999), Turner syndrome X,O (Hulcrantz et al. 2000), and deafwaddler (Konrad-Martin et al. 2001) mouse strains. Moreover, other inbred mouse strains were also investigated for their susceptibility or resistance to acoustic injury, including the 129/SvEv (Yoshida et al. 2000), B6.CAST (Vazquez et al. 2004), the WB/ReJ (WB), BALB/cByJ (BALB), and C57 (Jimenez et al. 2001), and the MOLF/Ei (Candreia et al. 2004) using DPOAEs.

5.3 Age-Related Hearing Loss

In humans, reduced TEOAE, SOAEs, and DPOAEs were shown in cases of aging or presbycusis in that emissions decrease with age over about 40–50 y (e.g., Cilento et al. 2003). However, using sophisticated statistical tests to separate the influence of reduced hearing sensitivity from aging effects, Stover and Norton (1993) established that the age-related threshold elevations caused the related reductions in emissions, and not aging in itself.

Although OAEs have been useful in studying the effects of aging in animal models such as the monkey (Torre and Fowler 2000), over the past decade there has been a proliferation of published studies of the effects of aging on DPOAEs in mouse models, in particular. For example, Parham (1997) was first to describe the progressive deterioration of DPOAE levels from high to low frequencies in the C57 strain, which represents the classic mouse model of early age-related hearing loss.

It is well known primarily from behavioral research that the mouse hears up to about 100 kHz, and optimally from ≈4–60 kHz (e.g., Ehret 1974). However, it is only relatively recently (Jimenez et al. 1999) that the majority of the mouse hearing range was tested with DPOAEs elicited by f_2's ranging from 6 to 54 kHz ($f_2/f_1 = 1.25$), thus representing the most sensitive frequency extent of mouse hearing. In the Jimenez et al. (1999) study, extensive serial measures from 2 to 15 months of age were obtained on both the C57 and CBA/CaJ (CBA) mouse strains, along with two other strains consisting of BALB and WB mice. These systematic data indicated that, whereas the C57 and CBA mice represented early and late-developing AHL models, respectively, the WB exhibited an age-related high-frequency absence of DPOAEs that was even earlier, at 2 months of age, than the C57 strain, and the BALB showed an intermediate aging effect. Thus, OHC pathology due to aging factors is strain specific and, across these particular mice, can be ordered from earliest to latest as WB > C57 > BALB > CBA (i.e., 2, 5, 8, and >15 months, respectively). Clearly, the OAE research on the effects of aging on OHC activity in progressively aging mouse ears will eventually contribute some significant information about the genetic basis of presbycusis, because of the wealth of available details on the mouse genome.

6. Conclusions

TEOAEs, SFOAEs, and SOAEs in animal models demonstrate extreme vulnerability to those traumas thought to affect the CA. The precise relationship between the generator of TEOAEs and SFOAEs and the generator of DPOAEs is unknown. Recent evidence (Shera and Guinan 1999; Chapter 8) suggests that they may reflect different aspects of the same process, presumably an aspect of CA function related to either OHC electromotility or stereociliary activity. Future studies that directly compare the effects of discrete manipulations of cochlear status on these OAEs may shed further light on the relationship of the various OAE phenomena.

The findings of both human and animal experiments indicate that appropriate use of OAEs allows the separation of cochlear from retrocochlear disorders, and further allows differential diagnosis of a specific sensory component of cochlear trauma, that is, dysfunction of the CA that acts to enhance sound-evoked basilar-membrane motion. This cochlear process appears to involve the OHCs and the stria vascularis, and may depend on other cochlear structures. In addition, the data demonstrate that OAEs have the potential to provide information regarding both the frequency pattern and time course of cochlear dysfunction. The findings of experimental studies undoubtedly aid in the design and interpretation of results of clinical tests utilizing OAEs to detect hearing impairments of various etiologies.

The available evidence on the relationship between the various OAE phenomena is consistent with the hypothesis by Shera and Guinan (1999) noted above that TEOAEs, SFOAEs, and the typical low-level SOAEs reflect the output of a common reflection mechanism responding to different stimuli, i.e., transient stimulation, tonal stimulation, and the positive feedback of OAE energy into the OAE-generator's input, respectively. The relationship of the extremely rare, high-level, atypical SOAEs observed in humans and other species to the other OAE phenomena is obscure. However, the atypical SOAEs tend to be associated with cochlear pathology, unlike TEOAEs, SFOAEs, and the typical SOAEs, which tend to be reduced or absent in the presence of cochlear pathology.

The earlier view that all OAEs are produced by the same nonlinear generator that is related, somehow, to the OHC's electromotiity (Brownell 1990) has been confronted in more recent times by the Shera and Guinan (1999) schema described above. However, our understanding of the relationship of DPOAEs to TEOAEs and SFOAEs remains confounded somewhat by the large differences in the relative levels of DPOAEs and TEOAEs/SFOAEs between primates and the common laboratory mammals. In any case, the conceptualization proposed by Shera and Guinan (1999) better explains the paradoxical observation that, whereas human and nonhuman primates tend to have many SOAES, large SFOAEs and TEOAEs, and small DPOAEs, rabbits, guinea pigs, chinchillas, mice, etc., have few SOAEs, small SFOAEs and TEOAEs, and large DPOAEs.

In summary, whereas the basic mechanisms and principles of OAE generation appear similar in humans, monkeys, and nonprimate mammals, there are some differences in the properties of OAEs between primates and the common laboratory mammals. While the overall similarities of OAEs across species indicate that it is appropriate to use animal models to study the effects of hearing pathologies on OAEs, such observed differences indicate that caution is required when extrapolating directly from animal models to the human situation. Critical unknowns still exist concerning the origins of the distinct families of DPOAEs, the pathway that propagates OAEs from their locus of generation along the cochlear partition to the stapes footplate, and the precise details concerning how the transduction process at the tips of the stereocilia contributes to the generation of emissions. Most certainly, experimental animal models will play a leading role in obtaining the objective evidence that will help solve these crucial issues.

Acknowledgments. This work was supported by grants from the Public Health Service (DC00613, DC03114). The authors thank Barden B. Stagner for assisting in the preparation of the figures.

References

Abdala C (2004) Distortion product otoacoustic emission ($2f_1 - f_2$) suppression in 3-month-old infants: evidence for postnatal maturation of human cochlear function. J Acoust Soc Am 116:3572–3580.

Abdala C, Sininger YS, Ekelid M, Zeng F-G (1996) Distortion product otoacoustic emission suppression tuning curves in human adults and neonates. Hear Res 98:38–53.

Anderson SD, Kemp, DT (1979) The evoked cochlear mechanical response in laboratory primates. Arch Otorhinolaryngol 224:47–54.

Avan P, Loth D, Menguy C, Teyssou M (1990) Evoked otoacoustic emissions in guinea pig: basic characteristics. Hear Res 44:151–160.

Avan P, Lemaire JJ, Dordain M, Chazal J, Buki B, Ribari O (1998) Otoacoustic emissions and monitoring of intracranial hypertension. J Audiolog Med 7:46–57.

Avan P, Bonfils P, Mom T (2001) Correlations among distortion product otoacoustic emissions, thresholds and sensory cell impairments. Noise Health 3:1–17.

Berlin CI, Hood LJ, Hurley A, Wen H (1994) Contralateral suppression of otoacoustic emissions: an index of the function of the medial olivocochlear system. Otolaryngol Head Neck Surg 110:3–21.

Brown AM (1987) Acoustic distortion from rodent ears: a comparison of responses from rats, guinea pigs and gerbils. Hear Res 31:25–38.

Brown AM, Gaskill SA (1990) Measurement of acoustic distortion reveals underlying similarities between human and rodent mechanical responses. J Acoust Soc Am 88: 840–849.

Brown AM, Kemp DT (1984) Suppressibility of the $2f_1 - f_2$ stimulated acoustic emissions in gerbil and man. Hear Res 13:29–37.

Brown MC, Nuttall AL (1984) Efferent control of cochlear inner hair cell responses in the guinea pig. J Physiol 354:625–646.

Brown AM, McDowell B, Forge A (1989) Acoustic distortion products can be used to monitor the effects of chronic gentamicin treatment. Hear Res 42:143–156.

Brown AM, Woodward S, Gaskill SJ (1990) Frequency variations in spontaneous sound emissions from guinea pig and human ears. Eur Arch Otorhinolaryngol 247:24–28.

Brownell WE (1990) Outer hair cell electromotility and otoacoustic emissions. Ear Hear 11:82–92.

Brownell WE, Bader CR, Bertrand D, De Ribaupierre Y (1985) Evoked mechanical responses of isolated cochlear hair cells. Science 227:194–196.

Candreia C, Martin GK, Stagner BB, Lonsbury-Martin BL (2004) Distortion product otoacoustic emissions show exceptional resistance to noise exposure in MOLF/Ei mice. Hear Res 194:109–117.

Canlon B, Borg E, Flock A (1988) Protection against noise trauma by pre-exposure to a low level acoustic stimulus. Hear Res 34:197–200.

Cappaert NL, Klis SF, Baretta AB, Juijser H, Smoorenburg GF (2000) Ethyl benzene-induced ototoxicity in rats: a dose-dependent mid-frequency hearing loss. J Assoc Res Otolaryngol 1:292–299.

Cevette MJ, Robinette MS, Carter J, Knops JL (1995) Otoacoustic emissions in sudden unilateral hearing loss associated with multiple sclerosis. J Am Acad Audiol 6:197–202.
Cilento B, Norton SJ, Gates GA (2003) The effects of aging and hearing loss on distortion product otoacoustic emissions. Otolaryngol Head Neck Surg 129:382–389.
Clark WW, Bohne BA (1978) Animal model for the 4-kHz tonal dip. Ann Otol Rhinol Laryngol 87 (Suppl 51):1–16.
Clark WW, Kim DO, Zurek PM, Bohne B.A. (1984) Spontaneous otoacoustic emissions in chinchilla ear canals: correlation with histopathology and suppression by external tones. Hear Res 16:299–314.
Cohn ES, Kelley PM, Fowler TW, Gorga MP, Lefkowitz DM, Kuehn JH, Schaefer GB, Gobar L, Hahn FJ, Harris DJ, Kimberling WJ (1999) Clinical studies of families with hearing loss attributable to mutations in the connexin 26 gene. Pediatrics 103:546–550.
Davis H (1983) An active process in cochlear mechanics. Hear Res 9:79–90.
de Kleine E, Wit HP, Avan P, van Dijk P (2001) The behavior of evoked otoacoustic emissions during and after postural changes. J Acoust Soc Am 220:973–980.
Dhar S, Talmadge CL, Long GR, Tubis A (2002) Multiple internal reflections in the cochlea and their effect on DPOAE fine structure. J Acoust Soc Am 112:2882–2897.
Dieler R, Shehata-Deiler WE, Brownell WE (1991) Concomitant salicylate-induced alterations of outer hair cell subsurface cisternae and electromotility. J Neurocytol 20: 637–653.
Dolan DF, Guo MH, Nuttall AF (1997) Frequency-dependent enhancement of basilar membrane velocity during olivocochlear bundle stimulation. J Acoust Soc Am 102:3587–3596.
Douek EE, Dodson HC, Bannister LH (1983) The effects of sodium salicylates on the cochlea of the guinea pig. J Laryngol Otol 93:793–799.
Ehret G (1974) Age-dependent hearing loss in normal hearing mice. Naturwissenschaften 11:506.
Engdahl B, Kemp DT (1996) The effect of noise exposure on the details of distortion product otoacoustic emissions in humans. J Acoust Soc Am 99:1573–1587.
Evans EF, Borerwe TA (1982) Ototoxic effects of salicylates on the responses of single cochlear nerve fibres. Br J Audiol 16:101–108.
Evans EF, Wilson JP, Borerwe TA (1981) Animal models of tinnitus. In: Evered D, Lawrenson G (eds) Tinnitus. London: Pitman Books, pp. 108–138.
Fahey PF, Stagner BB, Lonsbury-Martin BL, Martin GK (2000) Nonlinear interactions that could explain distortion product interference response areas. J Acoust Soc Am 108:1786–1802.
Franklin DJ, Lonsbury-Martin BL, Stagner BB, Martin GK (1991) Altered susceptibility of $2f_1 - f_2$ acoustic distortion products to the effects of repeated noise exposure in rabbits. Hear Res 53:185–208.
Gates GA, Karzon RK, Garcia P, Peterein J, Storandt M, Morris JC, Miller P (1995) Auditory dysfunction in aging and senile dementia of the Alzheimer's type. Arch Neurol 52:626–634.
Gorga MA, Neely ST, Bergman BM, Beauchaine KL, Kaminski JR, Peters J, Schulte L, Jesteadt W (1993) A comparison of transient-evoked and distortion product otoacoustic emissions in normal-hearing and hearing-impaired subjects. J Acoust Soc Am 94: 2639–2648.
Grewe TSD, Danhauer LJ, Danhauer KJ, Thornton ARD (1994) Clinical use of otoacoustic emissions in children with autism. Int J Ped Otorhinolaryngol 30:133–142.

Guinan JJ Jr (1986) Effect of efferent neural activity on cochlear mechanics. Scand Audiol Suppl 25:53–62.

Guinan JJ Jr (2006) Olivocochlear efferents: anatomy, physiology, function, and the measurement of efferent effects in humans. Ear Hear 27:589–607.

Harding GW, Bohn BA, Ahmad M (2002) DPOAE level shifts and ABR threshold shifts compared to detailed analysis of histopathological damage from noise. Hear Res 174:158–171.

Harris FP, Probst R (1992) Transiently evoked otoacoustic emissions in patients with Ménière's disease. Acta Otolaryngol 112:36–44.

Harris FP, Lonsbury-Martin BL, Stagner BB, Coats AC,. Martin GK (1989) Acoustic distortion products in humans: systematic changes in amplitude as a function of f_2/f_1 ratio. J Acoust Soc Am 85:220–229.

Hatzopoulos S, Di Stefano M, Campbell KC, Falgione D, Ricci D, Rosignoli M, Finesso M, Albertin A, Previati M, Capitani S, Martini A (2001) Cisplatin ototoxicity in the Sprague-Dawley rat evaluated by distortion product otoacoustic emissions. Audiology 40:253–264.

Hatzopoulos S, Petruccelli J, Laurell G, Avan P, Finesso M, Martini A (2002) Ototoxic effects of cisplatin in a Sprague-Dawley rat model as revealed by ABR and transiently evoked otoacoustic emission measures. Hear Res 170:70–82.

Hauser R, Probst R, Harris FP, Frei F (1992) Influence of general anesthesia on transiently evoked otoacoustic emissions in humans. Ann Otol Rhinol Laryngol 101:994–999.

Hayashida T, Hiel H, Dulon D, Erre J-P, Guilhaume A, Aran J-M (1989) Dynamic changes following combined treatment with gentamicin and ethacrynic acid with and without acoustic stimulation. Acta Otolaryngol 108:404–413.

He N-J, Schmiedt RA (1993) Fine structure of the $2f_1 - \mathrm{f}_2$ acoustic distortion product: Changes with primary level. J Acoust Soc Am 94:2659–2669.

Henley CM, Owings MH, Stagner BB, Martin GK, Lonsbury-Martin BL (1989) Postnatal development of $2f_1 - f_2$ otoacoustic emissions in pigmented rat. Hear Res 43:141–148.

Henley C, Wheatherly RA, Martin GK, Lonsbury-Martin B (1996) Sensitive developmental periods for kanamycin ototoxic effects on distortion-product otoacoustic emissions. Hear Res 98:93–103.

Horner KC, Lenoir M, Bock GR (1985) Distortion product otoacoustic emissions in hearing-impaired mutant mice. J Acoust Soc Am 78:1603–1611.

Howard MA, Stagner BB, Lonsbury-Martin BL, Martin GK (2002) Effects of reversible noise exposure on the suppression tuning of rabbit distortion-product otoacoustic emissions. J Acoust Soc Am 111:285–296.

Howard MA, Stagner BB, Foster PK, Lonsbury-Martin BL, Martin GK (2003) Suppression tuning in noise-exposed rabbits. J Acoust Soc Am 114:279–293.

Huang J-M, Money MK, Berlin CI, Keats BJB (1995) Auditory phenotyping of heterozygous sound-responsive (+/dn) and deafness (dn/dn) mice. Hear Res 88:61–64.

Hultcrantz M, Stenberg AE, Fransson A, Canlon B (2000) Characterization of hearing in an X,O "Turner mouse." Hear Res 143:182–188.

Jimenez AM, Stagner BB, Martin GK, Lonsbury-Martin BL (1999) Age-related loss of distortion-product otoacoustic emissions in four mouse strains. Hear Res 138:91–105.

Jimenez AM, Stagner BB, Martin GK, Lonsbury-Martin BL (2001) Susceptibility of DPOAEs to sound overexposure in inbred mice with AHL. J Assoc Res Otolaryngol 2:233–245.

Johnson A-C, Canlon, B (1994) Toluene exposure affects the functional activity of the outer hair cells. Hear Res 72:189–196.

Kalluri R, Shera CA (2001) Distortion-product source unmixing: a test of the two-mechanisms model for DPOAE generation. J Acoust Soc Am 109:622–637.

Katbamna B, Homnick DN, Marks JH (1999) Effects of chronic tobramycin treatment on distortion product otoacoustic emissions. Ear Hear 20:393–402.

Kemp DT (1978) Stimulated acoustic emissions from within the human auditory system. J Acoust Soc Am 64:1386–1391.

Kemp DT (1979) Evidence of mechanical nonlinearity and frequency selective wave amplification in the cochlea. Arch Otorhinolaryngol 224:37–45.

Kemp DT (1981) Physiologically active cochlear micromechanics-one source of tinnitus. In: Evered D, Lawrenson G (eds) Tinnitus. London: Pitman Books Ltd., pp. 54–81.

Kemp DT, Brown AM (1983) A comparison of mechanical nonlinearities in the cochleae of man and gerbil from ear canal measurements. In: Klinke R, Hartmann R (eds), Hearing—Physiological Bases and Psychophysics. Berlin: Springer-Verlag, pp. 82–88.

Kemp DT, Brown AM (1984) Ear canal acoustic and round window correlates of $2f_1 - f_2$ distortion generated in the cochlea. Hear Res 13:39–46.

Kemp DT, Chum R (1980) Properties of the generator of stimulated acoustic emissions. Hear Res 2:213–232.

Kemp DT, Bray P, Alexander L, Brown AM (1986) Acoustic emission cochleography—practical aspects. In: Cianfrone G, Grandori F (eds), Cochlear Mechanics and Otoacoustic Emissions. Scand Audiol Suppl 25, pp. 71–96.

Kim DO (1980) Cochlear mechanics: implications of electrophysiological and acoustical observations. Hear Res 2:297–317.

Kim DO, Molnar CE, Matthews JW (1980) Cochlear mechanics: nonlinear behavior in two-tone responses as reflected in cochlear-nerve-fiber responses and in ear-canal sound pressure. J Acoust Soc Am 67:1704–1721.

Knight RD, Kemp DT (2001) Wave and place fixed DPOAE maps of the human ear. J Acoust Soc Am 109:1513–1525.

Konrad-Martin D, Norton SJ, Mascher KE, Tempel BL (2001) Effects of PMCA2 mutation on DPOAE amplitudes and latencies in deafwaddler mice. Hear Res 151: 205–220.

Kössl M, Vater M (2000) Consequences of outer hair cell damage for otoacoustic emissions and audio-vocal feedback in the mustached bat. J Assoc Res Otolaryngology 1:300–314.

Kujawa SG, Fallon M, Bobbin RP (1992) Intracochlear salicylate reduces low-intensity acoustic and cochlear microphonic distortion products. Hear Res 64:73–80.

Laskey RE, Widholm JJ, Crofton KM, Schantz SL (2002) Prenatal exposure to Arocolor 1254 impairs distortion product otoacoustic emissions (DPOAEs) in rats. Toxicol Sci 68:458–464.

Le Calvez S, Avan P, Gilain L, Romand R (1998) CD1 hearing-impaired mice. I. Distortion product otoacoustic emission levels, cochlear function and morphology. Hear Res 120:37–50.

Li D, Henley CM, O'Malley BW Jr (1999) Distortion product otoacoustic emissions and outer hair cell defects in the hyt/hyt mutant mouse. Hear Res 138:65–72.

Liberman MC, Zuo J, Guinan JJ Jr (2004) Otoacoustic emissions without somatic motility: can stereocilia mechanics drive the mammalian cochlea? J Acoust Soc Am 116: 1649–1655.

Liu XZ, Newton VE (1997) Distortion product emissions in normal-hearing and low-frequency hearing loss carriers of genes for Waardenburg's syndrome. Ann Otol Rhinol Laryngol 106:220–225.

Long GR, Tubis A (1988) Modification of spontaneous and evoked otoacoustic emissions and associated psychoacoustic microstructure by aspirin consumption. J Acoust Soc Am 84:1343–1353.

Lonsbury-Martin BL, Martin GK (1988) Incidence of spontaneous otoacoustic emissions in macaque monkeys: a replication. Hear Res 34:13–317.

Lonsbury-Martin BL, Martin GK, Probst R, Coats AC (1988) Spontaneous otoacoustic emissions in a nonhuman primate: II. Cochlear anatomy. Hear Res 33:69–93.

Lonsbury-Martin BL, Harris FP, Hawkins MD., Stagner BB, Martin GK (1990) Distortion-product emissions in humans: I. Basic properties in normally hearing subjects. Ann Otol Rhinol Laryngol, Suppl 236:3–13.

Lonsbury-Martin BL, Whitehead ML, Martin GK (1991) Clinical applications of otoacoustic emissions. J Speech Hear Res 34:964–981.

Lonsbury-Martin BL, Martin GK, McCoy MJ, Whitehead ML (1994) Testing young children with otoacoustic emissions: middle-ear influences. Am J Otol 15, Suppl 1:13–20.

Maison SF, Liberman MC (2000) Predicting vulnerability to acoustic injury with a noninvasive assay of olivocochlear reflex strength. J Neurosci 20:4701–4707.

Manley GA (1983) Frequency spacing of acoustic emissions: A possible explanation. In: Webster WR, Aitkin LM (eds), Mechanisms of Hearing. Clayton, Australia: Monash University Press, pp. 36–39.

Martin GK, Shaw DWW, Dobie RA, Lonsbury-Martin BL (1983) Endolymphatic hydrops in the rabbit: auditory brainstem responses and cochlear morphology. Hear Res 12:65–87.

Martin GK, Probst R, Scheinin SA, Coats AC, Lonsbury-Martin BL (1987) Acoustic distortion products in rabbits. II. Sites of origin revealed by suppression and pure-tone exposures. Hear Res 28:191–208.

Martin GK, Lonsbury-Martin BL, Probst R, Coats AC (1988) Spontaneous otoacoustic emissions in a nonhuman primate. I. Basic features and relations to other emissions. Hear Res 33:49–68.

Martin GK, Stagner BB, Coats AC, Lonsbury-Martin BL (1989) Endolymphatic hydrops in rabbits: behavioral thresholds, acoustic distortion products, and cochlear pathology. In: Nadol JB (ed) Ménière's Disease: Pathogenesis, Pathophysiology, Diagnosis and Treatment, Amsterdam: Kugler and Ghedini Publications, pp. 205–219.

Martin GK, Franklin DJ, Harris FP, Ohlms LA, Lonsbury-Martin BL (1990) Distortion-product emissions in humans: III. Influence of hearing pathology. Ann Otol Rhinol Laryngol (Suppl) 236:29–44.

Martin GK, Jassir D, Stagner BB, Whitehead ML, Lonsbury-Martin BL (1998a) Locus of generation for the $2f_1 - f_2$ vs $2f_2 - f_1$ distortion-product otoacoustic emissions in normal-hearing humans revealed by suppression tuning, onset latencies, and amplitude correlations. J Acoust Soc Am 103:1957–1971.

Martin GK, Jassir D, Stagner BB, Lonsbury-Martin BL (1998b) Effects of loop diureics on the suppression tuning of distortion-product otoacoustic emissions in rabbits. J Acoust Soc Am 104:972–983.

Martin GK, Stagner BB, Jassir D, Telischi FF, Lonsbury-Martin BL (1999) Suppression and enhancement of distortion-product otoacoustic emissions by interference tones above f_2: I. Basic findings in rabbits. Hear Res 136: 105–123.

Martin GK, Candreia C, Bohne BA, Harding GW, Stagner BB, Lonsbury-Martin BL (2002) An augmented acoustic environment delays age-related hearing loss in C57BL/6J mice as revealed by DPOAEs and cochlear histopathology. Assoc Res Otolaryngol Abstr 25:265, pp. 69–70.

Martin GK, De La Garza AN, Stagner BB, Lonsbury-Martin BL (2005) Detailed DPOAE level/phase maps in normal and noise-damaged rabbit ears: insights into generation processes. Assoc Res Otolaryngol Abstr 28:141, p. 52.

Martin GK, Stagner BB, Lonsbury-Martin BL (2006) Assessment of cochlear function in mice: distortion-product otoacoustic emissions. Curr Protocols Neurosci 34:8.21C. 1–8.21C.18.

Mathis A, Probst R, De Min N, Hauser R (1991) A child with an unusually high-level spontaneous otoacoustic emission. Arch Otolaryngol Head Neck Surg 117:674–676.

McFadden D (2002) Masculinization effects in the auditory system. Arch Sex Behav 31:99–111.

McFadden D, Pasanen EG (1994) Otoacoustic emissions and quinine sulfate. J Acoust Soc Am 95:3460–3474.

McFadden D, Plattsmier HS, Pasanen E (1984) Aspirin-induced hearing loss as a model of sensorineural hearing loss. Hear Res 16:251–260.

McWilliams ML, Chen GD, Fechter LD (2000) Low-level toluene disrupts auditory function in guinea pigs. Toxicol Appl Pharmacol 167:18–29.

Mensh BD, Patterson MC, Whitehead ML, Lonsbury-Martin BL, Martin GK (1993) Distortion-product otoacoustic emissions in rabbit: I. Altered susceptibility to repeated pure-tone exposures. Hear Res 70:50–64.

Mills DM (1998) Interpretation of distortion product otoacoustic emission measurements. II. Estimating tuning characteristics using three stimulus tones. J Acoust Soc Am 103:507–523.

Mills DM, Norton SJ, Rubel EW (1993) Vulnerability and adaptation of distortion-product otoacoustic emissions to endocochlear potential variation. J Acoust Soc Am 94:2108–2122.

Moulin A (2000) Influence of primary frequencies ratio on distortion product otoacoustic emissions amplitude. II. Interactions between multicomponent DPOAEs, tone-burst-evoked OAEs, and spontaneous OAEs. J Acoust Soc Am 107:1471–1481.

Moulin A, Kemp DT (1996) Multicomponent acoustic distortion product otoacoustic emission phase in humans. II. Implications for distortion product otoacoustic emissions generation. J Acoust Soc Am 100:1640–1662.

Mountain DC (1980) Changes in endolymphatic potential and crossed olivocochlear bundle stimulation alter cochlear mechanics. Science 210:71–72.

Murphy WJ, Johnson JL (1999) Spontaneous otoacoustic emissions in chinchillas. Assoc Res Otolaryngol Abstr 22:372, p. 94.

Naeve SL, Margolis RH, Levine SC, Fournier EM (1992) Effect of ear-canal air pressure on evoked otoacoustic emissions. J Acoust Soc Am 91:2091–2095.

Norton SJ, Rubel EW (1990) Active and passive ADP components in mammalian and avian ears. In: Dallos P, Geisler CD, Matthews JW, Ruggero MA, Steele CR (eds) Mechanics and Biophysics of Hearing. New York: Springer-Verlag, pp. 219–226.

Ohyama K, Wada H, Kobayashi T, Takasaka T (1991) Spontaneous otoacoustic emissions in the guinea pig. Hear Res 56:111–121.

O Mahoney CF, Kemp DT (1995) Distortion product otoacoustic emission delay measurement in human ears. J Acoust Soc Am 97:3721–3735.

Parham K (1997) Distortion product otoacoustic emissions in the C57BL/6J mouse model of age-related hearing loss. Hear Res 112:216–234.

Pasic TR, Dobie RA (1991) Cis-platinum ototoxicity in children. Laryngoscope 101: 985–991.

Penner MJ, Zhang T (1997) Prevalence of spontaneous otoacoustic emissions in adults revisited. Hear Res 103:28–34.

Perez N, Espinosa JM, Fernandez S, Garcia-Tapia R (1997) Use of distortion-product otoacoustic emissions for auditory evaluation in Ménière's disease. Eur Arch Otorhinolaryngol 254:329–342.

Porter CA, Martin GK, Stagner BB, Lonsbury-Martin BL (2006) Distortion-product otoacoustic emission suppression growth in normal and noise-exposed rabbits. J Acoust Soc Am 120:884–900.

Pouyatos B, Campo P, Lataye R (2002) Use of DPOAEs for assessing hearing loss caused by styrene in the rat. Hear Res 165:156–164.

Probst R, Lonsbury-Martin BL, Martin GK, Coats AC (1987). Otoacoustic emissions in ears with hearing loss. Am J Otol 8:73–81.

Puel J-L, Bledsoe SC Jr, Bobbin RP, Caesar G, Fallon M. (1989) Comparative actions of salicylate on the amphibian lateral line and guinea pig cochlea. Comp Biochem Physiol 93C:73–80.

Rebillard G, Lavigne-Rebillard M (1992) Effect of reversible hypoxia on the compared time courses of endocochlear potential and $2f_1 - f_2$ distortion products. Hear Res 62:142–148.

Ress BD, Sridhar KS, Balkany TJ, Waxman GM, Stagner BB, Lonsbury-Martin BL (1999) Effects of cis-platinum chemotherapy on otoacoustic emissions. The development of an objective screening protocol. Otolaryngol Head Neck Surg 121:693–701.

Robertson D, Johnstone BM, McGill TJ (1980) Effects of loud tones on the inner ear: a combined electrophysiological and ultrastructural study. Hear Res 2:39–43.

Robles L, Ruggero MA (2001) Mechanics of the mammalian cochlea. Physiol Rev 81:1305–1352.

Ruggero MA, Kramek B, Rich NC (1984) Spontaneous otoacoustic emissions in a dog. Hear Res 13:293–296.

Rybak LP (1986) Drug ototoxicity. Ann Rev Pharmacol Toxicol 26:79–99.

Sakashita T, Minowa Y, Hachikawa K, Kubo T, Nakai Y (1991) Evoked otoacoustic emissions from ears with idiopathic sudden deafness. Acta Otolaryngol Suppl 486: 66–72.

Schmiedt RA (1986) Acoustic distortion in the ear canal: I. Cubic difference tones: Effects of acute noise injury. J Acoust Soc Am 79:1481–1490.

Schmiedt RA, Adams JC (1981) Stimulated acoustic emissions in the ear canal of the gerbil. Hear Res 5:295–305.

Schrott A, Puel J-L, Rebillard G (1991) Cochlear origin of $2f_1 - f_2$ distortion products assessed by using 2 types of mutant mice. Hear Res 52:245–254.

Shehata WE, Brownell WE, Dieler R (1991) Effects of salicylate on shape, electromotility and membrane characteristics of isolated outer hair cells from guinea pig cochlea. Acta Otolaryngol 111:707–718.

Shera CA (2004) Mechanisms of mammalian otoacoustic emission and their implications for the clinical utility of otoacoustic emissions. Ear Hear 25:86–97.

Shera CA, Guinan JJ Jr (1999) Evoked otoacoustic emissions arise by two fundamentally different mechanisms: a taxonomy for mammalian OAEs. J Acoust Soc Am 105: 782–798.

Shera CA, Tubis A, Talmadge CL (2004) Do forward- and backward-travel waves occur with the cochlea? Countering the critique of Nobili et al. J Assoc Res Otolaryngol 5:349–359.

Siegel JH, Cerka AJ, Recio-Spinoso A, Temchin N, van Dijk P, Ruggero MA (2005) Delays of stimulus-frequency otoacoustic emissions and cochlear vibrations contradict the theory of coherent reflection filtering. J Acoust Soc Am 118:2434–2443.

Stover L, Norton SJ (1993) The effects of aging on otoacoustic emissions. J Acoust Soc Am 94:2670–2681.

Stypulkowski PH (1990) Mechanisms of salicylate ototoxicity. Hear Res 46:113–146.

Subramaniam M, Salvi RJ, Spongr VP, Henderson D, Powers NL (1994) Changes in distortion-product otoacoustic emissions and outer hair cells following interrupted noise exposures. Hear Res 74:204–216.

Sutton LA, Lonsbury-Martin B.L, Martin GK, Whitehead ML (1994) Sensitivity of distortion-product otoacoustic emissions in humans to tonal over-exposure: time course of recovery and effects of lowering L_2. Hear Res 75:161–174.

Telischi FF, Roth J, Lonsbury-Martin BL, Balkany TJ (1995) Patterns of evoked otoacoustic emissions associated with acoustic neuromas. Laryngoscope 105:675–682, 1995.

Telischi FF, Mom T, Agrama M, Stagner BB, Ozdamar O, Bustillo A, Martin GK (1999) Comparison of the auditory-evoked brainstem response wave I to distortion-product otoacoustic emissions resulting from changes to inner ear blood flow. Laryngoscope 109:186–191.

Torre P 3rd, Fowler CG (2000) Age-related changes in auditory function of rhesus monkeys (*Macaca mulatta*). Hear Res 142:131–140.

Ueda H, Hattori T, Sawaki M, Niwa H, Yanagita, N (1992) The effect of furosemide on evoked otoacoustic emissions in guinea pigs. Hear Res 62:199–205.

Vazquez AE, Jimenez AM, Martin GK, Luebke AE, Lonsbury-Martin BL (2004) Evaluating cochlear function and the effects of noise exposure in the B6.CAST+ahl mouse with distortion product otoacoustic emissions. Hear Res 194:87–96.

West BA, Brummett RE, Himes DL (1973) Interaction of kanamycin and ethacrynic acid. Arch Otolaryngol 98:32–37.

Whitehead ML, Lonsbury-Martin BL, Martin GK (1991) Effects of the crossed acoustic reflex on distortion-product otoacoustic emissions in rabbits. Hear Res 51:55–72.

Whitehead ML, Lonsbury-Martin BL, Martin GK (1992a) Evidence for two discrete sources of $2f_1 - f_2$ distortion-product otoacoustic emission in rabbit: I. Differential dependence on stimulus parameters. J Acoust Soc Am 91:1587–1607.

Whitehead ML, Lonsbury-Martin BL, Martin GK (1992b) Evidence for two discrete sources of $2f_1 - f_2$ distortion-product otoacoustic emission in rabbit: II. Differential physiological vulnerability. J Acoust Soc Am 92:2662–2682.

Whitehead ML, McCoy MJ, Martin GK, Lonsbury-Martin BL (1993) Click-evoked and distortion-product otoacoustic emissions in adults: detection of high-frequency sensorineural hearing loss. Assoc Res Otolaryngol Abstr 16:397, p. 100.

Whitehead ML, Stagner BB, Martin GK, Lonsbury-Martin BL (1996) Visualization of the onset of distortion-product otoacoustic emissions, and measurement of their latency. J Acoust Soc Am 100:1663–1679.

Wier CC, Pasanen EG, McFadden D (1988) Partial dissociation of spontaneous otoacoustic emissions and distortion products during aspirin use in humans. J Acoust Soc Am 84:230–237.

Wilson JP (1980) Model for cochlear echoes and tinnitus based on an observed electrical correlate. Hear Res 2:527–532.

Wilson JP, Evans EF (1983) Effects of furosemide, flaxedil, noise and tone overstimulation on the evoked otoacoustic emission in cat. Proc Cong Int Union Phys Sci 15:100.

Wilson JP, Sutton GJ (1983) "A family with high-tonal objective tinnitus"—An update. In: Klinke R, Hartmann R (eds) Hearing—Physiological Bases and Psychophysics. Berlin: Springer-Verlag, pp. 97–103.

Wit HP, Ritsma RJ (1980) Evoked acoustical responses from the human ear: some experimental results. Hear Res 2:253–261.

Withnell RH, McKinley S (2005) Delay dependence for the origin of the nonlinear derived transient evoked otoacoustic emission. J Acoust Soc Am 117:281–291.

Withnell RH, Yates GK, Kirk DL (2000) Changes to low-frequency components of the TEOAE following acoustic trauma to the base of the cochlea. Hear Res 139:1–12.

Yates GK, Kirk DL (1997) Electrically evoked otoacoustic emissions—Implications for the reverse transduction process. In: Palmer A, Rees A, Summerfield A, Meddis R (eds) Psychophysical and Physiological Advances in Hearing. London: Whurr Publishers, pp. 39–45.

Yoshida N, Liberman MC (2000) Sound conditioning reduces noise-induced permanent threshold shift in mice. Hear Res 148:213–219.

Yoshida N, Hequembourg SJ, Atencio CA, Rosowski JJ, Liberman MC (2000) Acoustic injury in mice: 129/SvEv is exceptionally resistant to noise-induced hearing loss. Hear Res 141:97–106.

Zheng J, Shen W, He DZ, Long KB, Madison LD, Dallos P (2000) Prestin is the motor protein of cochlear outer hair cells. Nature 405:149–155.

Zurek PM, Clark WW (1981) Narrow-band acoustic signals emitted by chinchilla ears after noise exposure. J Acoust Soc Am 70:446–450.

Zwicker E, Manley G (1981) Acoustical responses and suppression-period patterns in guinea pigs. Hear Res 4:43–52.

9
Mechanisms of Mammalian Otoacoustic Emission

Christopher A. Shera and John J. Guinan, Jr.

1. Introduction

This chapter is not a survey of the literature. Neither is it particularly comprehensive, nor especially up to date. Rather, the aim is to provide an accessible introduction to a way of thinking about otoacoustic emissions that we have found productive. We show that a few general principles, systematically applied, can account for a remarkable variety of otoacoustic phenomena. Paramount among these principles is the fact—undisputed, underappreciated, and often utterly ignored—that evoked otoacoustic emissions arise from spatially distributed regions within the cochlea, not from discrete, pointlike sources. Interference among the many wavelets originating within these distributed source regions contributes essentially to shaping the sounds emitted by the ear.

2. Sources of Reverse Waves

Mammalian otoacoustic emissions (OAEs) involve the production of reverse-propagating waves within the cochlea. Reverse waves carry energy back toward the stapes, where it influences the motion of the ossicles; some fraction of the wave energy is thereby transmitted through the middle ear and appears in the ear canal as sound. Although reverse transmission through the middle ear has considerable influence on OAE amplitudes (e.g., Keefe 2002; Puria 2003), our focus here is entirely on mechanisms of emission generation within the cochlea.

2.1 Theoretical Source Types

Theoretically, one might hypothesize that reverse waves could originate from at least two different source types: (1) wave scattering off preexisting mechanical perturbations that exist independently of the stimulus wave ("reflection sources") and (2) wave generation by sources (or perturbations) induced by the stimulus itself via harmonic or intermodulation distortion ("distortion sources"). Reflection and distortion sources are a more mechanistic nomenclature for the emission sources Kemp and Brown (1983) identified as "place-fixed" and "wave-fixed."

2.1.1 Scattering by Preexisting Mechanical Perturbations

Preexisting mechanical perturbations can create reverse waves by disturbing the otherwise smooth forward flow of stimulus energy, a process equivalent to "scattering" the incoming wave. At the micromechanical level, the properties of the cochlear partition presumably vary somewhat irregularly with position (e.g., due to spatial variations in the number, geometry, or mechanical characteristics of the outer hair cells). In primates, for example, anatomical studies suggest that "general irregularity" and "cellular disorganization" characterize the arrangement of outer hair cells in the apical turns of the cochlea (Engström et al.1966; Bredberg1968; 1984; Lonsbury-Martin et al. 1988). These micromechanical perturbations (or "roughness") arise from the discrete cellular architecture of the organ of Corti and appear superimposed on the more gradual variation of parameters responsible for the position–frequency map. Mechanical perturbations can occur in both the passive and the active mechanics. For example, they may result from spatial variations in the density of radial fibers in the BM, or from cell-to-cell variations in the forces produced by hair cells, perhaps due to differences in the density of prestin molecules or the geometry of the hair bundle.

2.1.2 Wave-Induced Sources or Perturbations

Alternatively, the wave sources (or the mechanical perturbations that scatter the wave) can be induced by the stimulus wave itself. For example, nonlinearities in the mechanics can induce regions of mechanical distortion that act in effect as sources of wave energy. Unlike preexisting perturbations, these perturbations/sources are induced by the stimulus wave, and they therefore move with the wave when the stimulus frequency is varied. Regions of mechanical distortion can both "scatter" the energy in the incident wave and, via harmonic and intermodulation distortion, create sources at frequencies not present in the stimulus. For example, if two different stimulus tones are played simultaneously, intermodulation distortion occurs in the region of wave overlap. The region of induced intermodulation distortion acts as a source of waves (e.g., at combination-tone frequencies) that travel away from the source region in both directions. The mechanical nonlinearities in this example may arise from any number of mechanisms, including the "reverse transduction" via somatic electromotility of distortion in OHC receptor potentials and/or the nonlinear gating of stereociliary transduction channels.

2.2 *Distinguishing Source Types Experimentally*

How might reverse waves originating from these two theoretical source types be distinguished experimentally? An answer to this question lies in the frequency-dependence of their phase (Kemp and Brown 1983; Shera and Guinan

1999). The approximate form of this frequency-dependence can be deduced from the local "translation-invariance" of the traveling wave.

2.2.1 Local Translation Invariance of the Traveling Wave

Figure 9.1 illustrates the local translation invariance of the traveling wave. Each of the two panels in the figure shows schematic snapshots of forward waves at two nearby frequencies. The bottom panels show corresponding plots of phase lag versus location in the cochlea. The figures illustrate that the total phase lag experienced by the traveling wave as it propagates inward from the stapes to its characteristic place is approximately independent of frequency. This frequency-independence of the phase lag follows from the approximate local scaling symmetry (e.g., Zweig 1976) manifested by basilar-membrane transfer functions (e.g., Rhode 1971) and neural tuning curves (e.g., Kiang and Moxon 1974; Liberman 1978). When the cochlear frequency–position map is logarithmic, scaling symmetry implies that traveling-wave excitation patterns are locally translation-invariant, with the number of wavelengths in the traveling wave remaining nearly independent of frequency. Compared to a wave of low frequency (black line), a wave of higher frequency (gray line) travels a shorter distance along the cochlea and requires less time to reach its peak. But both waves travel the same number of wavelengths, and their total phase accumulation is therefore the same. Models indicate that similar conclusions apply to reverse waves.

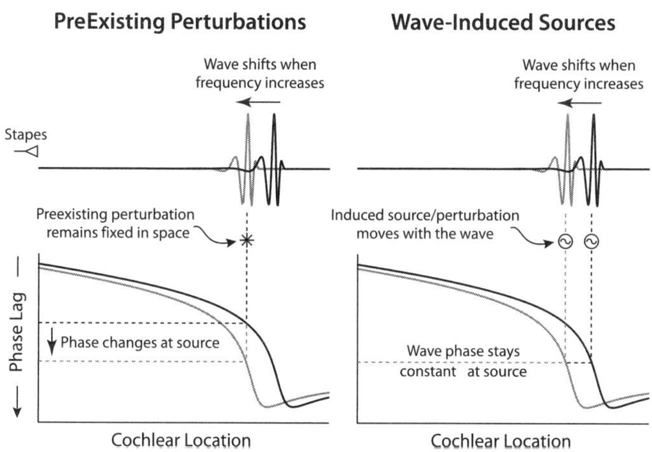

FIGURE 9.1. Schematic showing how the phase of the traveling wave changes at sites of pre-existing (*left*) and wave-induced (*right*) mechanical perturbations and/or sources. The black lines in each panel show a snapshot of the traveling wave (*top*) and the corresponding phase lag (*bottom*) versus distance from the stapes produced by a pure-tone stimulus. The *gray lines* show how the wave and phase profiles change after the stimulus frequency is increased. Although actual perturbations/sources are thought to be densely distributed along the cochlear partition, the figure illustrates the argument using isolated point sources for simplicity. (Adapted with permission from Kalluri and Shera 2001.)

2.2.2 Consequences for Emission Phase

We now consider the implications of this translation-invariance for the phase of the emissions generated by each of the two source types (for a full account, see Shera and Guinan 1999). In the left-hand panel, the stimulus wave encounters a preexisting perturbation; in the right-hand panel, the wave induces a source near the peak of the wave envelope. (For simplicity, the figure shows only single point sources; the arguments are readily generalized to more densely distributed and realistic spatial distributions.) In each case the phase of the reverse wave created by the source/perturbation depends on the phase of the incident wave at the source location (i.e., at the site of reverse wave generation). First consider the encounter with a preexisting perturbation. Because the preexisting perturbation is fixed in space, the phase of the stimulus wave incident on the perturbation varies when the stimulus frequency is changed. For example, when the frequency is increased, the wave pattern shifts basally along the partition, and the phase lag at the perturbation increases. As a result, the phase of the wave back-scattered from the perturbation varies with frequency. The situation is very different for a perturbation or source induced by the stimulus. Because the induced source moves with the stimulus wave, the phase of the incident wave at the source remains almost constant when the stimulus frequency is varied. As a result, the phase of the reverse wave generated by the source is nearly independent of frequency. Note that we assume here that in the process of generating the reverse wave, the induced source does not itself introduce substantial frequency-dependent phase shifts; for a theoretical justification of this assumption, again based on scaling, see the argument of the "demon emitter" (Shera and Guinan 1999).

The argument summarized in Figure 9.1 predicts that emissions evoked from the two theoretical source types will manifest very different phase-versus-frequency functions. Illustrated here for the simplest case (a single pure tone), the argument generalizes to any "frequency-scaled" stimulus. A frequency-scaled stimulus is one in which the corresponding wave pattern produced in the cochlea translates almost unchanged along the partition as the frequency components in the stimulus are varied. Suppose, for example, that the two waves shown in Figure 9.1 are presented simultaneously (so that the gray line corresponds to the f_2, or higher frequency wave, and the black line to the f_1, or lower-frequency wave). When the cochlear map is exponential, so that an octave corresponds to a constant distance along the basilar membrane, sweeping the primaries with the frequency ratio f_2/f_1 held fixed translates the stimulus wave pattern along the cochlear partition.

2.3 Source Types Operating in the Mammalian Ear

Using emission phase as the distinguishing experimental signature, Figure 9.2 demonstrates that both reflection and distortion sources operate in the mammalian ear. The figure shows the SFOAEs and $2f_1 - f_2$ DPOAEs evoked by frequency-scaled stimuli in the same ear. Whereas SFOAE phase varies

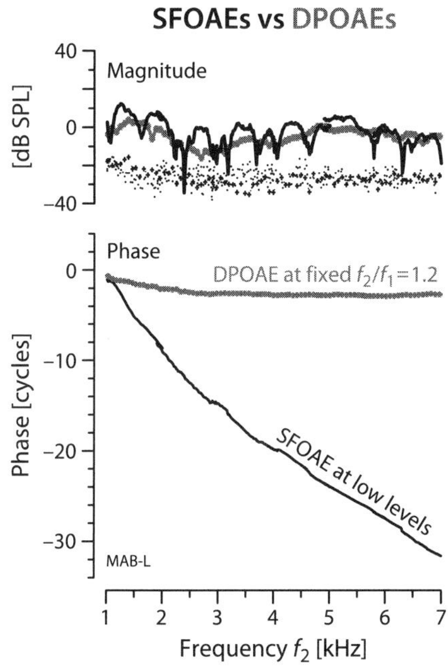

FIGURE 9.2. Experimental evidence that both preexisting and wave-induced emission sources operate in the mammalian ear. The *black lines* show SFOAE amplitude and phase versus frequency measured in a human subject at a probe level of 40 dB SPL. The *gray lines* show the $2f_1 - f_2$ DPOAE measured in the same subject using frequency-scaled stimuli ($f_2/f_1 = 1.2$). Corresponding primary levels were $\{L_1, L_2\} = \{50, 40\}$ dB SPL. To reduce unwanted reflection-source components of the DPOAE, responses were measured in the presence of a 55-dB SPL suppressor tone with frequency near $2f_1 - f_2$. The dots in the *upper panel* show the approximate measurement noise floor. (Adapted with permission from Shera and Guinan 1999.)

by more than 30 cycles, DPOAE phase is nearly independent of frequency, varying by less than one-half cycle over the almost three-octave range of the figure. (Although not shown here, transient-evoked OAEs measured at low and moderate levels manifest the same rapid phase rotation found for SFOAEs.) The phase behaviors shown here match those expected from the two theoretical source types. In particular, the rapid phase variation seen with SFOAEs (and TEOAEs) suggests that these OAEs originate via scattering from preexisting, place-fixed perturbations; likewise, the near constancy of DPOAE phase suggests an origin via sources induced by the stimulus.

2.4 A Puzzle to Ponder

The comparison shown in Figure 9.2 raises an intriguing puzzle: *Why are the results so different for the two stimulus paradigms?* After all, the stimuli used to evoke the DPOAEs and SFOAEs are really quite similar—both consist of one or two frequency-scaled pure tones of comparable intensity. Nevertheless, the resulting emissions are evidently generated by completely different source types. Why is this? Why aren't SFOAEs produced by the same induced nonlinear sources responsible for DPOAEs at comparable stimulus levels? More generally, for any given OAE, what determines the source type that generates it?

2.4.1 The Resolution in a Nutshell

We devote the bulk of this chapter to resolving this puzzle and to fleshing out the concepts its solution entails. In a nutshell, we argue that the resolution hinges on the spatially distributed nature of OAE generation. Although OAEs are often regarded as originating from discrete, punctate locations (indeed, we invoked this same approximation earlier, harmlessly in that case, in order to simplify the discussion of emission phase), OAEs do not emanate from isolated, pointlike sources. Rather, they are a collective phenomenon involving interference among multiple wavelets arising over a relatively broad region of the cochlea. As we review below, these interactions contribute essentially to determining OAE characteristics and apparent source types.

3. Wavelet Summation from Distributed Sources

To pursue the mechanisms of emission generation, we consider waves and distributed sources of waves in the context of the one-dimensional transmission-line model of cochlear mechanics. The presentation here is heuristic, emphasizing important concepts at the expense of rigor and precision. The approach has the advantage of allowing us to "derive" results with a minimum of fuss and overhead, thereby rendering the material more intuitively accessible. But the approach also has the unfortunate drawback of seeming to pull many things magically from a hat without full logical justification; as a result, the basic modeling framework may appear almost ad hoc, with no firm physical basis. Despite possible appearances, the key results we present below are not a priori postulates—although it might have simplified our task, we have refrained from making this stuff up according to whim—but physical consequences of Newton's laws and a few not-unreasonable assumptions applied to cochlear mechanics (e.g., Zweig and Shera 1995; Talmadge et al. 1998, 2000; Shera et al. 2004a).

3.1 Basis Waves and Sources

3.1.1 A Uniform Transmission Line

Before grappling with the complexities encountered in the cochlea, we first consider a simpler system that supports bidirectional wave propagation: a uniform electrical transmission line, infinite in both directions, arrayed along the x-axis.

Figure 9.3 shows the section of the transmission line located between two observation points x_0 and x_1. [Later on, when we apply these concepts to the cochlea, two especially useful observation points will be the stapes and the characteristic place of the stimulus tone.] When driven at angular frequency ω, the transmission line supports voltage waves of the form $\overrightarrow{V}(x) = V_0 e^{-i(kx-\omega t)}$ and $\overleftarrow{V}(x) = V_0 e^{+i(kx+\omega t)}$. The diacritical arrows indicate the direction of wave propagation; waves with rightward-pointing arrows move in the "forward" direction of increasing x. The waves $V^{\rightleftarrows}(x)$ represent plane waves of constant amplitude and wavelength traveling in the forward and reverse directions, respectively. The wave number, k, is equal to $2\pi/\lambda$, where λ is the wavelength. When rendered dimensionless and normalized to unit amplitude at $x = 0$, the forward and reverse waves on a particular transmission line are called its "basis waves." The basis waves for the uniform transmission line, denoted by $w^{\rightleftarrows}(x)$, are given by $w^{\rightleftarrows}(x) = e^{\mp ikx}$. (Here and hereafter the time-dependence of the form $e^{i\omega t}$ has been omitted to simplify the notation.)

Now suppose that we introduce an oscillating source of effective strength V_{src} at some point x' between the observation points x_0 and x_1. In the steady state, the source launches waves $V^{\rightleftarrows}(x) = V_{\text{src}} e^{\mp ik(x-x')}$ traveling away from the source in both directions. What are the total forward and reverse waves at the two observation points? The total reverse wave measured at x_0 is

$$\overleftarrow{V}(x_0) = V_{\text{src}} e^{+ik(x_0-x')} = \left[V_{\text{src}} e^{+ikx_0}\right] e^{-ikx'}. \tag{9.1}$$

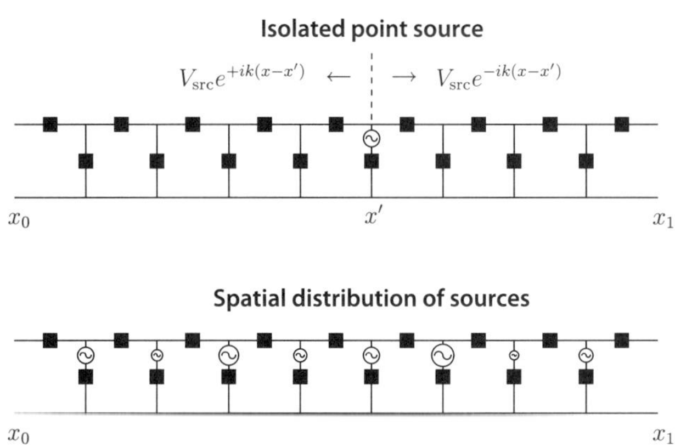

FIGURE 9.3. Waves and sources of waves in a uniform transmission line. The figure shows a segment of an infinite, one-dimensional transmission line consisting of an isotropic ladder network of series impedances and shunt admittances represented by black boxes. In the top panel an oscillating point source at x' launches waves of amplitude V_{src} that travel away from the source in both directions. In the bottom panel the transmission line contains a distribution of sources whose strengths vary somewhat irregularly with position.

Note that the result is proportional to $e^{-ikx'}$, which is simply the forward basis wave, $\overrightarrow{w}(x)$, evaluated at the source location, x'. Thus, apart from a constant factor (e^{+ikx_0}) that depends on the observation point but not on the source, the total *reverse* wave at x_0 is obtained by multiplying the source strength by the value of the *forward* basis wave at the source [i.e., $\overleftarrow{V}(x_0) \propto V_{\text{src}} \overrightarrow{w}(x')$, where $\propto$ means "is proportional to"]. Similarly, the forward wave at x_1 is

$$\overrightarrow{V}(x_1) = V_{\text{src}} e^{-ik(x_1 - x')} = \left[V_{\text{src}} e^{-ikx_1}\right] e^{+ikx'}; \tag{9.2}$$

in other words, the *forward* wave measured at x_1 is proportional to the source strength times the *reverse* basis wave at the source [i.e., $\overrightarrow{V}(x_1) \propto V_{\text{src}} \overleftarrow{w}(x')$].

These results are easily generalized to an arbitrary spatial distribution of sources. Let $v_{\text{src}}(x)$ be the source-strength density (i.e., source strength per unit length Δx). The total *reverse* wave at x_0 is obtained by taking each source element of length Δx located at $x > x_0$; multiplying its respective source strength, $v_{\text{src}}(x)\Delta x$, by the local value of the *forward* basis wave, $\overrightarrow{w}(x)$; and then adding up the wavelets produced by all such sources. In the continuum limit the result is the integral

$$\overleftarrow{V}(x_0) \propto \int_{x_0}^{\infty} v_{\text{src}}(x) \overrightarrow{w}(x) dx. \tag{9.3}$$

Similarly, the total *forward* wave at x_1 is obtained by integrating the product of the source-strength density and the *reverse* basis wave:

$$\overrightarrow{V}(x_1) \propto \int_{-\infty}^{x_1} v_{\text{src}}(x) \overleftarrow{w}(x) dx. \tag{9.4}$$

3.1.2 Generalization to a Cochlear Transmission Line

In the cochlea, the situation is somewhat more complicated because the system's mechanical properties (e.g., the stiffness of the partition and the acoustic mass of the fluids in the scalae) vary with position; the corresponding hydromechanical transmission-line analogue is therefore nonuniform and anisotropic (see Fig. 9.4). As a result, the forward and reverse basis waves that propagate within the structure are no longer simply plane waves with constant amplitude and wavelength (e.g., Shera and Zweig 1991; Talmadge et al. 1998). We denote these more general basis waves by $W^{\rightleftarrows}(x)$; they describe the propagation of transpartition pressure (or *pressure-difference*) waves in the hydromechanical transmission line (e.g., Peterson and Bogert 1950). As before, the dimensionless basis waves are normalized to unity at $x = 0$.

Despite the more complicated basis waves, expressions for the waves produced by a distribution of sources within the cochlea are exactly analogous to Eqs. (9.3) and (9.4) for the uniform transmission line. For example, the total reverse pressure-difference wave at the stapes, $\overleftarrow{P}(0)$, is obtained by multiplying the

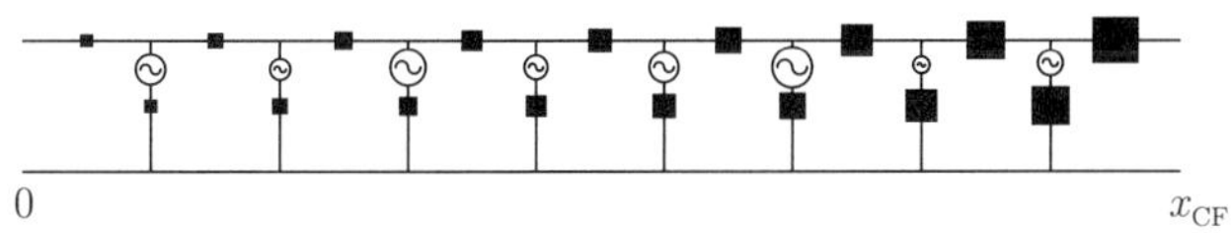

FIGURE 9.4. Simple transmission-line analogue of the cochlea. The figure represents the cochlear region between the stapes ($x = 0$) and the characteristic place (x_{CF}) for a pure tone of angular frequency ω as a nonuniform, hydromechanical transmission line in which the series impedance and shunt admittance vary with position. The transmission line includes a distribution of sources whose strengths are described by a complex-valued source density.

effective source strength, $\varrho(x)$, by the local value of the forward basis wave, $\overrightarrow{W}(x)$, and adding up the resulting wavelets for sources located throughout the cochlea:

$$\overleftarrow{P}(0) \propto \int_0^\infty \varrho(x)\overrightarrow{W}(x)dx. \tag{9.5}$$

Rather than introduce a special symbol for the length of the cochlea, we keep the upper limit of integration as ∞ simply to indicate that the summation extends along the entire cochlea. Similarly, the total forward pressure-difference wave at the characteristic place (x_{CF}) is found to be

$$\overrightarrow{P}(x_{CF}) \propto \int_0^{x_{CF}} \varrho(x)\overleftarrow{W}(x)dx. \tag{9.6}$$

Readers unsatisfied with these heuristic presentations can find more formal derivations elsewhere (Shera and Zweig 1991; Talmadge et al. 1998). Perhaps the most straightforward and intuitive derivation involves the explicit construction of the "traveling-wave Green's function" (Shera et al. 2005a). As a historical aside, we note that the complex addition of multiple wavelets produced by a field of distributed OAE generators was pioneered by Zwicker (1980, 1981), who called the process "vector summation" based on the geometric interpretation of complex numbers.

3.2 Importance of Phase Coherence

Our expressions for the emission pressures (Eqs. ⌊9.5⌋ and ⌊9.6⌋) all involve integrals of the form

$$I(\gamma) = \int A(x;\gamma)e^{i\phi(x;\gamma)}dx, \tag{9.7}$$

where $A(x;\gamma)$ and $\phi(x;\gamma)$ are the amplitude and phase of the integrand; both are assumed to depend on some parameter, here denoted by γ. (For example, in the case of distortion products, γ is the ratio f_2/f_1.) A useful but not inviolate mathematical rule of thumb states that the maximum values of $|I(\gamma)|$ (e.g., the largest emission amplitudes) occur near that value of γ for which the amplitude $A(x;\gamma)$—considered as a function of x—has a tall, broad maximum and, *simultaneously,* the phase $\phi(x;\gamma)$ remains nearly constant over the region of that maximum. (In practice, such values of γ do often exist.) A maximum in $A(x;\gamma)$ can arise when the sources are localized to a region (e.g., about the peak of the stimulus wave). The constant- or "stationary-phase" condition then guarantees that sources located within the peak region combine constructively (i.e., nearly in phase with one another), rather than canceling each other out through mutually destructive interference.

Figure 9.5 illustrates the power of phase-coherent summation with a simple example. The figure displays values of $|I(x;\gamma)|$, where

$$I(x;\gamma) = \int_0^x A(x')e^{i\phi(x';\gamma)}dx', \tag{9.8}$$

for two different values of the parameter γ as x varies continuously from 0 to 1. In both cases, the amplitude $A(x)$ is a bell-shaped curve peaked at $x = \frac{1}{2}$. In this example, the parameter γ controls the slope of the phase near the amplitude peak. When γ is 0 (solid line), the phase $\phi(x;\gamma = 0)$ varies linearly, changing by about one full cycle over the peak. Because of the rapidly rotating phase, the summation represented by the integral is largely incoherent, and the value of $I(x;\gamma = 0)$ near $x = 1$ remains small due to destructive interference among contributions originating at different spatial locations within the peak region. Cancellation is evident in the nonmonotonic behavior of $I(x;\gamma = 0)$, which increases to a maximum just beyond $x = \frac{1}{2}$ and then declines. By contrast, when γ is 1 (dashed line), the phase $\phi(x;\gamma = 1)$ flattens out somewhat about its

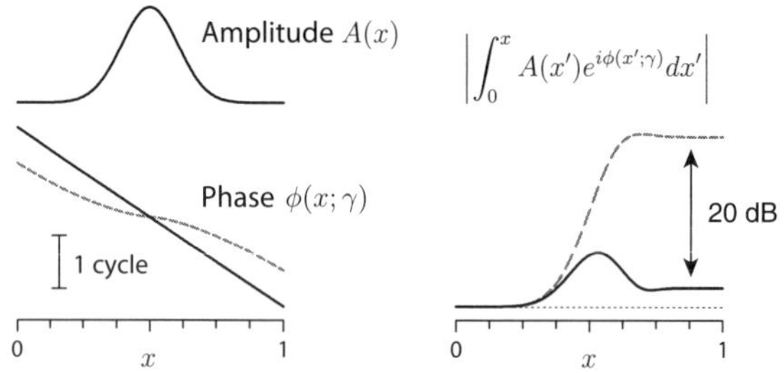

FIGURE 9.5. Importance of phase coherence. The panel on the *right* plots the magnitude of the integral $I(x;\gamma)$ versus x for two different values of γ. The *thin dotted line* shows the location of the 0 along the ordinate. The panel on the *left* plots the Gaussian amplitude $A(x)$ and phase $\phi(x;\gamma)$ of the integrand. The parameter γ controls the slope of the phase near the amplitude peak. The phase is linear for $\gamma = 0$ (*solid line*) but flattens out near the peak of $A(x)$ when $\gamma = 1$ (*dashed line*). The reduction in phase slope increases the magnitude of the integral at $x = 1$ by more than an order of magnitude.

central value at $x = \frac{1}{2}$. The reduction in phase slope, and the resulting increase in coherent summation across the peak region, dramatically boosts the value of $I(x; \gamma = 1)$ near $x = 1$, in this case by more than 20 dB. For another way of looking at this result, note that when $\gamma = 1$ (flattened phase), the amplitude $A(x)$ could everywhere be reduced by an order of magnitude and the integral would still yield a larger magnitude than it does when the original amplitude with $\gamma = 0$ (sloping phase) is used. In their collective effect, a distribution of relatively weak sources arrayed to combine nearly in phase with one another can easily come to dominate much stronger sources that sum incoherently.

3.3 Emergence of Source Directionality

Although Eqs. (9.5) and (9.6) for the emission pressures at the stapes and x_{CF} both have the same general form as integral (9.7) for $I(\gamma)$, their integrands are not identical. In particular, although both expressions contain the same effective source strength, $\varrho(x)$, they involve different basis waves: Whereas the net reverse pressure wave depends on the forward basis wave, the net forward pressure wave depends on the reverse basis wave. Physically, this means that wavelets generated by the distributed source $\varrho(x)$ combine and interfere with one another differently in the two directions. When the phase of the source distribution $\varrho(x)$ depends on x, these interference effects can result in the emergence of a pronounced "directionality" to the waves radiated by the source distribution.

To illustrate how this directionality comes about, we consider an artificial but illuminating example involving the uniform electrical transmission line discussed in Section 3.1.1. Assume that between 0 and L the source density $v_{\text{src}}(x)$ has the form $v_{\text{src}}(x) = v_0 e^{-ikx}$. [This is just the sort of source density function expected when the sources $v_{\text{src}}(x)$ are induced by a stimulus wave propagating along the line in the direction of increasing x.] Using Eqs. (9.3) and (9.4) with basis waves $w^{\rightleftarrows} = e^{\mp ikx}$ to calculate the net voltages at 0 and L resulting from the distribution of sources yields

$$\overleftarrow{V}(0) \propto \int_0^L v_0 e^{-2ikx} dx = \frac{v_0}{2ik}\left(1 - e^{-2ikL}\right) \tag{9.9}$$

and

$$\overrightarrow{V}(L) \propto \int_0^L v_0 dx = v_0 L. \tag{9.10}$$

Figure 9.6 shows the magnitudes of the voltage waves $\overleftarrow{V}(0)$ and $\overrightarrow{V}(L)$ produced by the distribution as a function of the length, L, of the source region. When the source region is small (i.e., much less than a wavelength in extent), the waves $\overleftarrow{V}(0)$ and $\overrightarrow{V}(L)$ emanating from the two ends of the region are nearly the same: the source region radiates equally in the two directions. But as L increases and the

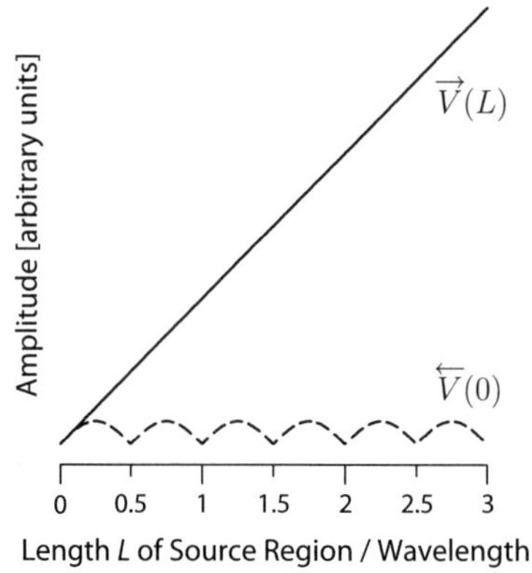

FIGURE 9.6. The emergence of source directionality. The figure shows the magnitudes of the net forward and reverse voltage waves [$\overrightarrow{V}(L)$ and $\overleftarrow{V}(0)$, respectively, computed from Eqs. (9.9) and (9.10)] that radiate from the two ends of the source distribution $v_{\text{src}}(x) = v_0 e^{-ikx}$. Wave magnitudes are plotted versus the assumed length of the source region, L, whose value is given relative to the wavelength of the basis waves for the transmission line. Except when the source region is much less than a wavelength in extent, the distributed source is highly directional, radiating much more in the forward direction than in the reverse.

source region grows in size, a pronounced asymmetry develops. In the forward direction, the wavelets always add in phase, and the amplitude of the net forward wave grows rapidly. But in the reverse direction, destructive interference among component wavelets causes strong cancellation within the source region, and the amplitude of the net reverse wave remains small. Indeed, whenever L matches an integral number of half-wavelengths, internal cancellation is complete and the net reverse wave is zero. Although each individual source within the distribution radiates equally in the two directions, collective interactions induce a strong directionality on the distribution as a whole.

4. Coherent Reflection from Preexisting Perturbations

We now apply the equations and concepts introduced in Section 3 to each of the emission source types. Our goal is to understand how the distributed nature of OAE generation underlies many OAE characteristics and provides the key to resolving the puzzle discussed in Section 2.4.

We begin by considering reverse waves created at the stimulus frequency by preexisting mechanical perturbations. The total reverse pressure wave (or SFOAE) at the stapes is given by Eq. (9.5):

$$\text{SFOAE at stapes} = \overleftarrow{P}(0) \propto \int_0^\infty \varrho(x)\overrightarrow{W}(x)dx, \tag{9.11}$$

an expression found by adding up wavelets created by perturbations located throughout the cochlea. The source-strength density, $\varrho(x)$, represents the local amplitude of the reverse wavelet created by a perturbation at point x. Since the perturbations act, in effect, by partially reflecting the stimulus wave, the source-strength density can be written in the form

$$\varrho(x) \propto r(x)\overrightarrow{W}(x), \tag{9.12}$$

where $\overrightarrow{W}(x)$ is the incident basis wave and $r(x)$ represents a local "reflectivity" (reflection coefficient density) that characterizes the effectiveness of the perturbations at scattering waves at the stimulus frequency (recall that the frequency-dependence has been suppressed in the notation). Note that whereas $\overrightarrow{W}(x)$ is "wave-fixed"—indeed, $\overrightarrow{W}(x)$ *is* the wave—the reflection coefficient $r(x)$ contains a substantial preexisting or "place-fixed" component.

4.1 An Optimal Spatial Frequency for Scattering

Just as a white noise signal can be regarded as consisting of many different temporal frequency components, an irregular reflectivity function can be represented as the sum of many different spatial frequencies. Although a wide range of spatial frequencies may contribute more or less equally to $r(x)$, not all of these spatial frequencies are equally effective at reflecting the traveling wave. Because high spatial frequencies involve more abrupt or discontinuous changes in the mechanics, one might naively expect that high spatial frequencies would reflect more strongly than low spatial frequencies. It turns out, however, that neither high nor low spatial frequencies make especially good scatterers. Because of interference effects, the most effective scattering components in $r(x)$ are those falling within a special *band* of spatial frequencies determined by the form of the traveling wave. Although some spatial frequencies reflect strongly, the vast majority hardly reflect at all, and the OAE ends up being dominated by wavelets scattered from a relatively narrow range of spatial frequencies.

To see what sets this special band of spatial frequencies, we consider the perturbation component $e^{i\kappa x}$ of spatial frequency κ. At the stapes, the contribution to the OAE from wavelets scattered by this spatial frequency is

$$\kappa \text{ contribution to OAE} \propto \int_0^\infty |\overrightarrow{W}(x)|^2 e^{i[\kappa x + 2\angle\overrightarrow{W}(x)]}dx, \tag{9.13}$$

an equation obtained from Eqs. (9.11) and (9.12) by taking $r(x) \propto e^{i\kappa x}$ and writing $\overrightarrow{W}(x)$ in terms of its amplitude, $|\overrightarrow{W}|$, and phase, $\angle\overrightarrow{W}$. Note that Eq. (9.13) has precisely the same form as $I(\gamma)$, with the spatial frequency κ playing the role of the parameter γ (see Eq. [9.7]). Since the amplitude $|\overrightarrow{W}(x)|^2$ is highly peaked, the strongest scattering would normally occur in the region about the characteristic place, unless phase cancellation precludes it. Contributions from neighboring spatial locations combine coherently when their phases are independent of position; reference to (13) shows that this requires that

$$\phi(x;\kappa) = \kappa x + 2\angle\vec{W}(x) \simeq \text{constant}. \tag{9.14}$$

As discussed in Section 3.2, we expect that the integral will be maximal when the region of stationary phase coincides with the region encompassing the amplitude peak. Whether or not these two regions coincide depends on the value of κ. Enforcing the correspondence by requiring that the phase be constant under the peak thus determines the "optimal" value of κ, denoted by κ_{opt}. By differentiating Eq. (9.14) and solving for κ we obtain

$$\kappa_{\text{opt}} = -2\frac{\partial\angle\vec{W}}{\partial x}\bigg|_{\text{peak}} = 2\frac{2\pi}{\lambda_{\text{peak}}}, \tag{9.15}$$

where λ_{peak} is the wavelength of the traveling wave at its maximum. Equation (9.15) sets the central value of the band of spatial frequencies that contribute most to the reflected wave; its value is determined via phase interference effects by the wavelength of the traveling wave.

The theory of coherent reflection briefly sketched out above indicates that the generation of realistic reflection-source OAEs hinges on three basic properties of the traveling wave. First, *the peak of the traveling wave is tall.* As a result, the scattering is effectively localized to the region about the peak. Second, *the peak of the traveling wave is broad relative to the distance between scattering centers* (e.g., between hair cells). The scattering region therefore contains many scattering centers arrayed at many different spatial frequencies. The many scattered wavelets combine and interfere with one another, both constructively and destructively. Finally, *the wavelength of the traveling wave is roughly constant over the scattering region.* Via an analogue of the Bragg scattering condition (Brillouin 1946; Zweig and Shera 1995), wavelets scattered by a particular spatial frequency (κ_{opt}) then combine coherently with one another, producing a large reverse wave.

4.2 Simulated Stimulus-Frequency Emissions

Figure 9.7 demonstrates that when introduced into a cochlear model that produces realistic traveling waves, an irregular array of mechanical perturbations of the sort considered above can indeed account for the major qualitative features of measured stimulus-frequency and transient-evoked OAEs. These features include a rapidly rotating phase and an amplitude spectrum whose leisurely noodling about is occasionally punctuated by deep notches (cf. Fig. 9.2). The perturbations used in the simulation were assumed to change randomly and discontinuously from hair cell to hair cell, and they therefore naturally contained a wide range of spatial frequencies. Although the underlying perturbations are irregular, only

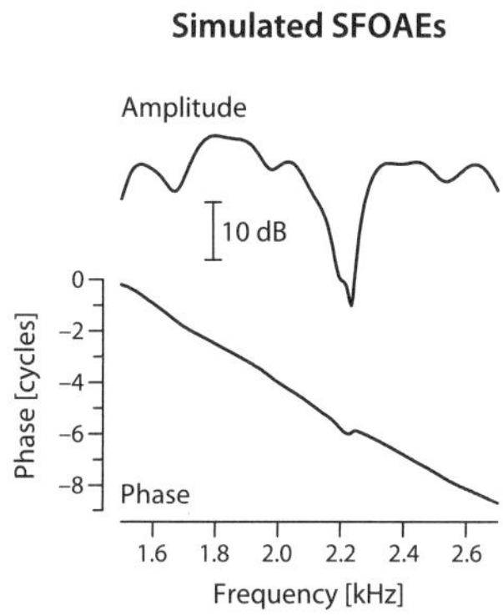

FIGURE 9.7. Simulated SFOAEs. The figure shows the amplitude and phase of stimulus-frequency OAEs versus frequency obtained from an active cochlear model supplemented with random perturbations in the mechanics (e.g., Zweig and Shera 1995; Talmadge et al. 1998; Shera et al. 2005a). Model parameters were adjusted to yield an SFOAE phase slope similar to that measured in human subjects. Detailed spectral features, such as the depths and locations of amplitude wobbles and notches, depend on the particular distribution of perturbations employed in the simulation.

a relatively small amount of this irregularity appears in the emission spectrum. Much of the randomness present in the distribution of perturbations has been "filtered out" through the dynamical action of the traveling wave. Thus, equivalent results would be obtained with much smoother patterns of micromechanical "roughness," so long as the perturbations are not specifically constructed to exclude spatial frequencies within the passband of the "spatial-frequency filter" centered at κ_{opt}.

A more detailed analysis than can be undertaken here shows that when combined with local scaling symmetry (Zweig 1976), the model predicts that mean SFOAE phase-gradient delay (τ_{SFOAE}) is set by the value of κ_{opt}. The delay τ_{SFOAE} is therefore approximately inversely proportional to λ_{peak} (e.g., Shera and Zweig 1993; Zweig and Shera 1995; Talmadge et al. 1998; Shera and Guinan 2003):

$$\tau_{SFOAE} f \backsimeq 2l/\lambda_{peak}, \tag{9.16}$$

where l is the "space constant" in the cochlear map (i.e., the distance over which the CF changes by a factor of e in the basal turns of the cochlea). Equation (9.16) provides a testable relation between SFOAE phase-gradient delay (proportional to the *frequency* derivative of the OAE phase) and the wavelength of the traveling wave (inversely proportional to the *spatial* derivative of the wave phase). Shera et al. (2005a) discuss corrections to this prediction that depend on the effective dimensionality of the fluid dynamics in the region of scattering.

5. Coherent Radiation from Induced Distortion Sources

5.1 Distortion Products at the Stapes

We now consider the generation of distortion products at the frequency f_{dp} in response to primary tones at f_1 and f_2. As always, the total reverse pressure wave at the stapes (our substitute for the DPOAE in the ear canal) is found using Eq. (9.5) by adding up appropriately weighted distortion-product wavelets created throughout the cochlea:

$$\text{DPOAE at stapes} = \overleftarrow{P}(0, f_{dp}) \propto \int_0^\infty \varrho(x, f_{dp}) \overrightarrow{W}(x, f_{dp}) dx. \tag{9.17}$$

We have explicitly included the dependence on f_{dp} as a reminder that multiple frequencies are now involved; our focus here is on a component not present in the original stimulus. Equation (9.17) assumes that distortion-product wavelets, once created, propagate independently of the primaries whose interactions produced them (e.g., the superposition involved ignores any suppressive interactions between the primaries and the distortion product). In effect, Eq. (9.17) treats distortion sources as perturbations in a linear model (e.g., Kanis and de Boer 1997). We expect this approximation (and the power-series approximation to the nonlinearity discussed below) to be best at low stimulus levels where the nonlinearity operates in its nonsaturating regime.

What, then, is the effective distortion-product source distribution $\varrho(x, f_{dp})$? For any particular distortion component the answer depends on micromechanical details not known with any certainty. For example, $\varrho(x, f_{dp})$ depends on the form and operating point of the cochlear nonlinearity (or nonlinearities) responsible for intracochlear distortion. Fortunately, we do not need to know these details to illustrate the general principles in play. Instead, we will focus on the cubic distortion product $2f_1 - f_2$ and approximate the nonlinearity (whatever it may be) by the first odd-order nonlinear term in a power-series expansion. When the primary levels and the frequency $f_{dp} = 2f_1 - f_2$ are held fixed, it then follows that the distortion-product source distribution has the general form (e.g., Talmadge et al. 2000)

$$\varrho(x, 2f_1 - f_2) \overset{\propto}{\sim} \overrightarrow{W}_1^2(x) \overrightarrow{W}_2^*(x), \tag{9.18}$$

where * represents complex conjugation, and we have adopted a notational shorthand in which wave frequencies are indicated using subscripts [e.g., $\overrightarrow{W}_1(x) \equiv \overrightarrow{W}(x, f_1)$]. Note that whereas the source distribution given in Eq. (9.12) has a significant place-fixed component [namely, $r(x)$], the distribution $\varrho(x, 2f_1 - f_2)$ consists entirely of wave-fixed factors. Substituting this source distribution into Eq. (9.17) and separating the amplitude and phase components of the integrand yields an integral in the canonical form $I(\gamma)$ defined by Eq. (19.7):

$$\overleftarrow{P}_{dp}(0; r) \propto \int_0^\infty |\overrightarrow{W}_1^2 \overrightarrow{W}_2 \overrightarrow{W}_{dp}| e^{i[2\angle \overrightarrow{W}_1 - \angle \overrightarrow{W}_2 + \angle \overrightarrow{W}_{dp}]} dx. \tag{9.19}$$

In this case, the role of the parameter γ is taken by the ratio $r = f_2/f_1$, which controls the spatial overlap between the responses at the three frequencies represented in the integral.

Knowing the qualitative form of the traveling wave, what can we deduce from Eq. (9.19) about the variation in emission amplitude as a function of the primary-frequency ratio, r? Applying our mathematical rule of thumb, we expect a maximum in the integral (i.e., maximal ear-canal DPOAE levels) when the integrand has an amplitude peak coincident with a region of stationary phase. Where is the amplitude peak at $\hat{x}$ located? Because of the sharp decrease in wave amplitude apical to the characteristic place, products containing wave amplitudes of different frequencies are generally maximal near the peak of the highest frequency wave in the product. In this case, the region of strongest distortion is therefore localized near the peak of the f_2 wave; thus, $\hat{x} \simeq x_2$. The stationary-phase condition requires

$$\frac{\partial}{\partial x}\left[2\angle\vec{W}_1(x) - \angle\vec{W}_2(x) + \angle\vec{W}_{\mathrm{dp}}(x)\right] = 0. \tag{9.20}$$

Since the spatial derivative of a wave's phase is proportional to its wave number (or, more precisely, to the real part of its wave number), (9.20) can be reexpressed as a condition on the wave numbers of the waves at the three relevant frequencies. Evaluating the result at $\hat{x} \approx x_2$ yields the condition

$$\sum k(x_2) \equiv 2k_1(x_2) - k_2(x_2) + k_{\mathrm{dp}}(x_2) = 0. \tag{9.21}$$

The mathematical rule of thumb thus states that we should expect a maximum in the ear-canal DPOAE at that value of r—call it r_{opt}, assuming it exists—for which the wave numbers $k_2(x)$, $k_1(x)$, and $k_{\mathrm{dp}}(x)$ satisfy (9.21) at $\hat{x} \simeq x_2$. Since the rule of thumb takes no account of variations with r in the *magnitude* (or height) of the peak in $|\vec{W}_1^2\vec{W}_2\vec{W}_{\mathrm{dp}}|$, this analysis is, at best, only approximate. Nevertheless, condition (9.21) yields results in excellent qualitative (and good quantitative) agreement with experiment: phase interference effects of precisely this sort appear to be the primary cause of the well-known "DP filter" (van Hengel 1996; Talmadge et al. 1998; Kemp and Knight 1999; Shera 2003b; Fahey et al. 2005).

How might one expect $\sum k(x_2)$—our shorthand for the sum of wave numbers in Eq. (9.21)—to vary as a function of r? First note that the three wave numbers are always positive, and are evaluated at the point x_2, the peak of the f_2 wave. This means that whereas the wave number k_2 is always evaluated at the peak of its *own* wave, the wave numbers k_1 and k_{dp} are not. Thus, although the value of $k_2(x_2)$ is nearly constant (by approximate translation-invariance), the values of $k_1(x_2)$ and $k_{\mathrm{dp}}(x_2)$ can vary substantially with r. (We assume that the frequency f_{dp} is fixed, so that both f_2 and f_1 depend on r.) Now, since $f_2 > f_1 > f_{\mathrm{dp}}$, the point x_2 is closer to the stapes than x_1 and x_{dp}, the respective peaks of the f_1 and f_{dp} waves. Recall that the wave number of a wave of given frequency increases with distance from the stapes (the wave number is inversely proportional to the wavelength, which decreases with x and decreases especially rapidly as the wave enters the so-called 'short-wavelength' region near its characteristic place). Therefore, $k_2(x_2) \geq k_1(x_2) \geq k_{dp}(x_2)$ at all values of r. When f_1 is roughly half an octave or more below f_2 (i.e., when $r > 1.4$), the f_1 and f_{dp} waves are in the "long-wavelength" region at x_2, so that $k_1(x_2)$ and $k_{dp}(x_2)$ are both small compared to $k_2(x_2)$, which is always evaluated in the short-wavelength region at the peak of the f_2 wave. Consequently, the $-k_2(x_2)$ term dominates the sum and $\sum k(x_2) < 0$; thus, condition (21) fails at large r, and the net reverse wave at f_{dp} is small. At the other extreme, when r is close to 1, all three waves bunch together and have very similar (and large) wave numbers at x_2. Consequently, $\sum k(x_2) \to 2k_{dp}(x_2) \gg 0$; thus, condition (21) fails even more dramatically as $r \to 1$. Since $\sum k(x_2)$ changes sign as r decreases from 1.4 toward 1, continuity requires that it pass through 0 at some intermediate value of r, thereby satisfying condition (21). This intermediate value of r—the value of r_{opt} defined above—leads to maximal DPOAE levels.

Figure 9.8 illustrates these ideas by plotting the amplitude and phase of the integrand of Eq. (9.19) versus x at three different values of r (large, optimal, and close to 1), all evaluated using a model of cochlear mechanics. At large values of r (top panel) the wave envelopes are widely separated along the BM, and the distortion sources are therefore relatively weak. Furthermore, the integrand phase changes considerably over the peak near x_2. Thus, at large ratios, an array of relatively weak sources combine incoherently, and the net DPOAE is small. As r decreases toward 1 (bottom panel), the three wave envelopes approach one another closely, and the value of $|\overrightarrow{W}_1^2\overrightarrow{W}_2\overrightarrow{W}_{dp}|$ at x_2 grows. But although the magnitude peak grows as $r \to 1$ (ignoring suppressive effects), the phase variation across that peak also increases. Since the width of the traveling wave is generally at least a wavelength in extent, the total phase variation across the peak is large, and significant cancellation occurs. For values of r close to 1, an array of strong sources combine incoherently, and the net DPOAE is again small. At values of $r \simeq r_{opt}$ (middle panel), however, conditions are right for producing maximal DPOAEs. Although the distortion sources are not as strong as they are when the waves completely overlap at $r \simeq 1$, the phase variation over the peak region is much smaller. The result is that sources of moderate strength combine coherently to yield a maximal ear-canal DPOAE.

5.2 Distortion Contributions to SFOAEs

A simple application of these ideas demonstrates that distortion contributions to SFOAEs can be strongly attenuated, even when the level of intracochlear distortion is relatively high. The distortion contribution to SFOAEs can be deduced from Eq. (9.19) by considering the limit $r \to 1$ in which the stimulus and emission frequencies coincide (see also Talmadge et al. 2000). Evaluating the integrand at $r = 1$ yields the relation

$$\text{Distortion component of SFOAE} \propto \int_0^\infty |\overrightarrow{W}(x)|^4 e^{i2\angle\overrightarrow{W}(x)}\,dx. \tag{9.22}$$

As discussed above, and illustrated in the bottom panel of Figure 9.8, this limit is characterized by rapid phase rotation over the integrand peak [$\sum k = 2k(x_{CF})$ is large], resulting in near total cancellation among the induced wavelets.

We thus have an answer to the puzzle posed earlier: Intracochlear distortion, even when strong, makes relatively little contribution to the SFOAE measured in the ear canal because the reverse distortion wave undergoes severe attenuation due to phase cancellation among the multiple distortion-source wavelets produced in the source region. This cancellation of distortion-source wavelets allows other mechanisms for producing reverse waves (i.e., coherent reflection from preexisting perturbations, as discussed in Section 4.) to dominate SFOAEs at low and intermediate stimulus levels (e.g., Goodman et al. 2003).

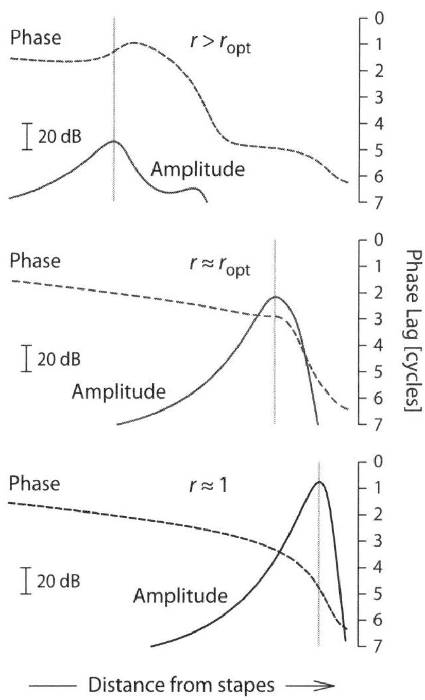

FIGURE 9.8. Coherent and incoherent summation of distortion sources. The figure shows the amplitude (*solid lines*) and phase (*dashed lines*) of the product $\overrightarrow{W}_1^2(x)\overrightarrow{W}_2^*(x)\overrightarrow{W}_{dp}(x)$ computed using a model of cochlear mechanics (Shera 2003b). This product of wave amplitudes appears as the integrand in Eq. (9.19) for $\overleftarrow{P}_{dp}(0; r)$. The top, middle, and bottom panels show results for three different values of the ratio $r = f_2/f_1$ (large, optimal, and close to 1, respectively). The primary levels and the distortion-product frequency $2f_1 - f_2$ were held fixed. The *gray vertical lines* mark the location $\hat{x} \cong x_2$. In the middle panel (the optimal case), the phase near $\hat{x}$ is approximately flat; as a result, the sources near the amplitude peak add coherently and produce the greatest reverse wave. In the top and bottom panels, by contrast, the phase near $\hat{x}$ varies significantly with position; the sources therefore largely cancel one another and do not produce a large reverse wave.

5.3 Distortion Products at their Characteristic Place

The mathematical framework developed above can also be used to compute the amplitude of the net forward wave produced by a distribution of distortion-product sources. Evaluated at its characteristic place (x_{dp}), the forward wave has the value

$$\text{DP at } x_{\mathrm{dp}} = \overrightarrow{P}_{\mathrm{dp}}(x_{\mathrm{dp}}; r) \propto \int_0^{x_{\mathrm{dp}}} |\overrightarrow{W}_1^2 \overrightarrow{W}_2 \overleftarrow{W}_{\mathrm{dp}}| e^{i[2\angle\overrightarrow{W}_1 - \angle\overrightarrow{W}_2 + \angle\overleftarrow{W}_{\mathrm{dp}}]} dx, \qquad (9.23)$$

an equation obtained by substituting Eq. (9.18) for $\varrho(x, f_{\mathrm{dp}})$ into Eq. (9.6) for $\overrightarrow{P}(x_{\mathrm{CF}})$. To deduce the major qualitative trends predicted by this equation, we make use of the approximation $\overleftarrow{W}(x) \simeq \overrightarrow{W}^{-1}(x)$ (Talmadge et al. 1998; Shera et al. 2005a); although this relation is satisfied exactly by the basis waves for a uniform transmission line, it is only approximate in cochlear models and other nonuniform media. As before, the amplitude of the integrand peaks near x_2 and increases to an overall maximum as r approaches 1. Arguments similar to those employed in the paragraphs following Eq. (9.19) then yield the stationary-phase condition

$$2k_1(x_2) - k_2(x_2) - k_{\mathrm{dp}}(x_2) \simeq 0. \qquad (9.24)$$

Note that this equation differs from Eq. (9.21) in the sign of the k_{dp} term; this sign flip dramatically alters the conclusions in the limit $r \to 1$. Whereas $\sum k$ for DPOAEs at the stapes was negative for large r, positive for r close to 1, and zero (indicating coherent summation) at intermediate r, the $\sum k$ for DPs at their characteristic place is negative at large and intermediate r but goes to zero (coherent summation) as $r \to 1$. Equation (24)—together with the increase in the peak magnitude of the integrand as the wave envelopes come together—thus implies that DPs measured at x_{dp} increase monotonically as $r \to 1$. In this case, the amplitude and phase variation, which oppose each other as $r \to 1$ for ear-canal DPOAEs, work together, both acting to increase the amplitude of the forward wave. At the DP place, strong sources combine coherently to produce a large response.

These predictions are consistent with psychophysical observations, which indicate an increase in DP amplitude at the DP place at close frequency ratios (e.g., Goldstein 1967; Wilson 1980a; Zwicker and Harris 1990). Agreement with studies of BM motion is less conclusive (reviewed in Robles and Ruggero 2001). Although some studies find a monotonic increase at close ratios (e.g., Robles et al. 1997), others do not (e.g., Cooper and Rhode 1997), perhaps because of suppressive effects not considered here. The most tightly controlled comparisons are those made using the "Allen–Fahey" paradigm (Allen and Fahey 1992), in which the ear-canal DPOAE is measured as a function of r while the DP response at x_{dp} is held constant (e.g., by monitoring the response of an auditory-nerve fiber tuned to f_{dp}). Studies performed using this paradigm all find that the ratio of ear-canal to intracochlear DPs falls sharply as $r \to 1$ (Allen and Fahey 1992; Shera and Guinan 1997, 2007; de Boer et al. 2005), just as predicted by the analysis above (Shera 2003b).

Interpreted using the framework outlined here, the simultaneous rise (in the intracochlear DP at x_{dp}) and fall (in the ear-canal DPOAE) as $r \to 1$ observed on opposite sides of the distortion-source region (which is centered near x_2) results from the emergence of a pronounced "directionality" to the waves radiated by the source region, as discussed in Section 3.3. The situation is roughly analogous to the example illustrated in Figure 9.6. At close ratios, the phase lag of the induced distortion sources increases significantly with x over the region of strong distortion, which is almost certainly a large fraction of a wavelength or more in extent. As a result, the distortion-source region "radiates" much more in the forward direction than it does in the reverse.

6. Source-Type Mixing

Our analysis has thus far considered reflection and distortion sources separately. But since actual cochleae are, to varying degrees, both nonlinear and mechanically irregular, both source types presumably operate simultaneously. Consequently, we expect sound stimuli to evoke reverse waves from both source types more or less independently. The reverse waves created in these different ways then combine to yield the total OAE measured in the ear canal.

6.1 Source-Type Mixing in DPOAEs

Figure 9.9 illustrates the source-type mixing expected during the generation of DPOAEs (e.g., Shera and Guinan 1999; Talmadge et al. 1999; Mauermann et al. 1999; Knight and Kemp 2001; Konrad-Martin et al. 2001). As discussed in Section 5, intermodulation distortion sources located in the overlap region of the primary traveling waves create wave energy at the frequency $2f_1 - f_2$. The resulting DP wavelets combine and travel away from the source region in both directions (Sections 5.1 and 5.3). The reverse wave that emanates from the distortion-source region propagates back to the stapes, where it is transmitted through the middle ear into the ear canal to become a so-called distortion-source OAE. By contrast, the forward wave that emanates from the source region propagates apically to its characteristic place, scattering off impedance perturbations encountered en route, as described in Section 4. The partial reflection caused by this scattering creates a second reverse wave that also makes its way to the ear canal (in this case, to appear as a "reflection-source" OAE). [Hoping to head off potential confusion, we note that although the reflection-source OAE originated from a forward wave generated by the distortion source, the actual *emission* (i.e., the reverse wave) was created by partial reflection.] The model thus predicts that the total DPOAE measured in the ear canal consists of two emission components, one from each source type.

Figure 9.10 shows that the predictions of the model schematized in Figure 9.9 are borne out experimentally. The solid lines show the total DPOAE; dashed lines show the emission components obtained by unmixing the total DPOAE into its

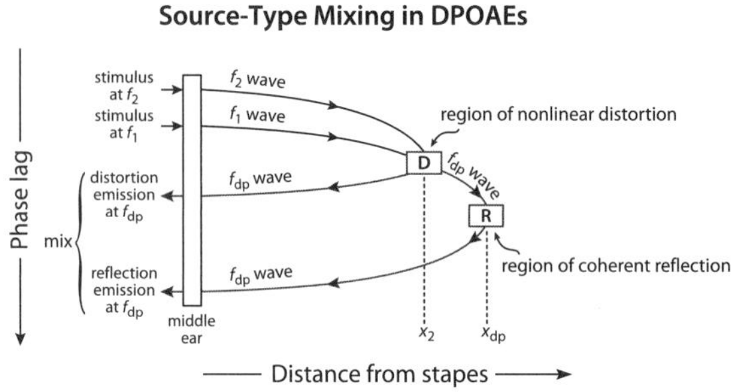

FIGURE 9.9. Source type mixing in DPOAEs. The figure panel shows phase lag relative to stimulus phase (lag increasing downward) of forward and reverse waves versus cochlear location. The primary traveling waves at frequencies f_1 and f_2 produce a region of nonlinear distortion (**D**), located near the peak of the f_2 wave (at x_2), that generates waves at the frequency f_{dp} traveling away from the source in both directions (here, $f_{dp} = 2f_1 - f_2$). The reverse wave propagates to the ear canal, where it appears as a distortion-source emission. The forward wave propagates to the region near its characteristic place at x_{dp}, where it undergoes partial reflection (**R**), generating a second reverse wave that propagates to the ear canal as a reflection-source emission. Emissions from the two source types combine to produce the DPOAE measured in the ear canal. For heuristic purposes, the relative phases of the primary tones have been specially adjusted so that the two curves intersect in region **D**. Phase shifts due to propagation through the middle ear are generally small and are not shown. (Adapted with permission from Shera and Guinan 1999.)

components (Kalluri and Shera 2001). The unmixing procedure employed in this example isolates the putative distortion-source component using a third stimulus tone with frequency near f_{dp} to suppress the amplitude of reflection-source wavelets that would otherwise scatter back from their characteristic place (e.g., Kemp and Brown 1983; Heitmann et al. 1998). As shown in the figure, the total DPOAE unmixes into components whose phase slopes and other characteristics are consistent with the source types predicted by the model. For example, the deep spectral notch and associated phase wiggle evident in the reflection-source component near 1.4 kHz are consistent with predictions of the coherent-reflection model (cf. Fig. 9.7). Finally—and crucially—the reflection-source component of the DPOAE closely matches the SFOAE (dotted lines) evoked at the same frequency under comparable stimulus conditions. Similar results are found in other subjects, demonstrating that DPOAEs typically comprise a mixture of emissions that arise from the two fundamentally different source types operating within the cochlea.

6.2 Changes in Distortion-Source Directionality

Our analysis suggests that the effective directionality of the distortion source depends on the primary-frequency ratio, r (cf. Sections 3.3 and 5.3). The phenomenon of DPOAE source-type mixing provides a simple way to test this

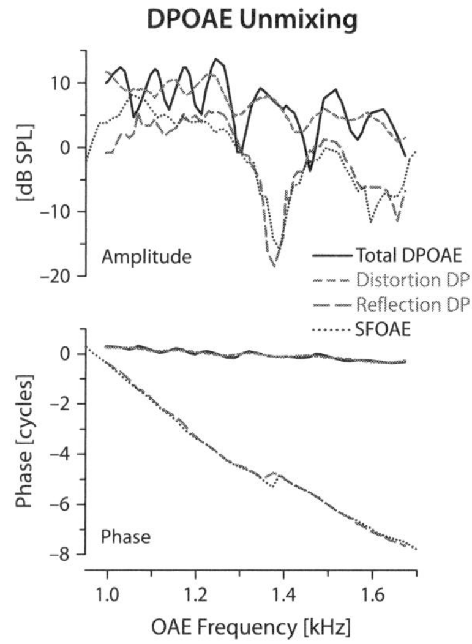

FIGURE 9.10. Distortion-product source unmixing. The figure shows the amplitude and phase of the total $2f_1 - f_2$ DPOAE (solid lines) along with the distortion-source (*short dashed lines*) and reflection-source (long-dashed lines) components obtained using suppression-based unmixing in a human subject ($\{L_1, L_2\} = \{60, 45\}$ dB SPL; $f_2/f_1 = 1.2$). For comparison, the dotted lines show the SFOAE measured in the same subject at a probe level of 40 dB SPL. To make the SFOAE and DPOAE stimulus conditions more comparable, the SFOAE was measured in the presence of an additional tone at the same frequency and level as the f_1 primary used to evoke the DPOAE. (Adapted with permission from Kalluri and Shera 2001.)

prediction. As illustrated in Figure 9.10, the mixing of distortion- and reflection-source OAEs manifests itself as an oscillatory interference pattern in the total DPOAE known as DPOAE fine structure. The interference pattern is created because the relative phase of the two components rotates with frequency. Peaks and valleys in the DPOAE amplitude occur as the relative sign of the components passes alternately through plus and minus one, creating bands of constructive and destructive interference. The magnitude of the oscillations depends on the relative amplitudes of the distortion- and reflection-source components. For example, the deepest interference notches occur when the two components have almost equal amplitudes but opposite phases, so that their sum is close to zero. Changes in the DPOAE interference pattern therefore signal changes in the relative amplitude of the two components.

Knight and Kemp (2000) painstakingly measured DPOAE interference patterns at a large number of ratios, r. Figure 9.11 reproduces some of the data from their experimental tour de force. To interpret the data, we assume that at any given frequency, region **R** in Figure 9.9 reflects back a constant fraction of the total forward DP wave. Changes in the interference pattern then provide a measure of changes in the directionality of the distortion-source region, **D**.

Furthermore, since the larger of the two components dominates the phase of the DPOAE, the overall slope of the DPOAE phase versus frequency function identifies the bigger component. The figure shows that as r decreases from near-optimal ratios, the data reveal precisely the pattern of changes predicted by the schematic model of source-type mixing shown in Figure 9.9 combined with the analysis of Sections 5.1 and 5.3 (see also Dhar et al. 2005). At near-optimal ratios ($r \simeq 1.23$), the shallow phase gradient and relatively small interference notch suggest that the distortion-source component dominates the total DPOAE, just as expected if the distortion-source region were radiating coherently back toward the stapes. At somewhat smaller ratios ($r \simeq 1.15$), the interference notch deepens and the phase jumps by about half a cycle across the notch, indicating that near the notch frequency the two components are approximately equal in amplitude but opposite in phase. This change in the relative amplitude of the two components is consistent with a change in source directionality, in particular, with a decrease in the amplitude of the wave radiated back toward the stapes and/or an increase in the wave radiated apically prior to reflection near the characteristic place. Finally, at the closest ratios ($r \simeq 1.08$), the steep phase gradient indicates that the reflection-source component dominates the total DPOAE. Once again, this is just as expected if the distortion source now radiates primarily in the forward direction so that its energy subsequently emerges from the cochlea primarily as a reflection-source OAE.

6.2.1 Complications Due to Changes in Amplification and Suppression

When interpreting the data in Figure 9.11 as evidence for changes in distortion-source directionality, we assumed for simplicity that the fraction of the incident DP wave reflected back from region **R** is independent of r. Although this assumption is surely false, a more refined analysis only strengthens our conclusions. To see this, note that as r decreases toward 1, the primaries draw closer to x_{dp}. As a result, the effective region of amplification for the forward f_{dp} wave (i.e., the region between x_2 and x_{dp}) shrinks (Allen and Fahey 1992) and its ability to amplify decreases because of increased suppression by the primaries. At close ratios both effects combine to reduce the amplification of f_{dp} waves traveling to and from the region **R** (Shera and Guinan 1997, 2007). [The amplitude of the wave radiated directly back from **D** is also affected, but to a lesser extent since the reverse wave from **D** never travels through regions of maximal f_{dp} gain and/or suppression, which are located on the apical side of **D** (i.e., near the peak of the f_{dp} wave).] We therefore expect a breakdown in the constant-fraction assumption, since the fraction of the DP energy initially radiated apically from **D** that is subsequently backscattered from **R** decreases as $r \to 1$. Returning now to Figure 9.11, note that if we had taken proper account of these effects when interpreting the DPOAE data (i.e., by factoring in the reduction in the size of the gain region and the increase in suppression near **R**), then our estimate of the relative amplitude of the forward DP wave radiated from **D** at close ratios would have *increased*. In other words, by employing the

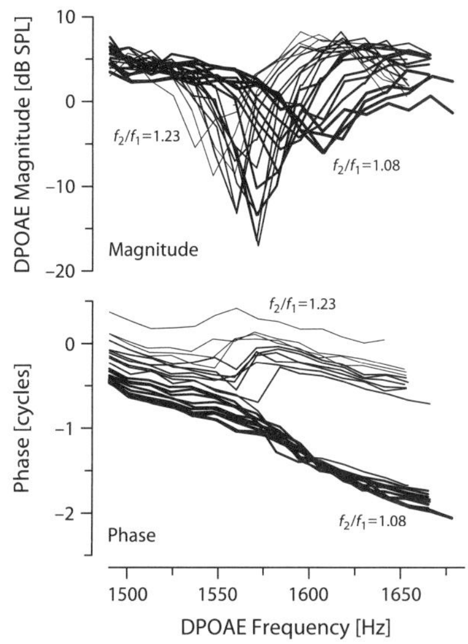

FIGURE 9.11. Evidence for changes in distortion-source directionality. The figure shows amplitude and phase of the $2f_1 - f_2$ DPOAE recorded in a human ear at 21 values of the primary frequency ratio $r = f_2/f_1$ ranging from $r \simeq 1.23$ (thinnest lines) to $r \simeq 1.08$ (thickest lines). The primary levels were fixed at $\{L_1, L_2\} = \{70, 70\}$ dB SPL during the sweeps. (The data are from Knight and Kemp 2000.)

constant-fraction assumption we *underestimated* the apparent change in source directionality.

7. Multiple Internal Reflection

Other than regarding the reverse pressure wave at the stapes as a convenient surrogate for the OAE pressure in the ear canal, our discussion has heretofore ignored what happens to these waves when they encounter the impedance mismatch at the cochlear boundary with the middle ear. Some of the wave energy is, of course, transmitted through the middle ear to the ear canal, but some of it is also reflected back into the cochlea. Like all forward waves, this wave reflected from the stapes is subject to "reemission" within the cochlea (e.g., via coherent reflection from random mechanical perturbations). The reemission process generates a new reverse wave, which is then partially reflected at the stapes, generating yet another forward wave subject to reemission, and so on ad infinitum (see Fig. 9.12). Although multiple internal reflection within the cochlea has many interesting consequences (e.g., Dhar et al. 2002), perhaps the most striking is the emergence of spontaneous otoacoustic emissions, or SOAEs (Kemp 1979; Talmadge et al. 1993 Zweig and Shera 1995; Allen et al. 1995; Shera 2003a).

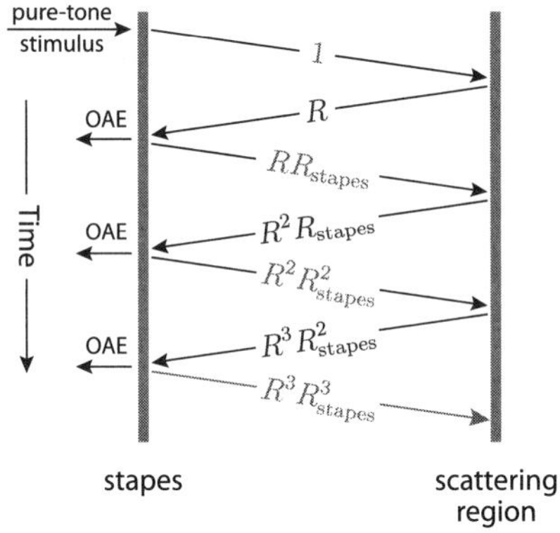

FIGURE 9.12. Spacetime diagram illustrating multiple internal reflection within the cochlea. Distance varies along the abscissa and time increases downward. R and R_{stapes} are the effective frequency-domain reflection coefficients at the stapes for forward waves (*gray lettering*) and reverse waves (*black lettering*), respectively. For simplicity, the forward and reverse waves are shown propagating at constant speed, and the distributed scattering of forward waves that occurs near the characteristic place has been represented as an encounter with a hard wall.

7.1 Standing Waves in the Cochlea

In the low-level linear regime, the effects of multiple internal reflection can be quantified as illustrated in Figure 9.12. We begin by introducing two effective complex-valued (frequency-domain) "reflection coefficients" evaluated near the stapes. The first, denoted by R, is defined looking apically into the cochlea from the stapes and characterizes the reflection (or reemission) of the forward wave within the cochlea; the second, R_{stapes}, is defined looking out of the cochlea from the stapes and characterizes the reflection of the reverse wave by the impedance mismatch with the middle ear. If we now launch an initial pressure wave of unit amplitude from the stapes, the resulting reflected waves have the amplitudes indicated in the figure. For example, the first reflected reverse wave has amplitude R; after reflection from the stapes, this wave, now traveling forward, has amplitude RR_{stapes}, and so on. The total reverse wave at the stapes is found by adding up the relevant contributions highlighted in black. The result is an infinite series whose sum is easily computed for $|RR_{stapes}| < 1$:

$$R\left[1 + RR_{stapes} + (RR_{stapes})^2 + \cdots\right] = R\sum_{n=0}^{\infty}(RR_{stapes})^n = \frac{R}{1 - RR_{stapes}}. \quad (9.25)$$

Note that whenever the product RR_{stapes} is positive real, the terms inside the square brackets all combine in phase. The multiple internal reflections then mutually reinforce one another, creating a significant standing-wave component in the cochlear pressure. As illustrated in Figure 9.13, the amplitude and bandwidth of this "standing-wave resonance" depends sensitively on the amplitude and phase of the product RR_{stapes}.

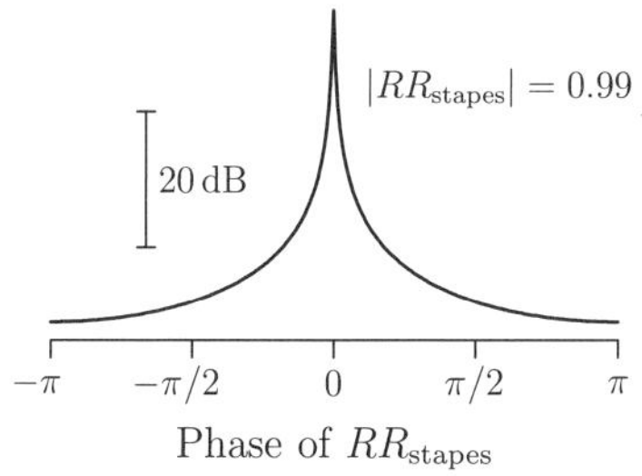

FIGURE 9.13. Standing-wave resonance resulting from the multiple internal reflection of traveling waves. The figure shows the relative standing-wave amplitude given by the infinite series $1 + RR_{stapes} + (RR_{stapes})^2 + \cdots$ (see Eq. [9.25]) versus $\angle\{RR_{stapes}\}$ for $|RR_{stapes}| = 0.99$. Note the sharp peak that occurs when RR_{stapes} has a magnitude close to 1 and a phase corresponding to an integral multiple of 2π.

This heuristic linear analysis suggests that if the round-trip traveling-wave power gain provided by the cochlear amplifier is sufficient to compensate for energy losses encountered en route (e.g., due to damping or acoustic radiation into the ear canal), then self-sustaining cochlear standing waves can form whenever the multiple forward waves all combine in phase with one another (i.e., whenever RR_{stapes} is positive real or, equivalently, $\angle\{RR_{stapes}\} = 2\pi n$). Although Eq. (9.25) predicts that the standing wave grows without bound as RR_{stapes} approaches 1, in the real cochlea, unconstrained growth is prevented by compressive nonlinearities that limit the energy produced. Standing-wave amplitudes grow until the nonlinearity reduces the round-trip power gain to match the net losses. Since the amount of amplification decreases as the wave amplitude grows, standing-wave amplitudes are stable against perturbations: If some random fluctuation increases the wave amplitude slightly, the round-trip gain decreases a little and the wave amplitude falls back down. Conversely, if the wave amplitude decreases, the total gain increases, pulling the oscillation amplitude back up. Once initiated—whether by sounds from the environment or by physiological noise—standing waves of this sort require no external sound for their maintenance. They would be manifest in the ear canal as spontaneous otoacoustic emissions.

7.2 Testing the Standing-Wave Model of SOAEs

The standing-wave model makes a number of predictions about SOAEs that can be tested by exploiting the observation that the phases of middle-ear transfer functions vary slowly compared to the phases of SFOAEs (e.g., Puria 2003). As a result, one can show (Shera 2003a) that

$$\theta(f) \equiv \angle\{RR_{stapes}\} \approx \angle P_{SFOAE}(f) + \text{constant}. \tag{9.26}$$

The lower panel of Figure 9.14 shows $\angle P_{SFOAE}(f)$ over a wide frequency range measured in a typical human subject. Equation (9.26) implies that to a good first

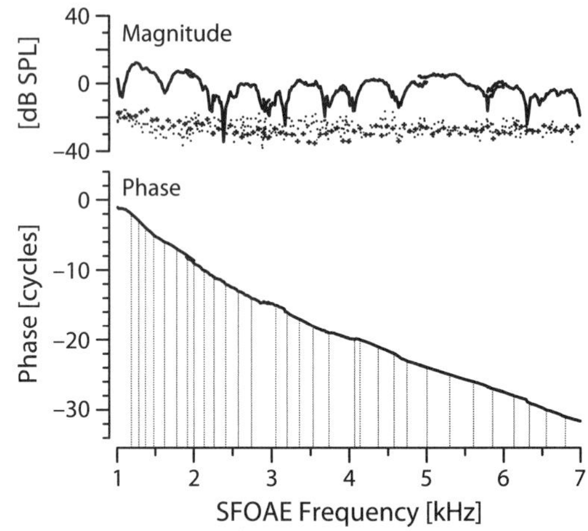

FIGURE 9.14. Potential SOAE frequencies determined by the standing-wave quantization condition. *Vertical gray lines* identifying frequencies at which $\angle P_{\mathrm{SFOAE}} = 2\pi n$ are shown superposed on the human SFOAEs reproduced from Figure 9.2. Note that the potential SOAE frequencies appear roughly quantized; adjacent SOAEs evince a characteristic minimum frequency spacing that both varies with SOAE frequency and manifests substantial variability about the mode.

approximation, $\theta(f)$ in this subject is just the curve shown in the figure shifted up or down by an unknown but constant amount. Recall now that the standing-wave model predicts that SOAE frequencies can occur only when $\angle\{RR_{\mathrm{stapes}}\}$ matches an integral multiple of 2π. In other words, SOAE frequencies must satisfy

$$\theta(f_{\mathrm{SOAE}}) = 2\pi n \qquad (n = 1, 2, \ldots), \tag{9.27}$$

an equation known as the *standing-wave quantization condition*. If we assume for purposes of illustration that the unknown constant in Eq. (9.26) is zero (it will, of course, vary from subject to subject), then SOAEs in this subject can occur whenever the corresponding SFOAE phase passes through an integral number of cycles. These potential SOAE frequencies are indicated on the graph using vertical gray lines.

Several important predictions of the standing-wave model can be deduced immediately by examining the spacing of the vertical lines:

1. Multiple SOAEs are possible in a single ear. This follows because SFOAE phase falls through many cycles, so that the standing-wave quantization condition (Eq. [9.27]) is satisfied at many different frequencies. Multiple SOAEs are in fact the rule. For example, Talmadge et al. (1993) report that 80 % of all emitting ears have more than one emission, with a median of five SOAEs per ear.
2. Adjacent SOAEs manifest a characteristic minimum frequency spacing determined principally by SFOAE phase slope. In other words, SOAE spacings

are roughly quantized. Quantization occurs because SFOAE phase varies approximately uniformly with frequency.
3. The characteristic minimum spacing increases with frequency in a way determined by SFOAE phase. The increased spacing occurs because SFOAE phase slope decreases at higher frequencies. (To convince yourself, hold a straightedge up to the phase curve in the figure.)
4. Frequent "dropouts" can occur. In other words, not all potential SOAEs (vertical lines) are realized as actual SOAEs. The model predicts that dropouts occur because the round-trip gain is not everywhere sufficient to maintain the emission. For example, SOAEs are unlikely to occur near the many dips in the SFOAE amplitude spectrum shown in the top panel. Because SFOAE amplitude changes relatively slowly with frequency compared to SFOAE phase, SOAEs are predicted to occur in irregularly spaced groups within which they appear separated by the local characteristic spacing.
5. SOAE spacings show considerable variation about the quantization mode. The variability occurs because SFOAE phase does not vary perfectly smoothly with frequency. According to the coherent reflection model, this variability in phase slope reflects the irregularity in the perturbations that scatter forward waves.

These qualitative predictions, and others not so immediately apparent from Figure 9.14, have been tested quantitatively (Shera 2003a). The results provide strong support for the standing-wave model and its suggestion that SOAEs are continuously self-evoking SFOAEs.

7.3 Analogy with a Laser Oscillator

The existence of amplitude-stabilized standing waves within the cochlea suggests that the cochlea is acting as a biological analogue of a laser oscillator (see Fig. 9.15). Reduced to its essentials, a laser oscillator consists of a resonant cavity enclosing a gain medium that supports coherent wave amplification. In the cochlea, the "resonant cavity" spans the region between the stapes and the scattering region near the peak of the traveling wave. At either end, cochlear

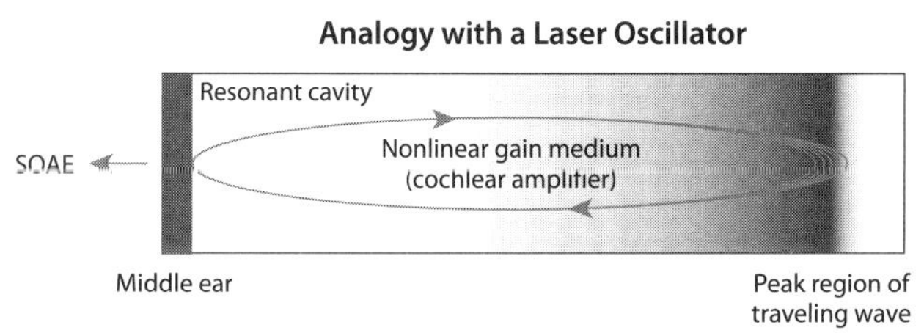

FIGURE 9.15. Schematic illustrating the analogy between the generation of SOAEs and the coherent emission of light by an optical laser. (Adapted with permission from Shera 2003a.)

traveling waves are partially reflected back into the cavity. The cochlear gain medium consists of the cellular force generators and attendant hydromechanical and electrical processes known collectively as the "cochlear amplifier." On each pass through the cavity, waves are amplified by their interaction with the gain medium. At certain special frequencies—namely, those for which the round-trip phase change is an integral multiple of 2π—multiple internal reflection creates standing waves. If the round-trip gain matches the losses, stable oscillations result that appear in the ear canal as SOAEs.

8. Recapitulation

In the introduction, we suggested that systematic application of a handful of general principles could account for a wide variety of otoacoustic phenomena. As a brief summary of the chapter, we now flesh out that suggestion.

8.1 A Handful of Simple Ingredients

Our recipe for otoacoustic emissions includes four basic ingredients. They are:

1. Two source types for reverse-propagating waves. The two source types are (1) scattering by preexisting mechanical perturbations and (2) the induction of sources or perturbations by the wave itself (e.g., via nonlinear distortion).
2. The form and symmetries of the traveling wave. In particular, at low and moderate sound levels, the peak of its envelope is both tall and broad and its wavelength remains approximately constant under the peak. At nearby frequencies the traveling wave manifests approximate translation-invariance.
3. Constructive and destructive interference among wavelets generated within a distributed source region.
4. Partial reflection of energy back into the cochlea by the impedance mismatch at the stapes.

8.2 A Gallimaufry of Otoacoustic Phenomena

With these four ingredients, and a little judicious mixing, we have explained:

1. How the characteristic features of SFOAEs and TEOAEs at low and moderate sound levels can arise by coherent reflection from random perturbations in the mechanics.
2. Why reflection from preexisting perturbations dominates nonlinear distortion in SFOAEs and TEOAEs at low and moderate sound levels (the puzzle of Section 2.4).
3. The shallow phase versus frequency function manifest by DPOAEs evoked by frequency-scaled stimuli.
4. The existence of an optimal f_2/f_1 ratio for DPOAEs measured in the ear canal.

5. The observation that distortion products measured on the BM (or psychophysically) generally increase as $f_2/f_1 \rightarrow 1$, in contrast to DPOAEs.
6. The mixing of distortion- and reflection-source components in DPOAEs and other emissions (e.g., DPOAE microstructure).
7. Systematic changes in the relative amplitudes of the reflection and distortion components of DPOAEs as the ratio f_2/f_1 is varied.
8. The implications for Allen–Fahey-type experiments arising from changes in the effective "directionality" of the distortion source (see also items 4, 5, and 7).
9. The existence and properties of SOAEs, which the model identifies as amplitude-stabilized standing waves produced by the cochlea acting as a biological, hydromechanical analogue of a laser oscillator.

Although the list of explained phenomena is already fairly extensive, the explanatory power of the recipe expands considerably with the admixture of just a single additional ingredient (namely, suppressive and other nonlinear interactions). Of course, we do not mean to suggest that this or any other simple recipe accounts for all otoacoustic phenomena. Nevertheless, given the simplicity of the four ingredients, the brew is quite satisfying.

9. Otoacoustic Emissions and Active Processes

Among the many ingredients not included in our recipe are biophysical details concerning the cochlear amplifier (e.g., the molecular workings of the "active process"). The standing-wave model of SOAEs, for example, presupposes the existence of a "nonlinear gain medium," but the Rube Goldbergian whirling and twirling of the gears and levers within the medium are of little concern. In this respect, the standing-wave model differs fundamentally from Gold's (1948) original conception of SOAEs and its intellectual descendants. Modern biophysical incarnations of Gold's local-oscillator model for SOAEs suppose that they originate via the autonomous motions of subcellular structures (e.g., hair bundles) within the cochlea (e.g., Martin and Hudspeth 1999, 2001; Jülicher et al.2003). In this view, SOAEs are regarded as direct macromechanical manifestations of events at the molecular level. SOAE frequencies, for example, are determined by local cellular properties such as hair-bundle stiffness and geometry, the gating compliance of the transduction channels, and the molecular kinetics of the adaptation and/or transduction processes.

But the standing-wave model and its laser analogy imply that none of these molecular and nanomechanical details are really central to the generation of SOAEs. So long as the gain medium amplifies traveling waves (and impedance perturbations reflect them), then the biophysical details underlying the operation of the cochlear amplifier remain effectively shielded from view and become largely irrelevant to the basic phenomenology. To use the evocative jargon of evolutionary biologists and condensed matter physicists, SOAEs are "emergent," a compelling example of the qualitatively different collective behavior that can occur when even a relatively small number of copies of relatively simple components couple together in relatively simple ways. In this view, surrounding context and coupling among the components are as crucial as the constituents themselves. As illustrated by the standing-wave model of SOAEs—in which SOAE frequencies are determined not by subcellular biophysics but by macromechanical (and emergent) properties of the cochlea such as round-trip traveling-wave phase shifts and middle-ear impedance—emergent phenomena often obey "macrolaws" formulated without reference to the detailed microlaws that govern the behavior of the components. As demonstrated here, many features of OAEs—and, we suspect, many cochlear responses relevant to its signal-processing function, including basic functional characteristics of the cochlear amplifier itself—are perhaps best understood in this way.

10. Epilogue

10.1 Coherent Reflection and Its Discontents

In recent years, the modeling framework reviewed here has been challenged on a number of fronts. These challenges include arguments that:

1. The wave equations underlying the analysis fundamentally misrepresent the hydrodynamics of the cochlea (Nobili et al. 1998, 2003a,b; Nobili 2000).
2. TEOAEs result from the nonlinear response of the cochlea to spectral irregularities in middle-ear transmission, not from scattering by micromechanical perturbations distributed along the cochlea (Nobili et al. 2003a).
3. Suppression experiments demonstrate that SFOAEs originate primarily from the tail of the traveling wave, not from the peak region as indicated by the model (Siegel et al. 2003, 2004).
4. Wave scattering cannot occur in the short-wave region near the peak of the traveling wave (Zwislocki 1983, 2002).
5. OAEs propagate back to the stapes via fast compressional (i.e., sound) waves rather than via slower, transverse pressure-difference waves (Wilson 1980b; Ren 2004; Ren et al. 2006; Ruggero 2004; Siegel et al. 2005).

Although this chapter is not the appropriate forum to delve into the nitty-gritty, the majority of these challenges (items 1, 2, and 4) have now been deflected (e.g., Shera et al. 2004a, 2005a; Kalluri and Shera 2007). For the remaining items (3 and 5), significant counter-arguments have been made (e.g., Cooper and Shera 2004; Shera et al. 2004b, 2005b, 2006, 2007), but as of this writing (spring 2005, references updated 2007), research and controversy continue.

To encourage those who might help resolve these issues or take the field in new directions, we offer some unsolicited but we hope not unhelpful advice. On the experimental front, remember that in addition to well-controlled measurements, compelling tests of any model require at least two things: (1) understanding what the model actually predicts and (2) recognizing that all models have, at best, only limited regions of validity; tests made outside these regions can guide the extension of the model into new regimes, but they cannot refute it. These minimal requirements are analogous to the expectation that those who use experimental data to support theoretical ideas or estimate model parameters need to understand the nature of the measurements and the magnitude of their uncertainties.

Apropos the first requirement, we note that the framework reviewed here indicates that its own predictions can be enormously complicated both by uncontrolled mixing of OAE source types and by multiple internal reflection within the cochlea. Failure to take proper account of these confounding effects can quickly lead to erroneous conclusions. As for the second requirement, we note that the model is based on concepts of cochlear function deduced from mechanical measurements made almost exclusively in the basal half of the cochlea. Mounting evidence suggests, however, that there are important mechanical differences between the base and the apex and that the middle may be most like the apex (e.g., Guinan et al. 2005). Indeed, strong evidence for base-to-apex variation comes from comparisons between otoacoustic and mechanical/neural delays motivated by the models outlined here (Shera and Guinan 2003; Siegel et al. 2005).

On the theoretical front, we note that many of the results discussed in this chapter follow from the analysis of Eqs. (9.5) and (9.6) applied to the various source types evidently operating in the mammalian cochlea. Although "derived" here by analogy with expressions for the waves in a simple, uniform transmission line, these results were first obtained by solving cochlear models in which the equations describing the cochlear partition were extended to include mechanical perturbations and/or sources of energy (Shera and Zweig 1993; Zweig and Shera 1995; Talmadge et al. 1998, 2000; Shera et al. 2005a). The key ideas that emerge from the analysis of these solutions—such as the propagation of otoacoustic energy via forward and reverse pressure-difference waves or the conclusion that coherent wave scattering occurs in a region about the wave peak—are not, therefore, mere postulates of prejudice or convenience. Rather, they are logical consequences of Newton's laws applied to cochlear mechanics in the context of a few reasonable—but not unassailable—assumptions. Our advice, then, is to focus not on the conclusions of the analysis but on the assumptions from which they follow. The challenge is to understand whether the predictions of Newton's laws change dramatically (and, if so, how) when certain ostensibly innocuous assumptions are properly relaxed.

10.2 Extension to Nonmammalian OAEs

Weinberg (1993) identifies two critical mistakes often made in the study of nature. The first mistake, not surprisingly, is taking models too seriously (this chapter?); but the second and perhaps more counterproductive mistake is failing to take them seriously enough. Although the title of our chapter explicitly limits its scope to mammalian OAEs, we close with the conjecture that many of the principles reviewed here will prove applicable to mechanisms of otoacoustic emission in nonmammalian species such as birds and lizards. Our conjecture might seem unlikely on its face, given the apparent absence of a clear analogue of the basilar-membrane traveling wave in many of these species. We caution, however, that readers not get too distracted by terminology or allow what may be superficial differences to mask a deeper underlying isomorphism. Although we make frequent use of the term "traveling wave" because of its demonstrated utility in the mammalian cochlea, the model reviewed here is not grounded on this concept. Rather, the model is ultimately defined by the equations used to describe and approximate the collective physical behavior of the system. Many aspects of this behavior are not unique to the mammalian ear, including active processes, significant phase shifts, and mechanical nonlinearity. Concepts such as constructive and destructive interference among responses generated over distributed source regions are thus likely to play important roles in nonmammalian ears, just as they evidently do in the ears of mammals. Only the ongoing dialog between theory and experiment will tell.

Acknowledgments. As the reference list only incompletely attests, the ideas reviewed here are the collective construction of many. We are especially indebted, however, to the work and insights of Egbert de Boer, David Kemp, Carrick Talmadge, Arnold Tubis, and George Zweig. We thank Christopher Bergevin, Paul Fahey, Radha Kalluri, and Robert Withnell for their helpful comments on the manuscript. This work was supported by grants R01 DC003687 (CAS) and R01 DC000235/R01 DC005977 (JJG) from the NIDCD, National Institutes of Health.

References

Allen JB, Fahey PF (1992) Using acoustic distortion products to measure the cochlear amplifier gain on the basilar membrane. J Acoust Soc Am 92:178–188.

Allen JB, Shaw G, Kimberley BP (1995) Characterization of the nonlinear ear canal impedance at low sound levels. Assoc Res Otolaryngol Abs 18:797.

Bredberg G (1968) Cellular patterns and nerve supply of the human organ of Corti. Acta Otolaryngol Suppl 236:1–135.

Brillouin L (1946) Wave Propagation in Periodic Structures. New York: McGraw-Hill.

Cooper NP, Rhode WS (1997) Mechanical responses to two-tone distortion products in the apical and basal turns of the mammalian cochlea. J Neurophysiol 78:261–270.

Cooper NP, Shera CA (2004) Backward traveling waves in the cochlea? Comparing basilar membrane vibrations and otoacoustic emissions from individual guinea-pig ears. Assoc Res Otolaryngol Abs 27:1008.

de Boer E, Nuttall AL, Hu N, Zou Y, Zheng J (2005) The Allen–Fahey experiment extended. J Acoust Soc Am 107:1260–1266.

Dhar S, Talmadge CL, Long GR, Tubis A (2002) Multiple internal reflections in the cochlea and their effect on DPOAE fine structure. J Acoust Soc Am 112:2882–2897.

Dhar S, Long GR, Talmadge CL, Tubis A (2005) The effect of stimulus-frequency ratio on distortion product otoacoustic emission components. J Acoust Soc Am 117:3766–3776.

Engström H, Ades HW, Andersson A (1966) Structural Pattern of the Organ of Corti. Baltimore: Williams & Wilkins.

Fahey PF, Stagner BB, Martin GK (2005) Mechanism responsible for DPOAE tuning. Assoc Res Otolaryngol Abs 28:319.

Gold T (1948) Hearing. II. The physical basis of the action of the cochlea. Proc R Soc B 135:492–498.

Goldstein JL (1967) Auditory nonlinearity. J Acoust Soc Am 41:676–689.

Goodman SS, Withnell RH, Shera CA (2003) The origin of SFOAE microstructure in the guinea pig. Hear Res 183:1–17.

Guinan JJ, Lin T, Cheng H (2005) Medial-olivocochlear-efferent inhibition of the first peak of auditory-nerve responses: Evidence for a new motion within the cochlea. J Acoust Soc Am 118:2421–2433.

Heitmann J, Waldman B, Schnitzler HU, Plinkert PK, Zenner H-P (1998) Suppression of distortion product otoacoustic emissions (DPOAE) near $2f_1 - f_2$ removes DP-gram fine structure—Evidence for a secondary generator. J Acoust Soc Am 103:1527–1531.

Jülicher F, Camalet S, Prost J, Duke TAJ (2003) Active amplification by critical oscillators. In: Gummer AW (ed), Biophysics of the Cochlea: From Molecules to Models. Singapore: World Scientific, pp. 16–27.

Kalluri R, Shera CA (2001) Distortion-product source unmixing: A test of the two-mechanism model for DPOAE generation. J Acoust Soc Am 109:622–637.

Kalluri R, Shera CA (2007) Near equivalence of human click-evoked and stimulus-frequency otoacoustic emissions. J Acoust soc Am 121: 2097–2110.

Kanis LJ, de Boer E (1997) Frequency dependence of acoustic distortion products in a locally active model of the cochlea. J Acoust Soc Am 101:1527–1531.

Keefe DH (2002) Spectral shapes of forward and reverse transfer functions between ear canal and cochlea estimated using DPOAE input/output functions. J Acoust Soc Am 111:249–260.

Kemp DT (1979) Evidence of mechanical nonlinearity and frequency selective wave amplification in the cochlea. Arch Otorhinolaryngol 224:37–45.

Kemp DT, Brown AM (1983) An integrated view of cochlear mechanical nonlinearities observable from the ear canal. In: de Boer E, Viergever MA (eds) Mechanics of Hearing. The Hague: Martinus Nijhoff, pp. 75–82.

Kemp DT, Knight RD (1999) Virtual DP reflector explains DPOAE "wave" and "place" fixed dichotomy. Assoc Res Otolaryngol Abs 22:396.

Kiang NYS, Moxon EC (1974) Tails of tuning curves of auditory-nerve fibers. J Acoust Soc Am 55:620–630.

Knight RD, Kemp DT (2000) Indications of different distortion product otoacoustic emission mechanisms from a detailed f_1, f_2 area study. J Acoust Soc Am 107:457–473.

Knight RD, Kemp DT (2001) Wave and place fixed DPOAE maps of the human ear. J Acoust Soc Am 109:1513–1525.

Konrad-Martin D, Neely ST, Keefe DH, Dorn PA, Gorga MP (2001) Sources of distortion product otoacoustic emissions revealed by suppression experiments and inverse fast Fourier transforms in normal ears. J Acoust Soc Am 109:2862–2869.

Liberman MC (1978) Auditory-nerve response from cats raised in a low-noise chamber. J Acoust Soc Am 63:442–455.

Lonsbury-Martin BL, Martin GK, Probst R, Coats AC (1988) Spontaneous otoacoustic emissions in the nonhuman primate. II. Cochlear anatomy. Hear Res 33:69–94.

Martin P, Hudspeth AJ (1999) Active hair-bundle movements can amplify a hair cell's response to oscillatory mechanical stimuli. Proc Natl Acad Sci USA 96:14306–14311.

Martin P, Hudspeth AJ (2001) Compressive nonlinearity in the hair bundles active response to mechanical stimulation. Proc Natl Acad Sci USA 98:14386–14391.

Mauermann M, Uppenkamp S, van Hengel PWJ, Kollmeier B (1999) Evidence for the distortion product frequency place as a source of distortion product otoacoustic emission

(DPOAE) fine structure in humans. I. Fine structure and higher-order DPOAE as a function of the frequency ratio f_2/f_1. J Acoust Soc Am 106:3473–3483.
Nobili R (2000) Otoacoustic emissions simulated by a realistic cochlear model. In: Wada H, Takasaka T, Ikeda K, Ohyama K, Koike T (eds) Recent Developments in Auditory Mechanics. Singapore: World Scientific, pp. 402–408.
Nobili R, Mammano F, Ashmore J (1998) How well do we understand the cochlea? Trends Neurosci 21:159–167.
Nobili R, Vetešník A, Turicchia L, Mammano F (2003a) Otoacoustic emissions from residual oscillations of the cochlear basilar membrane in a human ear model. J Assoc Res Otolaryngol 4:478–494.
Nobili R, Vetešník A, Turicchia L, Mammano F (2003b) Otoacoustic emissions simulated in the time domain by a hydrodynamic model of the human cochlea. In: Gummer AW (ed) Biophysics of the Cochlea: From Molecules to Models. Singapore: World Scientific, pp. 524–530.
Peterson LC, Bogert BP (1950) A dynamical theory of the cochlea. J Acoust Soc Am 22:369–381.
Puria S (2003) Measurements of human middle ear forward and reverse acoustics: implications for otoacoustic emissions. J Acoust Soc Am 113:2773–2789.
Ren T (2004) Reverse propagation of sound in the gerbil cochlea. Nat Neurosci 7: 333–334.
Ren T, He W, Nuttall AL (2006) Backward propagation of otoacoustic emissions in the cochlea. In: Nuttall AL, Ren T, Gillespie P, Grosh K, de Boer E (eds) Auditory Mechanisms: Processes and Models. Singapore: World Scientific, pp. 79–85.
Rhode WS (1971) Observations of the vibration of the basilar membrane in squirrel monkeys using the Mössbauer technique. J Acoust Soc Am 49:1218–1231.
Robles L, Ruggero MA (2001) Mechanics of the mammalian cochlea. Physiol Rev 81:1305–1352.
Robles L, Ruggero MA, Rich NC (1997) Two-tone distortion on the basilar membrane of the chinchilla cochlea. J Neurophysiol 77:2385–2399.
Ruggero MA (2004) Comparison of group delays of $2f_1 - f_2$ distortion product otoacoustic emissions and cochlear travel times. Acoust Res Lett Online 5:143–147.
Shera CA (2003a) Mammalian spontaneous otoacoustic emissions are amplitude-stabilized cochlear standing waves. J Acoust Soc Am 114:244–262.
Shera CA (2003b) Wave interference in the generation of reflection- and distortion-source emissions. In: Gummer AW (ed) Biophysics of the Cochlea: From Molecules to Models. Singapore: World Scientific, pp. 439–453.
Shera CA, Guinan JJ (1997) Measuring cochlear amplification and nonlinearity using distortion-product otoacoustic emissions as a calibrated intracochlear sound source. Assoc Res Otolaryngol Abs 20:51.
Shera CA, Guinan JJ (1999) Evoked otoacoustic emissions arise by two fundamentally different mechanisms: a taxonomy for mammalian OAEs. J Acoust Soc Am 105:782–798.
Shera CA, Guinan JJ (2003) Stimulus-frequency-emission group delay: a test of coherent reflection filtering and a window on cochlear tuning. J Acoust Soc Am 113:2762–2772.
Shera CA, Guinan JJ (2007) Cochlear travelling-wave amplification, suppression, and beamforming probed using noninvasive calibration of intracochlear distortion sources. J Acoust Soc Am 121:1003–1016.
Shera CA, Tubis A, Talmadge Cl, de Boer E, Fahey PF, Guinan JJ (2007) Allen-Fahey and related experiments support the predominance of cochlear slow-wave otoacoustic emissions. J Acoust soc Am 121:1564–1575.

Shera CA, Zweig G (1991) Reflection of retrograde waves within the cochlea and at the stapes. J Acoust Soc Am 89:1290–1305.

Shera CA, Zweig G (1993) Order from chaos: Resolving the paradox of periodicity in evoked otoacoustic emission. In: Duifhuis H, Horst JW, van Dijk P, van Netten SM (eds) Biophysics of Hair Cell Sensory Systems. Singapore: World Scientific, pp. 54–63.

Shera CA, Tubis A, Talmadge CL (2004a) Are there forward and reverse traveling waves in the cochlea? Countering the critique of Nobili et al. J Assoc Res Otolaryngol 5:349–359. [A version correcting JARO's numerous typesetting errors is available from the authors upon request (see also http://mit.edu/apg/).]

Shera CA, Tubis A, Talmadge CL, Guinan JJ (2004b) The dual effect of "suppressor" tones on stimulus-frequency otoacoustic emissions. Assoc Res Otolaryngol Abs 27:538.

Shera CA, Tubis A, Talmadge CL (2005a) Coherent reflection in a two-dimensional cochlea: Short-wave versus long-wave scattering in the generation of reflection-source otoacoustic emissions. J Acoust Soc Am 118:287–313.

Shera CA, Tubis A, Talmadge CL (2005b) Coherent-reflection models of reflection-source OAEs with and without slow transverse retrograde waves. Assoc Res Otolaryngol Abs 28:657.

Shera CA, Tubis A, Talmadge CL (2006) Four counter-arguments for slow-wave OAEs. In: Nuttall AL, Ren T, Gillespie P, Grosh K, de Boer E (eds) Auditory Mechanisms: Processes and Models. Singapore: World Scientific, pp. 449–457.

Siegel JH, Temchin AN, Ruggero MA (2003) Empirical estimates of the spatial origin of stimulus-frequency otoacoustic emissions. Assoc Res Otolaryngol Abs 26:679.

Siegel JH, Cerka AJ, Temchin AN, Ruggero MA (2004) Similar two-tone suppression patterns in SFOAEs and the cochlear microphonics indicate comparable spatial summation of underlying generators. Assoc Res Otolaryngol Abs 27:539.

Siegel JH, Cerka AJ, Recio-Spinoso A, Temchin AN, van Dijk P, Ruggero MA (2005) Delays of stimulus-frequency otoacoustic emissions and cochlear vibrations contradict the theory of coherent reflection filtering. J Acoust Soc Am 118:2434–2443.

Talmadge CL, Tubis A (1993) On modeling the connection between spontaneous and evoked otoacoustic emissions. In: Duifhuis H, Horst JW, van Dijk P, van Netten SM (eds) Biophysics of Hair Cell Sensory Systems. Singapore: World Scientific, pp. 25–32.

Talmadge CL, Long GR, Murphy WJ, Tubis A (1993) New off-line method for detecting spontaneous otoacoustic emission in human subjects. Hear Res 71:170–182.

Talmadge CL, Long GR, Tubis A, Dhar S (1999) Experimental confirmation of the two-source interference model for the fine structure of distortion product otoacoustic emissions. J Acoust Soc Am 105:275–292.

Talmadge CL, Tubis A, Long GR, Piskorski P (1998) Modeling otoacoustic emission and hearing threshold fine structures. J Acoust Soc Am 104:1517–1543.

Talmadge CL, Tubis A, Long GR, Tong C (2000) Modeling the combined effects of basilar membrane nonlinearity and roughness on stimulus frequency otoacoustic emission fine structure. J Acoust Soc Am 108:2911–2932.

van Hengel PWJ (1996) Emissions from cochlear modelling. PhD thesis, Rijksuniversiteit Groningen.

Weinberg S (1993) The First Three Minutes. New York: Perseus Books.

Wilson JP (1980a) The combination tone, $2f_1 - f_2$, in psychophysics and ear-canal recording. In: van den Brink G, Bilsen FA (eds), Psychophysical, Physiological and Behavioural Studies in Hearing. Delft: Delft University Press, pp. 43–50.

Wilson JP (1980b) Model for cochlear echoes and tinnitus based on an observed electrical correlate. Hear Res 2:527–532.

Wright AA (1984) Dimensions of the cochlear stereocilia in man and in guinea pig. Hear Res 13:89–98.

Zweig G (1976) Basilar membrane motion. In: Cold Spring Harbor Symposia on Quantitative Biology, Vol. XL. Cold Spring Harbor: Cold Spring Harbor Laboratory Press, pp. 619–633.

Zweig G, Shera CA (1995) The origin of periodicity in the spectrum of evoked otoacoustic emissions. J Acoust Soc Am 98:2018–2047.

Zwicker E (1980) Nonmonotonic behavior of $(2f_1 - f_2)$ explained by a saturation-feedback model. Hear Res 2:513–518.

Zwicker E (1981) Dependence of level and phase of the $(2f_1 - f_2)$-cancellation tone on frequency range, frequency difference, level of primaries, and subject. J Acoust Soc Am 70:1277–1288.

Zwicker E, Harris FP (1990) Psychoacoustical and ear canal cancellation of $(2f_1 - f_2)$-distortion products. J Acoust Soc Am 87:2583–2591.

Zwislocki JJ (1983) Sharp vibration maximum in the cochlea without wave reflection. Hear Res 9:103–111.

Zwislocki JJ (2002) Auditory Sound Transmission: An Autobiographical Perspective. Mahwah, NJ: Lawrence Erlbaum Associates.

10
Cellular and Molecular Mechanisms in the Efferent Control of Cochlear Nonlinearities

Ian J. Russell and Andrei N. Lukashkin

1. Introduction

With notable exceptions such as the hair cells of the bullfrog amphibian papilla (Flock and Flock 1966), outer hair cells of the cochlea of the horseshoe bats (*Rhinolophidae*) and *Hipposideros* (Bruns and Schmieszek 1980; Vater et al. 1992), and the entire outer hair-cell region of the mole rat (*Spalax ehrenbergi*, Raphael et al. 1991), the mechanosensitive hair cells of the lateral line, vestibular system, and cochlea receive direct efferent innervation. The structure and synaptic physiology of the efferent synapses in each of these sensory systems, where the hair cells subserve different sensory modalities, are similar in many respects. Through the action of the efferent fibers on the hair cells, the central nervous system can control the outflow of information from these sensory epithelia to the central nervous system.

This review is concerned with the cellular and molecular physiology of the postsynaptic action of the efferent synapses on hair cells and how these might interact with the inherently nonlinear cellular properties of the mechanotransducer conductance and the cochlear amplifier. The second part of the review discusses how efferent-induced changes in the nonlinear properties of hair cells may influence the mechanics of the cochlea, including their manifestation in distortion product otoacoustic emissions (DPOAEs). The cochlea efferent system has been extensively reviewed in recent years in this series (Guinan 1996; Smith and Spirou 2002), and this current review will deal with research more recent than this and with earlier research that relates to recent discoveries and concepts.

2. Functional Organization of the Organ of Corti

The organ of Corti contains two sensory cell types, the inner and outer hair cells (IHCs and OHCs), which are sandwiched between two sheets of extracellular matrix, the basilar and tectorial membranes (BM and TM). Sound-induced vibrations of the BM are detected and transduced by the hair cells, and the resultant electrical signals are transmitted to the brainstem by auditory nerve fibers. The electromotile OHCs are located over the BM, and their sensory hair bundles are embedded in the lower surface of the TM (Lim 1986). OHCs act as sensory-effector cells that detect and amplify the motion of the BM at low sound-pressure levels and compress it at high levels. IHCs are extensively innervated by afferent fibers (Spoendlin 1985) and act primarily as receptors. They are located near the limbal edge of the BM, and their hair bundles are not directly coupled to the underside of the TM (Lim 1986).

A division of labor therefore exists in the organ of Corti between the IHCs and OHCs, and each hair cell type is primarily associated with a particular efferent system. The unmyelinated lateral efferent neurons, with somata located predominantly in the ipsilateral superior olivary nucleus, form postsynaptic connections on the afferent synapses of type 1 ganglion cells that innervate the inner hair cells (see Guinan 1996 for a review). Thus the lateral efferent system is concerned with modulating directly the sensory outflow from the organ of Corti, and will not be discussed further in this review. The myelinated axons of the medial olivocochlear (MOC) efferent system with somata located bilaterally in the ventral nucleus of the trapezoid body and rostral periolivary regions terminate in large, impressive synaptic boutons on the basal poles of OHCs in the cochlea (reviewed by Guinan 1996). Early in development, the MOC efferents form transient synaptic connections with IHCs (Sobkowicz and Slapnick 1992).

It is now well established that activation of the MOC efferent system suppresses the sensory output of the cochlea by reducing its mechanical gain (see Guinan 1996 for a review). Activation of the MOC efferents alters, usually by suppression, micromechanical events within the cochlear partition in response to acoustic stimulation (Mountain 1980; Kemp and Souter 1988; Murugasu and Russell 1996a; Nuttal et al. 1997; Cooper and Guinan 2003). The action of the MOC efferent system on the nonlinear properties of the cochlea is the subject of this review.

3. Efferent Inhibition

Electrical activation of the MOC efferents has similar effects on neural, IHC, and BM responses in that it is frequency-dependent with suppression being greatest at the peak (characteristic, CF, or best frequency) of the tuning curve (see Guinan 1996 for a review). It has been shown that efferent inhibition acts on two different time scales: a “fast effect,” which develops over tens of milliseconds, and a “slow effect,” which develops over tens of seconds (Sridhar et al. 1995).

3.1 Inhibition of Afferent Activity Through Postsynaptic Action on Hair Cells

To date, the postsynaptic effects of the electrical activation of efferent fibers on intracellularly recorded receptor potentials from hair cells have been recorded in vivo only from the lateral line system of the freshwater cod *Lota lota* (Flock and Russell 1973, 1976). Inhibitory postsynaptic potentials (IPSPs) and currents (IPSCs) have been recorded in situ from the frog vestibular system (Ashmore and Russell 1982), the turtle cochlea (Art et al. 1984, 1985) (Fig. 10.1A), and in whole-cell patch-clamp recordings from cochlear IHCs (Goutman et al. 2005, Fig. 4C).

Low-frequency receptor potentials recorded from lateral-line hair cells (reviewed by Russell 1976) and receptor potentials recorded at frequencies more than half an octave below the resonance frequency of turtle hair cells are augmented following electrical stimulation of efferent fibers (Fig. 10.1A), largely as a consequence of the strong voltage sensitivity of the hair cell's basolateral membrane resistance (Art et al. 1985). Augmentation of the receptor potential is progressively reduced when the IPSPs become larger and the increased basal synaptic conductance overrides the resistance increase due to the efferent-induced hyperpolarization. Receptor potentials of turtle hair cells in the octave band

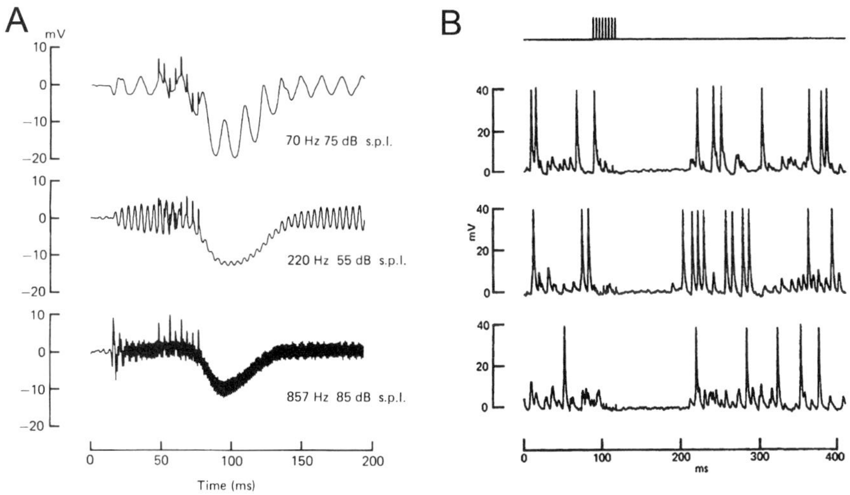

FIGURE 10.1. (**A**) Intracellular records showing efferent modulation of hair cell receptor potentials and suppression of afferent activity. Intracellular records from a hair cell showing efferent modulation of the receptor potential for three frequencies of acoustic stimulation. Sound pressure and frequency of the tone bursts are given beside each of the averaged responses. The efferents were stimulated with eight shocks, causing an 11-mV hyperpolarization, and the timing of the shocks is indicated by the capacitative artifacts. Ordinates are membrane potential relative to the resting potential (−45 mV). Note that the efferent effect on the receptor potential amplitude depends on stimulation frequency; for the middle trace, this was close to the characteristic frequency of the cell (200 Hz). (**B**) Intracellular records from an afferent nerve terminal showing inhibition of spontaneous activity following a train of eight efferent shocks whose timing is given in the top trace. Ordinate is membrane potential relative to resting potential −65 mV. (**A** reprinted with permission from Art et al. 1985. **B** from Art et al. 1984.)

around the characteristic frequency are reduced, partly as a consequence of the hyperpolarization, but also as a consequence of the shunting action of the efferent synapse, an idea first proposed by Fex (1962), which has been widely adopted (see Russell 1976 and Roberts and Meredith 1990 for reviews). From intracellular measurements in the lateral line and turtle basilar papilla it appears that the postsynaptic action of efferents on hair cells causes inhibition of both spontaneous and mechanically evoked postsynaptic responses in the afferent nerve fibers (Flock and Russell 1973, 1976; reviewed by Russell 1976; Art et al. 1984) (Fig. 10.1B). Hair cells in these sense organs are presynaptic to afferent fibers and postsynaptic to efferent fibers, so that efferent fibers can exert directly their influence on the sensory outflow from these organs.

3.2 Inhibition Through Remote Control: Efferent Modulation of the OHCs

The situation is different in the mammalian cochlea, where medial efferent inhibition of type I afferent activity in the auditory nerve is due largely, if not entirely, to modification of the mechanical input to the IHCs as a consequence of the postsynaptic action of the MOC efferents on the OHCs. Presumably, medial efferent inhibition can control, through presynaptic inhibition, the sensory outflow from the type-II afferent fibers of currently unknown function that innervate OHCs (Benson and Brown 2004). A similar situation arises in the avian basilar papilla, where, remarkably and apparently uniquely, the short hair cells, which constitute up to 35 % of the total number of hair cells, completely lack afferent fibers (Fischer 1994) and are innervated only by the large synaptic boutons of efferent fibers. The efferent innervation of the short hair cells can therefore influence only indirectly the sensory outflow of the basilar papilla (see Köppl et al. 2000 for a review).

So how do the MOC efferents exert remote control over the sensory outflow in the auditory nerve fibers? The primary and perhaps exclusive answer is that they do so by modulating the various forms of reverse transduction that take place in OHCs and short hair cells that amplify low-level and compress high-level responses of the cochlea and avian basilar papilla. Data from the vast majority of studies of the way the MOC efferents modulate the responses of the cochlea support this contention. Three observations in particular appear to contradict this view. Two of these observations are concerned with measurements that show electrical excitation of the MOC efferents to suppress cochlear responses at frequencies and levels outside of the range where BM displacements are normally influenced by OHC activity (see Robles and Ruggero 2001 for a review). Thus, medial efferent activation suppresses neural and basilar membrane responses at particular frequencies in the tail region of the tuning curve (Russell and Murugasu 1998; Stankovic and Guinan 1999) and at moderate to high levels at and above the characteristic frequency (Guinan and Stankovic 1996; Russell and Murugasu 1997).

The other observation is that spontaneous neural activity of auditory nerve fibers can be inhibited by electrical stimulation of the brainstem in experiments directed at exciting the MOC efferents (Mulders and Robertson 2005; see Guinan 1996 for a review). This is unexpected, because MOC efferents terminate exclusively on the OHCs. These apparently anomalous findings can still be accommodated within the proposal that modulation of cochlear responses by excitation of the MOC efferents is a consequence of the postsynaptic action of the MOC efferents on the OHCs.

4. Efferent Modulation of Cochlear Amplification

4.1 OHCs as the Source of Cochlear Amplification

Low-level signals at up to ultrasonic frequencies are amplified a thousandfold at low sound levels and compressed at high levels to provide the cochlea's enormous dynamic range. Currently, the choice for cochlear amplification is between fast, voltage-dependent, somatic motility, which can cause changes in the length and stiffness of OHC soma (see Santos-Sacchi 2003 for a review) and calcium-dependent fast adaptation of the hair-cell transducer, which can cause rapid changes in hair-bundle displacement and stiffness (see Hudspeth 1997; Hudspeth et al. 2000; Fettiplace et al. 2001 for reviews). The effects of medial efferent activation on DPOAEs (Mountain 1980; Siegel and Kim 1982) and tone- and noise-evoked BM displacement (Murugasu and Russell 1996a; Nuttal et al. 1997; Dolan et al. 1997; Cooper and Guinan 2003) provide direct confirming evidence that OHCs are the source of cochlear amplification and reveal that efferent fibers exert their influence on the sensory outflow from the cochlea through mechanical changes in the properties of OHCs. Electrical stimulation of the MOC efferents and perfusion of the scala tympani with acetylcholine (ACh) suppresses, by 20 dB or more, and linearizes BM responses to frequencies at the tip of the tuning curve (Murugasu and Russell 1996a,b) (Fig. 10.2) although BM responses to tones at frequencies above the tuning-curve tip are augmented (Dolan et al. 1997).

The effectiveness of the efferent suppression of neural and BM responses has been measured as a lateral shift (ΔL, Fig. 10.2A) of their level functions to more-insensitive regions along the sound-level axis. According to this measure, MOC efferent excitation is most effective in suppressing moderate-to-high-level responses to tones at and above the characteristic frequency (Guinan and Stankovic 1996; Russell and Murugasu 1997; Fig. 10.2B). This is largely because at and above the characteristic frequency, activation of the efferents causes the compressive region of the level functions to undergo changes in slope, and indeed shift to lower magnitudes or firing rates, at moderate to high sound levels. To describe the action of the efferents, therefore, as causing a lateral shift of the level function at these sound levels and frequencies is probably misleading. Efferent action also depends on the ambient state of the cochlea, and whether it is the fast or slow efferent effect that is being activated.

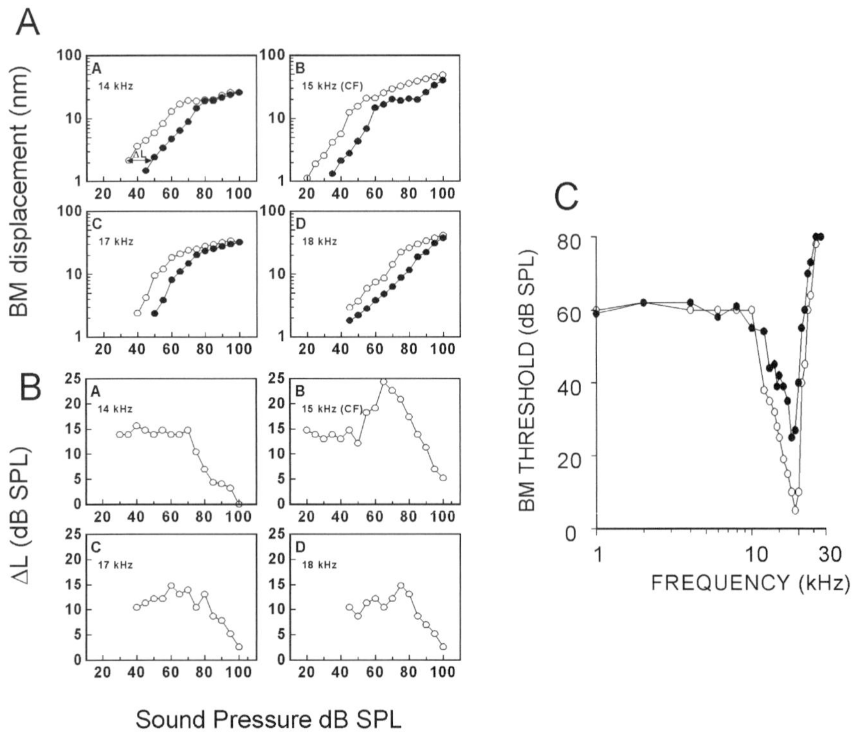

FIGURE 10.2. Suppression and augmentation of BM responses following activation of the MOC efferents. (**A**) The low-level slopes of BM displacement I/O functions (○) become steeper and level shifted (ΔL) during MOC stimulation (•). Measurements from a single location on the BM with a best frequency of 15 kHz. (**B**) As a consequence of MOC-induced changes in the compression of the level functions, the shift in level (ΔL) is greatest for moderately loud tones at CF. (**C**) Isoresponse tuning curve for 2-nm BM displacements for the tone presented alone (○) and during MOC stimulation (•) from a location tuned to 18 kHz. (**A, B** modified with permission from Russell and Murugasu 1997 **C** from Murugasu and Russell 1996a.)

4.2 Fast and Slow Effects Efferent Effects

In experiments designed to isolate the fast and slow efferent effects, Cooper and Guinan (2003) showed that the mechanical phase changes evoked by the fast and slow effects usually occur in opposite directions. Fast inhibition was associated with phase leads of between 20 and 70 degrees, while slow inhibition was associated with more-variable phase lags of between 10 and 50 degrees. Oppositely directed fast and slow phase changes were observed for most tones close to the characteristic frequency. The frequency and intensity dependence of fast and slow effects differed in at least two ways. First, while slow effects always involved phase lags, fast effects reversed from phase lags at low-to-moderate levels, $\frac{1}{4}$ to $\frac{1}{2}$ octave below characteristic frequency, to involving phase

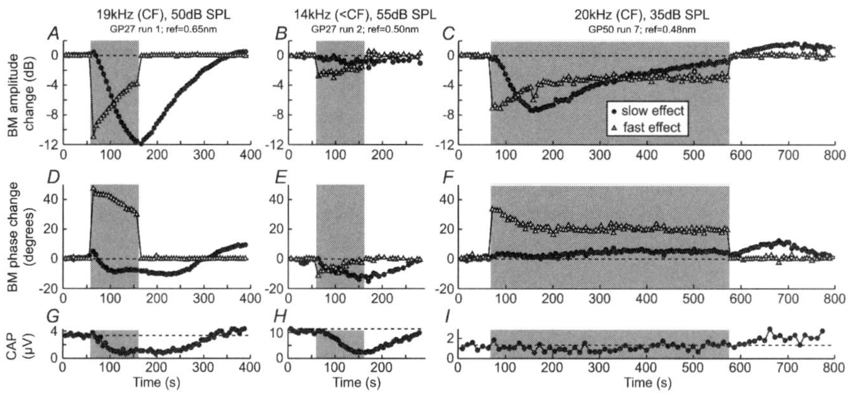

FIGURE 10.3. Fast and slow effects vary with tone frequency and intensity, and with the time course of efferent stimulation. BM vibration amplitude (**A–C**) and phase (**D–F**) changes attributed to fast (Δ) and slow (•) effects are shown for three runs of the single-tone paradigm (cf. Fig. 10.1**B**). Stimulus details and reference vibration levels for 0-dB amplitude changes are indicated at the top. *Shading* indicates the overall period of efferent stimulation (the test period; cf. Fig. 10.1**A**). **G–I**, amplitudes of simultaneously recorded CAPs, illustrating the slow effects of the efferent stimulation on the auditory nerve. (From Cooper and Guinan 2003, with permission.)

leads at higher levels and/or frequencies (Fig. 10.3). The different phase changes produced by the fast and slow effects of the efferent system led Cooper and Guinan to suggest that the efferent excitation could affect BM motion in two different ways. The phase changes associated with the fast effects may indicate increased damping of BM motion as a consequence of a decrease in the gain of the cochlea amplifier. The phase lags associated with the slow efferent effects are compatible with a decrease in the stiffness of the basilar membrane. These phase changes may indicate that efferent stimulation has altered the timing of OHC feedback to the cochlear partition. Timing of OHC feedback to the cochlea is crucial for cochlear amplification (e.g., Geisler and Sang 1995; Markin and Hudspeth 1995; Russell and Nilsen 2000), and is probably responsible for the amplification of cochlear responses at low stimulus levels and compression at moderate to high levels (Russell and Kössl 1992). It may also account for the efferent-induced augmentation, rather than suppression, of BM responses that Dolan et al. (1997) observed in response to tones at frequencies above the CF.

5. Cellular and Molecular Basis of Efferent Inhibition

5.1 Fast Efferent Effect

5.1.1 Fast Efferent Inhibition Involves SK2 Channels

The postsynaptic action of efferent fibers on hair cells is highly conserved throughout the lateral-line, vestibular, and auditory sense organs of vertebrates.

A similar sequence of postsynaptic events in response to electrical stimulation of efferent nerve fibers has been recorded in hair cells of sense organs as diverse as the lateral-line system of fish and amphibia (Russell 1976; Dawkins et al. 2005), the frog sacculus (Ashmore and Russell 1982), the turtle cochlea (Art et al. 1984; Fig. 10.1), and neonatal inner hair cells of the developing rat cochlea (Goutman et al. 2005; Fig. 10.4C). The primary efferent transmitter, ACh, acts through a nicotinic ionotropic receptor (nAChR) containing alpha 9 and 10 subunits (Elgoyhen et al. 1994, 2001; Marcotti et al. 2004) to hyperpolarize the hair cell (Flock and Russell 1976; Art et al. 1984; Housley and Ashmore 1991; Holt et al. 2001). Calcium influx through the nAChR-gated channel activates calcium-activated potassium channels to produce hyperpolarization within a few milliseconds (Evans 1996; Fig. 10.4B). The hyperpolarizing efferent potential is thought to be due to the gating of small-conductance (SK) calcium-activated potassium channels. SK2 channels are localized in the basolateral membrane of OHCs at the efferent terminals (Oliver et al. 2000). These findings are summarized in Figure 10.4.

5.1.2 The Efferent Synaptic Cistern: A Tightly Coupled Calcium Store to Convert Rapid Nicotinic Excitation into Inhibition

Crucial to the rapid and powerful postsynaptic action of the fast efferent effect is the close apposition of the synaptic cistern to the active zone of the efferent synapse (Fig. 10.4A). The synaptic cistern is believed to act as a rapid-release calcium store that can swiftly amplify the influx of calcium through the nAChR and hence the calcium levels in the vicinity of the SK channels through calcium-induced calcium release (CICR) from the synaptic cistern. The cistern can also quickly remove calcium from the postsynaptic membrane region of the efferent synapse.

These proposals are based on the finding that the fast efferent effect is sensitive to ryanodine. Calcium release can be both boosted and blocked by ryanodine, depending upon concentration and endogenous activity (Rousseau et al. 1987). When applied at 100 μM, (Evans et al. 2000) and the postsynaptic action of efferent fibers on hair cells of the *Xenopus* lateral-line system (Dawkins et al. 2005). In whole-cell recordings from OHCs in the apical turns of the excised rat organ of Corti, Lioudyno et al. (2004) showed that 1 μM ryanodine facilitated, and 100 μM ryanodine antagonized, ACh-evoked potassium currents and potassium-induced efferent stimulation. They also demonstrated the expression of type 1 ryanodine receptors close to the synaptic pole of OHCs but not IHCs, which is in the region of the synaptic cisternae of the OHCs.

5.1.3 Synaptic Facilitation of Efferent Inhibition

It is necessary to stimulate electrically efferent nerve fibres with pulse trains at relatively high frequencies to observe clear and consistent suppression of EPSPs and spike activity in afferent nerve fibres (Fex 1962; Russell 1976; Art et al. 1984; Guinan 1996). Studies in turtle hair cells (Art et al. 1984) and in

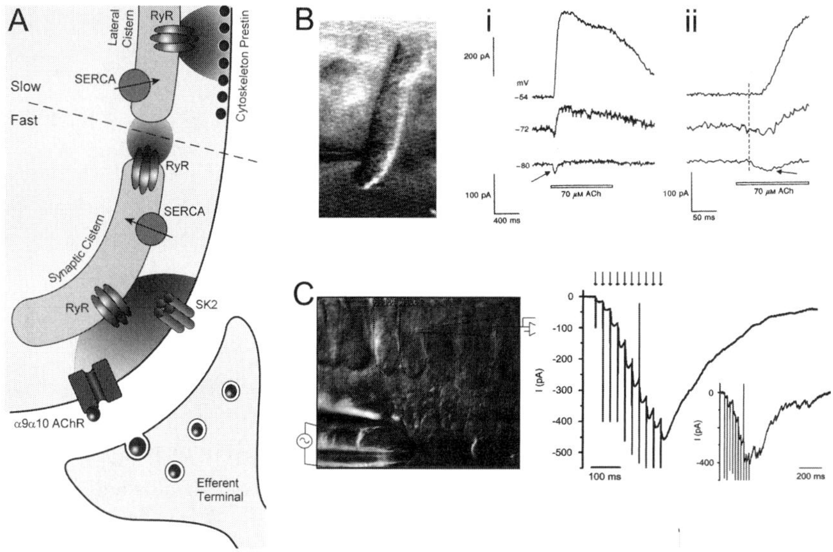

FIGURE 10.4. Fast and slow MOC efferent effects. (**A**) Scheme for generating fast and slow MOC efferent effects in OHCs based on in situ measurements reported in Section 5.1 and in vivo measurements reported in Section 5.2. *Gray shading* represents calcium release. (**B**) Acetylcholine activates two currents in guinea-pig outer hair cells. *Left*: Whole-cell patch of OHC. Perfusion pipette located close to basal, synaptic pole of OHC. *Right*: ACh produces a biphasic response at potentials close to −70 mV. (*i*) Average response to ACh in voltage clamp at the voltages shown beside the records. (*ii*) The onset of responses to the ACh applications shown on a faster time scale. The *vertical dashed line* indicates the start of the inward current at −80 mV. *Horizontal bars* indicate timing of the pressure pulse to eject the ACh solution. (**C**) *Left*: Inhibitory postsynaptic currents (IPSCs) evoked in neonatal rat apical IHCs by electrical stimulation of cholinergic efferent axons. For bipolar electrical stimulation of efferent fibers some supporting cells were removed and a theta glass micropipette, 10–15 μm in diameter, was positioned ≈ 10–20 μm below the base of the IHC. Scale bar 10 μm. Positions of two IHCs and recording pipette outlined. Repetitive stimulation of IHC efferents resulted in large compound IPSCs due to summation and facilitation. Right: Response of an IHC at a holding potential of −90 mV to a 10-pulse, 25-ms interval, train. The pulse train was presented at a 20-s interval and repeated 30–50 times; the average compound IPSC is shown. *Inset*: An individual record showing marked fluctuations on the falling phase of the compound IPSC. (**B** from Evans 1996; **C** from Goutman et al. 2005, with permission.)

rat IHCs (Goutman et al. 2005) show the resting probability of release at the efferent synapse is low, but facilitated strongly during repetitive stimulation of the efferent axons (Fig. 10.4C). Goutman et al. concluded that to be effective, efferent inhibition had to be repetitive and at sufficiently high frequencies for facilitation of transmitter release to occur. They proposed that efferent control of the cochlea operates in a form of fail-safe mode, being effective only when strongly driven, so that spontaneous, inadvertent activity would tend not to alter cochlear function. This may be so, but strong facilitation may also enable the postsynaptic effect of the efferent fibers to be closely titrated to the levels of the signals they are designed to suppress. The firing rate of the lateral line efferent neurons increases with active body movement and with that of the motor neurons that innervate the white voluntary muscles but not the red muscles. The stronger the activation of the white muscle, the greater the relative shear between body surface and surrounding water and stimulation of the lateral-line organs as the animal is propelled swiftly through the water, and the stronger the efferent inhibition of the lateral-line hair cells (Russell 1976).

5.1.4 MOC Efferents Are Essential to the Development of IHC Responses

The transient medial efferent innervation of IHCs that occurs between 7 and 21 days after birth in rats hyperpolarizes the IHCs with postsynaptic mechanisms that are very similar to those that have been demonstrated in OHCs (Glowatzki and Fuchs 2000). At this stage of development the IHCs generate spontaneous action potentials (Kros et al. 1998; Marcotti et al. 2003, 2004), the shape and frequency of which are believed to play an important role in the development and maturation of synaptic connections, as has been proposed for the developing retina (reviewed by Shatz 1996), and the activation of the efferent synapses modulates this activity (Marcotti et al. 2004; Goutman et al. 2005). The MOC efferents play an essential but as yet unknown role in the maturation of the cochlea. When the efferent innervation of the cochlea was sectioned, kittens failed to develop normal hearing (Walsh et al. 1998).

5.2 The Slow Efferent Effect

5.2.1 The Slow Efferent Effect Is Due to CICR from the Lateral Cisternae

The slow efferent effect is induced when the efferent synapse is excited repeatedly over several seconds (Sridhar et al. 1995, 1997) and can be mimicked by the perfusion of ACh through the scala tympani (Murugasu and Russell 1996b). When the scala tympani of the cochlea is perfused with artificial perilymph containing 100 μM ACh, BM responses are strongly attenuated at the CF, and the peak of the tuning curve is shifted to lower frequencies (Fig. 10.5A–C). Cyclopiazonic acid, which inhibits calcium uptake into endoplasmic reticulum-like structures, can enhance the slow effects of efferent stimulation on cochlear potentials (Sridhar et al. 1997) and BM responses to scala tympani perfusion of ACh (Fig. 10.5D) (Murugasu and Russell 1996b) in the guinea pig cochlea. Perfusion of 100 μM ryanodine through the scala tympani blocks ACh-induced changes in tone-evoked BM displacements (Murugasu and Russell 1996b), and at 30 μM concentration, ryanodine enhances the slow effect of the MOC efferents on cochlear potentials (Sridhar et al. 1997). On the basis of these observations it has been proposed that the slow efferent effect also depends on CICR from intracellular stores. In this scheme, CICR from the synaptic cistern builds up to a level that promotes calcium sparking to

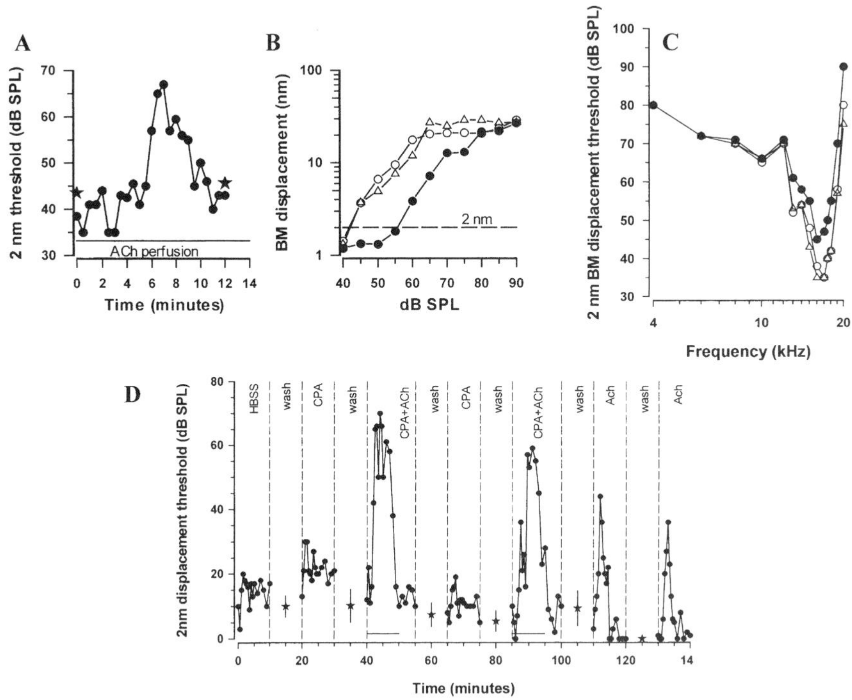

FIGURE 10.5. The thresholds of BM displacements to characteristic frequency tones are increased by, and then become desensitized to, continuous perfusion of the scala tympani with ACh in artificial perilymph (HBSS). Sensitization to ACh perfusion is restored following perfusion with HBSS. (**A**) 2-nm BM displacement threshold at the CF of the measuring point on the BM (15 kHz) as a function of time during continuous perfusion of the scala tympani with 100 μM ACh and 20 μM physostigmine in HBSS. (**B**) Shows the method of deriving the 2-nm BM displacement threshold from BM displacement I/O functions measured before (*open circle*), during (filled circle), and after (open triangle) ACh perfusion of the scala tympani. (**C**) The tip of the isoresponse BM tuning curve measured from close to the 18-kHz region of a cochlea is desensitized, broadened, and shifted downward in frequency by ACh perfusion of the scala tympani. HBSS perfusion (*open circles*), perfusion with 100 μM ACh and 20 μM physostigmine in HBSS (*filled circles*), and following washout with HBSS (*open triangles*). (**D**) Cyclopiazonic acid (CPA) augments and prolongs ACh suppression of BM responses. The effect on the 2-nm BM displacement threshold measured at the 18-kHz location of successive 10-minute perfusions of the scala tympani with HBSS (wash), 100 μM ACh, and ACh together with 10 μM (CPA). The *horizontal lines* beneath the CPA traces signify the 10-minute perfusion period. The *stars* represent the means and standard deviations of ten measurements of the threshold made during the washing periods. (Reprinted with permission from Murugasu and Russell 1996b.)

the neighboring lateral cisternae, thereby increasing the calcium levels adjacent to the lateral membranes of the OHCs (Fig. 10.4A).

5.2.2 The Slow Effect is Due to a Reduction in OHC Axial Stiffness as a Consequence of Protein Phosphorylation

The ACh-induced changes in the sensitivity and frequency tuning of the BM are interpreted as being due to a reduction in the stiffness of the cochlear partition, which was attributed to the OHCs (Murugasu and Russell 1996b), which accords with the conclusions of Cooper and Guinan (2003), who attribute the slow efferent effect to a reduction in BM stiffness. Dallos et al. (1997) discovered that the principal cause of the increased electromotile response of single OHCs in response to perfusion of the basal synaptic pole with ACh was a decreased global axial stiffness. Dallos and his colleagues and others have also shown that ACh-induced reduction in OHC axial stiffness occurs as a consequence of changes in protein phosphorylation of the OHC cytoskeleton, and possibly the motor protein prestin itself, via second messenger cascades mediated by calcium and cyclic nucleotides (see Frolenkov et al. 2000, 2003; Deak et al. 2005). Frolenkov et al. (2000, 2003) proposed that two distinct calcium-dependent signaling pathways regulate OHC motility. The first pathway, which causes OHCs to elongate (Dulon et al. 1990), reduces cytoskeletal stiffness. The second pathway shifts the voltage sensitivity of the OHC electromotile mechanism and may be activated by the release of calcium from intracellular stores located in the proximity of the lateral plasma membrane (Fig. 10.4A). The involvement, through activation of the "slow" efferent effect, of second-messenger pathways probably accounts for why the "slow effect" is indeed slow.

Thus, cholinergic inhibition of hair cells in organs as diverse as the lateral line and cochlea provides evidence of a role for calcium-activated calcium release in both fast and slow efferent effects. MOC efferents also appear to have a transient, but as yet unexplained, role in the development of afferent synaptic transmission in the cochlea.

6. Mechanisms

Through what mechanisms might the MOC efferents exert their influence on the cochlear amplifier?

6.1 Shunting

Shunting, as a consequence of increased OHC membrane conductance due to MOC efferents, was first proposed by Fex (1967). The shunting of current through the increased conductance of the hair cell membrane has profound effects on the electrical tuning in hair cells of the turtle cochlea, which to a first approximation is described as an increase in damping of a parallel resonant circuit caused by a decrease in the resistive element (Fig. 10.6A) (Art et al. 1985). OHCs do not share the strongly resonant properties of turtle, avian, and amphibian hair cells, but the membrane conductance of OHCs is strongly voltage sensitive (Kros 1996), and depending on the strength of the postsynaptic hyperpolarization, it is likely that efferent effects on OHC receptor potentials due to a voltage-dependent resistance increase will be balanced against the shunting effect. At high frequencies, the postsynaptic shunting action of the efferent synapse becomes ineffective in reducing the receptor potential because most of the transducer current flows across the membrane capacitance (Fig. 10.6) (Art et al. 1985; Russell and Kössl 1991; Dallos and Evans 1995; Murugasu and Russell 1996a; Chan and Hudspeth 2005), which means that shunting of the transducer current should be effective in modulating voltage-sensitive OHC motility only at frequencies below the electrical time constant of the OHC membrane. At higher stimulus levels, OHCs generate receptor potentials with a dc component that increases with increasing level (Russell et al. 1986). This component would be expected to be shunted by an efferent-induced increase in OHC membrane conductance and therefore could influence OHC and hence cochlear mechanical properties at moderate to high stimulus levels. This might be why the compression, which has been measured in BM and neural responses at moderate to high stimulus levels, is reduced by medial efferent stimulation (Guinan and Stankovic 1996; Russell and Murugasu 1997).

The shunting of the transducer current may, however, modulate feedback to the cochlear partition by altering the timing of the feedback so that it becomes less effective in pumping tone-evoked vibrations of the BM. Fex (1967) was the first to observe an efferent-induced augmentation of the CM, which he interpreted as the increased flow of receptor current resulting from the increase in OHC membrane conductance across resistive elements in the cochlear partition. The flow of transducer current across extracellular resistive elements in the immediate vicinity of OHCs has been proposed to be the source of driving voltage for OHC motility at high frequencies (Dallos and Evans 1995; Kolston 1995). Indeed, Kössl and Russell (1992) observed that for high-frequency tones the extracellular voltage changes in the immediate vicinity of the OHCs could be larger than those recorded intracellularly. If shunting not only augments but also alters the phase of this current, then this might be expected to alter the timing of OHC feedback, which might oppose rather than pump the swing of the BM.

6.2 Changing the Operating Point of the Mechanoelectrical Transducer

Activation of the efferent innervation can cause hair cells to be strongly hyperpolarized, by more than 20 mV in the case of frog sacculus and turtle cochlea (Ashmore and Russell 1982; Art et al. 1984). Similar potential changes occurring in OHCs might be expected to have significant effects on the highly nonlinear OHC mechanosensory transducer function. At low to moderate stimulus levels, OHCs in the basal turn of the guinea pig cochlea operate around a point close to

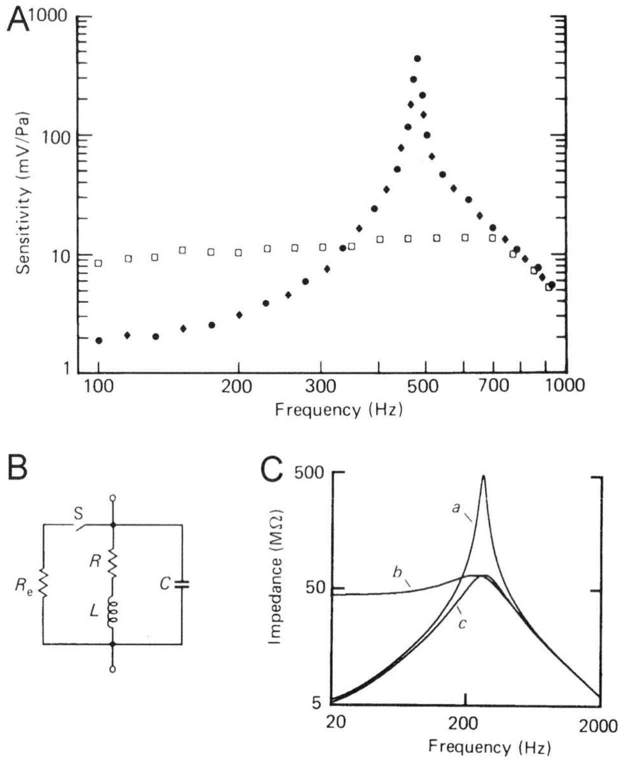

FIGURE 10.6. Shunting, due to efferent-induced increase in hair-cell membrane current suppresses receptor potential at the characteristic frequency, but can augment it at low frequencies. (**A**) Acoustic tuning curves a hair cell in the absence (*filled symbols*) and presence (*open symbols*) of continual efferent stimulation with single shocks at 50/s. Acoustic sensitivity (rms. voltage/sound pressure) is plotted against the sound frequency. Sound pressure 60 dB s.p.l. Characteristic frequency and quality factor of tuning for the hair cell: 488 Hz, 52. (**B**) Equivalent circuit used to describe electrical resonance in a hair cell. With the switch, s, open the circuit is tuned to a resonant frequency, f_0, of $(4\pi^2 LC)^{-1/2}$ with a quality factor of tuning, Q, of (L/R^2C). The circuit may be damped by increasing R, or by switching in R_e, identified with the efferent synaptic resistance. (**C**) Changes in the theoretical tuning curve of the circuit produced by the two kinds of damping. *a*, Impedance of the equivalent circuit versus frequency in the undamped condition with $f_0 = 263$Hz, Q = 10.9. Circuit parameters: L: 27.8 kH; $C = 13.2$pF; R = 4.2MΩ: (b): the Q of the circuit is reduced to 1.0 by increasing R to 44.5 MΩ. Note the increase in low-frequency impedance: (c): the Q is reduced from 10.9 to 1.5 by switching in a synaptic resistance R_e of 78.5 MΩ. Note that the impedance is reduced at all frequencies. The two conditions of damping were chosen to yield the same loss at the resonant frequency. (Reprinted with permission from Art et al. 1985.)

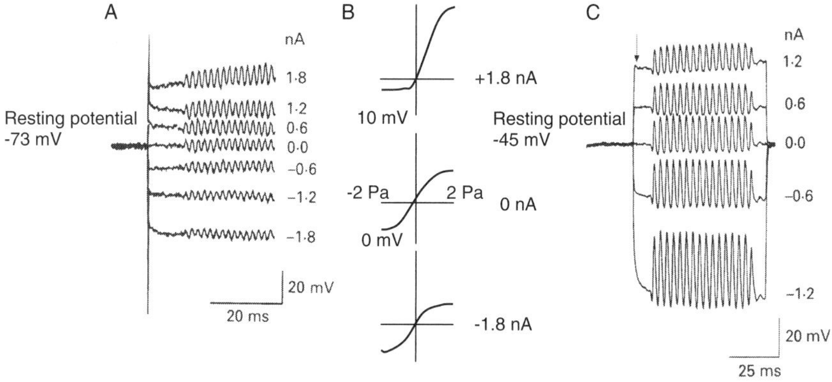

FIGURE 10.7. Intracellular current injection changes the symmetry of OHC but not IHC receptor potentials. (**A**) Receptor potentials recorded from an outer hair cell when stimulated with 600-Hz tones at 80 dB SPL and during current injection. (**B**) Transfer functions derived from receptor potentials of the outer hair cell shown and from a further four outer hair cells in two other preparations during the injection of 1.8, 0, and −1.8 nA of current. (**C**) Responses of an inner hair cell to positive and negative current injection, voltage responses to 300-Hz tones at 90 dB SPL. (Modified with permission from Russell and Kössl 1991.)

the maximum slope of their mechanosensory transducer function, where the DC component is minimal (Russell et al. 1986; Russell and Kössl 1992). Displacements of the BM either toward or away from the scala tympani reduce the sensitivity and frequency selectivity of the cochlea presumably by displacing the OHC transducer conductance away from its optimal operating position (Patuzzi et al. 1984). On the basis of calculations, however, Patuzzi and Rajan (1990) speculated that the shifts in operating position would be too small to influence cochlear amplification. In fact, when single OHCs in the basal turn of the guinea pig cochlea are injected in vivo through an intracellular electrode with current while the cochlea is stimulated with low-frequency tones, the OHC receptor potentials are decreased in amplitude and made more symmetrical with negative current (Fig. 10.7A,B) (which hyperpolarizes OHCs and presumably causes them to elongate).

The OHCs' receptor potentials are increased in amplitude and made asymmetrical in the depolarizing direction with positive current (which depolarizes OHCs and presumably causes them to shorten) (Russell and Kössl 1991). It is likely that these changes in the shape and magnitude of the receptor potential are due to interaction between the OHC hair bundles and the TM to which they are attached. Similar intracellular current injections into IHCs, which do not have hair bundles attached to the TM, do not strongly alter the symmetry of the waveforms of low-frequency receptor potentials, but they do alter the magnitudes (Fig. 10.7C). Hyperpolarizing currents increase and depolarizing currents decrease the magnitude of the receptor potentials, as might be expected when the driving voltages for the receptor current are increased and decreased respectively, and when voltage-dependent conductances in the basolateral membranes are activated (see Kros 1996 for a review). Efferent-induced length changes in the soma of OHCs have the potential to change the operating point of the OHC mechanosensory transducer function and thereby alter the gain of the cochlear amplifier.

6.3 Changing the Operating Point of the Electromechanical Transducer

Hyperpolarization of the OHC membrane potential through activation of MOC efferents might be expected to change the physiological properties of the OHCs and, consequently, the gain of the voltage-dependent motility or motility function. When the membrane potential becomes hyperpolarized, the axial forces acting on the OHCs will be altered as a consequence of voltage-dependent length and stiffness changes and subsequent mechanical interaction between the OHCs and adjacent cellular and noncellular structures in the organ of Corti. Changes in OHC membrane potential evoke an immediate shift in the voltage-dependence of the OHC capacitance with corresponding changes in the gain of the OHC motility.

Over a slower time scale of seconds, probably involving viscoelastic elements within the motor itself and or the lipid bilayer, OHC hyperpolarization shifts the voltage-dependence of the OHC capacitance (the electrical signature of the voltage-dependent motility or motility function) to more depolarizing levels. A hyperpolarizing shift in resting potential augmented by a depolarizing shift in the motility function should decrease the mechanical gain of the OHC motility. This reduction in gain may be either augmented or opposed by axial forces that are determined by the state of tension in the OHCs at the time of efferent activation and also by changes in OHC tension as a consequence of efferent activation. This is because the motility function can be effectively reduced by exerting tension on the plasma membrane. A reduction in membrane tension increases the mechanical gain at the normal resting potential, by shifting the most sensitive region of the motility function along the voltage axis toward the resting potential value. Activation of both fast and slow effects of the MOC efferents has the potential, therefore, to alter the gain of the OHC voltage-dependent motility and amplification in the cochlea (reviewed by Santos-Sacchi 2003).

6.4 Rapid Changes in Stiffness of the Cochlear Partition

Measurements of the axial stiffness of OHCs reveal that voltage-dependent stiffness contributes about 50% to the total axial stiffness of OHCs (He et al. 2003). The stiffness is related directly to the voltage-dependent motility and can therefore change rapidly through OHC transmembrane potential differences. Depolarization decreases and hyperpolarization increases axial stiffness. Accordingly, if the primary action of the fast effect of the MOC efferents is to hyperpolarize OHCs and cause them to lengthen and stiffen, then this action might be expected to cause a change in the resonance frequency of the cochlear partition and an upward shift in characteristic frequency when the contribution of OHCs to the mechanical properties is most effective. Gross electrical activation of the fast efferent system apparently causes little or no changes in the characteristic frequency of tuning curves measured from BM, IHC, and neural responses (Guinan 1996; Murugasu and Russell 1996a). However, electrical stimulation may mask subtle interplay among a number of factors including the initial state of tension of the OHCs and the relationship between the stimulus tone and the site of efferent activation along the length of the cochlea when the efferent system is activated naturally through frequency-selective descending pathways (Mulders and Robertson 2005).

The state of OHC tension at the time of efferent activation may be a significant factor in determining the consequences of the postsynaptic action of the efferent system as revealed in a study by Parthasarathi et al. (2003), who measured tone-evoked BM vibrations while injecting dc current into the fluid spaces of the cochlea. They observed that current that was predicted to hyperpolarize the OHCs caused them to shorten, rather than lengthen, when the OHCs were under tension due to the dominance of voltage-dependent stiffness change. If the medial efferents are activated when OHCs are under tension, as might be expected when the cochlea is excited by tones of moderate to high intensities, then efferent-induced hyperpolarization of the soma will result in a reduction in axial stiffness and shortening of the OHC soma, resulting in a reduction in BM stiffness at the characteristic frequency and possibly a shift to lower frequencies.

6.5 Rapid Changes in the Mechanical Properties of Hair Bundles

On the basis of research carried out on nonmammalian species for which prestin is not a candidate for the motor protein in cochlear amplification, force generation by the hair bundles has been proposed as a means of boosting the mechanical stimulus to the sensory epithelium (Hudspeth 1997; Fettiplace et al. 2001). Measurements of hair-bundle motion from IHCs and OHCs in the excised organ of Corti in response to mechanical and electrical stimuli have led to the conclusion that calcium-driven hair-bundle motility (Chan and Hudspeth 2005) and rapid force generation by the hair bundles, linked to rapid adaptation of the mechanotransducer channels in mammalian cochlear hair cells (Kennedy et al. 2005), can provide the amplification required to enhance the sensitivity and frequency selectivity of the mammalian cochlea. These conclusions are, however, disputed by Jia and He (2005), who find that OHC voltage motility powers the hair-bundle movements that can drive horizontal shearing movements of the tectorial membrane. Efferent activation might, via OHC electromotility and perhaps slow adaptation, control the operating point of the mechanotransducer, and thereby the gain of the cochlear amplifier (Chan and Hudspeth 2005).

Fast adaptation is calcium-dependent (Kennedy et al. 2003), and adaptation is slowed and reduced when the calcium current through the transducer channels is reduced, by reducing extracellular calcium. Reducing the extracellular calcium concentration from 1.5 to 0.02 mM shifted the mechanotransducer open probability displacement function by 200 nm in the negative direction, decreased the bundle force by 21 %, and doubled the passive stiffness of the hair bundle in measurements from the rat cochlea (Kennedy et al. 2005). Similar changes might be expected to occur if the concentration gradient for the influx of calcium were reduced as a consequence of an increase in intracellular free calcium during the slow efferent effect. However, it remains to be seen whether efferent activation has either a direct or indirect effect on the forces generated by the hair bundles of cochlear hair cells.

6.6 Do Fast Medial Efferents Exert Their Influence on Hair Cells and Afferent Synapses Through the Extracellular MOC Potential?

The MOC potential, an efferent-evoked voltage change that can be measured as a reduction of the endocochlear potential (EP) or as a positive potential within the organ of Corti (Fex 1967; see Guinan 1996 for a review), might modify the driving voltage of transducer currents of OHCs and/or IHCs. This possibility has been considered on several occasions (see Guinan 1996 for a review) with similar conclusions that the 2–3 mV reduction in EP resulting from the MOC efferents potential would have negligible effects (< 0.3 dB) either on the cochlear amplification or on IHC responses. The direct action of the MOC potential and the CM on the afferent postsynaptic terminals provide other possible ways in which MOC efferents acting on OHCs can indirectly suppress afferent activity. These possibilities, and the one above, could be discounted if stimulation of the MOC efferent system had no effect on the spontaneous discharge of afferent fibers. Suppression of spontaneous activity has been described with activation of the olivocochlear system (see Guinan 1996 for a review). However, Mulders and Robertson et al. (2005) conclude that the suppression of spontaneous activity as a consequence of electrical stimulation of the floor of the fourth ventricle may inadvertently result in the excitation of unmyelinated axons from the lateral efferent system, which can both increase and decrease responses of the auditory nerve without effects on cochlear responses dominated by the OHCs (Groff and Liberman 2003).

There are, therefore, several potential ways in which the fast and slow efferent activation can suppress neural responses in the auditory nerve via mechanical changes in the cochlear partition. The majority of the largely groundbreaking research on the MOC efferent system to date can provide only a limited view of the precise postsynaptic action of efferent fibers in the cochlea. Gross electrical stimulation of the MOC efferents is nonselective in its action on the cochlea, and measurement of the BM provides a limited view of these mechanical changes. The real action is within the organ of Corti itself. The efferent effects are manifested through interaction between the OHC and IHC stereocilia, TM, cuticular plate, and fluid movements in the subtectorial space. Direct in vivo observation of the way these structures respond to fast efferent activation is not yet possible. Can anything be deduced about efferent-induced changes in these structures through monitoring distortion-product otoacoustic emissions, which are determined largely by the strongest nonlinearity in the cochlea, namely the OHC mechanoelectrical transducer conductance?

7. Identification of Efferent-Dependent Mechanical Changes in the Cochlea Using Otoacoustic Emissions

The recording of otoacoustic emissions has proved to be particularly appealing as a relatively noninvasive technique for analyzing efferent-mediated conformational changes in the cochlea. This is because these conformational changes affect OHC-dependent cochlear amplification. As a first approximation, the output sound pressure recorded in the ear canal at the emission frequencies is proportional to the force generated by the active nonlinearity of the cochlea at these frequencies (Bian et al. 2004). Therefore, efferent-mediated changes in force production within the cochlea should lead to changes in the acoustical responses recorded in the ear canal. Indeed, Siegel and Kim (1982) described changes in the DPOAE amplitude during direct electrical stimulation of the MOC efferents. Contralateral acoustical stimulation, which activates the ipsilateral efferents (Fex 1959; Kiang 1970), suppresses mammalian spontaneous otoacoustic emissions (Grose 1983; Schloth and Zwicker 1983) and transient-evoked otoacoustic emissions (Collet et al. 1990; Ryan et al. 1991). The inherently nonlinear DPOAEs, however, have an additional advantage for monitoring efferent-dependent mechanical changes within the cochlea. This is because DPOAEs can be used to derive (Bian et al. 2002, 2004) and monitor efferent-mediated changes in the cochlear nonlinearity that are associated with the active amplified responses of the cochlea (see Robles and Ruggero 2001 for review).

Below it is shown that amplitude and phase patterns of the distortion products at the output of a nonlinear system are specific for distortion products of different orders and strongly depend on the type of nonlinearity. By observing changes in DPOAEs of different orders, small modifications in the nonlinearity, responsible for generating DPOAEs, can be detected (Kirk and Johnstone 1993; Frank and Kössl 1996, 1997). However, as is frequently the case with the DPOAE data, interpretation of these DPOAE changes should be made with care because of the complex dependence of DPOAEs on the stimulus parameters including the levels and frequencies of the primaries and the frequency ratio of the primaries. Care should also be taken in the interpretation of the DPOAE data because DPOAEs recorded in the ear canal could be a result of vector summation between DPOAEs of different origins, including those due to reemission (Kemp 1980), multiple reflections within the cochlea (Stover et al. 1996), or the generation of emissions over extended regions of the cochlea (Shera 2003).

7.1 DPOAE Changes Induced by Electrical Stimulation of the MOC Efferents

Electrical stimulation of the MOC efferents causes small changes in the DPOAEs (Mountain 1980; Siegel and Kim 1982). These pioneering works showed two distinctive consequences of MOC efferent stimulation. First, changes in DPOAE amplitude were observed over a limited range of the primary levels, at least for the $2f_1 - f_2$ component. Second, depending on the parameters of stimulation, both suppression and augmentation of the amplitude were seen for both $2f_1 - f_2$ and $f_2 - f_1$ DPOAEs. These amplitude changes were associated with postsynaptic action of the MOC efferents on the OHCs because they disappear after perfusion of the cochlea with curare, a known antagonist of nAChRs in the postsynaptic membrane of the neuromuscular junction (Katz and Miledy 1972).

7.2 DPOAE Changes During Prolonged Acoustical Stimulation

Somewhat different responses of the DPOAEs were observed when the acoustical stimulation was prolonged for several minutes. These responses depended on whether ipsi- or contralateral stimulation was applied, on the depth of anesthesia, and on the type of anesthetic agent. Changes in the $2f_1 - f_2$ component were very small or absent during continuous presentation of the primaries to the ipsilateral ear (Brown 1988; Whitehead et al. 1991; Kirk and Johnstone 1993). The largest reported increase of the $2f_1 - f_2$ amplitude did not exceed 1 dB (Kujawa et al. 1995). At the same time, most authors reported significant and frequently bidirectional changes of $\mathrm{f}_2 - f_1$ during continuous ipsilateral stimulation (Brown 1988; Whitehead et al. 1991; Kirk and Johnstone 1993; Kujawa et al. 1995; Chang and Norton 1997). Initial increase of the $f_2 - f_1$ amplitude by a few decibels over the first 2 minutes of stimulation was followed by a slower but more pronounced decrease. During this latter phase, the amplitude asymptotically declined to a new steady level and, depending on the stimulus parameters, could disappear below the noise floor. The amplitude decline continued for some time after the end of stimulation (off-effect, Kujawa et al. 1995), but the DPOAE eventually recovered to its prestimulation level. This recovery could be slowed down by contralateral acoustical stimulation when measured in guinea pigs and gerbils (Brown 1988) but not in rabbits (Whitehead et al. 1991). Suppression of the $f_2 - f_1$ amplitude showed strong dependence on the depth of anesthesia. Much smaller suppression was observed with deepening of the anesthesia (Brown 1988; Kujawa et al. 1995). At the same time, section of the OCB at the floor of the IVth ventricle did not change the time course and degree of the ipsilaterally induced changes in the $f_2 - f_1$ (Kujawa et al. 1995).

Data on the pharmacological sensitivity of the ipsilaterally induced $f_2 - f_1$ changes over a period of a few minutes are contradictory. Two antagonists of nAChR either did not block the effect of the prolonged ipsilateral stimulation (strychnine administrated intraperitoneally, Kirk and Johnstone 1993) or produced only a partial block (cochlear perfusion with curare, Kujawa et al. 1995). Kujawa et al. (1995) reported only a slight decrease in the effects of the ipsilateral stimulation during perfusion of the cochlea with a γ-aminobutyric acid (GABA) antagonist, bicuculline. However, in similar experiments, Kirk and Johnstone (1993) observed either complete blockade of the ipsilaterally induced suppression or augmentation of the $f_2 - f_1$ component amplitude during this stimulation. Cochlear perfusion with tetrodotoxin (TTX), in concentrations that completely blocked any neural activity in the cochlea, also produced apparently contradictory results. Ipsilaterally induced suppression vanished during TTX cochlear perfusion in experiments by Kirk and Johnstone (1993), but it was reported unchanged by Kujawa et al. (1995) The lack of the effects of TTX and OCB section on the ipsilaterally induced changes in the DPOAEs led Kujawa et al. (1995) to suggest that the DPOAE changes in their experiments, while signaling conformational alterations within the cochlea, were not due to primary efferent effects.

The time course and extent of the contralaterally induced suppression of the amplitude of the $f_2 - f_1$ component over a period of a few minutes is similar to changes induced by ipsilateral acoustic stimulation (Kirk and Johnstone 1993; Kujawa et al. 1995; Chang and Norton 1997). Moreover, Kirk and Johnstone (1993) reported superposition of the effects of ipsi- and contralateral stimulation. At the same time, Chang and Norton (1997) observed not only suppression but also augmentation of the $f_2 - f_1$ amplitude for some stimulating parameters. In contrast to ipsilateral stimulation, contralateral acoustical stimulation significantly suppresses the $2f_1 - f_2$ amplitude in lightly anaesthetized animals (Puel and Rebillard 1990). It is indicative that Chang and Norton (1997) observed contralateral suppression of $2f_1 - f_2$ and $3f_1 - 2f_2$ components only during the animal's recovery from ketamine-xylazine anesthesia. During the recovery, suppression of the $f_2 - f_1$ DPOAE became an "immediate" effect. It became fully developed (often with a significant overshoot) within a few seconds of the stimulus onset instead of a "delayed" decline during a few minutes observed under deep anesthesia (Chang and Norton 1997). This observation led Chang and Norton to conclude that xylazine anesthesia might modulate release of ACh, the known neurotransmitter of the MOC system. Nevertheless, contralaterally induced changes in the DPOAEs seemed to be mediated by the cochlear efferents because they were eliminated by section of the OCB at the midline of the IVth ventricle (Kujawa et al. 1995) and by intraperitoneal injection of, or cochlear perfusion with, ACh and GABA antagonists (Kirk and Johnstone 1993; Kujawa et al. 1995). These changes also disappeared after complete block of the cochlear neural activity with TTX (Kirk and Johnstone 1993; Kujawa et al. 1995). Further evidence in favor of contralaterally induced suppression being mediated by the cochlear efferents is the tonotopicity of the effect (Chang and Norton 1997). When pure tones were used for contralateral stimulation, the maximum suppression was observed with the tone's frequency about a half-octave below the f_2. It is consistent with findings by Guinan et al. (1984) that the MOC efferents project ipsilaterally to their characteristic frequency location on the BM, but to a position more basal than this in the contralateral cochlea.

Differences in the sensitivity of the odd- and even-order DPOAEs to acoustical and electrical stimulation might suggest that generators of the even and odd DPOAEs are dissociated in the cochlea (Whitehead et al. 1991). However, Kirk and Johnstone (1993) pointed out that from a theoretical point of view, a shift of the operating point of the OHC transfer function, for example due to a change in the resting position of the OHC stereocilia, might result in relatively large changes in even distortion products without affecting odd-order distortion. This phenomenon is illustrated in Figure 10.8, where changes in amplitude of the $f_2 - f_1$ component are much larger and opposite in sign to those observed for the $2f_1 - f_2$ component when the operating point of the OHC mechanoelectrical transducer is shifted from position C to position D. The possibility that different DPOAE components can be changed independently of each other received direct support from experiments by Frank and Kössl (1996, 1997). Frank and Kössl biased the cochlear partition acoustically and electrically so as to shift the OHC operating point around its resting position. It is possible that the corresponding shift of the OHC mechanoelectrical transducer operating point develops due to calcium-mediated, slow viscoelastic efferent effects associated with a decrease in OHC stiffness (Santos-Sacchi et al. 1998; Frolenkov et al. 2003). Independent changes of the $f_2 - f_1$ and $2f_1 - f_2$ amplitudes, however, might be observed even without the OHC operating point shift. It will be shown below that a more commonly recognized efferent effect, suppression of the cochlear amplifier (Guinan, 1996), might cause similar DPOAE amplitude changes.

7.3 DPOAE Changes Due to the Fast Effect of the MOC Excitation

More recently, Puria et al. (1996) reported changes in the $2f_1 - f_2$ component on a much shorter time scale of a few hundred milliseconds following the onset of the ipsilateral stimulation. Liberman et al. (1996) noted that the time constant of these relatively fast onset changes is similar to the time constant of efferent mediated effects recorded at different levels of the auditory system (see review by Guinan 1996 and the current chapter). These fast postonset changes in the DPOAE are considerably reduced following section of the OCB (Liberman et al. 1996; Kujawa and Liberman 2001) and intravenous injection of strychnine (Lukashkin et al. 2002).

An intriguing feature of the postonset changes is their bidirectionality. The DPOAE amplitude can either decrease or increase after the onset depending on the frequencies and intensities of the primary tones. Kim et al. (2001) have also observed biphasic postonset changes when a rapid reduction in DPOAE amplitude is followed by DPOAE growth. The fast postonset changes in DPOAE have been shown for a relatively narrow range of stimulus parameters. Kujawa and Liberman (2001) reported that these changes depended on the presence of amplitude minima (notches) routinely observed at the DPOAE I/O functions.

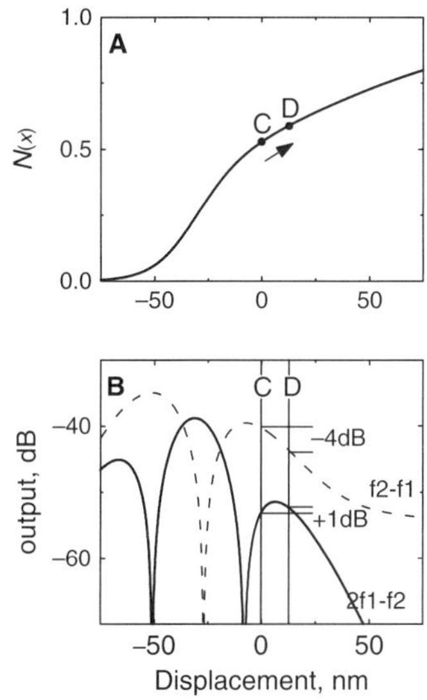

FIGURE 10.8. Changes in distortion amplitude due to a shift of the operating point of the OHC mechanoelectrical transducer conductance. (**A**) nonlinear I/O function, $N(x)$, of the outer-hair-cell mechanoelectrical transducer is approximated by a Boltzmann function with parameters used by Geleoc et al. (1997) to fit their experimental data. The nonlinearity was normalized for the maximum transducer conductance and modified to have about 50% of the transducer conductance activated at rest ($N(0) \approx 0.5$), which corresponds to the experimentally measured value (Russell and Kössl 1992; Legan et al. 2000). (**B**) dependence of the $f_2 - f_1$ and $2f_1 - f_2$ amplitudes on displacement of the operating point for the nonlinearity shown in **A**. Distortion amplitudes are calculated for constant levels of the input signals. $2f_1 - f_2$ component is moved upward by 10 dB. *Vertical lines C* and *D* indicate amplitude of the distortion products for two positions of the operating point shown in **A** with the same letters. Shift of the operating point from *C* to *D* results in different changes of the $f_2 - f_1$ and $2f_1 - f_2$ components indicated in the figure. (Modified from Fig. 1, Frank and Kössl 1996, with permission from Elsevier.)

Indeed, it was shown (Lukashkin et al. 2002) that all of the different fast postonset changes of the DPOAE amplitude (growth, decline, biphasic change) were associated with a single phenomenon, i.e., with the shift of the amplitude notch to higher levels of the primaries during the onset of stimulation (Fig. 10.9). This shift of the notch to higher levels increases for lower f_2/f_1 ratios but does not exceed about 2 dB. However, the relatively small shift of the notch is responsible for much larger changes in the DPOAE amplitude in the vicinity of the notch.

An increase in the DPOAE is observed for levels of primaries generating emissions on the left slope of *the initial notch* (Fig. 10.9, top panel, solid line). In contrast, the DPOAE decreases for levels of primaries that generate emissions on the right slope of *the shifted notch* (dashed line). A biphasic DPOAE response (i.e., a decrease followed by an increase; Fig. 10.9, bottom middle panel) takes place for any primary levels generating DPOAEs between

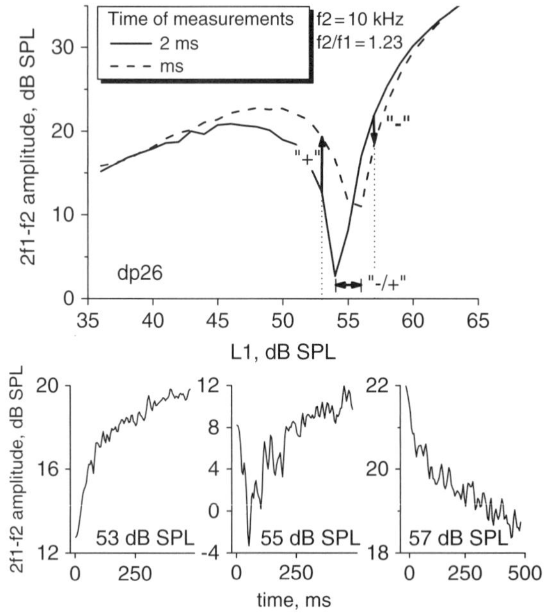

FIGURE 10.9. Dependence of the postonset changes of the DPOAE on the levels of the primaries. *Top panel* shows the response presented in Fig. 10.1B in the vicinity of the amplitude notch. Solid line shows DPOAE response at the beginning of the prolonged acoustical stimulation. This response does not change after the injection of strychnine. *Dashed curve* represents the DPOAE responses 480 ms after the onset of the stimulation. *Vertical arrows* indicate changes of the DPOAE amplitude for particular pairs of the primary levels; $L_1/L_2 = 53/43$ dB SPL (*left arrow*) and $L_1/L_2 = 57/47$ dB SPL (*right arrow*). Temporal courses of these changes are shown in the *bottom left* and *right panels*, respectively. *Middle panel of the bottom row* shows biphasic time-dependence of the DPOAE for the L_1 range indicated by the horizontal arrow in the top panel. f_2 and f_2/f_1 are indicated at the *top panel*. L_1 is indicated within each of the bottom panels. L_1 step is 1 dB. L_2 is 10 dB below L_1. (Used with permission from Lukashkin et al. 2002. Copyright 2002, Acoustical Society of America.)

the initial and steady-state notch positions. If the DPOAE level is far from the notch in its growth function, i.e., the levels of the primaries are more than about 10 dB higher or lower than those at the tip of the notch, then postonset changes in the DPOAE were not observed. In this respect, Siegel and Kim (1982) did not record changes in $2f_1 - f_2$ components for high- and low-level primaries but observed both increases and decreases in the $2f_1 - f_2$ components for midlevel primaries. Hence, the DPOAE changes they observed could also be a result of a shift in the amplitude notch.

Kujawa and Liberman (2001) speculated that changes in the notch position were observed because the OCB activation affected interaction between the multiple components of the DPOAE, which produced an amplitude notch due to phase cancellation between them. However, amplitude notches in the growth functions of the distortion component can be observed at the output of a single saturating nonlinearity (Lukashkin and Russell 1998, 1999). The notches originate in zeros of the higher-order derivatives of the nonlinearity (Bian et al. 2002; Lukashkin and Russell 2005). This genesis of the amplitude notch is attractive because it explains the specific DPOAE behavior seen for various experimental procedures (Lukashkin and Russell 2001; Lukashkin et al. 2002; Bian et al. 2004). Figure 10.10 shows that two different processes can generate an apparent shift, ΔL, of the notch to higher primary levels during the onset of the ipsilateral acoustical stimulation. First, this shift of the notch is possible if activation of the cochlear efferents move the OHC operating point in the positive direction by Δx. This mechanism is essentially the same as one suggested as an explanation for the slow efferent effect (Fig. 10.8). However, a similar notch shift is predicted to occur for a stable position of the OHC operating point (vertical line C in Fig. 10.10) when the cochlear amplifier is suppressed during the activation of the fast efferent effect. In fact, the notch shift could be due to a combination of a change in amplification and a change in the operating point of the mechanoelectrical transducer, although the DPOAE amplitude changes due to these two mechanisms may have different time scales.

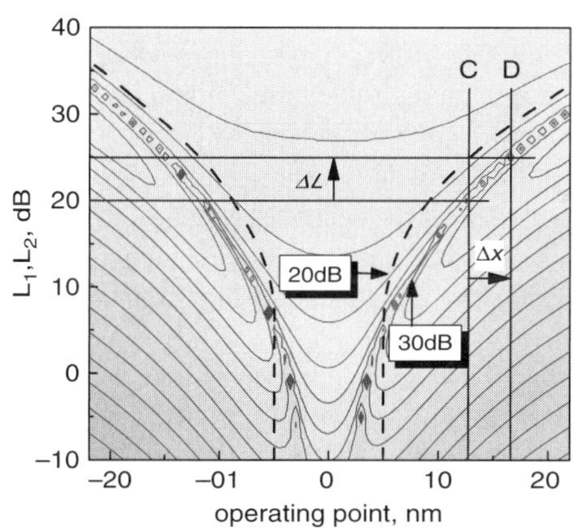

FIGURE 10.10. Two possible mechanisms for the amplitude notch shift. The contour plot shows dependence of the $2f_1 - f_2$ amplitude on the level of the primaries (*vertical axis*) and position of the operating point of the OHC mechanoelectrical transducer relative to the point of inflection of the transducer function (*horizontal axis*). The mechanoelectrical transducer is approximated by the same Boltzmann function as in Fig. 10.1. L_1 and L_2 are expressed in dB re 0.1 nm. The cochlear amplifier in this figure is modelled by a positive feedback loop (Yates et al. 1990). The overall closed-loop gain of the system is 30 dB. An *arrow* with the corresponding label points out at the amplitude notch for this level of the gain. Two *dashed curves* indicate new position of the amplitude notches when the system gain is 20 dB. The amplitude notch occurs for L_1 and L_2 of 20 dB when the system gain is 30 dB and the operating point is situated at position *C*. Apparent shift ΔL of the notch to L_1 and L_2 of 25 dB is observed when either the operating point moves to position *D* or the system gain drops to 20 dB.

8. Sound and Cortically Evoked Efferent Effects on Hair Cells and Mechanics of the Mustached Bat Cochlea

Probably more is understood about the neural basis of echolocation in the *Pteronotus parnellii* (the mustached bat) auditory system than about neural processing of auditory stimuli in any other mammalian species. This is largely through the considerable attention that has been paid to the behavioral and functional significance of the information contained in the echolocation calls and the types of information that can be extracted from the echoes. A consequence of these studies is that it is possible to selectively excite the cochlear efferent system of *P. parnellii* with behaviorally significant stimuli to reveal how efferent induced changes in the mechanical properties of OHCs interplay with sound-induced mechanical changes in the cochlear partition.

The echolocation calls of *P. parnellii* are adapted for hunting insects in dense vegetation and consist of a long constant-frequency (CF) component ending with a short downward frequency modulation (FM). The bats are able to detect the prey's wing beat as slight frequency and amplitude modulations in the CF echo, so-called glints (reviews by Neuweiler 1990, 2003 and or is this the whole, book). The CF component of the echo is analyzed by a massive expansion of the 54–70 kHz region of the cochlea, known as the acoustic fovea (see Kössl and Vater 1995). Within this fovea, a densely innervated and strongly resonant region processes the dominant echolocation frequency at 61 kHz. The frequency tuning of this resonance is incredibly narrow, with Q_{3dB} values in excess of 900 (Kössl and Vater 1995). Relative movements between bat and prey during echolocation cause Doppler shifts in echo frequency, which are compensated by lowering the frequency of the emitted call or indeed increasing the frequency of the resonance, so that the echo falls within the narrow "sweet spot" of the acoustic fovea.

The enhanced tuning is generated by hydromechanical specializations that involve morphological discontinuities of the BM and TM and result in the strong mechanical resonance (Russell and Kössl 1999; Russell et al. 2003). This resonance, which is very close to instability and can break down into sustained spontaneous otoacoustic emissions (SOAEs), is responsible for the extraordinarily loud OAEs and long ringing of delayed otoacoustic emissions (DEOAEs) and CM (Fig 10.11), (reviewed by Kössl and Vater 1995). Another striking feature of this remarkable mechanical analyzer is the density of the efferent innervation of the OHCs. Although each OHC receives only a single efferent terminal, which decreases in size from the first to the third row of OHCs, each OHC in the acoustic fovea is innervated by a single efferent neuron, an indication that cochlear mechanics of this region are finely controlled (Xie et al. 1993), a situation that contrasts with its Old-World counterpart, the horseshoe bat (Vater et al. 1992), whose OHCs do not receive efferent innervation. The auditory foveae of horseshoe bats (*Rhinolophidae*) are also finely tuned to the dominant echo frequency, but without the strong resonance displayed by the acoustic fovea of *P. parnellii* (reviewed by Neuweiler 2003). Perhaps the role

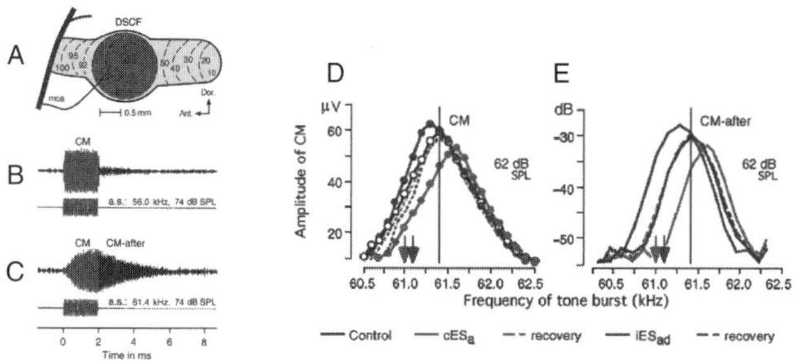

FIGURE 10.11. Corticofugal efferent control of cochlear tuning and sensitivity. (**A**–**C**) Cortical overrepresentation of, and sharply tuned hair cell responses to, a 61-kHz sound. (**A**) The primary auditory cortex. The dashed lines and numbers indicate iso-best frequency lines. The Doppler-shifted constant frequency (DSCF) area (*dark gray*) overrepresents a sound between 60.6 and 62.3 kHz. mca-middle cerebral artery. (**B**, **C**) Cochlear microphonic responses (CM) evoked by 2.0-ms tone bursts. Frequency and amplitude of an acoustic stimulus (a.s.) are indicated in kHz and dB SPL, respectively. The CM and CM-after are indicated in gray and black, respectively. The data in (**B**) and (**C**) were obtained from a single cochlea. (**D**, **E**) Corticofugal modulation of cochlear microphonic (CM) responses. Changes in CM (**D**) and CM-after (**E**) evoked by electric stimulation of cortical DSCF neurons (ESa). (**D**) The frequency-response curves of the CM measured with tone bursts at 62 dB SPL. (**E**) The amplitude spectrum of CM-after evoked by a 62-dB SPL tone burst at the BF. *Black curves*, control condition. *Solid pale gray curves*, "shifted" condition evoked by electric stimulation of contralateral DSCF neurons tuned to 61.00 kHz (cESa, *red arrows*). *Dashed pale gray curves*, recovered condition from cESa. *Solid dark gray curves*, shifted condition evoked by electric stimulation of "dorsal" ipsilateral DSCF neurons tuned to 61.10 kHz (iESad, *dark gray arrows*). *Dashed dark gray curves*, recovered condition from iESad. (Modified from Figs. 1, 2, Xiao and Suga 2002, with permission.)

of the OHCs and their efferent innervation in the *P. parnellii* cochlea is to control the stability of the sharply tuned resonance and fine-tune its frequency characteristics.

Studies by Henson et al. (1995), Goldberg and Henson (1998), and Xie and Henson (1998) showed that excitation of the medial efferent system through contralateral acoustic stimulation suppresses cochlear responses, as indicated by a shortening and frequency shift of the ringing of the acoustic fovea. Two studies confirm and extend Henson's pioneering studies and show that the efferent system finely controls the mechanical responses of the acoustic fovea of *P. parnellii*. A prerequisite for this is that the efferent fibres be themselves finely tuned. This property has yet to be demonstrated in the bat cochlea, but measurements from efferent fibers in other mammals reveal that their tuning can be similar to that of the afferent fibers (Cody and Johnstone 1982; Robertson 1984).

A recent and exciting focus of research into the workings of bat echolocation is the study of the corticofugal system by Suga and his colleagues (review by Suga et al. 2000, 2002), which projects through multiple descending loops to subcortical regions of the brain and the cochlea. They showed that the corticofugal system in *P. parnellii* adjusts and improves signal processing in the subcortical regions of the echolocation pathway, including the OHCs. Xiao and Suga (2002) found the OHCs are modulated through stimulation of the corticofugal system in a highly specific way that depends on the relationship between location of the neurons in the contra- and ipsilateral parts of the cortex concerned with analyzing the Doppler-shifted echoes to the constant-frequency component of the bats' echolocation calls (the so-called DSCF area; Fig. 10.11), the best frequencies (BFs) of the cortical neurons, and the OHCs. It appears that the way the OHCs are modulated, as measured from the CM, depends on whether the BFs of CM and cortical neurons are the same or different. When the BFs are matched, excitation of the contralateral DSCF causes the CM to be increased in amplitude and sharpened in frequency tuning. Unmatched CM is decreased in amplitude, sharpened, and centrifugally shifted (i.e., toward the BF of the cortical neuron) in frequency tuning. By contrast, stimulation of neurons in the ipsilateral DSCF area cause an increase in magnitude, broadening, and centripetal shift in CM frequency tuning that depends on the mismatch in the BFs of the neurons and the CM (Fig. 10.11). These changes in CM can last for several minutes following excitation of the DSCF neurons. Inactivation of either or both the ipsi- and contralateral DSCF areas through application of the $GABA_A$ agonist muscimol reduces the BF and amplitude of the CM and broadens the CM-frequency tuning curve, an indication that the frequency tuning of the cochlear hair cells is tonically influenced by the corticofugal system.

The other study by Drexl and Kössl (2003), which also builds on Henson's work, provides further indication that cochlear efferent neurons are selectively responsive to particular, biologically relevant, stimuli and that the postsynaptic influence of the efferent system on the cochlea depends on the ambient acoustic environment and hence the tonic state of the OHCs, which is probably determined through the interplay of the acoustic stimulus and fast and slow efferent effects. The most important function of the efferent system in *P. parnellii* is to optimize the stability and frequency tuning of the acoustic fovea and thereby the performance of a crucial piece of sensory equipment for the hunting of insect prey in a densely cluttered acoustic environment.

Drexl and Kössl found that MOC efferents can be excited by echolocation calls and broadband noise, but not pure tones below levels that would excite the middle-ear reflex, delivered to the contralateral ear (e.g., see Maison et al. 2000). Drexl and Kössl found that white noise delivered to the contralateral ear caused an upward shift in frequency of SFOAEs. Weak contralateral noise enhanced OAEs, and loud contralateral noise caused a decrease in the amplitude of OAEs. In accordance with efferent effects on BM and neural responses in other mammals, the efferent effects on the mechanical properties of the acoustic fovea of *P. parnellii*, as measured in DPOAEs, is greatest for sounds of moderate to loud intensity. The effect of contralateral noise on the level of tuning of the resonance was found to depend on the initial tuning, or Q values, of the resonance. If the Q was lower than the mean of about 500, then contralateral white noise increased the Q, so that it approached this mean value. If the initial Q value was above the mean, then contralateral white noise reduced the Q value. That these changes in the mechanical properties of the acoustic fovea had their counterparts in the electrical responses of the hair cells was revealed in simultaneous measurements of CM and DEOAEs, in that the Q values of the CM and DEOAEs were altered by similar amounts and in the same direction. There is therefore in the *P. parnellii* cochlea a direct correspondence between electrical events in the OHCs and mechanical changes in the organ of Corti. Significantly, decreases in Q associated with the presentation of white noise at the contralateral ear immediately reverted to initial values when the white noise ceased. By contrast, increases in Q did not return to original values within several hours, an indication perhaps that the increase in Q is due to the slow efferent effect, first describe by Sridhar et al. (1995).

Xie and Henson (1998) describe a tonic reduction of the CM decay time (reduction in resonance) and suppression of CM during contralateral presentation of *P. parnellii* echolocation and social calls that lasted from 20 minutes to 2 hours after stimulus offset. Similar tonic reductions in CM decay time were observed in bats that were just removed from their colony, probably caused by the broad band (6–120 kHz), intense (< 120 dB SPL) acoustic background of the colony, and after extensive echolocation during flight, most likely caused by their own echolocation calls. These results partially contrasted with those of Drexl and Kössl, who also found that changes in the cochlear resonance (presumably due to OHC-mediated stiffness changes in the cochlea) could occur on different time scales and independently of suppressive and augmenting effects (presumably due to changes in cochlear amplification).

The functions of the olivocochlear efferent fibers remain enigmatic and unresolved (see Guinan 1996 and Smith and Spirou 2002 for reviews). In the mustached bat, the BF of the CM increases with flight (Huffman and Henson 1993) and changes in an apparently unpredictable way with vocalization (Goldberg and Henson 1998). Perhaps, as with the lateral-line efferent system (Russell 1976), activity of the olivocochlear efferents is associated with voluntary motor activity. Xiao and Suga (2002) interpret their findings that cortical neurons of the DSCF area modulate frequency tuning of cochlear hair cells as indicating that the olivocochlear efferents are involved in selective attention and increase of the signal-to-noise ratio.

9. Summary

A similar sequence of postsynaptic events in response to electrical stimulation of efferent nerve fibers has been recorded in hair cells of sense organs as diverse as the lateral-line system, the vestibular system, and the cochlea, but not yet from OHCs of the mammalian cochlea. The primary efferent transmitter, ACh, acts through an nAChR containing alpha 9 and 10 subunits to hyperpolarize the hair cell. Calcium influx through the nAChR-gated channel rapidly activates calcium-activated potassium channels (SK2) in the close vicinity of the synapse to produce hyperpolarization within a few milliseconds. This process is focused at the efferent synapse and amplified by CICR from the subsynaptic cistern. Following repetitive excitation of the MOC efferent system, a second "slow" effect of the efferent system is manifested, again through the action of calcium. This time, calcium triggers a second messenger cascade, which leads to the phosphorylation of cytoskeletal proteins associated with the lateral walls of the OHCs, thereby changing the OHC's axial stiffness. The slow effect is thought to involve calcium "sparking" from the synaptic cistern to the lateral cisternae and CICR from these cisternae, which are located adjacent to the lateral walls of the OHCs. The mechanism of inhibition in the mammalian cochlea remains unresolved and depends very much on the mechanism of cochlear amplification. From direct measurements of BM vibrations to acoustic stimulation, and from the measurement of DPOAEs, the fast effect appears to attenuate cochlear amplification. It may do this through a variety of mechanisms including shunting, and resetting the operating points of the hair-cell mechanoelectrical transducer and the OHC voltage-dependent motility. Rapid hair-bundle adaptation, which is thought to provide amplification, at least in nonmammalian auditory organs, depends on the calcium gradient across the mechanoelectrical transducer. Perhaps efferent stimulation changes this gradient as a result of CICR from intracellular stores, which is perhaps why the OHC equivalents of reptilian and avian cochleae are dish-shaped, so that the hair bundle is closer to the efferent synapse than in other hair cells. The slow effect appears to change the stiffness of the cochlear partition and thereby the coupling of the hair cells to the acoustic stimulus. Changes in the stiffness, in turn, could lead to conformational changes within the partition as indicated by alterations in the cochlear nonlinearity as revealed by DPOAEs.

Selective excitation of the efferent system through corticofugal pathways and contralateral acoustic stimulation reveals that the efferents can finely tune the cochlear and the acoustic fovea of *P. parnellii* in particular. This is seen in changes in cochlear amplification, the shifting of the characteristic frequencies of the CM and SOAEs up and down in frequency, and increases and decreases of the Q of cochlear responses. These changes may come about through changes in OHC-mediated amplification and/or changes in the axial stiffness of the OHCs, which, in turn, will depend on the state of excitation and hence the axial tension of the OHCs at the time of efferent activation. Electrical stimulation of the MOC efferents provides a technique for understanding their postsynaptic action on the electrical and mechanical properties of the OHCs. Selective and behaviorally significant excitation of the efferent pathway reveals how efferent-induced changes in the mechanical properties of OHCs interplay with sound-induced mechanical changes in the cochlear partition. Through recording the nonlinear acoustical responses of the cochlea, DPOAEs provide a sensitive and subtle indicator of efferent-induced reduction in the amplification and changes in the mechanical properties of the cochlear partition. The utility of DPOAEs as a means for disclosing the details of efferent-dependent functional changes in the cochlea should increase in the future when the mechanisms for their generation are fully appreciated.

Acknowledgments. This work is supported by the MRC. We thank Corné Kros, Markus Drexl, and Arthur Popper for reading earlier drafts of the manuscript and for helpful advice on ways of improving it. We thank Stuart Johnson for revising and preparing Figure 10.4A.

References

Art JJ, Fettiplace R (1984) Efferent desensitization of auditory-nerve fiber responses in the cochlea of the turtle *Pseudemys scripta* elegans. J Physiol (Lond) 356:507–520.

Art JJ, Fettiplace R, Fuchs PA (1984) Synaptic hyperpolarization and inhibition of turtle cochlear hair-cells. J Physiol (Lond) 356:525–550.

Art JJ, Crawford AC, Fettiplace R, Fuchs PA (1985) Efferent modulation of hair cell tuning in the cochlea of the turtle. J Physiol (Lond) 360:397–421.

Ashmore JF, Russell IJ (1982) Effect of efferent nerve-stimulation on hair-cells of the frog sacculus. J Physiol (Lond) 329:P25–P26.

Benson TE, Brown MC (2004) Postsynaptic targets of type II auditory nerve fibers in the cochlear nucleus. JARO 5:111–125.

Bian L, Chertoff ME, Miller E (2002) Deriving a cochlear transducer function from low-frequency modulation of distortion product otoacoustic emissions. J Acoust Soc Am 112:198–210.

Bian L, Linhardt EE, Chertoff ME (2004) Cochlear hysteresis: observation with low-frequency modulated distortion product otoacoustic emissions. J Acoust Soc Am 115:2159–2172.

Brown AM (1988) Continuous low level sound alters cochlear mechanics: an efferent effect? Hear Res 34:27–38.

Bruns V, Schmieszek E (1980) Cochlear innervation in the greater horseshoe bat: demonstration of an acoustic fovea. Hear Res 3:27–43.

Chan DK, Hudspeth AJ (2005) Ca^{2+} current-driven nonlinear amplification by the mammalian cochlea in vitro. Nat Neurosci 8:149–155.

Chang KW, Norton SJ (1997) Efferently mediated changes in the quadratic distortion product ($f_2 - f_1$). J Acoust Soc Am 102:1719–1733.

Cody AR, Johnstone BM (1982) Acoustically evoked activity of single efferent neurons in the guinea-pig cochlea. J Acoust Soc Am 72:280–282.

Collet L, Kemp DT, Veuillet E, Duclaux R, Moulin A, Morgon A (1990) Effect of contralateral auditory-stimuli on active cochlear mechanics in human-subjects. Hear Res 43:251–261.

Cooper NP, Guinan JJ (2003) Separate mechanical processes underlie fast and slow effects of medial olivocochlear efferent activity. J Physiol (Lond) 548:307–312.

Dallos P, Evans BN (1995) High-frequency motility of outer hair-cells and the cochlear amplifier. Science 267:2006–2009.

Dallos P, He DZ, Lin X, Sziklai I, Mehta S, Evans BN (1997) Acetylcholine, outer hair cell electromotility, and the cochlear amplifier. J Neurosci 17:2212–2226.

Dawkins R, Keller SL, Sewell WF (2005) Pharmacology of acetylcholine-mediated cell signaling in the lateral line organ following efferent stimulation. J Neurophysiol 93:2541–2551.

Deak L, Zheng J, Orem A, Du GG, Aguinaga S, Matsuda K, Dallos P (2005) Effects of cyclic nucleoticles on the function of prestin. J Physiol (Lond) 563:483–496.

Dolan DF, Guo MH, Nuttall AL (1997) Frequency-dependent enhancement of basilar membrane velocity during olivocochlear bundle stimulation. J Acoust Soc Am 102:3587–96.

Drexl M, Kössl M (2003) Sound-evoked efferent effects on cochlear mechanics of the mustached bat. Hear Res 184:61–74.

Dulon D, Zajic G, Schacht J (1990) Increasing intracellular free calcium induces circumferential contractions in isolated cochlear outer hair-cells. J Neurosci 10: 1388–1397.

Elgoyhen AB, Johnson DS, Boulter J, Vetter DE, Heinemann S (1994) Alpha-9: An acetylcholine-receptor with novel pharmacological properties expressed in rat cochlear hair-cells. Cell 79:705–715.

Elgoyhen AB, Vetter DE, Katz E, Rothlin CV, Heinemann SF, Boulter J (2001) alpha 10: A determinant of nicotinic cholinergic receptor function in mammalian vestibular and cochlear mechanosensory hair cells. Proc Natl Acad Sci USA 98:3501–3506.

Evans MG (1996) Acetylcholine activates two currents in guinea-pig outer hair cells. J Physiol (Lond) 491:563–578.

Evans MG, Lagostena L, Darbon P, Mammano F (2000) Cholinergic control of membrane conductance and intracellular free Ca^{2+} in outer hair cells of the guinea pig cochlea. Cell Calcium 28:195–203.

Fettiplace R, Ricci AJ, Hackney CM (2001) Clues to the cochlear amplifier from the turtle ear. Trends Neurosci 24:169–75.

Fex J (1959) Augmentation of cochlear microphonics by stimulation of efferent fibres to the cochlea. Acta Otolaryngol 50:540–541.

Fex J (1962) Auditory activity in centrifugal and centripetal cochlear fibres in cat—a study of a feedback system. Acta Physiol Scand 55:1–68.

Fex J (1967) Efferent inhibition in cochlea related to hair-cell dc activity—study of postsynaptic activity of crossed olivocochlear fibres in cat. J Acoust Soc Am 41: 666–675.

Fischer FP (1994) General pattern and morphological specializations of the avian cochlea. Scan Microsc 8:351–364.

Flock Å, Flock B (1966) Ultrastructure of the amphibian papilla in the bullfrog. J Acoust Soc Am 40:1262.

Flock Å, Russell IJ (1973) Efferent nerve fibers—postsynaptic action on hair cells. Nature—New Biol 243:89–91.

Flock Å, Russell IJ (1976) Inhibition by efferent nerve-fibers—action on hair cells and afferent synaptic transmission in lateral line canal organ of burbot *Lota lota*. J Physiol (Lond) 257:45–62.

Frank G, Kössl M (1996) The acoustic two-tone distortions $2f1 - f2$ and $f2 - f1$ and their possible relation to changes in the gain and the operating point of the cochlear amplifier. Hear Res 98:104–115.

Frank G, Kössl M (1997) Acoustical and electrical biasing of the cochlear partition. Effects on the acoustic two tone distortions $f_2 - f_1$ and $2f_1 - f_2$. Hear Res 113:57–68.

Frolenkov GI, Mammano F, Belyantseva IA, Coling D, Kachar B (2000) Two distinct Ca^{2+}-dependent signaling pathways regulate the motor output of cochlear outer hair cells. J Neurosci 20:5940–5948.

Frolenkov GI, Mammano F, Kachar B (2003) Regulation of outer hair cell cytoskeletal stiffness by intracellular Ca^{2+}: underlying mechanism and implications for cochlear mechanics. Cell Calcium 33:185–195.

Geisler CD, Sang CN (1995) A cochlear model using feed-forward outer-hair-cell forces. Hear Res 86:132–146.

Glowatzki E, Fuchs PA (2000) Cholinergic synaptic inhibition of inner hair cells in the neonatal mammalian cochlea. Science 288:2366–2368.

Goldberg RL, Henson OW Jr. (1998) Changes in cochlear mechanics during vocalization: evidence for a phasic medial efferent effect. Hear Res 122:71–81.

Goutman, JD, Paul Albert Fuchs PA, Glowatzki E (2005) Facilitating efferent inhibition of inner hair cells in the cochlea of the neonatal rat. J Physiol 566:49–59.

Groff JA, Liberman MC (2003) Modulation of cochlear afferent response by the lateral olivocochlear system: activation via electrical stimulation of the inferior colliculus. J Neurophysiol 90:3178–3200.

Grose GH (1983) The effect of contralateral stimulation on spontaneous acoustic emissions. J Acoust Soc Am Suppl 1 74:S38.

Guinan JJ (1996) Physiology of olivocochlear efferents. In: Dallos P, Popper AN, Fay RR (eds) The Cochlea. New York: Springer-Verlag, pp. 435–502.

Guinan JJ, Stankovic KM (1996) Medial efferent inhibition produces the largest equivalent attenuations at moderate to high sound levels in cat auditory-nerve fibers. J Acoust Soc Am 100:1680–1690.

Guinan JJ, Warr WB, Norris BE (1984) Topographic organization of the olivocochlear projections from the lateral and medial zones of the superior olivary complex. J Comp Neurol 226:21–27.

He DZZ, Jia S, Dallos P (2003) Prestin and the dynamic stiffness of cochlear outer hair cells. J Neurosci 23:9089–9096.

Henson OW, Xie DH, Keating AW, Henson MM (1995) The Effect of contralateral stimulation on cochlear resonance and damping in the moustached bat—the role of the medial efferent system. Hear Res 86:111–124.

Holt JC, Lioudyno M, Athas G, Garcia MM, Perin P, Guth PS (2001) The effect of proteolytic enzymes on the alpha 9-nicotinic receptor-mediated response in isolated frog vestibular hair cells. Hear Res 152:25–42.

Housley GD, Ashmore JF (1991) Direct measurement of the action of acetylcholine on isolated outer hair-cells of the guinea-pig cochlea. Proc R Soc Lond B 244:161–167.

Hudspeth AJ (1997) Mechanical amplification of stimuli by hair cells. Curr Opin Neurobiol 7:480–486.

Hudspeth AJ, Choe Y, Mehta AD, Martin P (2000) Putting ion channels to work: Mechanoelectrical transduction, adaptation, and amplification by hair cells. Proc Natl Acad Sci USA 97:11765–11772.

Huffman RF, Henson OW Jr (1993) Labile cochlear tuning in the mustached bat. Concomitant shifts in biosonar emission activity. J Comp Physiol A 171:725–734.

Jia S, He DZ (2005) Motility-associated hair-bundle motion in mammalian outer hair cells. Nature Neurosci 8:1028–1034.

Katz B, Miledy R (1972) The statistical nature of the acetylcholine potential and its molecular components. J Physiol (Lond) 224:665–699.

Kemp DT (1980) Towards a model for the origin of cochlear echoes. Hear Res 2:533–548.

Kemp DT, Souter M (1988) A new rapid component in the cochlear response to brief electrical efferent stimulation: CM and otoacoustic observations. Hear Res 34:49–62.

Kennedy HJ, Evans MG, Crawford AC, Fettiplace R (2003) Fast adaptation of mechanoelectrical transducer channels in mammalian cochlear hair cells. Nat Neurosci 6: 832–836.

Kennedy HJ, Crawford AC, Fettiplace R (2005) Force generation by mammalian hair bundles supports a role in cochlear amplification. Nature 433:880–883.

Kiang NY-S, Moxon EC, Levine RA (1970) Auditory-nerve activity in cats with normal and abnormal cochleas. In: Wolstenhome GEW, Knight J (eds) Sensorineural Hearing Loss. CIBA Foundation Symposium. London: Churchill, pp. 241–268.

Kim DO, Dorn PA, Neely ST, Gorga MP (2001) Adaptation of distortion product otoacoustic emission in humans. JARO 2:31–40.

Kirk DL, Johnstone BM (1993) Modulation of $f_2 - f_1$: evidence for a GABA-ergic efferent system in apical cochlea of the guinea pig. Hear Res 67:20–34.

Kolston PJ (1995) A faster transduction mechanism for the cochlear amplifier? Trends Neurosci 18:427–429.

Köppl C, Manley GA, Konishi M (2000) Auditory processing in birds. Curr Opin Neurobiol 10:474–481.

Kössl M, Russell IJ (1992) The phase and magnitude of hair cell-receptor potentials and frequency tuning in the guinea-pig cochlea. J Neurosci 12:1575–1586.

Kössl M, Russell IJ (1995) Basilar-membrane resonance in the cochlea of the moustached bat. Proc Natl Acad Sci USA 92:276–279.

Kössl M, Vater M (1995) Cochlear structure and function in bats. In: Popper AN, Fay RR (eds) Hearing by Bats. New York: Spinger-Verlag, pp. 191–234.

Kros CJ (1996) Physiology of mammalian cochlear hair cells. In: Dallos P, Popper AN, Fay RR (eds) The Cochlea. New York: Springer-Verlag, pp. 318–385.

Kros CJ, Ruppersberg JP, Rüsch A (1998) Expression of a potassium current in inner hair cells during development of hearing in mice. Nature 394:281–284.

Kujawa SG, Liberman MC (2001) Effect of olivocochlear feedback on distortion product otoacoustic emission. JARO 2:268–278.

Kujawa SG, Fallon M, Bobbin RP (1995) Time-varying alterations in the $f_2 - f_1$ DPOAE response to continuous primary stimulation. I: Response characterization and contribution of the olivocochlear efferents. Hear Res 85:142–154.

Liberman MC, Puria S, Guinan JJ (1996) The ipsilaterally evoked olivocochlear reflex causes rapid adaptation of the $2f_1 - f_2$ distortion product otoacoustic emission. J Acoust Soc Am 99:3572–3584.

Lim DJ (1986) Functional structure of the organ of Corti—a review. Hear Res 22: 117–146.

Lioudyno M, Hiel H, Kong JH, Katz E, Waldman E, Parameshwaran-Iyer S, Glowatzki E, Fuchs PA (2004) A "synaptoplasmic cistern" mediates rapid inhibition of cochlear hair cells. J Neurosci 24:11160–11164.

Lukashkin AN, Russell IJ (1998) A descriptive model of the receptor potential nonlinearities generated by the hair cell mechanoelectrical transducer. J Acoust Soc Am 103:973–980.

Lukashkin AN, Russell IJ (1999) Analysis of the $f2 - f1$ and $2f1 - f2$ distortion components generated by the hair cell mechanoelectrical transducer: dependence on the amplitudes of the primaries and feedback gain. J Acoust Soc Am 106:2661–2668.

Lukashkin AN, Russell IJ (2001) Origin of the bell-like dependence of the DPOAE amplitude on primary frequency ratio. J Acoust Soc Am 110:3097–3106.

Lukashkin AN, Russell IJ (2002) Modifications of a single saturating non-linearity account for post-onset changes in $2f1 - f2$ distortion product otoacoustic emission. J Acoust Soc Am 112:1561–1568.

Lukashkin AN, Russell IJ (2005) Dependence of the DPOAE amplitude pattern on acoustical biasing of the cochlear partition. Hear Res 203:45–53.

Lukashkin AN, Lukashkina VA, Russell IJ (2002) One source for distortion product otoacoustic emissions generated by low- and high-level primaries. J Acoust Soc Am 111:2740–2748.

Maison S, Micheyl C, Andeol G, Gallego S, Collet L (2000) Activation of medial olivocochlear efferent system in humans: influence of stimulus bandwidth. Hear Res 140:111–25.

Marcotti W, Johnson SL, Kros CJ (2003) Sodium and calcium currents shape action potentials in immature mouse inner hair cells. J Physiol (Lond) 552:743–761.

Marcotti W, Johnson SL, Kros CJ (2004) Effects of intracellular stores and extracellular Ca^{2+} on Ca^{2+}- activated K^+ currents in mature mouse inner hair cells. J Physiol (Lond) 557:613–633.

Markin VS, Hudspeth AJ (1995) Modelling the active process of the cochlea: phase relations, amplification, and spontaneous oscillation. Biophys J 69:138–147.

Mountain DC (1980) Changes in endolymphatic potential and crossed olivocochlear bundle stimulation alter cochlear mechanics. Science 210:71–72.

Mulders W, Robertson D (2005) Diverse responses of single auditory afferent fibres to electrical stimulation of the inferior colliculus in guinea-pig. Exp Brain Res 160: 235–244.

Murugasu E, Russell IJ (1996a) The effect of efferent stimulation on basilar membrane displacement in the basal turn of the guinea pig cochlea. J Neurosci 16:325–332.

Murugasu E, Russell IJ (1996b) The role of calcium on the effects of intracochlear acetylcholine perfusion on basilar membrane displacement in the basal turn of the guinea pig cochlea. Aud Neurosci 2:363–376.

Neuweiler G (1990) Auditory adaptations for prey capture in echolocating bats. Physiol Rev 70:615–641.

Neuweiler G (2003) Evolutionary aspects of bat echolocation. J Comp Physiol A 189: 245–256.

Nuttall AL, Guo M, Ren T, Dolan DF (1997) Basilar membrane velocity noise. Hear Res 114:35–42.

Oliver D, Klocker N, Schuck J, Baukrowitz T, Ruppersberg JP, Fakler B (2000) Gating of Ca^{2+}-activated K^+ channels controls fast inhibitory synaptic transmission at auditory outer hair cells. Neuron 26:595–601.

Parthasarathi AA, Grosh K, Zheng J, Nuttall AL (2003) Effect of current stimulus on in vivo cochlear mechanics. J Acoust Soc Am 113:442–452.

Patuzzi R, Rajan R (1990) Does electrical stimulation of the crossed olivo-cochlear bundle produce movement of the organ of Corti? Hear Res 45:15–32.

Patuzzi R, Sellick PM, Johnstone BM (1984) The modulation of the sensitivity of the mammalian cochlea by low frequency tones. III. Basilar membrane motion. Hear Res 13:19–27.

Puel JL, Rebillard G (1990) Effect of contralateral sound stimulation on the distortion product $2F_1 - F_2$: Evidence that the medial efferent system is involved. J Acoust Soc Am 87:1630–1635.

Puria S, Guinan JJ, Liberman MC (1996) Olivocochlear reflex assays: Effects of contralateral sound on compound action potentials versus ear-canal distortion products. J Acoust Soc Am 99:500–507.

Raphael Y, Lenoir M, Wroblewski R, Pujol R (1991) The sensory epithelium and its innervation in the mole rat cochlea. J Comp Neurol 314:367–382.

Roberts BL, Meredith GE (1990) The efferent innervation of the ear: variations on an enigma. In: Webster DB, Fay RR, Popper AN (eds) The Evolutionary Biology of Hearing. New York: Springer-Verlag, pp. 185–210.

Robertson D (1984) Horseradish peroxidase injection of physiologically characterized afferent and efferent neurones in the guinea pig spiral ganglion. Hear Res 15:113–121.

Robles L, Ruggero MA (2001) Mechanics of the mammalian cochlea. Physiol Rev 81:1305–1352.

Rousseau E, Smith JS, Meissner G (1987) Ryanodine modifies conductance and gating behavior of single Ca^{2+} release channel. Am J Physiol 253:C364–C368.

Russell IJ (1976) Amphibian lateral line detectors. In: Llinas R, Precht W (eds) Frog Neurobiology. Berlin: Springer-Verlag, pp. 513–550.

Russell IJ, Kössl M (1991) The voltage responses of hair-cells in the basal turn of the guinea-pig cochlea. J Physiol (Lond) 435:493–511.

Russell IJ, Kössl M (1992) Sensory transduction and frequency-selectivity in the basal turn of the guinea-pig cochlea. Philos Trans R Soc Lond, Ser B 336:317–324.

Russell IJ, Kössl M (1999) Micromechanical responses to tones in the auditory fovea of the greater mustached bat's cochlea. J Neurophysiol 82:676–686.

Russell IJ, Murugasu E (1997) Medial efferent inhibition suppresses basilar membrane responses to near characteristic frequency tones of moderate to high intensities. J Acoust Soc Am 102:1734–1738.

Russell IJ, Murugasu E (1998) Efferent of the basilar membrane vibration depends on tone frequency and level: implications for the active control of basilar membrane mechanics. In: Palmer AR, Summerfield AQ, Meddis R (eds) Psychophysical and Physiological Advances in Hearing. London: Whurr, pp. 19–25.

Russell IJ, Nilsen KE (2000) The spatial and temporal representation of a tone on the guinea pig basilar membrane. Proc Natl Acad Sci USA 97:11751–11758.

Russell IJ, Cody AR, Richardson GP (1986) The responses of inner and outer hair-cells in the basal turn of the guinea-pig cochlea and in the mouse cochlea grown in vitro. Hear Res 22:199–216.

Russell IJ, Drexl M, Foeller E, Vater M, Kössl M (2003) Synchronization of a nonlinear oscillator: processing the CF component of the echo-response signal in the cochlea of the mustached bat. J Neurosci. 23:9508–9518.

Ryan S, Kemp DT, Hinchliffe R (1991) The influence of contralateral acoustic stimulation on click-evoked otoacoustic emissions in humans. Br J Audiol 25:391–397.

Santos-Sacchi J (2003) New tunes from Corti's organ: the outer hair cell boogie rules. Curr Opin Neurobiol 13:459–468.

Santos-Sacchi J, Kakehata S, Takahashi S. (1998) Effects of membrane potential on the voltage dependence of motility-related charge in outer hair cells of the guinea-pig. J Physiol (Lond) 510:225–235.

Schloth E, Zwicker E (1983) Mechanical and acoustical influences on spontaneous otoacoustic emission. Hear Res 11:285–293.

Shatz CJ (1996) Emergence of order in visual system development. Proc Natl Acad Sci USA 93:602–608.

Shera CA (2003) Wave interference in the generation of reflection- and distortion-source emissions. In: Gummer AW, Dalhoff E, Nowotny M, Scherer MP (eds) Biophysics of the Cochlea: From Molecules to Models. Singapore: World Scientific, pp. 439–453.

Siegel JH, Kim DO (1982) Efferent neural control of cochlear mechanics? Olivocochlear bundle stimulation affects cochlear biomechanical nonlinearity. Hear Res 6:171–182.

Smith PH, Spirou GA (2002) From the cochlea to the cortex and back. In: Oertel D, Popper AN, Fay RR (eds) Integrative Functions in the Mammalian Auditory Pathway. New York: Springer-Verlag, pp. 6–71.

Sobkowicz HM, Slapnick SM (1992) Neuronal sprouting and synapse formation in response to injury in the mouse organ of Corti in culture. Int J Dev Neurosci 10: 545–566.

Spoendlin H (1985) Anatomy of cochlear innervation. Am J Otolaryngol 6:453–467.

Sridhar TS, Liberman MC, Brown MC, Sewell WF (1995) A novel cholinergic slow effect of efferent stimulation on cochlear potentials in the guinea-pig. J Neurosci 15:3667–3678.

Sridhar TS, Brown MC, Sewell WF (1997) Unique postsynaptic signaling at the hair cell efferent synapse permits calcium to evoke changes on two time scales. J Neurosci 17:428–437.

Stankovic KM, Guinan JJ (1999) Medial efferent effects on auditory-nerve responses to tail-frequency tones. I. Rate reduction. J Acoust Soc Am 106:857–869.

Stover LJ, Neely ST, Gorga M P (1996) Latency and multiple sources of distortion product otoacoustic emissions. J Acoust Soc Am 99:1016–1024.

Suga N, Gao E, Zhang Y, Ma X, Olsen JF (2000) The corticofugal system for hearing: recent progress. Proc Natl Acad Sci USA 97:11807–11814.

Suga N, Xiao Z, Ma X, Ji W (2002) Plasticity and corticofugal modulation for hearing in adult animals. Neuron 36:9–18.

Vater M, Lenoir M, Pujol R (1992) Ultrastructure of the horseshoe bat's organ of Corti. II. Transmission electron microscopy. J Comp Neurol 318:380–391.

Walsh EJ, McGee J, McFadden SL, Liberman MC (1998) Long-term effects of sectioning the olivocochlear bundle in neonatal cats. J Neurosci 18:3859–3869.

Whitehead ML, Lonsbury-Martin BL, Martin GK (1991) Slow variation of the amplitude of acoustic distortion at $f_2 - f_1$ in awake rabbits. Hear Res 51:293–300.

Xie DH, Henson OW (1998) Tonic efferent-induced cochlear damping in roosting and echolocating mustached bats. Hear Res 124:60–68.

Xiao ZJ, Suga N (2002) Modulation of cochlear hair cells by the auditory cortex in the mustached bat. Nat Neurosci 5:57–63.

Xie DH, Henson MM, Bishop AL, Henson OW (1993) Efferent terminals in the cochlea of the moustached bat—quantitative data. Hear Res 66:81–90.

11
Cochlear Models Incorporating Active Processes

Stephen T. Neely and Duck O. Kim

1. Introduction

Although it was suggested more than 50 years ago that active elements might have a significant role in cochlear mechanics (Gold 1948), it was not until (1) direct observations of basilar membrane (BM) motion revealed clear differences between in vivo and postmortem responses (Rhode 1973; Cooper et al., Chapter 2) and (2) otoacoustic emissions (OAE) were convincingly recorded in human ear canals (Kemp 1978; Kemp, Chapter 1) that the need to consider the influence of active elements on BM mechanics became compelling. The first active model of cochlear mechanics (Kim et al. 1980b) used negative damping elements as an internal source of mechanical energy and was able to explain the higher sensitivity and sharper tuning characteristic of in vivo responses in contrast to the more broadly tuned, less sensitive postmortem responses.

1.1 Characteristics of Active Models

Lumped elements are commonly used to model continuous mechanical systems. Such a model may contain stiffness elements (e.g., springs), inertial elements (e.g., masses), and damping elements (e.g., friction or viscosity). A *passive* mechanical system has no internal sources of energy. Any lumped model representing a passive system will contain only positive damping elements. A mechanical system is said to be *active* when it contains an internal energy source capable of introducing *signal* vibrations into the system. A cochlear model with negative damping elements is an example of an *active* model; however, negative damping elements are not the only way that an active mechanical system can be modeled.

A passive mechanical system is always *stable* in the sense that its response to external excitation eventually decreases in amplitude when the excitation is removed. An active mechanical system, on the other hand, may be *unstable*, with responses that grow with unbounded amplitude even in the absence of external excitation. There are two intermediate stability situations that are possible in an active, nonlinear system such as the cochlea. (1) It is possible for an active system to have responses that always decrease with time, just like a passive system. (2) A nonlinear, active system can be *conditionally stable* if one of its vibration modes continues to oscillate indefinitely, even in the absence of external excitation, with amplitude that never exceeds a certain maximum value.

Prevailing theories of OAEs (Shera and Guinan 1999; Lonsbury-Martin and Martin, Chapter 8) suggest that an active mechanical system is required for their production. Nonlinearity is essential for the production of some types of OAEs, while the influence of cochlear nonlinearity can be observed to influence all types of OAEs under certain conditions.

2. The Need for an Active Model

Development of the concept of *active mechanics* of the cochlea is intimately tied to the findings of *nonlinearity* and *physiological vulnerability*, two major characteristics present in both the *mechanics* and the *physiology* of the cochlea. The possibility that the mechanics and the physiology of the cochlea may be intimately tied together was a foreign concept in early years (i.e., 1960s and 1970s) of research on the cochlea (see Cooper et al., Chapter 2). In particular, the pioneering postmortem measurements of BM motion in human ears (von Békésy 1960) were widely accepted as closely resembling mechanical response in vivo.

2.1 Nonlinearity, Physiological Vulnerability, and Neural Control of Cochlear Mechanics

The discovery of nonlinearity in BM motion (Rhode 1971) was inconsistent with then-current notions about cochlear mechanics and was met with strong skepticism. For example, measurements of BM motion by other investigators (Evans and Wilson 1975; Wilson and Johnstone 1975) indicated that the cochlear mechanical response was linear, in direct opposition to Rhode's observations (Rhode 1971; Rhode and Robles 1974). The nonlinearity controversy lasted for some time.

While studies of the cochlear mechanics were plagued with a conflict in the experimental data from multiple investigators, neurophysiological studies of the cochlea did not suffer from such a conflict in the experimental data. In particular, studies of single cochlear nerve fibers uniformly described sharp frequency tuning (Kiang et al. 1965; Evans and Wilson 1975). The contrast between the mechanical and neurophysiological studies regarding the presence or absence of controversies is related to the fact that cochlear mechanisms are vulnerable to insults (Robertson 1974; Robertson and Manley 1974; Robertson and Johnstone 1979; Siegel et al. 1982) and that measurements of basilar-membrane motion involve procedures invasive to the cochlea, whereas recordings of cochlear nerve fibers commonly leave the cochlea intact.

In the present context of tracing the rationale for the introduction of active elements into cochlear models, it is noteworthy that neurophysiological observations in the 1960s and 1970s revealed nonlinear behavior of the cochlea. By measuring responses of single cochlear nerve fibers to click pairs, Goblick and Pfeiffer (1969) attempted to determine the impulse responses of the cochlea, anticipating that the response should be linear. Instead, they found that the cochlear response, observed through the nerve fibers, was nonlinear. Goblick and Pfeiffer could not distinguish whether the cochlear nonlinearity was mechanical or electrophysiological in nature. Another study, which used a method of recording responses of populations of single cochlear nerve fibers in individual animals, revealed that the spatial distributions of the phases of distortion products along the length of the cochlear partition were the same as the natural pure tones of matching frequency (Pfeiffer and Kim 1975; Kim et al. 1979, 1980a). These observations provided strong evidence that nonlinear distortion products are mechanically propagated along the cochlear partition.

By the late 1970s, a number of observations thus showed that basilar-membrane motion was nonlinear and that various features of cochlear mechanics, i.e., nonlinearity, frequency-tuning, and sensitivity, were all physiologically vulnerable (Rhode 1973). Rhode's direct mechanical observations, however, were not convincing enough by themselves, because other mechanical studies were contradictory. Neurophysiological and otoacoustic studies (Kemp 1978; Kim et al. 1979, 1980a; Mountain 1980; Kim 1980; Siegel et al. 1982) provided important support for the above view. The first active cochlear model (Kim et al. 1980b) was introduced in this setting. Not only were active elements of the cochlea thought to be a means of accounting for the sharp tuning and high sensitivity of basilar-membrane motion, but they were also thought to be intimately associated with the physiological vulnerability and nonlinearity of the cochlea observed mechanically and neurally (Kim 1980; Kim et al. 1980a; Siegel et al. 1982). It is noteworthy that these developments preceded, by several years, the discovery of the "somatic motility" of the mammalian outer hair cells (OHC) (Brownell et al. 1985) and the "hair-bundle motility" of the turtle auditory hair cells (Art et al. 1986).

Control of cochlear mechanics by cellular mechanisms was further demonstrated in the observations that distortion-product otoacoustic emissions (DPOAE) were affected by neural activities of the efferent olivocochlear bundle (Mountain 1980; Siegel et al. 1982; Guinan 1986). These observations suggested that the brain can modulate the mechanical properties of the cochlea and perhaps control the gain of the mechanical amplification.

2.2 Sharp Tuning in the Presence of Viscous Damping

During the era when von Békésy's measurements were considered representative of cochlear mechanics, it was also known from direct measurements that neural tuning was considerably sharper than the BM tuning that he observed. The popular resolution for this inconsistency was the so-called *second filter*, which, being interposed between the BM and neural recording site, had the ability to transform the broad BM tuning into the sharp tuning necessary for the exquisite frequency selectivity achieved by the central auditory system (e.g., Evans and Wilson 1975).

An alternative theory, proposed earlier by Thomas Gold and not appreciated until much later, was that the cochlea contained a biological vibration amplifier (Gold 1948). He argued, by analogy to microwave radio receivers, that signal amplification needed to occur prior to detection in order to provide improved signal-to-noise ratio. In the case of the cochlea, where vibrations were inevitably damped by viscous fluid, this meant that amplification needed to occur in cochlear mechanics, prior to detection of vibrations by sensory hair cells. Unfortunately, it was not until sharp tuning was observed in BM measurements (Rhode 1978; Sellick et al. 1982; Ruggero et al. 1997) that Gold's idea started to be taken seriously by other hearing researchers.

3. Development of Active Cochlear Models

After the need to consider active elements in cochlear mechanics was established on the basis of experimental observations and optimal receiver principles, skepticism remained regarding the possibility of introducing vibration amplifiers into cochlear models while maintaining mechanical stability. Results from cochlear models with active elements demonstrated that stable amplification was possible (e.g., Neely and Kim 1983).

Attempts to reproduce the sharp tuning observed in BM measurements with conventional models motivated inverse solutions of cochlear mechanics. The usual approach to cochlear modeling is to specify the mechanical impedance of the BM and solve for BM vibration patterns that result from driving the stapes. Specifying the BM vibration pattern and solving for the BM impedance needed to produce such a pattern is considered an inverse solution. When applied to BM measurements, inverse solutions indicated the need for negative damping elements (Neely 1980; Zweig 1991; de Boer 1995a,b). This result was discounted at first as being unreasonable. Later, it became apparent that negative damping elements could provide a feasible model for the vibration amplifiers that were needed to explain observed features of cochlear mechanics.

3.1 Basilar-Membrane Tuning in Vivo and post Mortem

The cochlear model that incorporated active elements (Kim et al. 1980b) was developed to provide an explanation for the change in BM tuning that Rhode observed when an animal died (Rhode 1973). This model specified typical BM impedance except that damping elements were given a negative value over a small region basal to the *characteristic place* where the traveling wave reaches its maximum amplitude. This negative damping region caused the traveling wave to be amplified, resulting in larger BM vibration amplitudes in a restricted portion of the excitation pattern. Assuming a logarithmic frequency-to-place map, the excitation pattern in the model with negative damping resembled BM measurements in vivo (Figure 11.1, curve 1). Removal of the negative damping elements produced an excitation pattern in the model that resembled BM measurements post mortem (Figure 11.1, curves 2 and 3).

Figure 11.2 describes profiles of power flux density in the active cochlear model, i.e., the time-averaged power per unit area, over the basal and cochlear-partition boundaries. Positive (or negative) power flux density here corresponds to power flow from the partition into the fluid (or from the fluid into the partition). The top panel of Figure 11.2 describes the region of the partition where energy was added by the partition into the fluid (i.e., the more basally located region of the partition having negative damping) and the region of the partition where energy was dissipated by the partition (i.e., the apically adjoining region having positive damping). The lower two panels of Figure 11.2 describe that when the internal energy source (i.e., negative damping) was removed, the total power delivered by the same stimulus to the cochlear partition was reduced

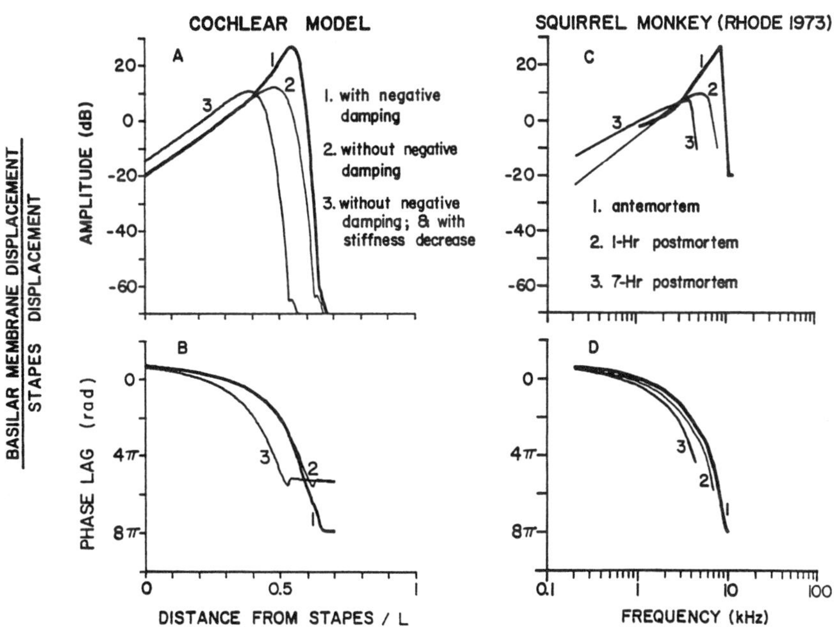

FIGURE 11.1. Comparison between a cochlear model with negative damping (Kim et al. 1980b) and measurements in squirrel monkey (Rhode 1973) of BM displacement relative to stapes displacement. (**A, B**) Amplitude and phase, respectively, of BM displacements in a cochlear model as a function of place for a stimulus frequency of 1550 Hz. The three conditions represent model parameter values representing BM stiffness and damping. (**C, D**) Amplitude and phase of BM displacements in a squirrel monkey as a function of frequency at one place. The three conditions represent physiological states of the cochlea. (Reprinted from Kim et al. 1980b.)

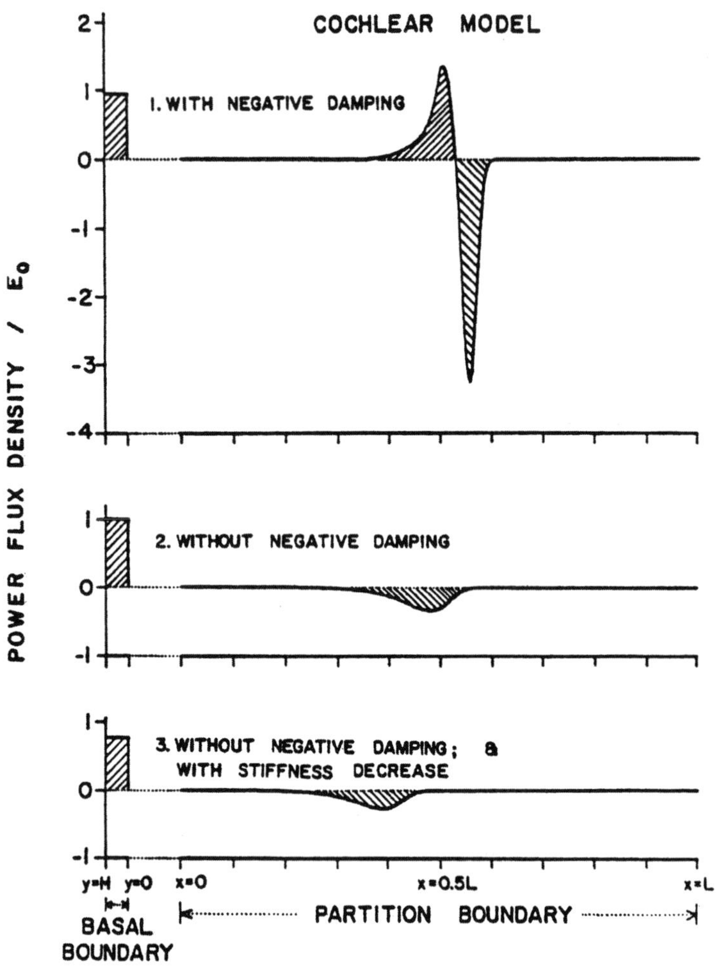

FIGURE 11.2. Power flux density in an active cochlear model, i.e., the time-averaged power per unit area, over the basal and cochlear-partition boundaries. (Reprinted from Kim et al. 1980b.)

in magnitude and became more broadly distributed. Thus, changes taking place in the power flux densities of the model described in Figure 11.2 underlie the changes taking place in the partition motion described in Figure 11.1.

Time-domain solutions of this linear, active model were also obtained to demonstrate stability despite the presence of negative damping elements; however, the model worked properly for only one frequency at a time because a different negative damping profile was required for each frequency.

The incorporation of a second mechanical degree of freedom into the cochlear model, analogous to the concept of the tectorial membrane and basilar membrane as two resonant systems (Zwislocki and Kletsky 1979; Allen 1980), allowed the damping profile to be frequency-dependent, so that a single BM impedance specification could provide stable amplification over a wide range of frequencies (Neely and Kim 1983, 1986). This linear, active cochlear model with two mechanical degrees of freedom in BM impedance provided a convincing explanation for Rhode's ante versus postmortem observations and the sharp frequency-tuning characteristics of single cochlear nerve fibers. The role of active elements in providing the cochlea with high sensitivity and sharp tuning became known as the *cochlear amplifier*, a term introduced by Davis (1983).

3.2 Alternative Cochlear Amplifier Models

Although the possibility of active elements in cochlear mechanics gained rapid acceptance after the introduction of the first cochlear models, theoretical arguments for the necessity of active elements continue to the present time. The modeling work of de Boer and Zweig based on inverse methods and physical principles (de Boer 1983a,b,1995a,b; Zweig 1991; Dimitriadis and Chadwick 1999) argued that observed BM responses were not possible unless the cochlea incorporated active elements.

Notable alternatives to the use of negative damping in active cochlear models include explicit representation of feedback forces and incorporation of limit-cycle oscillators. Even before OHC motility was observed experimentally, it was conjectured that OHC feedback forces, generated by electromechanical transduction, were the sources of energy that powered the cochlear amplifier (Mountain et al. 1983; Geisler 1991). One variation of this feedback idea is called feed-forward, because the place along the BM where OHCs apply their force is slightly different from the place where OHCs sense BM displacement (Geisler and Sang 1995; Fukazawa and Tanaka 1996).

Some early active cochlear models incorporated limit-cycle oscillators, such as van der Pol oscillators (Duifhuis et al. 1986). Limit-cycle oscillators are conditionally stable active elements that continue to vibrate in the absence of external excitation, but the amplitude of this vibration is limited by nonlinear parameters. More recent oscillator models conjecture that hair bundles are on the verge of local instability by being conditioned to operate near a critical point called a Hopf bifurcation (see Duke and Jülichr, Chapter 3).

4. Models of Otoacoustic Emissions

A complete model of OAEs requires not only a representation of cochlear mechanics that incorporates either nonlinear or active elements, but also a coupled representation of the middle ear that is capable of propagating acoustic signals in both forward and reverse directions.

4.1 Hardware Model

Zwicker (1986) described results of a hardware model of the cochlea with nonlinear feedback elements. This model could simulate simultaneous evoked as well as delayed evoked emissions with many similarities to human OAEs. Hardware models became less attractive as computational models as computer processing power increased.

4.2 Spontaneous Emissions

Spontaneous otoacoustic emissions (SOAE) provide important evidence for the existence of active elements in cochlear mechanics. Not only were they predicted on theoretical grounds, but due to the unstable nature of positive feedback amplifiers (Gold 1948), measurements of SOAEs have features that are characteristic of limit-cycle oscillators. (1) The distribution of instantaneous amplitudes of SOAE waveforms is bimodal, which is characteristic of limit-cycle oscillators and differs from the Gaussian shape that is characteristic of the distribution of instantaneous amplitudes of filtered noise (Bialek and Wit 1984; van Dijk 1990; Shera 2003). (2) Multiple SOAEs have correlated amplitudes suggesting that they are produced by coupled oscillators (Wit 1990). And (3) SOAEs can be entrained by presentation of an external tone at a nearby frequency (Long et al. 1991). All of these features have also been demonstrated for SOAE in nonmammals (see Manley and van Dijk, Chapter 7).

4.3 Electrically Evoked Otoacoustic Emissions

Active elements in the cochlea derive their energy from the endocochlear potential, an electrical voltage that exists across the BM. The physical implementation of the active element contains a mechanoelectrical transducer at the hair bundle of the OHC and electromechanical transducer, which is either in the hair bundle or in the lateral wall of the OHC. In laboratory preparations, electrical signals can be introduced into the cochlea to stimulate electromechanical transducers in the absence of (or in conjunction with) acoustic stimulation. The vibrations induced by electrical stimulation can be measured acoustically at the eardrum, where they are called electrically evoked otoacoustic emissions (EEOAE). Comparisons between measurements and models of EEOAEs can provide insights into OHC transduction processes that are not obtainable using acoustic stimulation alone (Yates and Kirk 1998; Nakajima et al. 2000).

4.4 Coherent Reflection and Fine Structure

The regular spacing of local maxima in evoked OAE spectra is evidence that forward-traveling waves are reflected within the cochlea. The measured delay associated with these reflections indicates that reflection occurs at the place where BM vibrations have their maximum amplitude. The most successful explanation for the origin of these reflections has them intimately linked to active elements. Forward-traveling waves reflect off random irregularities in the micromechanics of the BM. The tall, broad peak of the BM excitation pattern defines a localized region in which the wavelength of the traveling wave is relatively constant.

Within this region, coherent scattering occurs off irregularities, resulting in *coherent reflection* (Zweig and Shera 1995). The retrograde waves produced by coherent reflection are thought to play a role, to some extent, in all types of OAEs (Shera and Guinan 1999).

The regular spacing of spectral maxima is the result of interaction between forward-traveling waves and retrograde waves (Shera and Guinan Chapter 9). The resulting spectral periodicity is also called *fine structure* because it appears to be superimposed onto broader spectral features. The fine structures of OAEs and behavioral thresholds have similar frequency spacing (Johnson et al., Chapter 12) that is determined by characteristics of apical reflectance at the tonotopic place and basal reflectance at the stapes. Because these fine structures require retrograde waves, which is thought to require coherent reflection, which, in turn, requires the presence of active elements, fine structure disappears when these elements are deactivated (e.g., Long et al. 1991).

4.5 Modeling DPOAE Growth

Perhaps the most important function of active elements in the cochlea is their ability to compress a large (100 dB) dynamic range of sound levels entering the ear and deliver a much smaller (60 dB or less) dynamic range of vibrations to the inner hair cells (Allen and Neely 1992). This dynamic range compression is accomplished by providing maximum amplification to low-level sounds and gradually reducing the amount of amplification as the level of the sound increases. This gain reduction can be modeled as a saturation of OHC feedback forces (Yates 1990; Neely and Stover 1993). The level- dependent gain causes the dynamic range of cochlear responses to be less than that of sound-pressure levels entering the ear.

The loss of OHC function eliminates the ability to hear low-level sounds (e.g., Robertson and Johnstone 1979) and also causes the loss of compression. The loss of compression of the cochlear response is thought to underlie the phenomenon of recruitment (e.g., Neely and Allen 1997). Cochlear compression is a significant factor in determining auditory performance limits on basic psychoacoustic tasks such as intensity discrimination and increment detection. An important contribution of cochlear models is their ability to tie together objective measures, such as OAEs, with psychoacoustic measures, such as loudness growth (Neely et al. 2000).

4.6 Modeling DPOAE Adaptation

A significant feature of OHCs is their innervation by efferent fibers from the medial olivocochlear (MOC) system. This connection provides a route for top-down control of active elements in the cochlea. MOC innervation serves a developmental role in configuration of the cochlear amplifier as it first becomes functional (Walsh et al. 1998). Stimulation of efferent fibers can reduce the gain of the cochlear amplifier in a fully developed cochlea (Murugasu and Russell

1996; Shre and Guinan, Chapter 9). DPOAEs exhibit adaptation (Liberman et al. 1996; Kim et al. 2001) as a result of the reflex activation of the MOC system. Models of cochlear mechanics that simulate efferent control of active elements can help illuminate the role of the MOC efferent system in normal auditory function. For example, Kim et al. (2003) found that a model incorporating a dynamic change of the operating point of the OHCs nonlinearity under the control of the MOC system provided a means of accounting for a variety of complex time courses of DPOAE level and phase for various stimulus conditions.

5. Roles of Multiple OHC Motilities

Two types of hair-cell motility have been observed: (1) *somatic* (Brownell et al. 1985) and (2) *hair-bundle* motility (Art et al. 1986). Somatic motility is mediated by a voltage-sensitive protein, prestin, which is located in the lateral wall of mammalian OHCs (Zheng et al. 2000; Dallos and Fakler 2002). Hair-bundle motility has been observed in both nonmammals (Martin et al. 2000; Ricci et al. 2002) and mammals (Chan and Hudspeth 2005; Kennedy et al. 2005) and is associated with the mechanoelectrical transduction channel located in the stereocilia of hair cells (see Duke and Jülichr, Chapter 3). Hair-bundle motility is highly vulnerable to insults, whereas somatic motility is highly resistant to insults. A prevalent current view is that active amplification of the cochlear response is mediated by the somatic motility of the OHCs in mammals (e.g., Dallos and Fakler 2002), whereas in nonmammals, it is mediated by the hair-bundle motility (e.g., Martin et al. 2000).

What are the respective roles of the hair-bundle and somatic motilities in mammalian outer hair cells? In 1986 it was proposed that in the mammalian cochlea, the hair-bundle motility of the OHC mediates amplification of the cochlear-partition motion and that the somatic motility mediates optimizing the operating point of the OHC embedded in the organ of Corti (Kim 1986). In this hypothesis, the MOC system is postulated to control the gain of the cochlear amplifier by modulating the OHC's membrane potential and thereby changing the cell's soma length and the operating point of the hair-bundle motility (Kim 1986; Kim et al. 2003). The studies of Chan and Hudspeth (2005) and Kennedy et al. (2005) are consistent with this hypothesis.

Although the recent in vitro observations of motilities of the cochlear hair cells offer insights, there remains a further need to improve the methods. For example, the amount of the amplification ($\approx$2.5 dB) due to hair-bundle motility that was observed in vitro was much lower than the value ($\approx$20–50 dB) expected in vivo. Somatic motility has not yet been ruled out as the source of amplification. A thorough understanding of the respective roles of the different types of hair-cell motilities represents an important goal of research in this field. Achieving the goal will require further studies.

Summary

Active elements in cochlear models represent the contribution of OHCs in providing mechanical feedback forces that amplify BM vibrations. The amplification allows lower-level sounds to be heard than would otherwise be possible and also generates retrograde waves that are measurable in the ear canal as OAEs. This amplification is level-dependent, which not only provides dynamic-range compression to the auditory system, but also generates distortion products that are measurable in the ear canal as DPOAEs. Cochlear models that incorporate nonlinear, active elements provide a means to conceptually tie together all of these important features of normal auditory function.

Acknowledgments. This work was partially supported by a grant from NIH-NIDCD.

References

Allen JB (1980) Cochlear micromechanics—a physical model of transduction. J Acoust Soc Am 1980 68:1660–1670.

Allen JB, Neely ST (1992) Micromechanical models of the cochlea. Physics Today July:40–47.

Art JJ, Crawford AC, Fettiplace R (1986) Electrical resonance and membrane currents in turtle cochlear hair cells. Hear Res 22:31–36.

Bialek W, Wit HP (1984) Quantum limits to oscillator stability: theory and experiments on acoustic emissions from the human ear. Phys Lett 104A:173–178.

Brownell WE, Bader CR, Bertrand D, Ribaupierre Y (1985) Evoked mechanical responses of isolated cochlear outer hair cells. Science 227:194–196.

Chan DK, Hudspeth AJ (2005) Ca^{2+} current-driven nonlinear amplification by the mammalian cochlea in vitro. Nat Neurosci 8:149–155.

Dallos P, Fakler B. (2002) Prestin, a new type of motor protein. Nat Rev Mol Cell Biol 3:104–111.

Davis H (1983) An active process in cochlear mechanics. Hear Res 9:79–90.

de Boer E (1983a) No sharpening? A challenge for cochlear mechanics. J Acoust Soc Am 73:567–573.

de Boer E (1983b) On active and passive cochlear models—toward a generalized analysis. J Acoust Soc Am 73:574–576.

de Boer E (1995a) The inverse problem solved for a three-dimensional model of the cochlea. I. Analysis. J Acoust Soc Am 98:896–903.

de Boer E (1995b) The inverse problem solved for a three-dimensional model of the cochlea. II. Application to experimental data sets. J Acoust Soc Am 98:904–910.

Dimitriadis EK, Chadwick RS (1999) Solution of the inverse problem for a linear cochlear model: a tonotopic cochlear amplifier. J Acoust Soc Am 106:1880–1892.

Duifhuis H, Hoogstraten HW, van Netten SM, Diependaal RJ, Bialek W (1986) Modelling the cochlear partition with coupled Van der Pol oscillators. In: Allen JB, Hall JL, Hubbard AE, Neely ST, Tubis A (eds) Peripheral Auditory Mechanisms. New York: Springer-Verlag, pp. 290–297.

Evans EF, Wilson JP (1975) Cochlear tuning properties: concurrent basilar membrane and single nerve fiber measurements. Science 19:1218–1221.

Fukazawa T, Tanaka Y (1996) Spontaneous otoacoustic emissions in an active feed-forward model of the cochlea. Hear Res 95:135–143.

Geisler CD (1991) A cochlear model using feedback from motile outer hair cells. Hear Res 54:105–117.

Geisler CD, Sang C (1995) A cochlear model using feed-forward outer-hair-cell forces. Hear Res 86:132–46.

Goblick TJ, Pfeiffer RR (1969) Time-domain measurements of cochlear nonlinearities using combination click stimuli. J Acoust Soc Am 46:924–938.

Gold T (1948) Hearing II. The physical basis of the action of the cochlea. Proc R Soc Lond B 135:492–498.

Guinan JJ (1986) Effect of efferent neural activity on cochlear mechanics. Scand Audiol Suppl 25:53–62.

Kemp DT (1978) Stimulated acoustic emission from the human auditory system. J Acoust Soc Am 64:1386–1391.

Kennedy HJ, Crawford AC, Fettiplace R (2005) Force generation by mammalian hair bundles supports a role in cochlear amplification. Nature 433: 880–883.

Kiang NYS, Watanabe T, Thomas EC, Clark LF (1965) Discharge Patterns of Single Fibers in the Cat's Auditory Nerve. Cambridge, MA: MIT Press, pp. 1–154.

Kim DO, Siegel JH, Molnar CE (1979) Cochlear nonlinear phenomena in two-tone responses. Scand Audiol (Suppl) 9:63–81.

Kim DO (1980) Cochlear mechanics: implications of electrophysiological and acoustical observations. Hear Res 2:297–317.

Kim DO (1986) Active and nonlinear cochlear biomechanics and the role of outer-hair-cell subsystem in the mammalian auditory system. Hear Res 22:105–114.

Kim DO, Molnar, CE, Matthews, JW (1980a) Cochlear mechanics: nonlinear behavior in two-tone responses as reflected in cochlear-nerve-fiber responses and in ear-canal sound pressure. J Acoust Soc Am 67:1704–1721.

Kim DO, Neely ST, Molnar, CE, Matthews, JW (1980b) An active cochlear model with negative damping in the partition: comparison with Rhode's ante- and post-mortem observations. In: van den Brink G, Bilsen FA (eds) Psychophysical, Physiological, and Behavioral Studies in Hearing. Delft: Delft University Press, pp. 7–14.

Kim DO, Dorn PA, Neely ST, Gorga (2001) Adaptation of distortion product otoacoustic emission in humans. J Assoc Res Otolar 2:31–40.

Kim DO, Yang XM, Neely ST (2003) Effects of the medial olivocochlear reflex on cochlear mechanics: experimental and modeling studies of DPOAE. In: Gummer AW (ed) Biophysics of the Cochlea: From Molecules to Models. Singapore: World Scientific, pp. 506–516.

Liberman MC, Puria S, Guinan JJ (1996) The ipsilaterally evoked olivocochlear reflex causes rapid adaptation of the 2f1-f2 distortion product otoacoustic emission. J Acoust Soc Am 99:3572–3584.

Long GR, Tubis A, Jones KL (1991) Modeling synchronization and suppression of spontaneous otoacoustic emissions using Van der Pol oscillators: effects of aspirin administration. J Acoust Soc Am 89:1201–1212.

Martin P, Mehta AD, Hudspeth AJ (2000) Negative hair-bundle stiffness betrays a mechanism for mechanical amplification by the hair cell. Proc Natl Acad Sci USA 97:12026–12031.

Mountain DC (1980) Changes in endolymphatic potential and crossed olivocochlear bundle stimulation alter cochlear mechanics. Science 210:71–72.

Mountain DC, Hubbard AE, McMullen TA (1983) Electromechanical processes in the cochlea. In: de Boer E, Viergever MA (eds) Mechanics of Hearing. The Hague: Martinus Nijhoff, pp. 119–126.

Murugasu E, Russell IJ (1996) The effect of efferent stimulation on basilar membrane displacement in the basal turn of the guinea gig cochlea. J Neurosci 16:325–332.

Nakajima HH, Hubbard AE, Mountain DC (2000) Effects of acoustic trauma on acoustic enhancement of electrically evoked otoacoustic emissions. J Acoust Soc Am 107: 2603–2614.

Neely ST (1980) Backward solution of a two-dimensional cochlear model. J Acoust Soc Am 67:S75.

Neely ST, Allen JB (1997) Relation between the rate of growth of loudness and the intensity DL. In: Jesteadt W (ed) Modeling Sensorineural Hearing Loss. Mahwah, NJ: Lawrence Erlbaum, pp. 213–222.

Neely ST, Kim DO (1983) An active cochlear model showing sharp tuning and high sensitivity. Hear Res 9:123–130.

Neely ST, Kim DO (1986) A model for active elements in cochlear biomechanics. J Acoust Soc Am 79:1472–1480.

Neely ST, Stover LJ (1993) Otoacoustic emissions from a nonlinear, active model of cochlear mechanics. In: Duifhuis H, Horst JW, van Dijk P, van Netten SM (eds) Biophysics of Hair Cell Sensory Systems. Singapore: World Scientific, pp. 64–71.

Neely ST, Gorga MP, Dorn PA (2000) Distortion product and loudness growth in an active, nonlinear model of cochlear mechanics. In: Wada H, Takasaka T, Ikeda K, Ohyama K, Koike T (eds) Recent Developments in Auditory Mechanics. Singapore: World Scientific, pp. 237–243.

Pfeiffer RR, Kim DO (1975) Cochlear nerve fiber responses: distribution along the cochlear partition. J Acoust Soc Am 58:867–869.

Rhode WS (1971) Observations of the vibration of the basilar membrane in squirrel monkeys using the Mössbauer technique. J Acoust Soc Am 49:1218–1231.

Rhode WS (1973) An investigation of post-mortem cochlear mechanics using the Mössbauer effect. In: Møller AR (ed) Basic Mechanisms in Hearing. New York: Academic Press, pp. 49–67.

Rhode WS (1978) Some observations on cochlear mechanics. J Acoust Soc Am 64: 158–176.

Rhode WS, Robles L (1974) Evidence from Mössbauer experiments for nonlinear vibration in the cochlea. J Acoust Soc Am 55:588–596.

Ricci AJ, Crawford AC, Fettiplace R (2002) Mechanisms of active hair bundle motion in auditory hair cells. J Neurosci 22:44–52.

Robertson D (1974) Cochlear neurons: frequency analysis altered by perilymph removal. Science 186:153–155.

Robertson D, Johnstone BM (1979) Aberrant tonotopic organization in the inner ear damaged by kanamycin. J Acoust Soc Am 66:466–469.

Robertson D, Manley GA (1974) Manipulation of frequency analysis in the cochlear ganglion of the guinea pig. J Comp Physiol 91:363–375.

Ruggero MA, Rich NC, Recio A, Narayan, Robles L (1997) Basilar-membrane responses to tones at the base of the chinchilla cochlea. J Acoust Soc Am 101:2151–2163.

Sellick PM, Patuzzi R, Johnstone BM (1982) Measurement of basilar membrane motion in the guinea pig using the Mossbauer technique. J Acoust Soc Am 72:131–141.

Shera CA (2003) Mammalian spontaneous otoacoustic emissions are amplitude-stabilized cochlear standing waves. J Acoust Soc Am 114:244–262.

Shera CA, Guinan JJ (1999) Evoked otoacoustic emissions arise by two fundamentally different mechanisms: a taxonomy for mammalian OAEs. J Acoust Soc Am 105: 782–798.
Siegel JH, Kim DO (1982) Efferent control of cochlear mechanics? Olivocochlear bundle stimulation affects cochlear biomechanical nonlinearity. Hear Res 6:171–182.
Siegel JH, Kim DO, Molnar CE (1982) Effects of altering organ of Corti on cochlear distortion products f2-f1 and 2f1-f2. J Neurophysiol 47:303–328.
van Dijk P, Wit HP (1990) Amplitude and frequency fluctuations of spontaneous otoacoustic emissions. J Acoust Soc Am 88:1779–1793.
von Békésy G (1960) Experiments in Hearing. New York: McGraw-Hill.
Walsh EJ, McGee J, McFadden SL, Liberman MC (1998) Long-term effects of sectioning the olivocochlear bundle in neonatal cats. J Neuroscience 18:3859–3869.
Wilson JP, Johnstone JR (1975) Basilar membrane and middle-ear vibration in guinea pig measured by capacitive probe. J Acoust Soc Am 57:705–723.
Wit HP (1990) Spontaneous otoacoustic emission generators behave like coupled oscillators. In: Dallos P, Geisler CD, Matthews JW, Ruggero MA, Steele CR (eds) The Mechanics and Biophysics of Hearing. Berlin: Springer-Verlag, pp. 259–268.
Yates GK (1990) Basilar membrane nonlinearity and its influence on auditory nerve rate-intensity functions. Hear Res 50:145–162.
Yates GK, Kirk DL (1998) Cochlear electrically evoked emissions modulated by mechanical transduction channels. J Neurosci 18:1996–2003.
Zheng J, Shen W, He DZ, Long KB, Madison LD, Dallos P (2000) Prestin is the motor protein of cochlear outer hair cells. Nature 11:149–55.
Zweig (1991) Finding the impedance of the organ of Corti. J Acoust Soc Am 89: 1229–1254.
Zweig G, Shera CA (1995) The origin of periodicity in the spectrum of evoked otoacoustic emissions. J Acoust Soc Am 98:2018–2047.
Zwicker E (1986) "Otoacoustic" emissions in a nonlinear cochlear hardware model with feedback. J Acoust Soc Am 80:154–162.
Zwislocki JJ, Kletsky EJ (1979) Tectorial membrane: a possible effect on frequency analysis in the cochlea. Science 204:639–641.

12
Relationships Between Otoacoustic and Psychophysical Measures of Cochlear Function

Tiffany A. Johnson, Michael P. Gorga, Stephen T. Neely, Andrew J. Oxenham, and Christopher A. Shera

1. Introduction

Both otoacoustic emission (OAE) and psychoacoustic measures provide insights into active cochlear processes in the human auditory system. This chapter provides an overview of the similarities in these two measures, one a physiological response generated within the cochlea (OAEs) and the other a behavioral measure reflecting processing in the periphery, but also at all other levels of the auditory system, including the auditory cortex. The major phenomena that will be reviewed here are measures of OAE and behavioral-threshold microstructure, estimates of cochlear-response growth, and estimates of frequency specificity in the cochlea, including frequency tuning and the upward spread of suppression/masking.

2. Threshold Microstructure

Considerable effort has been devoted to relating OAE responses to behavioral thresholds for pure-tone stimuli. Currently, the primary clinical application of OAEs is screening for hearing loss, a task that at some level requires knowledge of the relation between OAE levels and behavioral thresholds. The reader is referred to Janssen and Miller, Chapter 13, for a discussion of this topic. The focus here will be on relating microstructure observed in threshold measures and OAE responses.

2.1 *Nature of OAE and Threshold Microstructure*

When behavioral thresholds are measured for closely spaced frequencies, quasiperiodic maxima and minima can be observed (e.g., Elliott 1958; Long 1984). These maxima and minima are commonly referred to as threshold microstructure or threshold fine structure. Similar periodic spacings of maxima and minima have been observed for various types of evoked OAE responses including stimulus–frequency OAEs (SFOAEs) (e.g., Lutman and Deeks 1999), transient OAEs (TEOAEs) (e.g., Zwicker and Peisl 1990), and distortion product OAEs (DPOAEs) (e.g., He and Schmiedt 1993; Mauermann et al.1999a,b). The frequency spacing of spontaneous OAEs (SOAEs) also has been observed to follow regular intervals similar to those observed in threshold microstructure (Zwicker and Schloth 1984; Long and Tubis 1988a,b). As will be discussed below, the underlying mechanisms responsible for these microstructure patterns vary among the various classes of OAEs. An example of behavioral threshold microstructure for an individual subject is shown in Figure 12.1 (reproduced from Zwicker and Schloth 1984). In this figure, behavioral thresholds (circles) are plotted as a function of frequency. The upward-pointing arrows and corresponding dB SPL values indicate the frequencies and levels at which SOAEs were observed. For this subject, SOAEs occurred at frequencies for which thresholds were lowest.

The characteristic frequency spacing between adjacent maxima and minima in the behavioral and OAE interference microstructure increases with increasing frequency (Zwicker and Schloth 1984). Many studies (e.g., Zwicker 1989; Zwicker and Peisl 1990; Russell 1992; He and Schmiedt 1993; Talmadge et al. 1993; Braun 1997) have suggested that the characteristic spacings correspond to a constant fraction of an octave (or, to what is nearly equivalent, a constant distance along the basilar membrane or a constant fraction of the psychophysical

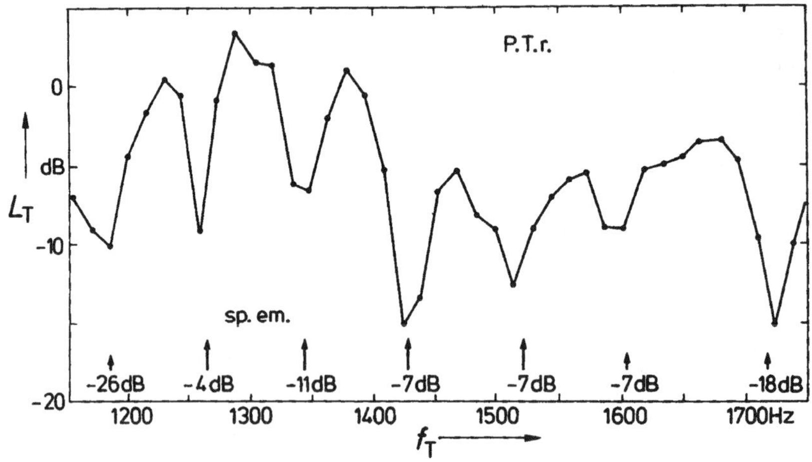

FIGURE 12.1. Behavioral thresholds (*circles*) as a function of frequency for an individual subject. The frequencies (and levels) at which spontaneous OAEs were observed are indicated by upward-pointing arrows (and corresponding dB values). (Reprinted with permission from Zwicker and Schloth 1984.)

critical band measured in bark).[1] For example, characteristic SOAE spacings are generally reported as 0.4 bark (Zwicker 1989). However, these correspondences are typically based on OAE or behavioral data from a limited frequency range and/or on histograms representing data pooled across frequency. An analysis of data from a wider frequency range shows that characteristic OAE spacings, and presumably behavioral spacings as well, are neither a constant fraction of an octave nor a constant distance along the basilar membrane; neither do they manifest any simple relationship to Zwicker's critical band rate scale (Shera 2003).

The depth of the microstructure (i.e., the level difference between adjacent maxima and minima) is typically 10–15 dB for behavioral thresholds, with some depths as large as 20 dB (e.g., Long 1984; Long and Tubis 1988a,b; Mauermann et al. 2004). The depths observed for evoked-OAE microstructure are similar to those observed in behavioral-threshold microstructure, although depths exceeding 20 dB also have been observed (e.g., He and Schmiedt 1993; Heitmann et al. 1998; Kalluri and Shera 2001).

2.2 Relationships Between Threshold and OAE Microstructure

A number of studies have demonstrated that SOAEs in humans occur at frequencies corresponding to behavioral-threshold minima (e.g., Zwicker and Schloth 1984; Zwicker 1986; Long and Tubis 1988a; Furst et al. 1992). Although most of the data in these studies are drawn from the 1- to 3-kHz region in which SOAEs are most prevalent, current understanding suggests that the correlation is likely to hold throughout the cochlea (see Section 2.3).

The frequency locations of evoked-OAE minima/maxima show more variability in relation to behavioral thresholds compared to the relationship between SOAE and behavioral-threshold minima. The alignment of maxima and minima for the evoked OAE and behavioral thresholds varies across emission type and perhaps across individuals. Zwicker and Schloth (1984) showed evidence of maxima in tone-burst-evoked OAEs aligning with minima in threshold microstructure in an individual subject. In contrast, Long (1984) reported minima in behavioral thresholds aligning with minima in synchronous-evoked OAE responses, also for a single subject. Furthermore, Lutman and Deeks (1999) reported a weak correspondence between behavioral-threshold and SFOAE microstructure in a group of 10 normal-hearing subjects, but no correspondence for DPOAE and behavioral-threshold microstructure.

---

[1]One bark corresponds to the width of one critical band and, therefore, its width in Hz increases with increasing frequency (Hartmann 1998). The Bark scale is sometimes referred to as the critical band rate scale.

2.3 Mechanisms Responsible for Threshold and OAE Microstructure

The similarity in microstructure for behavioral thresholds and some OAEs suggest commonalities in the cochlear mechanisms responsible for the microstructure. Talmadge et al. (1998) described a class of cochlear models that accounts for many of the empirical observations of both behavioral threshold and OAE microstructure. Their models, which draw from the work of Zweig and Shera (1995), attribute the microstructure primarily to the existence of reflections due to impedance roughness in the tall and broad peak region of a traveling wave (see also Shera and Guinan, Chapter 9). These reflections at the peak of the traveling wave have varying phase and interact with forward-propagating waves to create alternating constructive and destructive interference patterns and subsequent microstructure. The models described by Talmadge et al. (1998) predict identical frequency spacing for SOAEs, TEOAEs, and behavioral-threshold microstructure and similar but different spacings for SFOAEs and DPOAEs. In their model, the difference in frequency spacing for these different emission types is a result of differences in the relative importance of contributions from cochlear standing waves set up by basal reflectance at the stapes footplate and, in the case of DPOAEs, interaction among multiple cochlear sources. While the details of this model are beyond the scope of this chapter, all types of microstructure depend on the existence of spatially distributed tonotopic reflections from the tall and broad peak region of the traveling wave, a cochlear excitation pattern that depends on cochlear-amplifier gain. This suggests that microstructure of all types depends, at least indirectly, on cochlear-amplifier gain.

Observations of changes in microstructure following experimental manipulations known to reduce cochlear-amplifier gain are consistent with many of the predictions of Talmadge et al. (1998). For example, Long and Tubis (1988a,b) recorded SOAEs, TEOAEs, and synchronous-evoked OAEs in a group of normal-hearing subjects who consumed large doses of aspirin over the course of 3–4 days. Aspirin is known to interfere with outer hair cell somatic motility and to provide a reversible reduction in cochlear-amplifier gain (Stypulkowski 1990; Shehata et al. 1991). At the end of the treatment with aspirin, previously observed SOAEs were absent and the microstructure observed in both the behavioral thresholds and the evoked OAEs was reduced.

In another example in which OAE and threshold microstructure covaried, Furst et al. (1992) exposed normal-hearing subjects to white noise at 100 dB SPL in order to produce temporary threshold shifts. SOAEs and threshold microstructure were evaluated prior to and following noise exposure. Their results indicated a reduction in SOAE level and a reduction in threshold microstructure following noise exposure, in general agreement with the findings of Long and Tubis (1988a,b).

There are also data to suggest that masking produced by the presence of SOAEs may play a role in producing the microstructure observed in behavioral-threshold measures. Long and Tubis (1988a,b) observed that SOAEs were eliminated in their subjects during the course of aspirin consumption prior to the point at which behavioral-threshold microstructure and evoked-OAE microstructure were reduced. They also noted that in regions of microstructure where SOAEs had been present but were now absent, the thresholds near microstructure maxima were actually improved temporarily (i.e., observed at lower SPLs) before returning to their original levels as the subject continued to ingest aspirin. This observation was interpreted as being consistent with the idea that in the absence of experimental treatment, SOAEs were producing masking in regions of the threshold maxima. Following this logic, release from masking (i.e., lower behavioral thresholds) occurs when the SOAEs are first eliminated, followed by a return of thresholds to maximum levels as the aspirin further affects the cochlear-amplification process.

Smurzynski and Probst (1998) reported changes in behavioral thresholds that covaried with changes in SOAE level in a single subject. In this subject, SOAEs were either absent or present at levels near the noise floor at the start of behavioral-threshold testing. Over the course of behavioral-threshold measurements, SOAEs appeared and grew in level. As the SOAE levels increased, the behavioral thresholds measured in the vicinity of the SOAEs also increased (i.e., the threshold was increasingly elevated), while the depth of the threshold microstructure was reduced. The mechanisms responsible for the consistent appearance of SOAEs in this subject during the course of behavioral testing are unclear. However, the results suggest that SOAEs produced masking, given the increase in behavioral threshold.

While DPOAE microstructure shows periodicity similar to that observed for other evoked OAEs and behavioral thresholds, the mechanisms responsible for the microstructure may be different. Current thinking (see Shera and Guinan, Chapter 9) suggests that, as typically measured in the ear canal, DPOAEs contain contributions from two cochlear sources: a nonlinear-distortion source and a coherent-reflection source (Talmadge et al.1998, 1999; Mauermann et al.1999a,b; Shera and Guinan1999; Stover et al.1999; Konrad-Martin et al.2001, 2002). These sources are believed to generate OAEs by two fundamentally different mechanisms (Shera and Guinan 1999). The distortion component arises as a by-product of nonlinearities in the cochlear processing of the two primaries close to the f_2 place. The coherent-reflection component is generated at the cochlear location corresponding to the $2f_1 - f_2$ frequency by the reflection mechanism described above. The reader is referred to Shera and Guinan (1999) and Shera (2004) for more details of this interpretation of DPOAE sources.

In the context of the present discussion, DPOAE microstructure is believed to arise from the alternating constructive and destructive interference of these two cochlear sources, one with a slowly varying phase (distortion component) and one with a rapidly varying phase (reflection component). The source of microstructure in the case of SFOAEs and TEOAEs also is believed to be due to the interference between two components, one with a rapidly varying phase and one with a slowly varying phase. However, in contrast to

DPOAEs, where the two components are generated by the ear, for SFOAEs and TEOAEs the component with the slowly varying phase is the stimulus (Shera and Guinan1999).

Experimental observations are consistent with the idea that DPOAE microstructure arises from the interference of these two cochlear sources. DPOAE microstructure can be reduced or eliminated by presenting a suppressor tone in conjunction with the primaries (Heitmann et al. 1998; Kalluri and Shera 2001; Konrad-Martin et al. 2001). The suppressor tone, which is chosen to have a frequency within several Hz of the $2f_1 - f_2$ frequency, reduces the reflection-source contribution. In addition, Mauermann et al. (1999b) observed DPOAE microstructure in ears with hearing loss only when normal or near-normal hearing existed at the $2f_1 - f_2$ frequency. If hearing loss was present at the $2f_1 - f_2$ frequency, it was still possible to record DPOAEs, but no microstructure was observed.

2.4 *Summary of Threshold Microstructure*

Both behavioral-threshold microstructure and OAE microstructure repeat in quasiperiodic intervals. SOAEs are correlated with behavioral-threshold minima, at least for mid-frequencies (between 1 and 3 kHz). The alignment between evoked-OAE minima/maxima and behavioral-threshold minima/maxima are variable across individuals, and the reasons for a lack of a strong correlation have not been well described. However, empirical observations suggest that both behavioral-threshold and OAE microstructure are by-products of a normal cochlear amplifier.

3. Response Growth

Direct measurements of basilar-membrane motion in nonhuman mammals have characterized cochlear-response growth as nonlinear and compressive (see Robles and Ruggero 2001 for a comprehensive review). The invasive nature of these measurements of basilar-membrane motion precludes making similar measurements in humans. It is therefore necessary in humans to use techniques that provide indirect measures of cochlear-response growth. This section focuses on growth of suppression (OAEs), growth of masking (psychoacoustics), estimates of compression derived from these measures, and experiments that provide a direct comparison of the form of loudness growth and OAE input/output functions. Here, the separation of the discussion into sections on growth of masking/suppression, estimates of compression, and comparison of loudness growth and OAE input/output functions is largely arbitrary and is used for purposes of illustration. It is assumed, perhaps too simplistically, that all of the measures relate to the underlying response growth of the basilar membrane.

3.1 Growth of OAE Suppression and Psychophysical Masking

OAE responses can be suppressed by the simultaneous presentation of a tone or noise. Information about underlying cochlear-response properties in humans can be inferred by examining the growth of suppression in OAE level for various frequency relationships between the OAE eliciting stimulus and the suppressor stimulus. When used in this manner, the OAE stimulus (the f_2 primary tone in the case of DPOAEs) can be viewed as a probe that provides information about a restricted cochlear region, particularly for low stimulus levels. This is analogous to the assumptions made during psychophysical masking experiments, in which a probe stimulus is used to provide information about cochlear processing for a localized region of the basilar membrane. All types of OAEs can be reduced in level by the presentation of a suppressor. This section, however, primarily describes data for DPOAEs, as the majority of the literature regarding growth of suppression for OAEs has involved DPOAEs.

Several investigators have recorded DPOAEs in a suppression paradigm (e.g., Brown and Kemp 1984; Harris et al. 1992; Kummer et al. 1995; Abdala 1998, 2001; Martin et al. 1998, 2003; Gorga et al. 2002a, 2003; Abdala and Fitzgerald 2003; Johnson et al. 2004). For purposes of illustration, the focus here will be on the data described by Gorga et al. (2002a). There is, however, consistency in the results across many of these studies.

Figure 12.2 plots the median DPOAE decrement as a function of suppressor level for $f_2 = 4$ kHz and L_2 ranging from 20 to 60 dB SPL (Gorga et al. 2002a). Thin lines represent suppressors with frequencies distant from f_2. The threshold of suppression (i.e., the lowest suppressor level at which DPOAE level is just reduced in the presence of a suppressor) increases as the suppressor moves away from the probe frequency (toward either higher or lower frequencies). The growth of suppression with suppressor level tends to be steepest (once threshold is exceeded) for suppressors lower in frequency than the probe, with more shallow slopes observed for suppressors higher in frequency than the probe. With some variation across probe level (L_2), the slope of these functions ranges from a maximum of 2.5 dB/dB for suppressors lower in frequency than the probe (f_2) to less than 1 dB/dB for suppressors higher in frequency than f_2. A slope of nearly 1 dB/dB is observed for the suppressor closest to f_2 (4.1 kHz). These general trends have been observed in other investigations of DPOAE suppression (e.g., Kummer et al. 1995; Abdala 1998, Abdala; Abdala and Fitzgerald 2003; Gorga et al. 2003) and in the suppression of tone-burst-evoked OAEs (Zettner and Folsom 2003). These observations have been interpreted as providing information about the growth of response to the suppressor at the f_2 place, and suggest that the response to suppressors lower than the probe frequency grows more rapidly than the response to suppressors at or above the probe frequency once the suppression threshold is reached.

Slopes of growth of masking (GOM) functions (probe level vs. masker level) for simultaneous masking data are similar to those reported above for the DPOAE growth of suppression data (e.g., Egan and Hake 1950; Nelson and Bilger 1974;

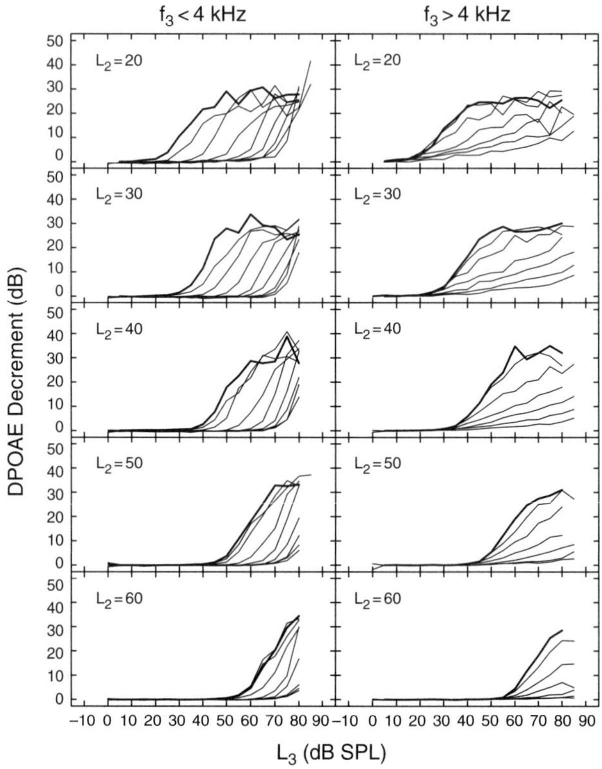

FIGURE 12.2. Median DPOAE decrement (suppression) as a function of suppressor level (L_3) for suppressor frequencies (f_3) lower (*left column*) and higher (*right column*) than f_2. The parameter in each panel is f_3, with the *heavy lines* representing f_3 closest to f_2. The level of f_2 is indicated in each panel. (Reprinted with permission from Gorga et al. 2002a.)

Widin and Viemeister 1979; Bacon and Viemeister 1985; Stelmachowicz et al. 1987; van der Heijden 1995; Bacon et al. 1999). Psychophysical GOM slopes are generally steepest (as much as 2–2.5 dB/dB) for maskers lower in frequency than the probe stimulus and decrease to < 1 dB/dB for maskers higher in frequency than the probe stimulus (reviewed in Stelmachowicz et al. 1987). As was true for DPOAE growth of suppression, these observations for psychophysical GOM are consistent with the idea that the growth of response to a low-frequency masker is more rapid at the cochlear location corresponding to the probe frequency, compared to the growth of response to the probe at the same place.

In addition to decrement-versus-suppressor-level functions, Gorga et al. (2002a) evaluated the suppressor level required to produce a criterion amount of decrement (suppression) as a function of L_2 for on- and low-frequency suppressors. A subset of these results is reproduced in Figure 12.3. Here, suppressor level (L_3) is plotted as a function of L_2 for $f_2 = 4$ kHz, which results

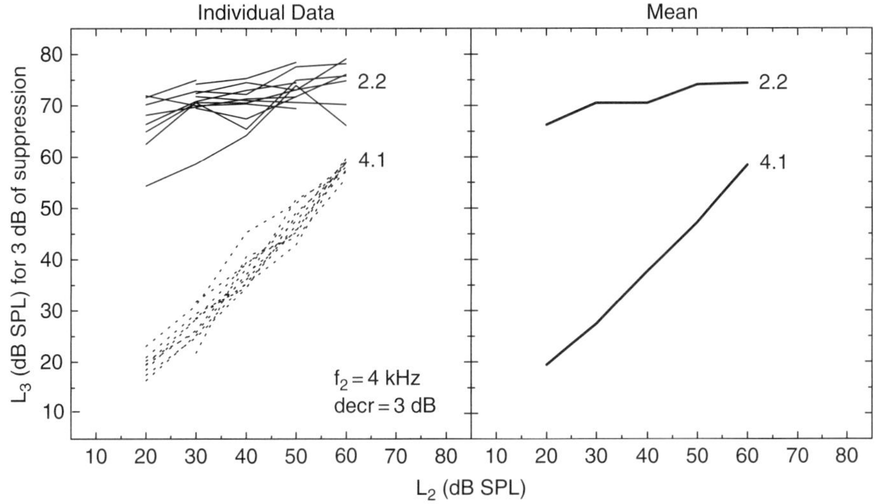

FIGURE 12.3. Suppressor level (L_3) required to produce 3 dB of suppression in DPOAE level when $f_2 = 4$ kHz. The parameter is suppressor frequency, as indicated in each panel in kHz. Data for individual subjects are plotted in the *left panel*, with mean data shown in the *right panel*. (Reprinted with permission from Gorga et al. 2002a.)

in slopes that are the reciprocal of the slopes reported when probe level is the dependent variable. The parameter is suppressor frequency, as indicated in each panel. These results indicate that there is linear growth of suppression for the on-frequency suppressor (4.1 kHz) and compressive growth (slope of 0.26 dB/dB) for the low-frequency suppressor (2.2 kHz). The off-frequency (2.2 kHz) growth of suppression results demonstrate the nonlinear growth of response to the probe (f_2). The seemingly contradictory observation that the growth of response to the on-frequency suppressor is linear while the off-frequency suppressor growth is compressive is a consequence of differences in response growth to the suppressor and probe near the probe-frequency place for on- and off-frequency stimuli. The on-frequency suppressor (4.1 kHz) is processed by the same nonlinearity as the probe stimulus ($f_2 = 4$ kHz). Therefore, even though the response to the probe grows nonlinearly, the response to the suppressor grows at the same rate, resulting in a linear growth in suppression. In contrast, a suppressor approximately an octave below f_2 will result in linear response growth at the f_2 place (see also Fig. 12.2). As a result, the relative growth of suppressor level with probe level for a constant amount of suppression will reflect the compressive growth of the probe response. This is essentially the same argument offered to explain similar observations in the psychophysical masking data to be reviewed below (e.g., Stelmachowicz et al. 1987; Oxenham and Plack 1997), and is consistent with direct observations of basilar-membrane motion at fixed cochlear locations when driven by best-frequency and low-frequency stimuli (e.g., Sellick et al. 1982; Ruggero and Rich 1991) and single-unit recordings from auditory-nerve fibers (Sachs and Abbas 1974). As will be discussed below (see Section 3.2), estimates of cochlear compression can be derived from data such as these.

Oxenham and Plack (1997) described a GOM paradigm, in which the probe was a brief-duration tone at 6 kHz presented at levels ranging from 40 to 90 dB SPL. A long-duration sinusoidal masker at either 3 (off-frequency) or 6 (on-frequency) kHz was presented in a forward-masking paradigm. Masker level was adjusted to mask the signal for each of several signal levels. Results for three normally hearing subjects are shown in Figure 12.4. The form of the on- and off-frequency GOM functions is similar to those shown in Figure 12.3 for DPOAE suppression data. The slope of the masker level versus signal level for the on-frequency condition (6-kHz signal, 6-kHz masker) revealed nearly linear growth of masking (slopes ranging from 0.96 to 1.08 across the three listeners). In contrast, the GOM for the off-frequency condition revealed compressive growth of masking, with a mean slope of 0.16 dB/dB.

While the on-versus-off-frequency trends were similar for psychophysical and DPOAE measures, the off-frequency GOM reported by Oxenham and Plack (0.16 dB/dB) is more compressive than the off-frequency DPOAE growth of suppression (0.26 dB/dB) (Gorga et al. 2002a). This may be due, in part, to additional suppressive effects present in the DPOAE stimulus paradigm (due to the presence of an f_1 tone) that were avoided in the psychoacoustic forward-masking paradigm. Data from direct mechanical measurements in nonhuman mammals suggest that the presentation of a low-frequency suppressor results in more linear growth of response to a stimulus presented at characteristic frequency (e.g., Ruggero et al. 1992). The use of simultaneous masking/suppression paradigms may, therefore, underestimate the compressive nonlinearity present in the cochlear response. At least partially due to this concern, there has been interest recently in evaluating psychophysical growth of masking using forward-masking paradigms (e.g., Oxenham and Plack 1997; Hicks and

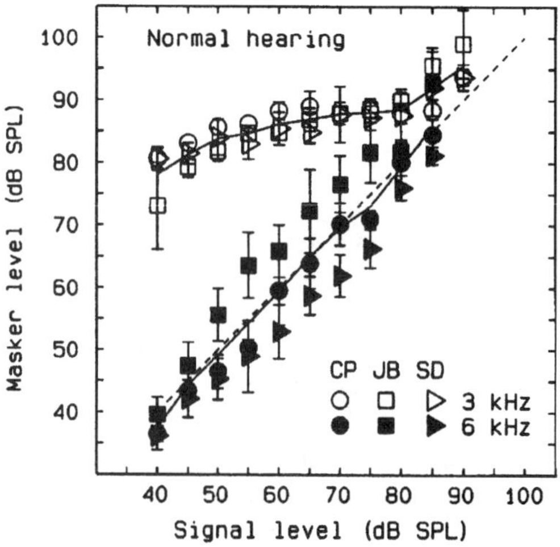

FIGURE 12.4. Masker level necessary to mask a 6-kHz probe (signal) as a function of signal level. Data for three normal-hearing subjects and two masker frequencies are plotted, as indicated on the figure. (Reprinted with permission from Oxenham and Plack 1997.)

Bacon 1999; Moore et al. 1999; Plack and Drga 2003; Rosengard et al. 2005), something that is not possible for DPOAE measurements.

3.2 Cochlear-Compression Estimates

Estimates of cochlear compression can be obtained from OAE data by (1) evaluating the slope of input/ouput (I/O) functions or (2) comparing shifts in the tails of suppression tuning curves (STCs) with shifts at the tips of STCs as a function of stimulus (probe) level (as shown in Fig. 12.3). In many ways, these two methods are variants of one another, and, as a result, produce similar estimates of compression in most cases.

Compression estimates based on the slope of OAE I/O functions have been derived for DPOAEs (Kummer et al. 1998; Dorn et al. 2001; Schairer et al. 2003), SFOAEs (Schairer et al. 2003), and tone-burst-evoked OAEs (Epstein et al. 2004). For f_2 frequencies ranging from 1 to 6 kHz, DPOAE minimum growth-rate estimates range from 0.16 dB/dB (output/input) to 0.34 dB/dB (Dorn et al. 2001; 2003). Kummer et al. (1998) reported slopes approximating 0.2 dB/dB over a broader range of frequencies (0.5 to 8 kHz). A similar growth rate of approximately 0.2 dB/dB was estimated from the slope of tone-burst-evoked OAEs at 1 kHz (Epstein et al. 2004). However, for reasons that are unclear, growth rates derived from a study using SFOAE data were less compressive. Schairer et al. (2003) reported growth rates of 0.44 dB/dB at 2 kHz and 0.61 dB/dB at 1 and 4 kHz for SFOAE I/O functions. Cochlear compression has been reported for other approaches to estimating compression from DPOAE data (i.e., comparison of tip-versus-tail shifts on DPOAE STCs; see Section 4.). The cochlear compression estimated from this alternate approach is similar to that observed for DPOAE I/O functions (see, e.g., Gorga et al. 2002a; Neely et al. 2003).

Growth rates can be transformed to compression ratios by taking their reciprocal. Compression ratios derived from DPOAEs and tone-burst-evoked OAEs range from approximately 3:1 to 6:1, with lower compression ratios predicted from SFOAE measures (1.6:1 to 2.3:1). Thus, DPOAE and tone-burst-evoked OAE compression estimates predict that for moderate-level stimuli, somewhere between a 30- and 60-dB change in input level is required to produce a 10-dB change in output level, while SFOAE data suggest that 16- to 23-dB changes in input level produce a 10-dB change in output.

Various methods have been used to estimate cochlear compression from psychophysical measures; these include GOM and temporal masking. Other techniques can be used, including estimates based on growth of loudness. A complete discussion of loudness growth is beyond the scope of this chapter, although recent efforts to make direct comparisons between loudness growth and OAE I/O functions will be briefly discussed in Section 3.3.

Compression estimates based on GOM data can be made by taking the ratio of the on- and off-frequency masking function slopes (see Fig. 12.4). It is necessary to take the ratio of these two slopes to account for any deviation from linear response growth in the on-frequency condition (this is required for psychoacoustic measures because the measured quantity does not necessarily depend linearly on the cochlear response to the probe tone). Using this technique, Oxenham and Plack (1997) reported compressive growth rates of 0.16 dB/dB for a 6-kHz signal. Other estimates of compressive growth using this technique are in the range of approximately 0.2 to 0.3 dB/dB for high-frequency signals (4–6 kHz) (Moore et al. 1999; Nelson et al. 2001; Plack and Drga 2003). Several studies using this GOM approach to estimate compression for a broader range of frequencies have suggested that compression decreases for more apical regions of the cochlea (e.g., Hicks and Bacon 1999; Moore et al. 1999), a view that may not agree with results from other techniques (e.g., Lopez-Poveda et al. 2003; Plack and Drga 2003; Nelson and Schroder 2004).

Nelson et al. (2001) used a temporal-masking paradigm to estimate cochlear compression. In this forward-masking paradigm, the probe level was fixed at a low level, while masker level was varied for various probe-masker intervals. Masked thresholds as a function of time delay were measured for both on- and off-frequency maskers. The ratio of the slopes of the on- and off-frequency functions was used to estimate cochlear-response-growth rate. Using temporal masking, growth rates between 0.1 and 0.3 dB/dB have been reported for signal frequencies ranging from 0.25 to 8 kHz (Nelson et al. 2001; Lopez-Poveda et al. 2003; Plack and Drga 2003; Nelson and Schroder 2004). These growth rates are similar to those reported for high frequencies with GOM, and indicate more compressive growth for low frequencies than some other estimates from GOM. This discrepancy has been explained, in part, by suggesting that the assumption of linear response growth for the off-frequency masker is not met for low signal frequencies. Data from direct measurements in nonhuman mammals suggest that cochlear nonlinearity is more distributed, i.e., less frequency-specific, in the apical cochlea than in the basal cochlea (e.g., Cooper and Rhode 1997). If this observation is correct, then when testing with a probe frequency of 0.5 kHz, for example, the response to an off-frequency masker at 0.25 kHz may not grow linearly at the 0.5-kHz place, whereas a 1-octave separation between probe and masker may be sufficient at higher frequencies. This would result in an underestimation of compression at the 0.5-kHz place, when taking the ratio of the off-and-on-frequency responses. Studies reporting equal compression along the cochlea with the temporal-masking technique typically used off-frequency functions from higher frequency regions to estimate compression at low frequencies (e.g., Lopez-Poveda et al. 2003; Plack and Drga 2003; Nelson and Schroder 2004).

The extent to which compression varies along the length of the cochlea remains controversial. In frequency regions where both OAE and psychophysical estimates of compression have been made (frequencies at and above 1 kHz), both techniques provide roughly similar estimates, with compression ratios between 3:1 and 5:1. A single OAE-based estimate below 1 kHz (Kummer et al. 1998) suggests similarly compressive growth, as compared to higher frequencies. More data are needed to clarify the extent to which compression differs as a function of frequency.

3.3 OAE Input/Output Functions and Loudness

Growth of loudness provides another method for estimating response growth psychophysically. Growth of loudness may be viewed as a behavioral measure of response growth in much the same way as an OAE I/O function might be viewed as a "mechanical" measure of response growth. Given the similar predictions of response growth for OAE-suppression and psychophysical-masking procedures, it is reasonable to predict a relationship between loudness growth and OAEs.

Neely et al. (2003) described the similarity in shape between median DPOAE I/O functions and the classic Fletcher–Munson (1933) loudness curve. Furthermore, they demonstrated similarity in cochlear-compression estimates derived from the Fletcher–Munson loudness data and DPOAE I/O functions, with both predicting maximum compression of about 4:1 for stimulus levels of 65–70 dB SPL. Despite the similarity between compression estimates from DPOAE I/O functions and loudness functions in the group data, Neely et al. were skeptical regarding the correspondence between the two estimates in individual ears because of the large variability observed in compression estimates from DPOAE data in individual normal-hearing ears. Similar variability may exist in individual estimates of loudness growth; however, Neely et al. used only the classic Fletcher–Munson loudness curve for comparison.

More recent data suggest that there may be a correlation between OAE I/O functions and loudness in individual ears. Müller and Janssen (2004) reported findings from a within-subject comparison of loudness growth (based on categorical loudness scaling) and DPOAE I/O functions in normal and impaired subjects. Despite uncertainties regarding how to normalize DPOAE I/O functions in order to compare them to loudness growth, compression estimates (defined as the reciprocal of the slope of the growth functions) from the two methods were similar. Compression was estimated at a probe level of 65 dB SPL for both measures. On average, the compression estimated from loudness growth (6.3:1) and the compression estimated from DPOAE I/O functions (5.8:1) was similar for normal ears. Epstein and Florentine (2005) reported a correlation of 0.65 between the slope of temporal-integration-based loudness estimates and tone-burst-evoked OAE I/O functions. A correlation of 0.87 was found for similar comparisons between pulsation thresholds and OAE I/O functions. These measures provide support for the view that estimates of response growth and compression derived from OAE and loudness-growth measures share, at least in part, the same underlying mechanisms.

3.4 Summary of Response Growth

Both OAEs and psychophysical measures provide similar estimates of response growth that depend on frequency in the same way. Cochlear compression estimates based on DPOAE measures predict less compression than similar estimates from forward-masking GOM or temporal masking because of suppression effects due to the stimuli (i.e., f_1 and f_2) used to elicit the response. However, the compression estimates derived from the various types of OAE and psychophysical measures are similar in magnitude.

4. Frequency Tuning

Both OAE and psychophysical measures have been used as noninvasive probes of the frequency-tuning or frequency-resolving capabilities of the human auditory system. One commonly used approach to estimating frequency resolution is through the construction of tuning curves in which the probe stimulus (f_2 in the case of DPOAEs, a single tone in the case of behavioral masking measurements) is fixed in frequency and level while maskers of varying frequencies are adjusted in level until the probe is masked (or suppressed in the case of OAEs). A variant on this method, in which the masker is a notched noise of varying notch widths, also has been used in psychoacoustics. Results for the notched-noise method will be discussed in Section 4.2.

One measure of tuning is the quality factor, Q, which is defined as the center frequency of the filter divided by its bandwidth. A high Q value indicates sharp tuning. In auditory research, two commonly used definitions are Q_{10} and Q_{ERB}. Q_{10} is calculated by dividing the characteristic frequency (CF) by the bandwidth of the tuning curve 10 dB above the most sensitive point (or tip), which occurs at the CF. Q_{ERB} is analogous to Q_{10} but is calculated by dividing the CF by the equivalent rectangular bandwidth (ERB). For any filter, the ERB corresponds to the bandwidth of a rectangular filter that has the same CF response and passes the same total power. Both Q_{10} and Q_{ERB} provide estimates of tuning around the tip of the tuning curve.

4.1 OAE Suppression Tuning Curves and Psychophysical Tuning Curves

Although both transient-evoked OAEs and SFOAEs have been used to estimate frequency selectivity (see, e.g., Neumann et al. 1997 and Shera and Guinan 2003), the focus in this section is on suppression tuning curves (STCs) derived from DPOAE data. The use of SFOAE group delay as an estimate of frequency tuning will be discussed in Section 4.2.

Several investigators have used DPOAE STCs to estimate frequency selectivity in human ears (e.g., Brown and Kemp 1984; Harris et al. 1992; Cianfrone et al. 1994; Kummer et al. 1995; Abdala 1998, 2003; Gorga et al. 2002a, 2003; Abdala and Fitzgerald 2003; Martin et al. 2003). DPOAE STCs typically are constructed using suppression criteria of either 3 or 6 dB, a criterion that represents only a partial reduction in the DPOAE response. This contrasts with the approach used to construct psychophysical tuning curves (PTCs) in which the probe is rendered inaudible by the masker. Johnson et al. (2004) described an approach for constructing DPOAE STCs from a completely suppressed response, which might be more directly analogous to the approach used in constructing PTCs. Despite this difference in methodology, DPOAE STCs have features that are at least qualitatively similar to many of the features observed in PTCs.

Figure 12.5 plots examples of DPOAE STCs for probe (f_2) levels ranging from 20 to 60 dB SPL when $f_2 = 4$ kHz (Gorga et al. 2002a). In this figure, the

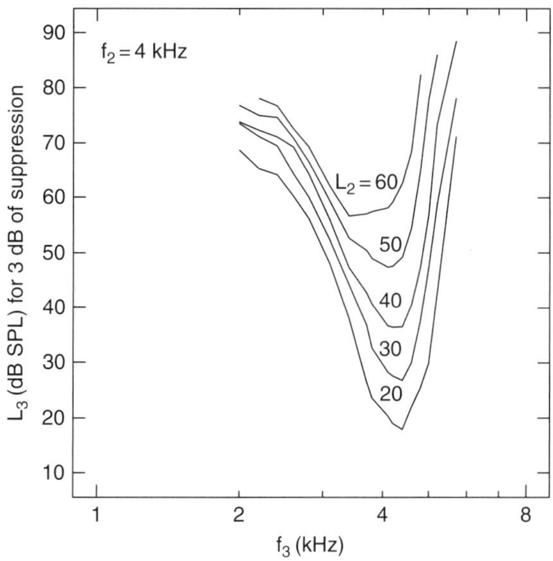

FIGURE 12.5. Suppressor level (L_3) required for 3-dB suppression of DPOAE level as a function of suppressor frequency (f_3) when $f_2 = 4$ kHz ($f_1 = 3.2$ kHz). The level of f_2 (L_2) is indicated for each DPOAE STC. (Reprinted with permission from Gorga et al. 2002a.)

STC becomes progressively broader as L_2 increases. Q_{10} values range from 3.5 when $L_2 = 20$ dB SPL to 2.0 when $L_2 = 60$ dB SPL. Comparison of reported Q_{10} values across various studies suggests that larger Q_{10} values (consistent with sharper tuning) are observed as primary levels decrease (e.g., Harris et al. 1992; Cianfrone et al. 1994; Kummer et al. 1995; Abdala 1998, 2001; Gorga et al. 2002a, 2003; Martin et al. 2003). Across these studies, Q_{10} ranges from 2.5 to 4 for probe levels between 20 and 40 dB SPL to between 1 and 3.3 for probe levels of 50 to 70 dB SPL. The effect of stimulus frequency is less systematic. Several studies have reported an increase in Q_{10} as f_2 increases from 1 to 2 or 4 kHz, followed by a decrease in Q_{10} for higher f_2 frequencies (Harris et al. 1992; Cianfrone et al. 1994; Kummer et al. 1995; Abdala 1998; Abdala and Fitzgerald 2003). In contrast, Martin et al. (2003) report observing the largest Q_{10} for moderate-level stimuli when $f_2 = 1$ kHz and the smallest Q_{10} when $f_2 = 2$ kHz ($f_2 = 3$ and 4 kHz were intermediate). The reasons for these nonmonotonicities and/or differences across studies are not clear, although the general trend is for lower-frequency stimuli to have lower Q_{10} values (approximately 2–3) than higher-frequency stimuli, at least through 3 to 4 kHz (approximately 2.5–4).

Many recent psychoacoustic studies of frequency selectivity have used forward-masking paradigms. This is primarily because estimates made using simultaneous masking are believed to include influences of both excitation and suppression in the cochlea, which complicates the interpretation. Tuning estimates based on simultaneous-masking data typically result in lower Q values than those observed with forward masking (e.g., Moore 1978; Vogten 1978; Moore et al. 1984, 1987; Oxenham and Shera 2003). However, in this section the focus will be on PTCs obtained with simultaneous masking, in order to facilitate comparison with DPOAE STCs, which, because of the nature of the response, require simultaneous presentation of both primary and suppressor stimuli.

The Q_{10} values observed for simultaneous-masking PTCs range from approximately 3 to 9 for probe frequencies between 0.5 and 6 kHz (Moore 1978; Moore et al. 1984; Bacon and Moore 1986). The probe levels at which these Q_{10} values were obtained ranged from 10 to 35 dB SL, corresponding to sound-pressure levels between 40 and 55 dB SPL. There is some indication that tuning sharpness increases as probe frequency increases (e.g., Moore et al. 1984). However, the change in tuning with frequency has been a matter of some debate, as ERB estimates from the notched-noise technique for estimating auditory-filter shape (e.g., Glasberg and Moore 1990) have suggested that tuning sharpness is relatively constant above 1 kHz, an observation that has been challenged by more-recent data (e.g., Shera et al. 2002; Oxenham and Shera 2003). This topic will be discussed further in Section 4.2. The Q_{10} values observed for PTCs are similar to those observed for DPOAE STCs, although they generally indicate sharper tuning than the DPOAE STCs.

Also apparent in Figure 12.5 is a systematic elevation in DPOAE STC tip threshold as level is increased from 20 to 60 dB SPL, with less change in tail threshold over the same level range. The differential shift in tip-versus-tail threshold is a consequence of the difference in response growth at a fixed place for on- and off-frequency stimuli as discussed in Section 3.1. This difference in threshold for the STC tip as compared to the STC tail (i.e., the tip-to-tail difference) has been described as a measure of cochlear-amplifier gain (Mills 1998; Pienkowski and Kunov 2001). The data shown in Figures 12.3 to 12.5 are consistent with the idea that cochlear-amplifier gain decreases as level increases. In Figure 12.5, for example, maximum gain of 45 to 50 dB is observed when $L_2 = 20$ dB SPL, with gain of 35 to 40 dB observed when $L_2 = 40$ dB SPL. These results are consistent with those reported by Pienkowski and Kunov (2001).

The same approach can be taken with PTC and psychophysical GOM data. Estimating gain at 1 kHz from the PTC reported by Moore et al. (1984) suggests cochlear-amplifier gain of 40 to 45 dB for probe levels of 45 dB SPL. Comparing the on- and off-frequency masked thresholds in Figure 12.4 also predicts 40 to 45 dB of gain for a probe frequency of 6 kHz (Oxenham and Plack 1997). Thus, the psychophysical measures predicted more gain for probe levels of 40 to 50 dB SPL than that predicted from DPOAE STCs.

4.2 SFOAE Group Delay and Psychophysical Forward-Masking Estimates

An alternate approach to estimating frequency tuning in humans with OAE data was recently proposed by Shera et al. (2002) and Shera and Guinan (2003). The method is based on the empirical correlation established in cats and guinea pigs between Q_{ERB} values measured physiologically in auditory-nerve fibers and SFOAE group delays measured in the same species. (SFOAE group delays, measured at 40 dB SPL, were defined as the negative of the slope of the phase-versus-frequency function.) The observed correlation can be understood in the context of filter theory as reflecting the inverse relationship between filter bandwidth and group delay. By assuming that the correlation observed in laboratory mammals also applies in humans, Shera et al. (2002) estimated that the human Q_{ERB} increases monotonically with frequency from a value near 10 at 1 kHz to about 30 at 8 kHz. These Q_{ERB} values are larger (by a factor of 2 or more) than predicted from either DPOAE or simultaneous masking tuning estimates reviewed in Section 4.1. These Q_{ERB} values also are greater than those observed in other mammals, which is a direct consequence of the longer OAE group delays measured in humans (Shera et al. 2002).

Shera et al. (2002) and Oxenham and Shera (2003) also described Q_{ERB} estimates using a technique believed to compensate for the limitations in estimating frequency tuning from either simultaneous masking or from many of the forward-masking approaches. Specifically, they used the notched-noise technique for estimating auditory-filter bandwidth (e.g., Patterson 1976; Glasberg and Moore 1990), a forward-masking paradigm, and a fixed-level probe stimulus (approximately 40 dB SPL) with varying masker level. The Q_{ERB} estimates derived from these behavioral-masking measurements ranged from approximately 10 at 1 kHz to 20 at 8 kHz. These values are similar to those derived from the SFOAE group delay and are larger than those derived from ERB values obtained from the notched-noise technique using simultaneous masking (e.g., Moore and Glasberg 1983; Dubno and Dirks 1989; Glasberg and Moore 1990; Moore et al. 1990; Shailer et al. 1990). These SFOAE and behavioral data also indicate that the sharpness of tuning increases with increasing stimulus frequency, at least for frequencies between 1 and 8 kHz. Although this predicted increase in the sharpness of human tuning at higher CFs disagrees with estimates obtained from simultaneous-masking notched-noise experiments (reviewed in Glasberg and Moore 1990), it agrees with the trend found in neural tuning curves measured in laboratory animals (Liberman 1990; Tsuji and Liberman 1997; Shera et al. 2002).

4.3 Summary of Frequency Tuning

Tuning estimates based on DPOAE data generally are more similar to those obtained from simultaneous-masking psychophysical data than from forward-masking psychophysical data. This trend is expected because of the necessity of presenting primary and suppressor stimuli simultaneously when making DPOAE measures. The data reviewed above are consistent with the idea that cochlear tuning broadens as stimulus level increases. The most recent data presented by Shera, Oxenham, and colleagues for both psychophysical and OAE measures (see Section 4.2) suggest sharper tuning than previously had been described in humans and other mammals, and indicate that cochlear tuning increases with stimulus frequency.

5. Upward Spread of Excitation

Direct observation of basilar membrane motion in nonhuman mammals, as well as measures of neural responses from auditory-nerve fibers, demonstrates upward spread of excitation such that a low-frequency stimulus can affect the response of a cochlear location with a higher CF (e.g., Pfeiffer and Kim 1975; Kim and Molnar 1979; Kim 1983; Kim et al. 1990; Ruggero et al. 1992; Rhode and Recio 2000; Robles and Ruggero 2001). In contrast, a stimulus higher in frequency than the CF will have little effect on the CF response except for frequencies close to the CF and/or for high-level stimuli. Evidence of this phenomenon in OAE and psychoacoustic measures can be seen in the data reviewed in Sections 3 and 4 above. In both OAE and psychoacoustic measures, low-frequency stimuli can mask or suppress a response to a high-frequency probe stimulus more effectively than a stimulus higher in frequency than the probe stimulus. This can be seen in the STCs shown in Figure 12.5 by noting that the low-frequency flank of the STC has a shallower slope than the high-frequency flank, a trend that has been observed across many different studies (e.g., Brown and Kemp 1984; Harris et al. 1992; Cianfrone et al. 1994; Kummer et al. 1995; Abdala 1998, 2001; Martin et al. 1998; Gorga et al. 2002a, 2003) and is also present in PTCs (e.g., Moore 1978; Moore et al. 1984; Bacon and Moore 1986; Vogten 1978). Plots of growth of suppression and masking (see Fig. 12.2) also show more rapid growth of suppression or masking for suppressors/maskers that are lower in frequency than the probe as compared to those higher in frequency than the probe (e.g., Egan and Hake 1950; Stelmachowicz et al. 1987; Kummer et al. 1995; Abdala 1998, 2001; Bacon et al. 1999; Gorga et al. 2002a, 2003; Abdala and Fitzgerald 2003).

In the data reviewed in Sections 3 and 4, the probe frequency is fixed and the suppressor or masker frequency is varied. While this approach provides a demonstration of the upward spread of excitation for the suppressing or masking stimuli, it also implicitly assumes that the cochlear response to the probe tone remains centered at its tonotopic place. This assumption probably is violated, at least at high-stimulus levels. A more direct demonstration of the upward spread of excitation/suppression might be to use a technique in which the suppressor or masker frequency is fixed and probe frequency is varied. In this manner, the influence of either the suppressor or masker on the response to probes of different frequencies can be assessed. This approach has been used both psychophysically (e.g., Wegel and Lane 1924; Egan and Hake 1950) and with OAEs (Gorga et al. 2002b).

Using this stimulation paradigm, Gorga et al. (2002b) recorded DPOAE responses for suppressor frequencies at both 2 and 4 kHz. Their results, shown in Figure 12.6, show asymmetric spread of suppression. The suppressor stimulus reduces the response to probe stimuli higher in frequency than the suppressor to a larger degree than probe stimuli on the low-frequency side of the suppressor, consistent with upward spread of excitation for the suppressor.

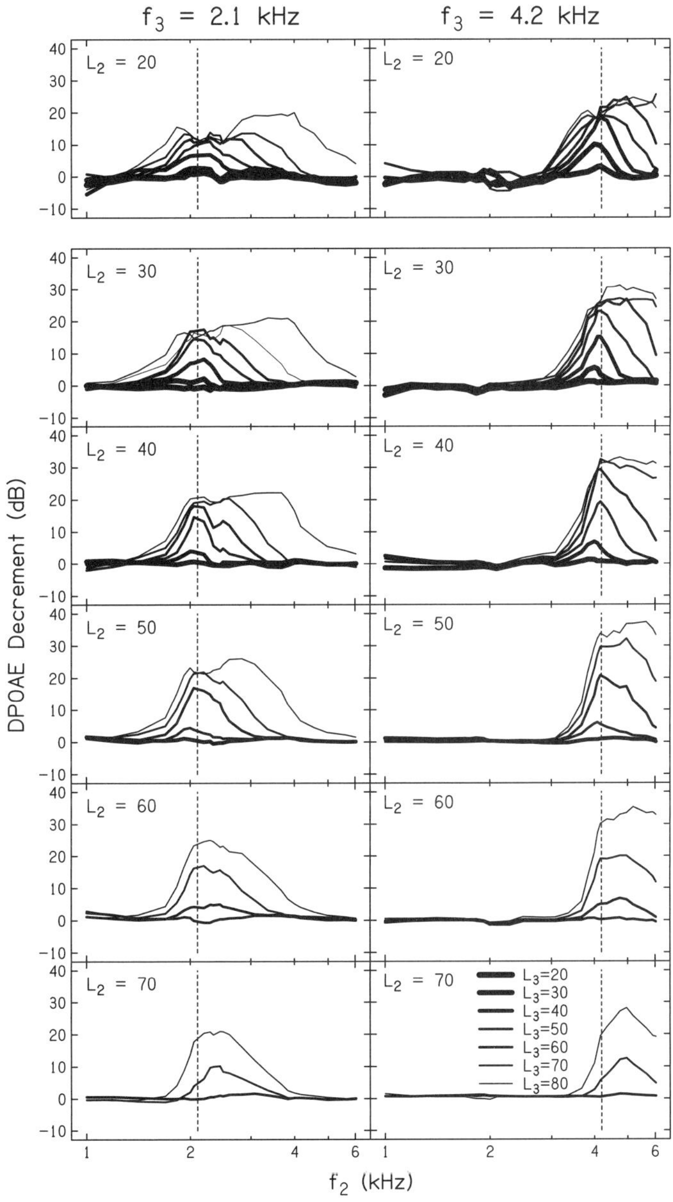

FIGURE 12.6. DPOAE decrement (suppression) as a function of f_2 when the suppressor frequency was either 2.1 kHz (*left column*) or 4.1 kHz (*right column*). Each row represents a different L_2. The parameter in each panel is suppressor level (L_3), with the *dashed vertical line* in each panel representing the suppressor frequency. (Reprinted with permission from Gorga et al. 2002b.)

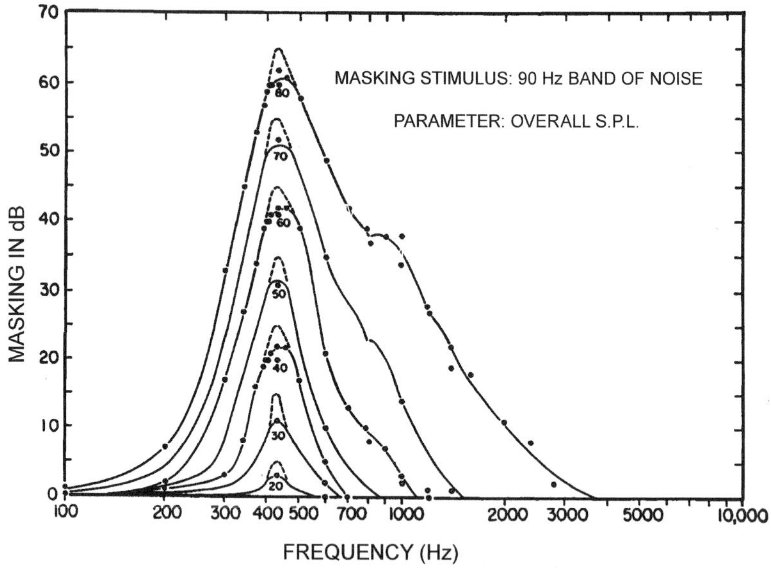

FIGURE 12.7. Amount of masking as a function of probe frequency for a 90-Hz band of noise centered at 410 Hz. The *continuous lines* represent a visual fit to the data. The dashed lines represent a 4.2-dB extension of the fit at the peak of the masking function, as described by Egan and Hake (1950). (Reprinted with permission from Egan and Hake 1950.)

The upward spread of suppression observed in the DPOAE responses and shown in Figure 12.6 is similar to the upward spread of masking reported in the classic psychoacoustic literature on masking patterns (e.g., Wegel and Lane 1924; Egan and Hake 1950). Figure 12.7 shows asymmetric spread of masking for a 400-Hz narrowband noise masker. These data were collected using a paradigm similar to that described above, where the masker is fixed in frequency and threshold is measured for various probe frequencies. More recent data from Oxenham and Plack (1998), using both simultaneous and nonsimultaneous masking techniques, also demonstrated upward spread of masking in both masking paradigms.

6. Summary and Future Directions

Both OAE and psychoacoustic measures provide a means to assess cochlear function noninvasively in humans. Although estimates of cochlear function obtained from OAEs and psychoacoustics are not in perfect agreement, they provide similar estimates in most cases. While there is relative agreement in many areas, the issue of cochlear function at the apex as compared to the base remains an area where additional data from OAE measures may help to clarify the extent to which cochlear compression is similar (or different) in these two locations. In addition, future work aimed at understanding the sources of the variability typically observed across individuals with normal hearing will help to clarify the extent to which these individuals can be expected to differ from one another and to quantify the contribution of cochlear function to any differences. Finally, although the data presented in this chapter focus exclusively on results from ears with normal hearing, efforts toward understanding the influence of hearing loss on cochlear function, as measured with OAEs and comparable psychoacoustical approaches, may help to improve clinical services offered to patients with cochlear hearing loss.

Acknowledgments. This work was supported by the NIH-NIDCD (F32 DC7536, R01 DC2251, R01 DC03909, and R01 DC03687). The authors would like to thank Connie Converse for her assistance with manuscript preparation.

References

Abdala C (1998) A developmental study of distortion product otoacoustic emission (2f1-f2) suppression in humans. Hear Res 121:125–138.

Abdala C (2001) Maturation of the human cochlear amplifier: distortion product otoacoustic emission suppression tuning curves recorded at low and high primary tone levels. J Acoust Soc Am 110:1465–1476.

Abdala C (2003) A longitudinal study of distortion product otoacoustic emission ipsilateral suppression and input/output characteristics in human neonates. J Acoust Soc Am 114:3239–3250.

Abdala C, Fitzgerald TS (2003) Ipsilateral distortion product otoacoustic emission (2f1−f2) suppression in children with sensorineural hearing loss. J Acoust Soc Am 114:919–931.

Bacon SP, Moore BCJ (1986) Temporal effects in masking and their influence on psychophysical tuning curves. J Acoust Soc Am 80:1638–1645.

Bacon SP, Viemeister NF (1985) Simultaneous masking by gating and continuous sinusoidal maskers. J Acoust Soc Am 78:1220–1230.

Bacon SP, Boden LN, Lee J, Repovsch JL (1999) Growth of simultaneous masking for $f_m < f_s$: effects of overall frequency and level. J Acoust Soc Am 106:341–350.

Braun M (1997) Frequency spacing of multiple spontaneous otoacoustic emissions shows relation to critical bands: a large-scale cumulative study. Hear Res 114:197–203.

Brown AM, Kemp DT (1984) Suppressibility of the $2f_1 - f_2$ stimulated acoustic emissions in gerbil and man. Hear Res 13:29–37.

Cianfrone G, Altissimi G, Cervellini M, Musacchio A, Turchetta R (1994) Suppression tuning characteristics of $2f_1 - f_2$ distortion product otoacoustic emissions. Br J Audiol 28:205–212.

Cooper NP, Rhode WS (1997) Mechanical responses to two-tone distortion products in the apical and basal turns of the mammalian cochlea. J Neurophysiol 78:261–270.

Dorn PA, Konrad-Martin D, Neely ST, Keefe DH, Cyr E, Gorga MP (2001) Distortion product otoacoustic emission input/output functions in normal-hearing and hearing-impaired human ears. J Acoust Soc Am 110:3119–3131.

Dubno JR, Dirks DD (1989) Auditory filter characteristics and constant recognition for hearing-impaired listeners. J Acoust Soc Am 85:1666–1675.

Egan JP, Hake HW (1950) On the masking pattern of a simple auditory stimulus. J Acoust Soc Am 22:622–630.

Elliott E (1958) A ripple effect in the audiogram. Nature 181:1076.

Epstein M, Florentine M (2005) Inferring basilar-membrane motion from tone-burst otoacoustic emissions and psychoacoustic measurements. J Acoust Soc Am 117: 263–274.

Epstein M, Buus S, Florentine M (2004) The effects of window delay, delinearization, and frequency on tone-burst otoacoustic emission input-output measurements. J Acoust Soc Am 116:1160–1167.

Fletcher H, Munson WA (1933) Loudness, its definition, measurement, and calculation. J Acoust Soc Am 5:82–108.

Furst M, Reshef I, Attias J (1992) Manifestations of intense noise stimulation on spontaneous otoacoustic emission and threshold microstructure: experiment and model. J Acoust Soc Am 91:1003–1014.

Glasberg BR, Moore BCJ (1990) Derivation of auditory filter shapes from notched-noise data. Hear Res 47:103–138.

Gorga MP, Neely ST, Dorn PA, Konrad-Martin D (2002a) The use of distortion product otoacoustic emission suppression as an estimate of response growth. J Acoust Soc Am 111:271–284.

Gorga MP, Neely ST, Dorn PA, Dierking D, Cyr E (2002b) Evidence of upward spread of suppression in DPOAE measurements. J Acoust Soc Am 112:2910–2920.

Gorga MP, Neely ST, Dierking DM, Dorn PA, Hoover BM, Fitzpatrick DF (2003) Distortion product otoacoustic emission suppression tuning curves in normal-hearing and hearing-impaired human ears. J Acoust Soc Am 114:263–278.

Harris FP, Probst R, Xu L (1992) Supression of the $2f_1 - f_2$ otoacoustic emission in humans. Hear Res 64:133–141.

Hartmann WM (1998) Signals, Sounds, and Sensation. New York: Springer-Verlag, pp. 251–252.

He N-j, Schmiedt RA (1993) Fine structure of the $2f_1 - f_2$ acoustic distortion product: changes with primary level. J Acoust Soc Am 94:2659–2669.

Heitmann J, Waldmann B, Schnitzler H-U, Plinkert PK, Zenner H-P (1998) Suppression of distortion product otoacoustic emissions (DPOAE) near $2f_1 - f_2$ removes DP-gram fine structure—Evidence for a secondary generator. J Acoust Soc Am 103:1527–1531.

Hicks ML, Bacon SP (1999) Psychophysical measures of auditory nonlinearities as a function of frequency in individuals with normal hearing. J Acoust Soc Am 105: 326–338.

Johnson TA, Neely ST, Dierking DM, Hoover BM, Gorga MP (2004) An alternate approach to constructing distortion product otoacoustic emission (DPOAE) suppression tuning curves. J Acoust Soc Am 116:3263–3266.

Kalluri R, Shera CA (2001) Distortion-product source unmixing: a test of the two-mechanism model for DPOAE generation. J Acoust Soc Am 109:622–637.

Kim DO (1983) Spatiotemporal response patterns in populations of cochlear nerve fibers: single- and two-tone studies. Ann NY Acad Sci 405:68–78.

Kim DO, Molnar CE (1979) A population study of cochlear nerve fibers: comparison of spatial distributions of average-rate and phase-locking measures of responses to single tones. J Neurophysiol 42:16–30.

Kim DO, Chang SO, Sirianni JG (1990) A population study of auditory-nerve fibers in unanesthetized decerebrate cats: responses to pure tones. J Acoust Soc Am 87: 1648–1655.

Konrad-Martin D, Neely ST, Keefe DH, Dorn PA, Gorga MP (2001) Sources of distortion product otoacoustic emissions revealed by suppression experiments and inverse fast Fourier transforms in normal ears. J Acoust Soc Am 109:2862–2879.

Konrad-Martin D, Neely ST, Keefe DH, Dorn PA, Cyr E, Gorga MP (2002) Sources of DPOAEs revealed by suppression experiments, inverse fast Fourier transforms, and SFOAEs in impaired ears. J Acoust Soc Am 111:1800–1809.

Kummer P, Janssen T, Arnold W (1995) Suppression tuning characteristics of the $2f_1 - f_2$ distortion-product otoacoustic emission in humans. J Acoust Soc Am 98:197–210.

Kummer P, Janssen T, Arnold W (1998) The level and growth behavior of the $2f_1 - f_2$ distortion product otoacoustic emission and its relationship to auditory sensitivity in normal hearing and cochlear hearing loss. J Acoust Soc Am 103:3431–3444.

Liberman MC (1990) Effects of chronic cochlear de-efferentation on auditory-nerve response. Hear Res 49:209–224.

Long GR (1984) The microstructure of quiet and masked thresholds. Hear Res 15:73–87.

Long GR, Tubis A (1988a) Investigations into the nature of the association between threshold microstructure and otoacoustic emissions. Hear Res 36:125–138.

Long GR, Tubis A (1988b) Modification of spontaneous and evoked otoacoustic emissions and associated psychoacoustic microstructure by aspirin consumption. J Acoust Soc Am 84:1343–1353.

Lopez-Poveda EA, Plack CJ, Meddis R (2003) Cochlear nonlinearity between 500 and 8000 Hz in listeners with normal hearing. J Acoust Soc Am 113:951–960.

Lutman ME, Deeks J (1999) Correspondence amongst microstructure patterns observed in otoacoustic emissions and Bekesy audiometry. Audiol 38:263–266.

Martin GK, Jassir D, Stagner BB, Whitehead ML, Lonsbury-Martin BL (1998) Locus of generation for the $2f_1 - f_2$ vs $2f_2 - f_1$ distortion-product otoacoustic emissions in normal-hearing humans revealed by suppression tuning, onset latencies, and amplitude correlations. J Acoust Soc Am 103:1957–1971.

Martin GK, Villasuso EI, Stagner BB, Lonsbury-Martin BL (2003) Suppression and enhancement of distortion-product otoacoustic emissions by interference tones above f_2. II. Findings in humans. Hear Res 177:111–122.

Mauermann M, Uppenkamp S, van Hengel PWJ, Kollmeier B (1999a) Evidence for the distortion product frequency place as a source of distortion product otoacoustic emission (DPOAE) fine structure in humans. I. Fine structure and higher-order DPOAE as a function of the frequency ratio f2/f1. J Acoust Soc Am 106:3473–3483.

Mauermann M, Uppenkamp S, van Hengel PWJ, Kollmeier B (1999b) Evidence for the distortion product frequency place as a source of distortion product otoacoustic emission (DPOAE) fine structure in humans. II. Fine structure for different shapes of cochlear hearing loss. J Acoust Soc Am 106:3484–3491.

Mauermann M, Long GR, Kollmeier B (2004) Fine structure of hearing threshold and loudness perception. J Acoust Soc Am 116:1066–1080.

Mills DM (1998) Interpretation of distortion product otoacoustic emission measurements. II. Estimating tuning characteristics using three stimulus tones. J Acoust Soc Am 103:507–523.

Moore BCJ (1978) Psychophysical tuning curves measured in simultaneous and forward masking. J Acoust Soc Am 63:524–532.

Moore BCJ, Glasberg BR (1983) Suggested formulae for calculating auditory-filter bandwidths and excitation patterns. J Acoust Soc Am 74:750–753.
Moore BCJ, Glasberg BR, Roberts B (1984) Refining the measurement of psychophysical tuning curves. J Acoust Soc Am 76:1057–1066.
Moore BCJ, Poon PWF, Bacon SP, Glasberg BR (1987) The temporal course of masking and the auditory filter shape. J Acoust Soc Am 81:1873–1880.
Moore BCJ, Peters RW, Glasberg BR (1990) Auditory filter shapes at low center frequencies. J Acoust Soc Am 88:132–140.
Moore BCJ, Vickers DA, Plack CJ, Oxenham AJ (1999) Inter-relationship between different psychoacoustic measures assumed to be related to the cochlear active mechanism. J Acoust Soc Am 106:2761–2778.
Müller J, Janssen T (2004) Similarity in loudness and distortion product otoacoustic emission input/output functions: implications for an objective hearing aid adjustment. J Acoust Soc Am 115:3081–3091.
Neely ST, Gorga MP, Dorn PA (2003) Cochlear compression estimates from measurements of distortion-product otoacoustic emissions. J Acoust Soc Am 114:1499–1507.
Nelson DA, Bilger RC (1974) Pure-tone octave masking in normal-hearing listeners. J Speech Hear Res 17:223–251.
Nelson DA, Schroder AC (2004) Peripheral compression as a function of stimulus level and frequency region in normal-hearing listeners. J Acoust Soc Am 115:2221–2233.
Nelson DA, Schroder AC, Wojtczak M (2001) A new procedure for measuring peripheral compression in normal-hearing and hearing-impaired listeners. J Acoust Soc Am 110:2045–2064.
Neumann J, Uppenkamp S, Kollmeier B (1997) Relations between notched-noise suppressed TEOAE and the psychoacoustical critical bandwidth. J Acoust Soc Am 101:2778–2788.
Oxenham AJ, Plack CJ (1997) A behavioral measure of basilar-membrane nonlinearity in listeners with normal and impaired hearing. J Acoust Soc Am 101:3666–3675.
Oxenham AJ, Plack CJ (1998) Suppression and the upward spread of masking. J Acoust Soc Am 104:3500–3510.
Oxenham AJ, Shera CA (2003) Estimates of human cochlear tuning at low levels using forward and simultaneous masking. J Assoc Res Otolaryngol 4:541–554.
Patterson RD (1976) Auditory filter shapes derived with noise stimuli. J Acoust Soc Am 59:640–654.
Pfeiffer RR, Kim DO (1975) Cochlear nerve fiber responses: distribution along the cochlear partition. J Acoust Soc Am 58:867–869.
Pienkowski M, Kunov H (2001) Suppression of distortion product otoacoustic emissions and hearing threshold. J Acoust Soc Am 109:1496–1502.
Plack CJ, Drga V (2003) Psychophysical evidence for auditory compression at low characteristic frequencies. J Acoust Soc Am 113:1574–1586.
Rhode WS, Recio A (2000) Study of mechanical motions in the basal region of the chinchilla cochlea. J Acoust Soc Am 107:3317–3332.
Robles L, Ruggero MA (2001) Mechanics of the mammilian cochlea. Physiol Rev 81:1305–1352
Rosengard PS, Oxenham AJ, Braida LD (2005) Comparing different estimates of cochlear compression in listeners with normal and impaired hearing. J Acoust Soc Am 117: 3028–3041.
Ruggero MA, Rich NC (1991) Furosemide alters organ of corti mechanics: evidence for feedback of outer hair cells upon the basilar membrane. J Neurosci 11:1057–1067.

Ruggero MA, Robles L, Rich NC (1992) Two-tone suppression in the basilar membrane of the cochlea: mechanical basis of auditory-nerve rate suppression. J Neurophysiol 68:1087–1099.

Russell AF (1992) Heritability of spontaneous otoacoustic emissions. Ph.D. thesis, University of Illinois.

Sachs MB, Abbas PJ (1974) Rate versus level functions for auditory-nerve fibers in cats: tone-burst stimuli. J Acoust Soc Am 56:1835–1847.

Schairer KS, Fitzpatrick D, Keefe DH (2003) Input-output functions for stimulus-frequency otoacoustic emissions in normal-hearing adult ears. J Acoust Soc Am 114:944–966.

Sellick PM, Patuzzi R, Johnstone BM (1982) Measurement of basilar membrane motion in the guinea pig using the Mossbauer technique. J Acoust Soc Am 72:131–141.

Shailer MJ, Moore BCJ, Glasberg BR, Watson N, Harris S (1990) Auditory filter shapes at 8 and 10 kHz. J Acoust Soc Am 88:141–148.

Shehata WE, Brownell WE, Dieler R (1991) Effects of salicylate on shape, electromotility and membrane characteristics of isolated outer hair cells from guinea pig cochlea. Acta Otolaryngol 111:707–718.

Shera CA (2003) Mammalian spontaneous otoacoustic emissions are amplitude-stabilized cochlear standing waves. J Acoust Soc Am 114:244–262.

Shera CA (2004) Mechanisms of mammalian otoacoustic emission and their implications for the clinical utility of otoacoustic emissions. Ear Hear 25:86–97.

Shera CA, Guinan JJ (1999) Evoked otoacoustic emissions arise by two fundamentally different mechanisms: a taxonomy for mammalian OAEs. J Acoust Soc Am 105: 782–798.

Shera CA, Guinan JJ (2003) Stimulus-frequency-emission group delay: a test of coherent reflection filtering and a window on cochlear tuning. J Acoust Soc Am 113:2762–2772.

Shera CA, Guinan JJ, Oxenham AJ (2002) Revised estimates of human cochlear tuning from otoacoustic and behavioral measurements. Proc Natl Acad Sci USA 99: 3318–3323.

Smurzynski J, Probst R (1998) The influence of disappearing and reappearing spontaneous otoacoustic emissions on one subject's threshold microstructure. Hear Res 115: 197–205.

Stelmachowicz PG, Lewis DE, Larson LL, Jesteadt W (1987) Growth of masking as a measure of response growth in hearing-impaired listeners. J Acoust Soc Am 81: 1881–1887.

Stover LJ, Neely ST, Gorga MP (1999) Cochlear generation of intermodulation distortion revealed by DPOAE frequency functions in normal and impaired ears. J Acoust Soc Am 106:2669–2678.

Stypulkowski PH (1990) Mechanisms of salicylate ototoxicity. Hear Res 46:113–145.

Talmadge CL, Long GR, Murphy WJ, Tubis A (1993) New off-line method for detecting spontaneous otoacoustic emission in human subjects. Hear Res 71:170–182.

Talmadge CL, Tubis A, Long GR, Piskorski P (1998) Modeling otoacoustic emission and hearing threshold fine structures. J Acoust Soc Am 104:1517–1543.

Talmadge CL, Long GR, Tubis A, Dhar S (1999) Experimental confirmation of the two-source interference model for the fine structure of distortion product otoacoustic emissions. J Acoust Soc Am 105:275–292.

Tsuji J, Liberman MC (1997) Intracellular labeling of auditory nerve fibers in guinea pig: central and peripheral projections. J Comp Neurol 381:188–202.

van der Heijden M, Kohlrausch A (1995) The role of envelope fluctuations in spectral masking. J Acoust Soc Am 97:1800–1807.

Vogten LLM (1978) Low-level pure-tone masking: a comparison of "tuning curves" obtained with simultaneous and forward masking. J Acoust Soc Am 63:1520–1527.

Wegel RL, Lane CE (1924) The auditory masking of one pure tone by another and its probable relation to the dynamics of the inner ear. Phy Rev 23:266–285.

Widin GP, Viemeister NF (1979) Intensive and temporal effects in pure-tone forward masking. J Acoust Soc Am 66:388–395.

Zettner EM, Folsom RC (2003) Transient emission suppression tuning curve attributes in relation to psychoacoustic threshold. J Acoust Soc Am 113:2031–2041.

Zweig G, Shera CA (1995) The origin of periodicity in the spectrum of evoked otoacoustic emissions. J Acoust Soc Am 98:2018–2047.

Zwicker E (1986) Spontaneous oto-acoustic emissions, threshold in quiet, and just noticeable amplitude at low levels. In: Moore BCJ, Patterson RD (eds) Auditory Frequency Selectivity. New York: Plenum Press, pp. 49–59.

Zwicker E (1989) Otoacoustic emissions and cochlear traveling waves. In: Wilson JP, Kemp DT (eds) Cochlear Mechanisms Structure, Function, and Models. New York: Plenum Press, pp. 359–366.

Zwicker E, Peisl W (1990) Cochlear preprocessing in analog models, in digital models and in human inner ear. Hear Res 44:209–216.

Zwicker E, Schloth E (1984) Interrelation of different oto-acoustic emissions. J Acoust Soc Am 75:1148–1154.

13 Otoacoustic Emissions as a Diagnostic Tool in a Clinical Context

THOMAS JANSSEN AND JÖRG MÜLLER

Ja so gar zwey Flutes douces geben, wenn man c" und das a" ' rein zusammen bläset, noch den dritten Klang, nemlich ein f, welches zu probieren stehet. (Georg Andreas Sorge, *Vorgemach der musikalischen Composition*, Lobenstein, Germany, 1745)

Georg Andreas Sorge was the first to describe the phenomenon of the hearing organ's cubic distortion.

1. Introduction

The discovery of otoacoustic emissions (OAEs) (Kemp 1978)—which were measured initially in humans and not in animals—has not only changed previous thinking about cochlear sound processing, but has also produced a totally new tool for diagnosing hearing problems. OAEs are the by-product of the nonlinear sound amplification process in the cochlea (Davis 1983; Dallos 1992), referred to as the cochlear amplifier (CA), and hence can serve as a measure for evaluating cochlear integrity. OAEs are the only audiological test to selectively assess cochlear dysfunction. Along with tympanometry and auditory brain stem responses, OAEs can differentiate between middle-ear, cochlear, and neural disorders.

Spontaneous OAEs (SOAEs) appear without any sound stimulation at a few frequencies in a healthy cochlea and seem to be a direct consequence of the cellular force generation of outer hair cells (OHCs) (Zwicker and Schloth 1984; Burns et al. 1998; Jülicher et al. 2003). Evoked OAEs are generated by external sounds, either by transient (clicks and tone bursts) or stationary stimuli (tones). Transiently evoked OAEs (TEOAEs) represent the sum of the pulse responses of OHCs along the cochlea, whereas distortion product OAEs (DPOAEs) represent cubic distortions of OHCs when stimulated simultaneously by two tones f_1 (lower frequency) and f_2 (higher frequency), its frequencies predicted by the term $(f_1 - f_2)^3$ (see Fig. 13.1). DPOAEs arise directly from the frequency-selective compressive nonlinearity of OHCs (Brownell et al. 1985; Kemp et al. 1986).

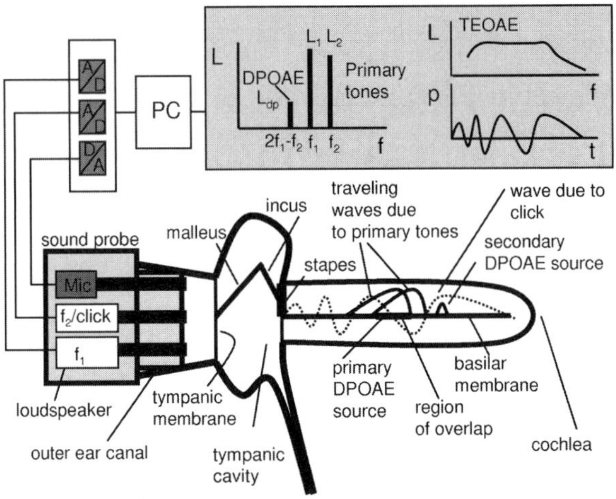

FIGURE 13.1. Schematic drawing of how TEOAEs and DPOAEs are evoked within the cochlea and subsequently measured in the outer ear canal. The sound probe consists of either one loudspeaker for applying a click stimulus (TEOAE) or two loudspeakers for applying two independently generated primary tones shown at *top right* with frequencies f_1 and f_2 and levels L_1 and L_2 (DPOAE), and one microphone (Mic) for measuring the acoustic response signal in the outer ear canal. TEOAEs represent OHC pulse responses along the basilar membrane. Basal responses appear at the beginning, and apical responses at the end of the TEOAE time function. DPOAEs are generated within the region of overlap of the traveling waves of the two primary tones and close to the f_2 place. In humans, the $2f_1 - f_2$ distortion component usually exhibits the highest amplitude. A/D = analog–digital converter, D/A = digital–analog converter, PC = personal computer, L = sound-pressure level, p = sound pressure, f = frequency, t = time.

The two tones interact in the cochlea close to the characteristic place of f_2. Thus, DPOAEs can be applied as a probe for frequency-specific assessment of cochlear dysfunction at the f_2 place. In humans, the $2f_1 - f_2$ distortion component yields the highest amplitude and is therefore primarily used in audiological diagnostics (Gorga et al. 2000a). Stimulus frequency OAEs (SFOAEs) are elicited by one continuous low-level sinusoidal signal. Recording of SFOAEs is difficult because stimulus and response superimpose.

SOAEs do not appear in each normally hearing subject and are present in about a half of the normally hearing population, with a distinctly higher prevalence in women than in men (Bilger et al. 1990; Penner et al. 1993; Penner and Zhang 1997). SOAEs are therefore not suited for audiological diagnostics. TEOAEs and DPOAEs are present in essentially every normally hearing subject. TEOAEs already disappear at mild hearing losses and are therefore a suitable tool for newborn hearing screening, which is a selection procedure for deciding whether further diagnostics are advised (Glattke and Robinette 2002). DPOAEs are reported to be measurable at a cochlear hearing loss of up to 40–50 dB HL, representing approximately the range of the cochlear amplifier (Davis 1983;

Ruggero et al. 1997). In sound-conductive hearing loss, where both the stimulus and the response amplitude are reduced, OAEs are difficult to measure, even with a mild hearing loss (Margolis 2002).

OHCs are reported to be impaired by sound overexposure, ototoxic drugs (e.g., therapeutic antibiotics), infections (e.g., meningitis, mumps, materno-fetal infection), and anoxia (e.g., birth trauma), or to be partly missing in genetic hearing loss. OHC impairment results in a loss of sensitivity and frequency selectivity of the hearing organ (Liberman and Dodds 1984). OAEs, as a by-product of cochlear nonlinear sound amplification, then appear with reduced amplitude or disappear (Mills and Rubel 1994). Even minute changes in the functioning of OHCs, caused, for example, by low-level noise exposure (Skellet et al. 1996), increased body temperature due to fever (O'Brien 1994), administration of salicylate (McFadden and Plattsmier 1984), or alteration of body posture (Frank et al. 2000) are known to affect the amplitude of OAEs.

There are four main OAE applications in clinical diagnostics: (1) newborn hearing screening, (2) proof of a cochlear hearing-loss along with tympanometry and auditory brainstem responses, (3) quantitative evaluation of hearing loss and recruitment, and (4) detecting beginning cochlear impairment during noise exposure or ototoxic drug administration. Providing hearing-aid fitting parameters and assessment of cochlear vulnerability to acoustic overexposure are potential applications in the future.

The primary purpose of this chapter is to show how OAEs work in a clinical context and what they are able to achieve in clinical diagnostics.

2. Capability and Limitation of OAE Measures

Whereas TEOAEs more qualitatively assess cochlear function, DPOAEs provide quantitative information about the range and operational characteristics of the CA, i.e., sensitivity, compression, and frequency selectivity. There are several DPOAE measures shown in Figure 13.2 that are used for assessing the functioning of the CA:

1. DPOAE-grams (Fig. 13.2A) plot the DPOAE level L_{dp} as a function of f_2 (the main DPOAE generation site) for a selected combination of the primary-tone levels L_1 and L_2. It should be emphasized that DPOAE-grams reflect CA sensitivity best when recorded at close-to-threshold stimulus levels (Janssen et al. 1998; Kummer et al. 1998; Dorn et al. 2001). DPOAE-grams when recorded with narrow frequency spacing of f_2 (DPOAE fine structure) may give information about the fine structure of the behavioral pure-tone threshold. However, due to the superposition of the second DPOAE source (He and Schmiedt 1993, 1996, 1997; Shera and Guinan 1999); the correlation between the two measures is not clear. In normal hearing (normal CA), DPOAE-grams are close to each other at high and more separated at low stimulus levels, reflecting cochlear nonlinear sound processing. In cochlear hearing loss (impaired CA), DPOAE-grams are more separated even at high stimulus

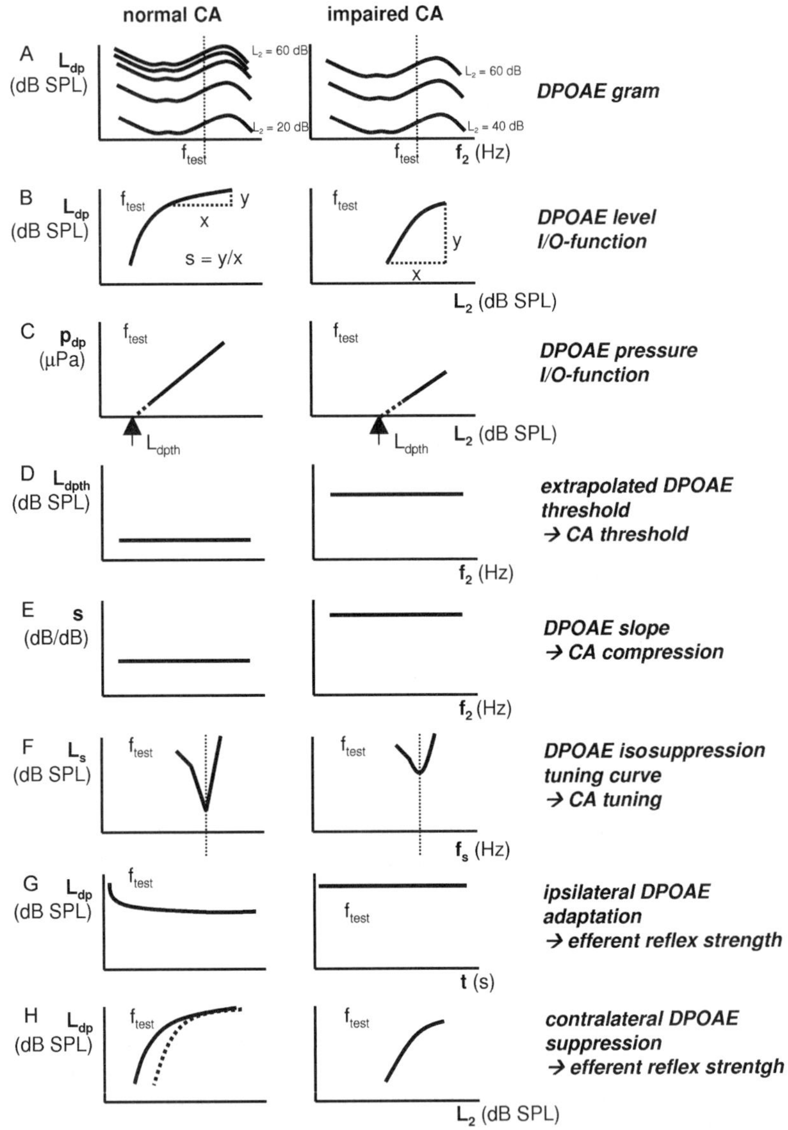

FIGURE 13.2. Schematic presentation of commonly used graphic renditions for DPOAE data assessment. All plots are shown for a normally hearing subject at *left* (normal CA) and a hearing-impaired subject at *right* (impaired CA). The term commonly used for the respective DPOAE measurement plot is given on the *right-hand side* of each row. The physiological parameters that can be derived from the DPOAE measures are indicated by an *arrow* at the respective term. For more detailed explanation of these plots, see the text.

levels, revealing loss of CA compression (Janssen et al. 1998; Kummer et al. 1998; Neely et al. 2003).

2. DPOAE level I/O functions (Fig. 13.2B) plot the DPOAE level L_{dp} as a function of primary-tone level L_2 for a selected f_2 (test frequency f_{test}) and thus reflect CA dynamics at the f_2 place in the cochlea (Dorn et al. 2001). In normal hearing, in response to low-level stimuli, DPOAE level I/O functions exhibit steep slopes, while at high stimulus levels, slopes decrease, thus mirroring the strong amplification at low and decreasing amplification (saturation) at moderate sound levels. However, this is true only when a stimulus setting is used, which accounts for the different compression of the primary tones at the f_2 place (Kummer et al. 2000).
3. DPOAE pressure I/O functions (Fig. 13.2C) plot the DPOAE pressure p_{dp} (instead of the DPOAE level L_{dp}) as a function of the primary-tone level L_2. Due to the logarithmic dependency of the DPOAE level on the primary-tone level, there is a linear dependency between DPOAE pressure p_{dp} and primary-tone level L_2 (Boege and Janssen 2002). Thus, DPOAE data can easily be fitted by linear regression analysis. The intersection point of the linear regression line with the L_2-axis at $p_{dp} = 0$ Pa can then serve as an estimate of the stimulus level at the DPOAE threshold, that is, L_{dpth} (Boege and Janssen 2002).
4. Estimated threshold level L_{dpth} (Fig. 13.2D) plotted across frequency f_2 provides a measure for estimating CA threshold at the f_2-place in the cochlea. Note that changes in DPOAE level with frequency do not have an influence on L_{dpth} if the separation of the DPOAE-grams is independent of frequency.
5. DPOAE slope (Fig. 13.2E), calculated from DPOAE level I/O functions (e.g., between stimulus levels L_2 of 40 and 60 dB SPL), indicates CA compression. When plotted across frequency, a slope profile can be established. In ears with cochlear hearing loss, the slope s of the DPOAE level I/O function increases with increasing hearing loss, indicating loss of CA compression (Janssen et al. 1998; Kummer et al. 1998).
6. DPOAE isosuppression tuning curves (Fig. 13.2F) mirror CA frequency selectivity at the f_2 place in the cochlea. They plot the level L_s of an ipsilateral suppressor tone at which the DPOAE level L_{dp} is decreased by a particular amount (e.g., 3 dB) across the frequency of the suppressor tone f_s. DPOAE isosuppression tuning curves exhibit the V-shape of neural tuning curves. In a cochlear hearing loss, the tip of the tuning curve disappears, revealing loss of both CA sensitivity and CA frequency selectivity (Kummer et al. 1995; Abdala et al. 1996; Abdala 1998; Gorga et al. 2003b).
7. Ipsilateral DPOAE adaptation graphs (Fig. 13.2G) plot the DPOAE level L_{dp} as a function of time at a selected f_2 beginning at the stimulus onset. The reduced DPOAE level at steady state indicates the reduction in OHC motility as controlled by medial olivocochlear (MOC) efferents and can serve as a measure for evaluating the reflex strength of the crossed MOC efferents (Liberman et al. 1996; Agrama et al. 1998; Kim et al. 2001; Bassim et al. 2003; Meinke et al. 2005).

8. Contralateral DPOAE suppression (Fig. 13.2H) is the difference between the DPOAE level L_{dp} measured in the absence and in the presence of a contralateral acoustic stimulus. White noise yields the most DPOAE suppression. The difference in DPOAE level increases with decreasing primary-tone level. Thus, the difference at the lowest L_2 reflects the reflex strength of the uncrossed MOC efferents best (Collet et al. 1990; Puel and Rebillard 1990; Moulin et al. 1993; Williams and Brown 1995; Puria et al. 1996; Maison et al. 2000; Janssen et al. 2003).

There are some limitations of DPOAEs in clinical applications. First, electric microphone noise, physiological noise (breathing, blood flow) and external acoustic noise do not allow OAE measurements at very low stimulus levels. Especially below 0.5 kHz, reliable OAE measurements are not possible even at high stimulus levels. Second, because of the limited frequency range of the sound probe's electroacoustic transducers, high-frequency OAE measurements are difficult without using specialized devices. Third, standing waves in the outer ear canal make a defined stimulus setting difficult to obtain. Fourth, besides the main DPOAE source at f_2, a secondary DPOAE source is present at the $2f_1 - f_2$ place, which interacts with the main source constructively or destructively at the f_2 place (Whitehead et al. 1992; Brown et al. 1996; Shera and Guinan 1999). Therefore, the DPOAE does not exactly reflect OHC function at the f_2 place. When all these factors are considered, it is easy to see how OAE measures can be misinterpreted.

Examples of current clinical applications of the DPOAE measures will be presented below. Also, the relationship between DPOAE and behavioral measures (e.g., hearing threshold, cochlear tuning, and loudness) and potential clinical applications of the DPOAE measures for providing hearing-aid fitting parameters or for predicting cochlear vulnerability to acoustic overexposure will be discussed.

3. Technical Aspects

3.1 Stimulus Calibration

Stimulus calibration is important to ensure proper data interpretation and comparability. There is a serious problem with stimulus calibration, because, due to standing waves in the ear canal, the sound pressure level at the eardrum cannot be accurately set from measurements at the tip of the sound probe.

The most commonly used calibration method is the in-the-ear calibration based on the measurement of the sound-pressure level at the ear probe microphone for constant voltage at the loudspeaker (Whitehead et al. 1995c). However, the ear probe microphone is located in the outer ear canal, while the relevant magnitude for the generation of OAEs is the actual sound-pressure level at the eardrum. Thus, dependent on ear canal length and middle-ear impedance, there is a frequency-dependent deviance of unknown quantity between the nominal sound-pressure level at the tip of the ear probe and the actual sound-pressure level at the eardrum due to standing-wave effects (Siegel 1994). This deviation is usually highest at frequencies corresponding to $d = \lambda/4$ and $\lambda/2$ (d: distance between ear probe and eardrum). Thus, problems become serious around 3 kHz and above 6 kHz in adults, but are less important in newborns and infants due to the smaller ear canal length (Keefe et al. 1993). Calibration errors may cause a change in the shape and thus the compression of DPOAE I/O functions. Further improvement of ear probe calibration is therefore necessary to enhance the accuracy of clinically relevant measures deduced from DPOAE recordings such as, for example, estimated CA threshold and compression.

In testing TEOAEs, calibration errors have less impact, because the wide-band stimulus is not influenced that strongly by standing-wave problems. Apart from this, TEOAEs are usually stimulated with a click level that is relatively high, that is, where cochlear compression already saturates cochlear motion. As opposed to DPOAEs, no level ratio between primary tones needs to be fulfilled. This results in TEOAEs being less susceptible to stimulus calibration errors.

3.2 Stimulus Setting

TEOAE responses can be evoked by two types of stimulus trains: (1) by a set of four clicks of equal magnitude (referred to as the linear protocol) or (2) by three clicks of positive polarity followed by a fourth click of an inverse polarity with a relative magnitude of 9.5 dB higher than the corresponding positive clicks (referred to as the nonlinear protocol) (Kemp et al. 1986; Bray 1989). Under the hypothesis that the TEOAE recordings originate from saturated cochlear generators, it is assumed that the nonlinear protocol removes stimulus artifacts of a linear nature, i.e., the stimulus itself, because sound signals increase linearly with the stimulus level, while sound signals increasing nonlinearly with the stimulus level, i.e., the emission emerging from the nonlinear operation of OHCs, remain. It is generally accepted that the nonlinear protocol is a practical compromise to maximize the reliability of a TEOAE recording (Kemp et al. 1990a,b; Grandori and Ravazzani 1993; von Specht et al. 2001; Hatzopoulos et al. 2003).

When using a primary-tone setting, which accounts for the different compression of the primary-tone traveling waves at the f_2 place (Whitehead et al. 1995a,b; Kummer et al. 2000; Boege and Janssen 2002), the DPOAE I/O function reflects compressive nonlinear sound processing (see Fig. 13.3A) known from direct measurements of basilar membrane displacements (Ruggero et al. 1997). Due to the steep slope of the traveling wave toward the cochlear apex, the maximum interaction site is close to the f_2 place in the cochlea, as indicated in Figure 13.3C. Thus, OHCs at the f_2 place contribute most to DPOAE generation. The number of OHCs contributing to DPOAE generation depends on the size of the overlapping region, which is determined by the primary-tone levels L_1 and L_2, respectively, and the frequency ratio f_2/f_1 of the primary tones f_1 and f_2. To preserve optimum overlap of the primary-tone traveling waves, the primary-tone

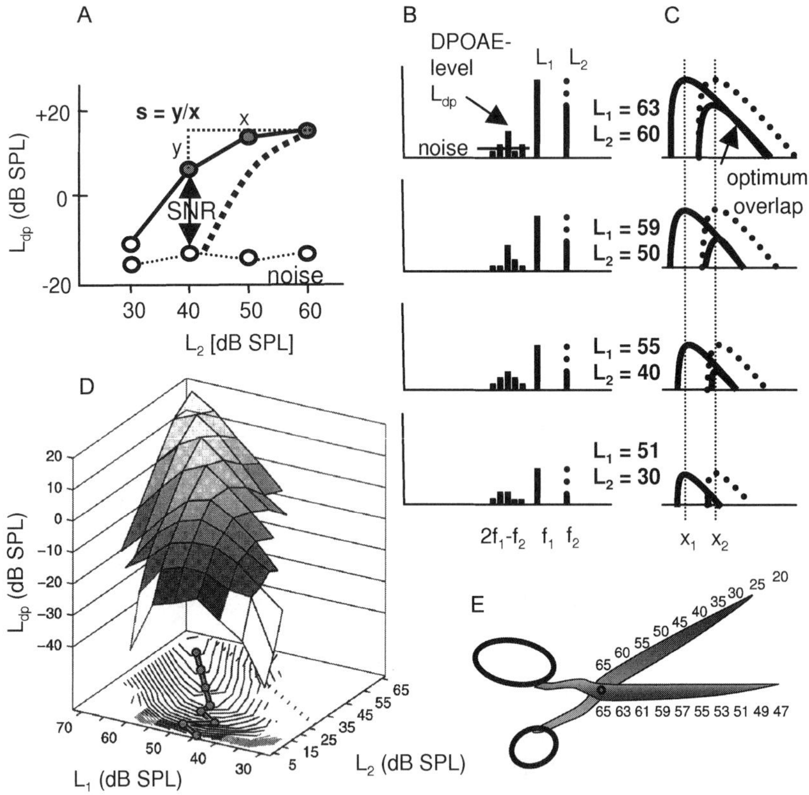

FIGURE 13.3. Schematic description of the influence of the primary-tone setting on the generation of DPOAE I/O functions. (**A**) DPOAE I/O functions elicited either with the scissor paradigm $L_1 = 0.4L_2 + 39$ dB SPL (*thick solid line*) or the $L_1 = L_2$ paradigm (*thick dotted line*). The thin dotted line along the abscissa represents the noise floor. (**B**) Spectral functions of the recorded response signal during two-tone stimulation. The plots show the DPOAE level L_{dp} and the noise floor together with the spectral lines of L_1 and L_2. The L_2 spectral line at f_2 is set either according to the scissor paradigm equation (*solid line*) or the $L_1 = L_2$ paradigm (*dotted line*). (**C**) Corresponding envelopes of the traveling waves of L_1 and L_2. (**D**) Typical DPOAE data recorded at $f_2 = 3955$ Hz from a normally hearing subject. L_{dp} is plotted against L_1 and L_2. The shaded area indicates valid L_{dp} with an SNR exceeding 6 dB. Contours of the data are projected into the L_1, L_2 stimulus level plane. The circles and line indicate optimal L_1 levels, where a local maximum of the DPOAE level occurred at a given L_2. This line continuously deviates with decreasing L_2 from the diagonal dashed line that indicates $L_1 = L_2$. (**D** adapted from Kummer et al. 2000.) (**E**) Graphical presentation of the "scissor" paradigm ($L_1 = 0.4L_2 + 39$ dB SPL) primary-tone levels. (Reprinted from Kummer et al. 2000, p. 49, Copyright 2000, with permission from Elsevier.)

level difference has to be increased with decreasing stimulus level, resulting in a $L_1|L_2$ setting described by $L_1 = 0.4L_2 + 39$ (Kummer et al. 2000). This "scissor" $L_1|L_2$ setting (Fig. 13.3E) can be derived from findings on the influence of the $L_1|L_2$ setting on the DPOAE level L_{dp}, as shown in Figure 13.3D. From the 3D plot shown here, it can be seen that it is not the $L_1 = L_2$ setting, but the "scissor" $L_1|L_2$ setting that yields compressive DPOAE growth reflecting nonlinear CA sound processing.

3.3 OAE Recording

OAEs are low-level sound emissions occurring with sound pressure levels from a maximum amounting to about 20 dB SPL down to the limiting noise-floor level. Therefore, the recording of OAEs requires the use of a highly sensitive low-noise microphone. Loudspeakers need to exhibit a low distortion factor to minimize technical distortion. For DPOAE recording, separate loudspeakers are commonly used for each primary tone in order to exclude technically generated distortion components. Unfortunately, nowadays, manufacturers do not provide sound probe loudspeakers that are able to apply sound-pressure levels exceeding 70 dB SPL without distortion. Especially for assessing sound-conductive hearing loss, where both the stimulus and the response are attenuated, a higher stimulus level would be desirable.

Both the microphone and the loudspeakers of the ear probe need to be miniaturized so that the ear probe is small enough (especially in newborns) to be placed inside the ear canal. A tight fit of the probe is essential for OAE recording. If there is leakage between the ear-tip and ear canal, low-frequency sound components cannot be recorded properly. Furthermore, the closure of the ear canal by the ear-tip excludes external sounds. In addition, it is essential to make sure that the ear canal is clean and that the ear probe ports are not blocked with cerumen. For clinical practice, it is important that ear probes be designed to offer easy access for cleaning ports or replacing clogged ear-tips.

To achieve low noise-floor levels, OAE measurements should be conducted in a sound-attenuating booth or any other quiet environment without continuous background noise. The measuring device should be portable, too, for bedside use in newborn hearing screening. Automated measuring and evaluation procedures, which are essential prerequisites for newborn hearing screening, guarantee test consistency and simplify the interpretation of OAE recordings in clinical diagnostics.

Determining the OAE spectrum by means of fast Fourier transform (FFT) computations from the time-domain signal allows for automatic evaluation of OAE signals. To improve measurement performance, that is, to minimize the influence of unwanted external signals, algorithms for noise reduction and artifact rejection are applied. In addition, the noise-floor level is reduced by time-domain averaging of the recorded signal. For example, in averaging 50 buffers recorded with a sample rate of 48 kHz and a buffer size of 4096 sample points, the noise-floor level decreases down to about 0 dB SPL at 500 Hz and about −30 dB

SPL at 4 kHz. The measuring time then amounts to 4 s. Averaging 200 sweeps of 20 ms each (which is common for TEOAE recording), the measuring time to get a valid TEOAE response also amounts to about 4 s. Recall that the TEOAE response "sees" almost the whole cochlea, whereas the DPOAE response reflects only a limited region of the cochlea. Therefore, TEOAEs give a rapid overview of cochlear function, whereas DPOAEs provide quantitative information about sound processing at distinct cochlear places.

The noise-floor level is usually higher at low frequencies due to microphone properties and low-frequency body sounds such as breathing. Artifact rejection can be performed by elimination of high noise level buffers sensitive to, for example, breathing, swallowing, or head movements of the subject. Furthermore, when recording TEOAEs, stimulus artifacts may generate signals being in phase in two averaging buffers, resulting in a pseudoresponse of high reproducibility. By means of windowing functions, the stimulus artifact can be excluded so that the reproducibility of the overall signal is restricted to the signal section of interest (Kemp et al. 1990a,b). Since the stimulus artifact always appears in the early recording period, the high-frequency TEOAE components get lost as a result of the windowing procedure.

There are several objective methods for separating the emission signal from the background noise and for automatically evaluating the validity of a recorded emission. For both TEOAEs and DPOAEs, there are approaches based on quantitative as well as statistical criteria. In the case of TEOAEs, clinical OAE systems yield mainly three signal evaluation approaches. The first method is based on the calculation of the buffer correlation of the time-domain averaged signals between two separate signal buffers *a* and *b* (Kemp et al. 1990a). If the two buffers are completely identical, the correlation coefficient is 1, and thus the reproducibility is 100 %. A signal is commonly accepted as valid for a reproducibility exceeding a minimum of 60 %. The second method relies on the computation of the spectral power ratio of the sum and the difference of the two signal buffers. This measure is mostly denoted as the signal-to-noise ratio (SNR). The pass criterion for a valid signal is typically set to an SNR of 6 dB. The third signal validation procedure is based on a binomial statistical test, which determines the statistical probability that an emission has been recorded. Binomial statistics reduce the recorded signal to binary events, and use knowledge on the expected distribution of these events (binomial distribution) (Giebel 2001). The pass criterion comprises the detection of eight significant peaks in the binomial waveform with alternating sign, with each peak corresponding to a significance of 99.7 %. In the case of DPOAEs, commercially available systems perform mainly two different data validation procedures. The first method is based on the calculation of the noise-floor level by averaging the levels of several adjoining frequency components around the DPOAE frequency component, with the SNR being indicated by the difference between the emission level and the noise-floor level. Here also, the SNR criterion is usually set to 6 dB. The second data validation procedure is based on a phase statistics method, which checks the coupling of the DPOAE component phase to the phase of the primary tones. The phase statistics average normalized phase vectors of the signal received at the known DPOAE frequency. Like the binomial statistics, the vector sum can be scaled in probability terms, providing defined and very high sensitivities. A typical level of significance exceeds 99 % per single frequency test.

4. OAE Characteristics

4.1 OAE Level and OAE Reproducibility

TEOAEs exhibit (like DPOAEs) a nonlinear relationship between stimulus and response, revealing the compressive sound processing of OHCs (Kemp 1979; Kemp and Chum 1980; Bray 1989). The measured OAE presssure in the outer ear canal depends on the ear canal volume. Because of the smaller ear canal volume, OAE pressure in newborns is higher compared to that of adults (Norton 1992; Lasky 1998a,b; Abdala 2000). As a consequence, OAEs in newborns are easier to measure.

The interindividual variance of the OAE level is high (standard deviation > 10 dB; Kemp et al. 1986; Probst et al. 1987; Bonfils and Uziel 1989; Smurzynski and Kim 1992). The intraindividual variance of the OAE level is much smaller (standard deviation < 2 dB) (Johnsen and Elberling 1982a,b; Harris et al. 1991). Recently, Janssen et al. (2005a) showed that repetitive DPOAE measurements with unchanged sound-probe position exhibited an exponentially increasing standard deviation of DPOAE level with increasing SNR. For example, at an SNR of 10 dB, the standard deviation amounts to 1.8 dB, at an SNR of 20 dB to 0.7 dB, and at an SNR of 40 dB to 0.1 dB. This means that the higher the SNR, the higher the reliability of the DPOAE measurement. This finding is important with respect to the evaluation of small DPOAE changes, which, for example, is the case during contralateral DPOAE suppression or ipsilateral DPOAE adaptation. For clinical practice, however, repetitive OAE measurements with changed sound-probe position are relevant. For example, the standard deviation when sound-probe position was changed amounted to about 1.6 dB (Müller et al. 2005).

DPOAEs are supposed to primarily reflect OHC activity at the f_2 place. However, there is evidence that DPOAEs are generated by two distinct cochlear sources (Whitehead et al. 1992; Brown et al. 1996; Shera and Guinan 1999). As already mentioned, the first source, the effect of which is actually intended to be measured, is located at the region of overlap of the traveling waves of the two primary tones near the f_2 place and is due to intermodulation distortion (see Figs. 13.1 and 13.3). The second source, which is unintentionally adding to the first source emission, is located at the characteristic frequency place of the emission at $2f_1 - f_2$ and is due to reflection of energy that has traveled apically from the overlap region near f_2 (see Fig. 13.1). Thus, energy from both interacting sources yield the composite DPOAE signal that is actually recorded in the outer ear canal. The influence of the second DPOAE source may be observed when monitoring the DPOAE level across frequency with narrow frequency spacing of f_2 (DPOAE fine structure). Because of either destructive or constructive superposition of the two DPOAE sources across frequency, a number of investigators have described a pattern of dips and peaks in the DPOAE fine structure in normal hearing subjects (He and Schmiedt 1993, 1996, 1997; Talmadge et al. 1999). However, for clinical evaluation of DPOAE I/O functions, the interference of the second DPOAE source makes the interpretability and the accuracy of deduced measures, such as DPOAE threshold and compression, difficult. There have been attempts to minimize the influence of the second source by presenting an additional suppressor tone near $2f_1 - f_2$ (Heitmann et al. 1998), or by using a time-windowing procedure, which is able to separate the two sources (Mauermann and Kollmeier 2004).

In newborns, click-evoked TEOAEs are larger when elicited in the right ear, whereas tone-evoked DPOAEs are larger in the left ear (Sininger and Cone-Wesson 2004). This stimulus-guided asymmetrical nature of OAEs in newborns suggests that at the early stages of auditory system development, initial sound processing in the auditory system at the level of the cochlea may serve to facilitate later development of hemispheric specialization for sound processing.

4.2 Age-Related Changes

DPOAEs of term-born neonates appear to a large extent to be adult-like. This is true for various parameters such as the shape of I/O functions, DPOAE phase delays (Eggermont et al. 1996), suppression tuning, and DPOAE growth (Abdala 1998, 2001). Especially at lower frequencies, the DPOAE levels in adults are reported to be lower than those in newborns (Lasky 1998a). DPOAE I/O functions also seem to be more linear in adults, with saturation at higher primary-tone levels (Lasky 1998b). However, premature neonates show non-adult-like behavior of DPOAE suppression growth and tuning (Abdala 2000, 2003; Abdala and Chatterjee 2003). These differences lead to the assumption that the CA matures close to term birth. Moreover, Mills and Rubel (1998) showed that the properties of DPOAE I/O functions measured in gerbils aged from 15 days after birth to adult suggested that CA gain increased with age at very high frequencies (> 40 kHz). Conclusions from these findings should not be drawn for cochlear maturation in human neonates, because of the high-frequency hearing capabilities of these experimental species. Although differences in cochlear mechanics may explain developmental differences in DPOAEs, developmental differences in the resonance characteristics of the outer and middle ear and/or calibration errors due to different ear-canal lengths may also be involved (Lasky 1998b).

Etiology and mechanisms of age-related hearing loss (presbyacusis) are conjectural at present, because age-related degeneration of the cochlea and/or nerve fibers cannot be differentiated from effects of extrinsic damage (noise, ototoxic drugs) and intrinsic disorders (e.g., systemic diseases). Findings based on the comparison of DPOAE I/O functions and audiometric thresholds support the hypothesis that strial dysfunction may be a substantive factor in cochlear aging (Gates et al. 2002).

4.3 DPOAE Suppression and Adaptation

The effect of contralateral (CAS) and ipsilateral acoustic stimulation (IAS) on DPOAEs has been extensively studied in animals and humans. The observed change in DPOAE level due to CAS was found to be more commonly a decreasing effect (suppression) than an increasing effect (enhancement) and amounted to only a few decibels, depending on the type of the contralateral stimulus and the primary-tone level setting (e.g., Collet et al. 1990; Veuillet et al. 1991). In general, with increasing contralateral stimulus bandwidth and with decreasing primary-tone level, the suppression effect increased (Chéry-Croze et al. 1993; Moulin et al. 1993; Janssen et al. 2003).

Ipsilateral DPOAE adaptation is reported to exhibit a decrease in the steady-state DPOAE level of up to 4 dB in cats (Liberman et al. 1996). In contrast, DPOAE adaptation was found on average to be lower than 1 dB in human subjects (Agrama et al. 1998; Kim et al. 2001; Bassim et al. 2003).

4.4 Middle-Ear Influence

Due to different middle-ear (linear) and cochlear (nonlinear) sound processing, sound-conductive and cochlear hearing losses cause different DPOAE growth behavior (Topolska et al. 2000; Gehr et al. 2004; Janssen et al. 2005b). A simple model, illustrated in Figure 13.4, that simulates middle-ear and cochlear hearing loss of different degrees, shows the resulting changes in DPOAE level and slope of DPOAE I/O functions in either condition. In the model, the DPOAE behavior in sound-conductive hearing loss (curves A, B, and C) is simulated by shifting the normal-hearing reference DPOAE I/O function on the L_2-axis (representing the damping of the primary-tone level) as well as on the L_{dp}-axis (representing the damping of the DPOAE level). The shift on the L_2-axis results in a change of the estimated DPOAE threshold level corresponding to a hearing loss of 10, 20, and 30 dB HL and in an increase in slope at a fixed L_2. The shift on the L_{dp}-axis also results in a significant decrease of L_{dp} at high primary-tone levels. For simulating the DPOAE behavior in cochlear hearing loss, I/O functions are adopted from real DPOAE I/O functions corresponding to a hearing loss of 10, 20, and 30 dB HL (curves *a*, *b*, and *c*). The slope in this case reflects the compression loss of the CA that increases with increasing hearing loss. From the model, it can be shown that in sound-conductive hearing loss, a considerable reduction in the absolute DPOAE level occurred and hence DPOAEs cannot be measured at a hearing loss exceeding 20 dB HL. When comparing the slope in both sound-conductive and cochlear hearing loss at a hearing loss of up to 20 dB HL, the magnitude of the slope is similar in both conditions. A differentiation between sound-conductive and cochlear hearing loss solely by means of the slope is therefore not possible. However, the difference between the estimated DPOAE threshold L_{dpth} and the DPOAE detection threshold (which is the lowest primary-tone level at which a valid DPOAE is measurable) is extremely different in both conditions and thus could serve as a measure for distinguishing between sound-conductive and cochlear hearing loss.

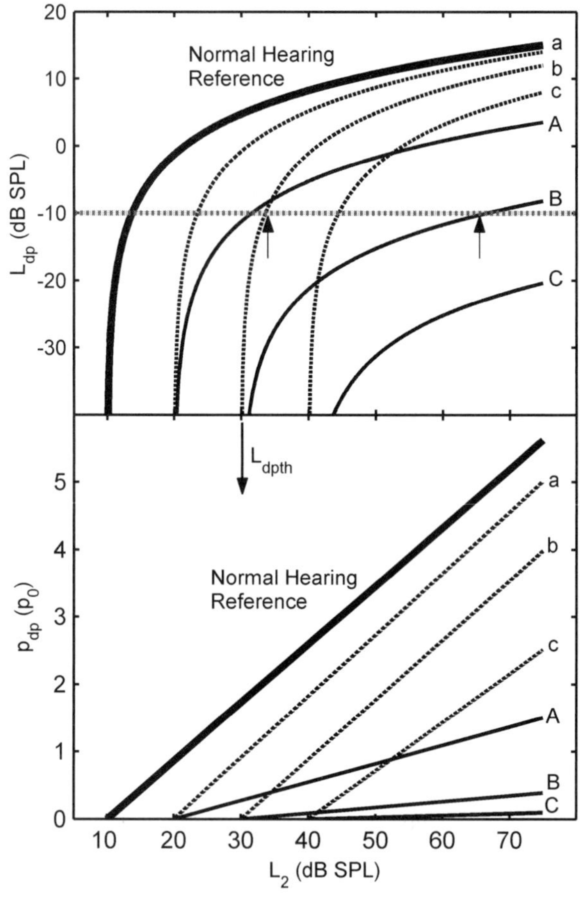

FIGURE 13.4. DPOAE model functions simulating cochlear and middle-ear disorders shown in logarithmic (*top*) and semilogarithmic scale (*bottom*). The *thick solid line* represents a normally hearing reference subject with an estimated DPOAE threshold level of 10 dB SPL (= 0 dB HL). *Solid curves A, B, C* represent 10, 20, 30 dB HL sound-conductive hearing loss. Dashed curves *a, b, c* represent 10, 20, 30 dB HL cochlear hearing loss. The horizontal line in the upper panel indicates the DPOAE detection threshold (noise level + 6 dB), which is typical for measurements in neonates under screening conditions at $f_2 = 4$ kHz. Thin upward arrows indicate the different DPOAE detection thresholds, the *single downward arrow* indicates the common estimated hearing threshold in the case of the 20 dB HL cochlear and sound-conductive hearing loss, respectively. (Reprinted from Janssen et al. 2005, p. 2974, Copyright 2005, with permission from Acoustical Society of America.)

4.5 Relationship Between DPOAE and Behavioral Measures

The relationship between OAE level and auditory threshold—or rather the lack of it—is strongly debated. Earlier, it was common to define confidence limits to determine the degree of certainty with which any measured response could be assigned to either normal or impaired hearing (Gorga et al. 1996, 2000a), or to define a "DPOAE detection threshold" as the stimulus level at which the response equaled the noise present in the instrument (Dorn et al. 2001). However, since the noise is of technical origin (e.g., microphone noise), the threshold evaluated in this way does not match the behavioral threshold. A more relevant measure is the intersection point between the extrapolated DPOAE I/O function and the primary-tone level axis at which the response's sound pressure is zero and hence at which OHCs are inactive (Boege and Janssen 2002; Gorga et al. 2003a). A linear dependency between the DPOAE sound-pressure and the primary-tone sound-pressure level is present (Boege and Janssen 2002) when the "scissor" paradigm is used for eliciting DPOAEs (Kummer et al. 2000). Because of the linear dependency, DPOAE data can be easily fitted by linear regression analysis (see Fig. 13.5) in a semilogarithmic plot, where the intersection point of the regression line with the L_2 primary-tone level axis at $p_{dp} = 0$ Pa can thus serve as an estimate of the DPOAE threshold. The estimated DPOAE threshold L_{dpth} is independent of noise and seems to be more closely related to behavioral threshold than the DPOAE detection threshold (Boege and Janssen 2002; Gorga et al. 2003a).

The DPOAE growth rate steepens when the auditory threshold is elevated (Janssen et al. 1998; Kummer et al. 1998; Boege and Janssen 2002; Gorga et al. 2003a; Neely et al. 2003) and differs significantly between hearing-loss classes (their width being 10 dB) (Janssen et al. 2005b). The slope of DPOAE I/O functions is reported to be related to the slope of the loudness functions (Neely et al. 2003; Müller and Janssen 2004). Thus, the slope s of DPOAE I/O functions is suggested to allow a quantitative assessment of CA compression and hence provide an objective recruitment test.

DPOAE I/O functions in newborns within the early postnatal period (mean age = 3.2 days), in follow-up newborns (4 weeks later), in normally hearing adults, and in patients suffering from cochlear hearing loss exhibit a different DPOAE threshold L_{dpth} and slope s of the DPOAE I/O function, as illustrated in Figure 13.5. In newborns and adults, Figure 13.5C shows that L_{dpth} across frequency is similar to that known from behavioral pure-tone thresholds, being lowest at $f_2 = 4$ kHz. Figure 13.5D indicates that the slope s also varies with frequency, being lowest around 4 kHz. Thus, s and L_{dpth} share similar features. It has been shown that the estimated threshold for normally hearing subjects is shifted by about 20 dB to higher levels relative to the auditory threshold in quiet. Also, L_{dpth} in newborns tends to be higher compared to that in older newborns and adults. Other work in our laboratory showed that s was higher in newborns compared to that in follow-up newborns and adults. Further, in cochlear-hearing-loss ears, L_{dpth} and s increased with increasing cochlear hearing loss. Moreover, L_{dpth} and s of newborns were similar to that of the cochlear-hearing-loss ears with hearing losses of 15–20 dB HL. In contrast, in cochlear-hearing-loss ears exhibiting hearing losses of 35–40 dB HL, L_{dpth} and s were considerably higher. Whether the higher L_{dpth} and the higher s in newborns is

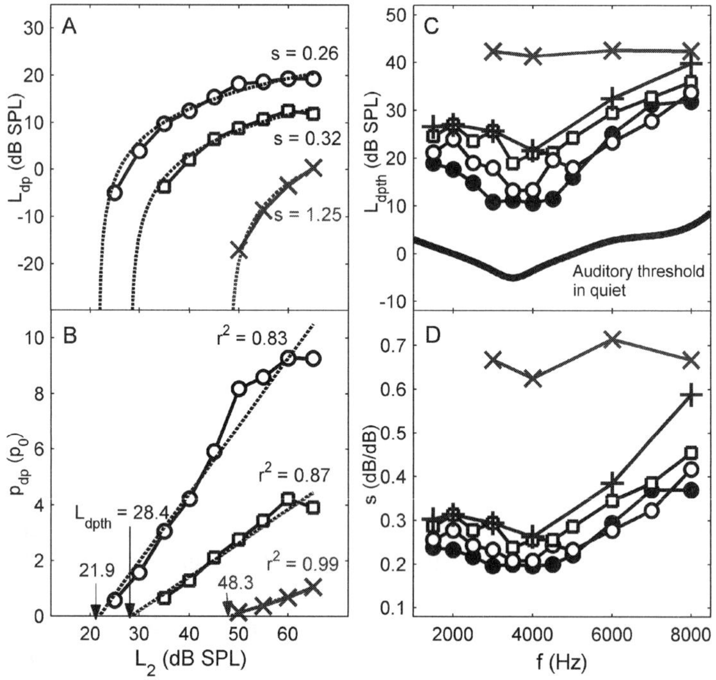

FIGURE 13.5. DPOAE I/O functions are shown on a logarithmic (**A**) and semilogarithmic (**B**) scale. In addition, the DPOAE threshold (**C**) and DPOAE slope (**D**) are shown for different subject groups. (**A, B**) Data from one neonate at $f_2 = 4$ kHz measured directly after birth (*open squares*) and about 4 weeks after birth (*open circles*). For comparison, data for an adult with a hearing loss of 40 dB HL at 6 kHz are shown (*gray* x symbols). In **A**, the DPOAE slope s is given beside each curve. In **B**, the DPOAE threshold L_{dpth} is given for each data set. Further on, r^2 represents the coefficient of determination for each linear extrapolation line. (**C, D**) L_{dpth} and s for averaged data. *Filled black circles* in **C** and **D** represent data of normally hearing adult subjects (hearing loss at a maximum of 10 dB HL), *open black circles* indicate data from early measurements in neonates, and *open squares* represent data for follow-up measurements in neonates. *Gray crosses* and x *symbols* indicate data from adults with hearing losses of 15–20 dB HL, and of 35–40 dB HL, respectively. In **C**, for comparison, auditory thresholds in quiet (*bold line*) are also shown. (Adapted Janssen et al. 2005, p. 2972, Copyright 2005, with permission from Acoustical Society of America.)

due to cochlear maturation or transitory middle-ear dysfunction (e.g., due to the presence of amniotic fluid or Eustachian tube dysfunction) remains unknown. However, from the simple model presented above (see Section 4.4), it can be surmised that the observed compression loss in newborns is due to a change in the middle-ear transfer function (compare Figs. 13.4A,B to Fig. 13.5A,B).

Human DPOAE isosuppression tuning curves (STCs) share similar features to neural tuning curves of other mammals in that the slopes of the high- and low-frequency flanks and the Q_{10} dB value of the tuning curve are similar

(Kummer et al. 1995). Also, in patients with cochlear hearing loss, the tip of the DPOAE STCs disappeared, reflecting loss of frequency selectivity at the respective cochlear place (Gorga et al. 2003b), similar to that found in animals when OHCs are impaired by ototoxic drugs (e.g., Liberman and Dodds 1984). Up to now, DPOAE STCs have not been used in clinical practice.

5. Clinical Applications

Today, nearly 30 years after the discovery of OAEs, OAE measurements are a standard part of the audiometric diagnostic test battery. OAEs, as an epiphenomenon of hearing, are a means of acquiring noninvasively information about disorders of an essential element of sound processing, i.e., the CA, and hence allow assessment of loss of sensitivity, compression, and frequency-selectivity of the hearing organ. Like tympanometry, which examines the status of the middle ear, OAEs are a fast and easy-to-handle method for examining cochlear function objectively. In fact, OAEs extend the objective audiometric test battery by filling a gap that existed before their discovery, that is, they allow a direct examination of cochlear function. Subjective tests are only able to assess disorders of sound processing as a whole. Tympanometry, OAEs, and auditory brainstem responses (ABRs), in combination, allow for a differentiation between sound-conductive, cochlear, and neural hearing loss. Minute changes in hearing capability are detectable by OAEs. They are therefore a suitable means for detecting beginning impairment, for example, caused by noise exposure or ototoxic drugs, as well as for monitoring hearing recovery, for example, after a sudden hearing loss. Due to the simple handling and due to the fact that, contrary to ABRs, OAEs are fully developed at birth (Abdala 2000), OAEs are the first choice in newborn hearing screening.

5.1 Newborn Hearing Screening

Screening is a selection procedure used to decide whether further diagnostics are advised (Wilson 1968). Consequently, a screening decision is binary: "pass" (negative finding, no diagnostics necessary), or "refer" (positive finding, follow-up diagnostics advised). The requirements for screening are different from those for diagnostics. Because screening is performed in large populations, the devices used typically provide automatic evaluation rather than rely on an expert's judgment. Therefore, a modern screening device will not have a graphic output showing raw data that have to be interpreted, but will instead produce a binary screening result with a defined error probability (Joint Comittee on Infant Hearing 2000). This can best be achieved by making use of signal statistics. It is desirable that a screening result discriminates between groups with normal hearing (no treatment necessary) and impaired hearing (treatment advised). The common agreement describes a threshold elevation of more than 30 dB in the speech-frequency region between 500 to 4000 Hz as a finding that, in time, requires treatment. The discriminative power of a screening procedure is described by the term "validity," which includes two psychometric aspects: sensitivity (defined as the percentage of correctly classified impaired individuals) and specificity (defined as the percentage of correctly classified normally hearing individuals). International recommendations set minimum standards for screening validity. While sensitivity should be close to 100 %, specificity is required to be above 96 % in order to avoid an unnecessarily large number of follow-up diagnostics, thus minimizing both parental concerns for their potentially impaired child and the costs of testing (National Institutes of Health 1993; Grandori and Lutman 1999).

A screening test should be performed as quickly as possible. Therefore, the respective methods must avoid long preparation times, and the test should stop automatically if the desired quality of the result is achieved. Otoacoustic emissions (TEOAE and DPOAE) are widely regarded as being suitable for screening in newborns and infants, since they are not present in the case of OHC dysfunction (e.g., Kemp and Ryan 1991; Gorga et al. 2000b; Norton et al. 2000a,b). The premise for this approach is that inner-ear hearing loss always includes OHC damage or malfunction. Additionally, significant conductive losses also cause "refer" results under screening conditions, mainly due to the attenuation of an existing OAE signal.

A major disadvantage of using OAEs in screening protocols is a lower validity as compared to ABR methods (Barker et al. 2000; Norton et al. 2000a). This is especially true in populations with a high prevalence of slight threshold elevation due to a temporary sound-conductive hearing loss, as is found in term neonates in the first 36 hours of life because of Eustachian tube dysfunction or amniotic fluid in the tympanic cavity or due to a persisting sensory hearing loss in premature and neonatal intensive-care-unit infants. To maintain a high sensitivity, the specificity may be reduced dramatically, making a screening procedure inefficient. To avoid high referral rates, OAE referrals should be followed up with an ABR screening before diagnostic assessment, that is, two-stage screening (Rhodes et al. 1999; Norton et al. 2000b). This procedure has proved to be very efficient with respect to saving costs and must therefore be regarded as the state of the art in screening programs.

TEOAE recordings (time and frequency domain) from two infants with "pass" and from one infant with "refer" response at newborn hearing screening are shown in Figures 13.6E–J. TEOAEs that were recorded in one of the "pass" babies at the age of 3 months are presented in Figures 13.6E and F, while those from the other "pass" baby were recorded 4 hours after birth and are shown in Figure 13.6G and H. The recording immediately after birth exhibited low-level TEOAEs. Despite the low level and due to the high reproducebility (88 %) of the response, the newborn hearing test was negative ("pass"). The low emission level can be attributed to a transitory sound-conductive hearing loss due to residual amniotic fluid in the middle-ear cavity or Eustachian tube dysfunction. In cases of severe amniotic fluid and/or Eustachian tube dysfunction, the screening device produces a "refer" response, so that the screening test is often negative only at the

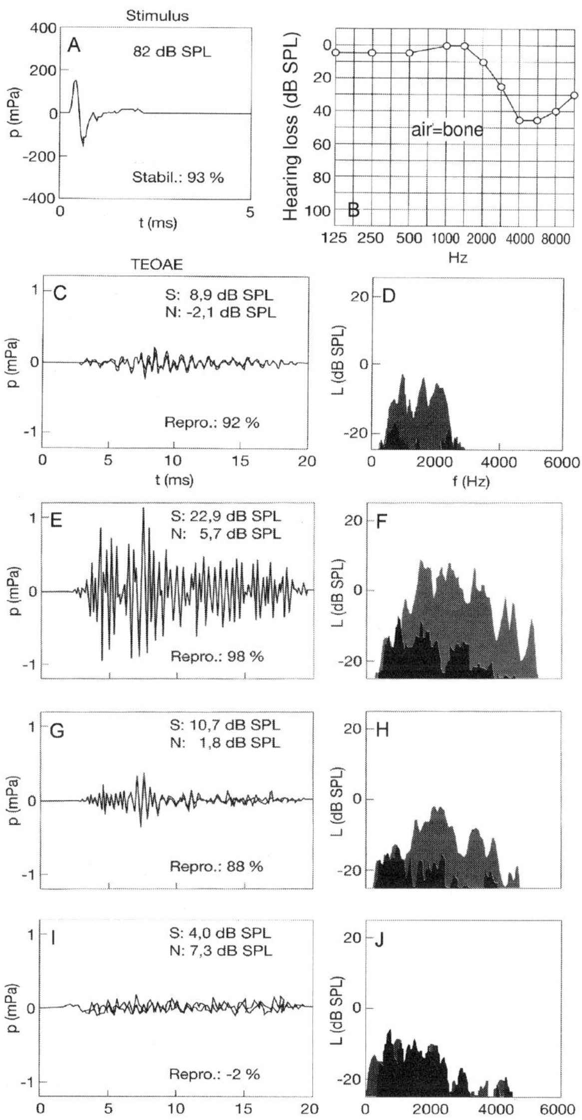

FIGURE 13.6. TEOAE patient case examples. (**A–D**) Data for an adult exhibiting a noise-induced cochlear hearing loss (see audiogram in **B**). (**A**) TEOAE stimulus in the time domain (stimulus stability: 93 %). (**C**) Corresponding TEOAE response in the time domain. The response signal level *S* and the noise-floor level *N* are given together with the response reproducibility (*Repro.*), which is derived from the buffer correlation (see Section 3.3). (**D**) TEOAE response in the frequency domain with the *black area* representing the stimulus and the *gray area* the response level. (**E/F**) TEOAE data in time and frequency domains, respectively, for a 3-month-old infant. (**G/H**) TEOAE data for a newborn (4 hours after birth). (**I/J**) TEOAE data for a 18-month-old infant. (Adapted from Janssen, 2000, pp. 95–96, Copyright 2001, with permission from Thieme.)

second screening stage, and hence no further audiological diagnostics are advised. To reduce the rate of false-positive screening responses at the first screening stage, additional measurement of ABRs (where only the stimulus and not the response is reduced by the sound-conductive hearing loss) or DPOAEs (which—compared to TEOAEs—are present at a higher hearing loss) is recommended.

Figures 13.6I and J show TEOAEs from an 18-month-old infant that exhibited a "refer" response during newborn hearing screening, indicating middle-ear or cochlear disorder. Tympanometry and ABRs revealed a sensorineural cochlear hearing loss of 50–60 dB HL. Due to the fact that TEOAEs were absent, OHC dysfunction must have been the cause of the hearing loss.

Today, there are hearing-screening instruments available that display more than a "refer" or a "pass." They show the calibration transfer function (see Figure 13.7A) and stop measuring if, for example, there is a leakage between the ear-tip and the sound probe. Due to sound leakage, low frequencies are

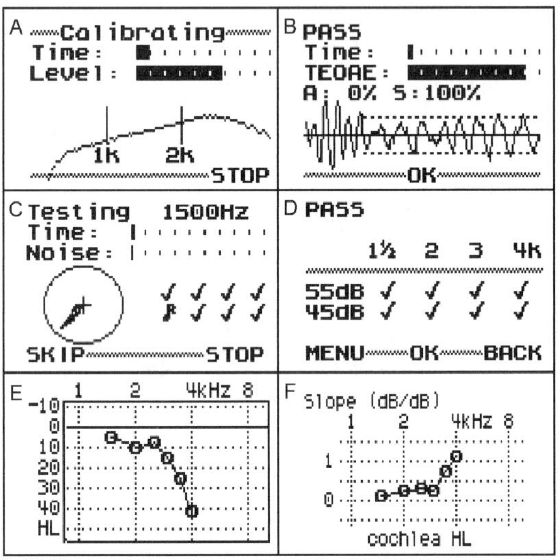

FIGURE 13.7. Screen shots from a newborn-hearing-screening device (Fischer-Zoth). (**A**) Calibration transfer function. The response level is shown as a horizontal bar. The time bar shows the time elapsed during calibration. A time-out occurs if within a given time, no valid calibration is possible. (**B**) Recorded TEOAE signal in the time domain. The resulting pass or refer decision and the elapsed time, the TEOAE signal level and the artifact rejection rate are also displayed. (**C**) DPOAE recording procedure. The picture shows the test frequency, the elapsed recording time, the current background noise level, and the phase statistics plot (see Section 3.3). (**D**) The graph displays the resulting DPOAE pass or refer decision on the basis of a multifrequency and multilevel test procedure. (**E/F**) Extended information about the estimated DPOAE threshold (**E**) and the DPOAE slope (**F**) derived from DPOAE I/O functions, shown for a cochlear-hearing-loss ear. On the basis of these DPOAE measures, a simple, preliminary diagnosis (here: "cochlea HL") is given.

missing in the transfer function, so that leakage can automatically be detected by the instrument. In this case, the user is informed by a relevant message being displayed on the screen (e.g., "leakage") and is asked to change the position of the sound probe. Also, the time function of the TEOAE response is shown (Figure 13.7B), from which high-frequency (basal OHC activity) and low-frequency components (apical OHC activity) can be traced.

DPOAE newborn hearing screening is usually performed in the mid-frequency region (e.g., between 1.5 and 4 kHz), as shown in Figure 13.7D. High-frequency testing is difficult, due to the standing-wave problem and the limitation of the electroacoustic transducer. Below 1.5 kHz, the SNR is insufficient, so that also low-frequency testing is unadvised. DPOAEs are elicited at different primary-tone levels (e.g., 55 and 45 dB SPL), so that a multifrequency and multilevel test procedure is provided. A rough estimation of hearing loss is possible using this procedure. In a new approach, DPOAE I/O functions are recorded in a wide primary-tone-level range (e.g., from 15 to 65 dB SPL) to get more-quantitative information about the hearing loss. Extrapolated DPOAE I/O functions are used to estimate the DPOAE threshold level (Boege and Janssen 2002) with which a DPOAE audiogram is established and displayed on the screen of the hearing-screening device, as shown in Figure 13.7E. Additionally, as illustrated in Figure 13.7F, the slope of the DPOAE I/O function is calculated to provide information about loss of CA compression at the test frequencies. This new technique may fill the gap between newborn hearing screening and audiological diagnostics, because it gives more-detailed information than either TEOAEs or ABRs.

5.2 Topological Diagnostics

A therapy for a hearing disorder can be developed only after knowing which stage of the auditory pathway is impaired. Psychoacoustic tests are able to differentiate between sound-conductive and sensorineural hearing loss by evaluating the difference between sound and bone-conductive pure-tone thresholds. The discrimination of a sensorineural hearing loss, that is, the differentiation between a sensory (cochlear) and a neural disorder with subjective testing, is unreliable, because the validity of these tests (Short Increment Sensitivity Index (SISI), Fowler, Carhart) is limited. Moreover, in uncooperative patients or infants, psychoacoustic tests cannot be performed. In such cases, only objective tests help in achieving the goal of determining end-organ integrity. Using tympanometry, OAEs, and ABRs, the site of the impairment in the auditory pathway can be ascertained with high reliability.

OAEs are used primarily as function detectors in clinical practice. As a general rule, if there is a suspicion of a hearing disorder, the measurement of OAEs should be used as the first audiological test. It is fast and helps to confirm normal middle-ear *and* cochlear function. This is the case if OAEs are present over a wide frequency range. If OAEs are absent, then middle-ear *or* cochlear (OHC) pathology is likely. OAEs then should be followed by tympanometry. If the tympanogram is normal and OAEs are absent, then a cochlear disorder is likely. Since OAEs directly reflect OHC dysfunction, they are therefore the method of choice in occupational medicine, where indisputable proof of a cochlear hearing loss is required. If the tympanogram is abnormal, a sound-conductive hearing loss is likely. If both the tympanogram and OAEs are normal, ABRs are needed to reveal whether there is a cochlear (inner hair cell) or neural pathology. For example, in auditory neuropathy, where synchronization of neural activity is malfunctioning (either due to inner hair cell synaptic or neural dysfunction), normal OAEs and abnormal ABRs occur (Starr et al. 1996; Doyle et al. 1998). OAEs are, of course, also a suitable means for testing individuals who are only simulating a hearing loss.

5.3 Quantitative Evaluation of Hearing Loss and Recruitment

Especially for hearing-aid adjustment in infants, a quantitative evaluation of the hearing loss and recruitment is necessary. When elicited by high stimulus levels (which is common in clinical practice), TEOAEs are absent at a cochlear hearing loss exceeding 20 dB HL, whereas DPOAEs are absent only at a cochlear hearing loss exceeding 40–50 dB HL. Thus, when TEOAEs *and* DPOAEs elicited at high stimulus levels are used, a rough estimate of the hearing loss is possible. For example, when TEOAEs are absent and DPOAEs are present, then the hearing loss is suggested to be not more than 20–40 dB HL.

In principle, both DPOAEs and TEOAEs allow acquisition of frequency-specific information about a hearing-loss problem. When stimulating the ear with a click or a tone burst, almost all OHCs along the cochlear partition, or a part of them (the site in the cochlea depending on the carrier frequency of the tone burst), are excited. Due to frequency dispersion in the cochlea, a specific component of the TEOAE response can be directly traced to a specific frequency component of the transient signal. As the basilar membrane at basal sites moves faster than at more apical sites, high-frequency TEOAE components stem from basal cochlear sites, whereas low-frequency TEOAE components come from more apical ones.

A patient case example presented in Figure 13.6 illustrates the frequency-specificity of TEOAEs. In a patient suffering from a high-frequency hearing loss (Fig. 13.6B), TEOAEs elicited by a broadband click (Fig. 13.6A) exhibit a TEOAE response in which the high-frequency components are missing in the time and spectral domains (Fig. 13.6C and D, respectively). However, due to the fact that the stimulus and the high-frequency TEOAE components superimpose (and therefore have to be canceled during TEOAE recording), TEOAEs fail to measure cochlear function above 4 kHz. In contrast, DPOAEs have the advantage of being capable of superior detection of a high-frequency hearing loss. This is due to the fact that stimulus (primary tones at f_2 and f_1) and response ($2f_1 - f_2$) do not superimpose. However, calibration errors due to standing waves in the outer-ear canal, as discussed above, can lead to misinterpretation of DPOAE results above 6 kHz.

The higher frequency specificity of DPOAEs is demonstrated in Figure 13.8 in a case example of a noise-induced hearing loss with a notch at 4 kHz. The high-resolution audiogram shows a maximum hearing loss of about 35 dB HL. Below 3 kHz and above 6 kHz, the audiogram shown in Figure 13.8A indicates normal hearing. In the TEOAE spectrum of Figure 13.8A, high-frequency components are missing, although hearing is normal above 6 kHz. In contrast, the DPOAE-grams of Figure 13.8B show normal DPOAE levels at 6 kHz. Moreover, the notch in the audiogram is exactly replicated in the DPOAE-grams, with the DPOAE level being lowest at 4 kHz. The precise correspondence between audiogram and DPOAE-gram gives evidence that the main DPOAE source is the f_2 place, and thus plotting the DPOAE level across f_2 should be the procedure of choice for a frequency-specific evaluation of the hearing loss.

DPOAE-grams, however, do not allow a precise quantitative assessment of the hearing loss. When looking at the DPOAE-grams in Figure 13.8B obtained by high stimulus levels ($L_2 = 60, 55, 50$ dB SPL; traces 1, 2, and 3), the difference between the DPOAE level at a frequency where hearing is normal (e.g., at 6 kHz) and that at 4 kHz amounted to only about 25 dB and hence does not reflect the actual hearing loss (which is about 35 dB HL). Due to the noise floor and the criteria used for accepting a DPOAE response as valid (which in this case is an SNR higher than 10 dB), DPOAEs are not measurable around 4 kHz when they are elicited at low stimulus levels ($L_2 = 45, 40, 35$ and 6 in Fig. 13.8B).

However, in evaluating the difference between the highest DPOAE level at normal hearing (at 6 kHz) and the noise-floor level at 4 kHz, then the difference matches the actual hearing loss. This occurrence is pure chance, because the noise-floor level is −30 dB SPL at 4 kHz. If the notch were at lower frequencies, or another DPOAE instrument with a higher noise floor had been used, a precise measurement of the hearing loss would not have been possible.

When the slope of the DPOAE I/O functions is plotted across test frequencies (f_2), as illustrated in Figure 13.8C, s increases with increasing hearing-loss, revealing an increasing loss of CA compression. Like the DPOAE-grams, the slope profile matches the audiometric threshold exactly with respect to the test frequency.

As already mentioned, extrapolated DPOAE I/O functions allow a more precise estimate of hearing loss, since the estimation is independent of the noise floor. This outcome is demonstrated in Figure 13.9 by three clinical cases. These plots show the audiometric threshold, the estimated hearing threshold (derived from extrapolated DPOAE I/O functions), the slope profile, and the DPOAE-grams obtained in two patients suffering from sensorineural hearing loss and in a newborn with a "refer" hearing-screening response. In the patients' ears, the DPOAE threshold estimated by means of extrapolated I/O functions (filled circles) was similar to the audiometric threshold (open circles in the audiogram). The estimation was possible up to a hearing loss of 50 dB HL (see Fig. 13.9A).

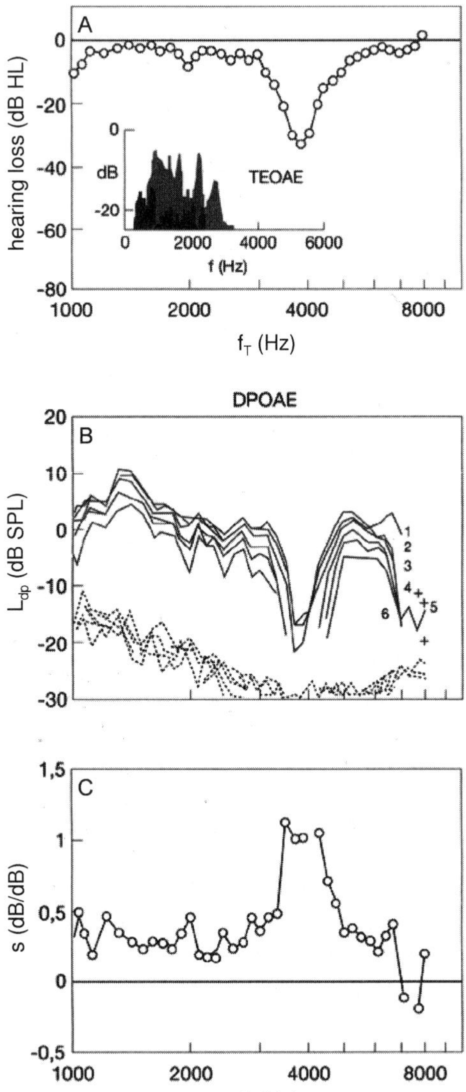

FIGURE 13.8. (**A**) High-resolution hearing threshold and TEOAE. (**B**) DPOAE-grams ($35 < L_2 < 60$ dB SPL; $L_1 = 0.4L_2 + 39$ dB SPL). (**C**) Slope profile of a hearing loss patient with unknown cause of disease. With increasing hearing loss, DPOAE level decreased and the slope calculated from DPOAE I/O functions increased. (Adapted from Janssen 2001, p. 97, Copyright 2001, with permission from Thieme.)

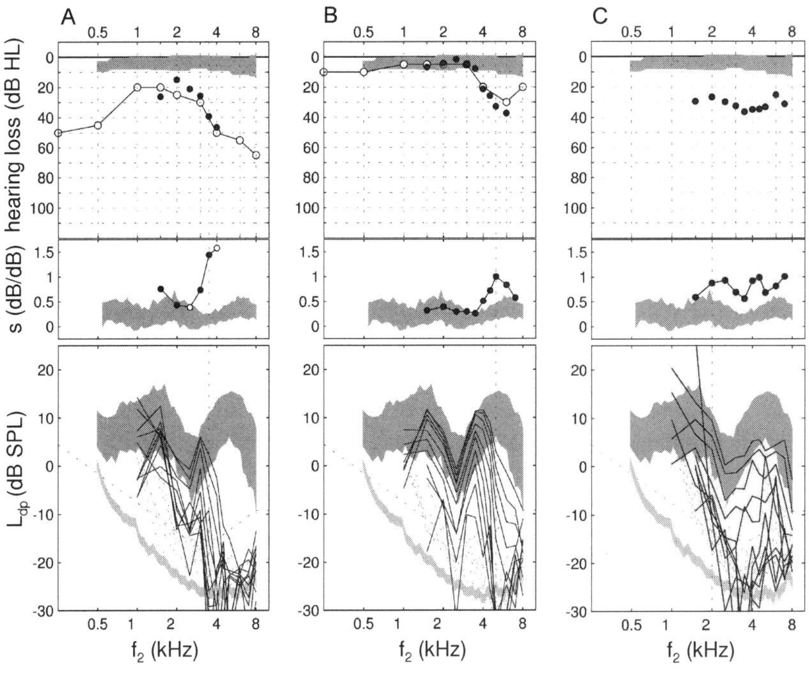

FIGURE 13.9. Measures deduced from DPOAE recordings in three different patients. Clinical audiogram (*open circles*) and reconstructed audiogram from extrapolated DPOAE I/O functions (*filled circles*) (*top*), slope profile (*middle*), and DPOAE-grams ($20 < L_2 < 65$ dB SPL; $L_1 = 0.4L_2 + 39$ dB SPL) (*bottom*) are shown. The *gray areas* constitute the respective range of mean ± standard deviation of normative data derived from a group of 30 normally hearing adults. (**A**, **B**) Data from two patients with differing cochlear hearing loss. (**C**) Data from a 3-days-old infant. In general, it can be seen that the objective DPOAE threshold data almost exactly reproduce the subjective audiogram hearing-threshold data. Further on, slope increases with increasing hearing threshold representing increasing loss of cochlear compression.

With increasing hearing loss, the slope of the DPOAE I/O function increased. In the case of the high-frequency hearing loss, the slope increased from about 0.3 dB/dB at 2 kHz to more than 1.5 dB/dB at 4 kHz (see Fig. 13.9A). In the case of the high-frequency notch, there was an increase from 0.25 dB/dB at 3 kHz to a maximum of 1 dB/dB around 6 kHz, and a decrease to 0.5 dB/dB at about 8 kHz (see Fig. 13.9B). This example also shows that the patients' hearing thresholds were replicated adequately in the slope of the DPOAE I/O functions.

It is important to note that an estimate of an accurate hearing threshold solely from the DPOAE-grams was not possible. Although the high-frequency hearing loss could be seen by missing DPOAEs, the degree of the hearing loss could not be estimated because the DPOAEs disappeared into the noise floor, being −25 dB SPL around 6 kHz. Moreover, the notch in the DPOAE-grams around 3 kHz (which was due to calibration errors) was not seen either in the estimated hearing threshold by means of extrapolated DPOAE I/O functions, nor in the slope of the DPOAE I/O functions. Therefore, the calibration error due to standing waves did not seem to affect either the estimated hearing threshold or the slope. Moreover, in the case of the audiogram notch at 6 kHz shown in Figure 13.9B, the absolute DPOAE level at frequencies exceeding 6 kHz decreased, even though the ear showed nearly normal hearing at 8 kHz. Thus, the estimated hearing threshold failed to match the audiogram at 8 kHz. In contrast, the audiogram notch is clearly reflected in the slope profile. The reason for this agreement is that the extrapolation method used demands at least three valid DPOAE data points (Boege and Janssen 2002). In this case, on the basis of the applied SNR criterion (>6 dB), only two data points were available at 8 kHz, and therefore, no hearing threshold estimate was possible at this frequency. In contrast, for a reliable estimate of the slope (i.e., CA compression), two consecutive data points are sufficient.

Figure 13.9C presents a case of a newborn in which no TEOAEs were measurable during the newborn hearing-screening session. The estimated hearing threshold was nearly independent of frequency, with the hearing loss being between 30 and 40 dB HL. DPOAE-grams were clearly separated, and hence the slope was increased over the entire frequency range, thus revealing a loss of CA compression. Due to the increased slope of the I/O functions and to the fact that DPOAEs were measurable even at moderate primary-tone levels, a sound-conductive hearing loss was excluded, so that a cochlear hearing loss was considered likely. Later, a cochlear hearing loss was confirmed by ABRs and tympanometry. With knowledge of the estimated hearing threshold and the slope of the DPOAE I/O functions, in this case, important information about loss of CA sensitivity and CA compression was obtained and resulted in an improved hearing-aid adjustment.

5.4 Monitoring Cochlear Function

OAE measures are stable through time in any particular individual and hence are capable of monitoring recovery from OHC impairment. Therapeutic drugs such as antibiotic (e.g., aminoglycosides) and antitumor chemotherapeutic (e.g., cisplatin) agents are reported to induce an irreversible hearing loss, which typically affects the highest frequencies first, with hearing loss systematically progressing to the lower frequencies (e.g., Kopelman et al. 1988; Fausti et al. 1994; Berg et al. 1999; Stavroulaki et al. 2001). Early detection of ototoxicity is important for providing effective management options such as substitution of medications, change of dosage, and mode of administration (Lonsbury-Martin and Martin 2001). Because TEOAEs are less effective above 4 kHz, DPOAEs are the test of first choice for detecting and monitoring OHC dysfunction due to ototoxic drugs. Moreover, DPOAEs have an additional advantage over TEOAEs in that they can give information about CA compression. If OHC function is disturbed during the toxic process, then not only DPOAE level, but also DPOAE growth, should be altered.

Like antibiotic and chemotherapeutic drugs, salicylate is also known to affect hearing sensitivity and to induce tinnitus (Myers and Bernstein 1965; McFadden and Plattsmier 1984; Wier et al. 1988; Long and Tubis 1988; Boettcher and Salvi 1991; Brown et al. 1993; McFadden and Pasanen 1994). However, most importantly, impairment due to salicylate toxicity is reversible.

Figure 13.10 illustrates the case of a young female who consumed an overdose of salicylate with suicidal intent (10 g) resulting in a high-frequency hearing loss of 50 dB HL at 6 kHz and strong tinnitus. Two days later, hearing was restored and tinnitus abolished. In the state of intoxication, DPOAEs elicited

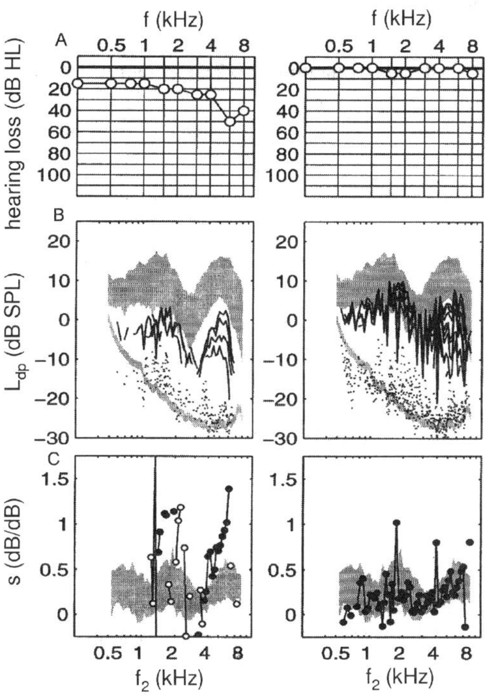

FIGURE 13.10. Audiograms (*top*), DPOAE-grams ($20 < L_2 < 65$ dB SPL; $L_1 = 0.4L_2 + 39$ dB SPL) (*middle*), and slope *s* of the DPOAE I/O functions (*lower*) in the right ear during an acute aspirin intoxication episode (left side) and after recovery (right side). Shaded areas indicate the respective range of mean ± standard deviation of the corresponding measures calculated for a normally hearing subject sample for the DPOAE level (at $L_2 = 65$ dB SPL only), and the slope *s*. *Dotted lines* in the middle panels indicate the noise floor for the different DPOAE-grams. The small lower shaded area represents the normal noise condition. In the lower panels, open circles show slope values calculated from only two $L_{dp}(L_2)$ values, while filled circles mark slope values calculated from three or more $L_{dp}(L_2)$ values. (Reprinted from Janssen et al. 2000 p. 1791, Copyright 2000, with permission from Acoustical Society of America.)

at high primary-tone levels appeared with nearly normal levels, despite the 50 dB HL hearing loss. However, the slope of the DPOAE I/O functions increased considerably, indicating loss of CA compression. The steep slope around the tinnitus frequency reveals that the low-level response was affected more than the high-level response of the CA, and hence suggests impairment of active cochlear processes at the cochlear place of the tinnitus. This interpretation is in accordance with the report of Wier et al. (1988), who also found steepened DPOAE I/O functions in volunteers who consumed salicylate over a period of 4 days (325 mg per day). In these cases, structural changes in OHCs appeared to be the most likely pathological correlate of tinnitus. For example, loss of stiffness of the OHC body or of its stereocilia and a corresponding alteration in coupling of stereocilia with the tectorial membrane may play a key role in this pathological process. Assuming that a loss of OHC stiffness is responsible for distortions within cochlear micromechanics, the corresponding tonic change at the inner hair cell activity may be one potential correlate of tinnitus (Janssen et al. 2000).

After recovery, as shown in the right panels of Figure 13.10, DPOAE level and slope were within the normal range, revealing normal CA sensitivity and compression. One explanation for the recovery is reversible impairment of cochlear micromechanics. It seems reasonable to assume that the effect of salicylate is to remove OHCs from their role as force feedback elements, in that the salicylate molecule partitions into the OHC membrane and alters the OHC motor protein prestin (Tunstall et al. 1995). The effect of salicylate-induced tinnitus on DPOAEs may better explain the underlying mechanisms of ototoxicity. Whereas TEOAEs give evidence only that something may have changed in the cochlea, DPOAEs are suggested to be a more useful and robust quantitative indicator of cochlear impairment.

A case example of a patient suffering from sudden hearing loss shown in Figure 13.11 demonstrates the ability of DPOAE I/O functions to monitor recovery from a loss of CA sensitivity and compression. When the patient was admitted, the hearing loss was 60 dB HL and DPOAEs were immeasurable. On the fourth day of therapy, recovery had begun and the hearing loss was now about 30 dB HL. Exemplary DPOAE I/O functions show the DPOAE behavior during complete and incomplete recovery between the fourth and eighth days of therapy at two distinct cochlear places, their characteristic frequencies being $f_2 = 1900$ Hz and 3320 Hz. At $f_2 = 1900$ Hz, where hearing loss completely recovered, DPOAE I/O functions changed their shape from linear to compressive. The slope of the I/O function calculated between $L_2 = 45$ and 65 dB SPL continuously decreased from about 1.1 dB/dB at the beginning to about 0.25 dB/dB on the eighth day of therapy. At $f_2 = 3320$ Hz, where hearing loss recovered only from 30 to 20 dB HL, DPOAE I/O functions exhibited a linear course (the slope being about 1 dB/dB), revealing a remaining loss of CA compression. These findings also imply that in sudden hearing loss, where the etiology is not fully understood, CA dysfunction is involved. It should be emphasized that in this case there was a complete recovery at least in one part of the cochlea. The fact

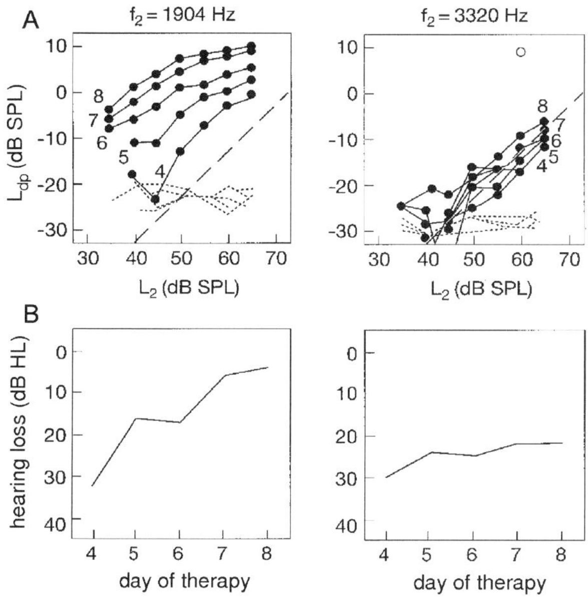

FIGURE 13.11. Examples of DPOAE I/O functions (**A**) and hearing threshold data (**B**) of a patient suffering from sudden hearing loss, on five consecutive days during recovery of hearing at $f_2 = 1904$ Hz, where hearing completely recovered (*left*) and at $f_2 = 3320$ Hz, where it showed limited recovery (*right*). Numbers at the individual I/O function curves refer to the day of therapy. *Filled circles* indicate DPOAEs, *dotted lines* the corresponding noise-floor levels. For comparison, the dashed diagonal line indicates unity slope. The lower **B** panels show the hearing loss measured with an audiometer at 50-Hz frequency resolution. (Adapted from Janssen, 2001, p. 101, Copyright 2001, with permission from Thieme.)

that DPOAEs in sudden hearing loss, salicylate intoxication (Fig. 13.10), and acoustic trauma (Janssen et al. 1998) behave similarly suggests the influence of similar underlying pathogenetic mechanisms.

In general, DPOAE I/O functions have the ability to monitor minute changes of CA function. It may be possible that DPOAE I/O functions can predict whether, and to what extent, cochlear dysfunction recovers (Hoth 2005).

6. Future Applications

6.1 Assessment of Cochlear Vulnerability

About 10 million workers in the United States are subject to job-related sound exposure in a potentially hazardous sound-pressure-level range (Environmental Protection Agency 1981). The vulnerability to noise-induced hearing loss has long been known to be highly variable across subjects. A screening tool that distinguishes between "tough" and "tender" ears would be useful, e.g., in occupational medicine, for preventing noise-induced hearing loss.

Due to the small contralateral DPOAE suppression and ipsilateral DPOAE adaptation effects commonly measured in humans (e.g., Collet et al. 1990; Kim et al. 2001), the clinical application of DPOAEs for investigating dysfunction of the MOC efferents seems to be limited. Maison and Liberman (2000), however, established an experimental approach for obtaining greater DPOAE adaptation effects. Specifically, they showed that in guinea pigs, presumably in notch regions of DPOAE I/O functions, a small primary-tone level change of 1 dB could yield a great change in adaptation magnitude, typically progressing from an increasing (enhancement) to a decreasing (suppression) postonset time course, with the difference being up to 30 dB.

In guinea pigs, the reflex strength of the medial olivocochlear (MOC) system, which can be assessed by means of contralateral DPOAE suppression and/or ipsilateral DPOAE adaptation, was found to be inversely correlated to the degree of noise-induced hearing loss (Maison and Liberman 2000; Luebke and Foster 2002). Recent findings in humans (Müller et al. 2005) have shown that similar bipolar changes occur for contralateral DPOAE suppression at frequencies that are located in dips of the DPOAE fine structure. In contrast, ipsilateral DPOAE adaptation did not exhibit such high bipolar effects and is, in addition, a much more difficult and time-consuming measurement procedure. Hence, contralateral DPOAE suppression seems to be the first choice in humans to serve as a clinical tool for detecting the efficiency of the efferent MOC system and, with that, possibly the vulnerability of the human cochlea to sound overexposure.

An approach for acquiring the efferent reflex strength is shown in Figure 13.12. First, as shown in Figure 13.12A, DPOAE fine structure was measured. The frequency where the deepest dip occured in the fine structure was then used for further contralateral DPOAE suppression measurements, which were conducted over a wider primary-tone stimulus level range (see Fig. 13.12B). As illustrated in Figure 13.12C, the slice through the three-dimensional DPOAE showed a huge bipolar effect. In 10 normally hearing subjects, the maximum bipolar effect (the ΔL_{dp} *range*) amounted to about 14 dB, as shown in Figure 13.12D. These findings suggest the bipolar effect is able to provide information about the MOC reflex strength and possibly about cochlear vulnerability to noise exposure.

6.2 Objective Hearing-Aid Fitting Parameters

DPOAE I/O and loudness functions are reported to have a close resemblance in their slope characteristics (Zhang and Zwislocki 1995, Neely et al. 2003; Müller and Janssen 2004). Thus, DPOAEs have the potential for determining characteristic quantities of the cochlear-impaired ear and might provide parameters for a noncooperative hearing-aid-adjustment situation (Müller and Janssen 2004). DPOAE and categorical loudness I/O functions exhibited a similar behavior, which was manifested in similar threshold level estimates and compression, suggesting that both measures are determined by the same source of nonlinearity.

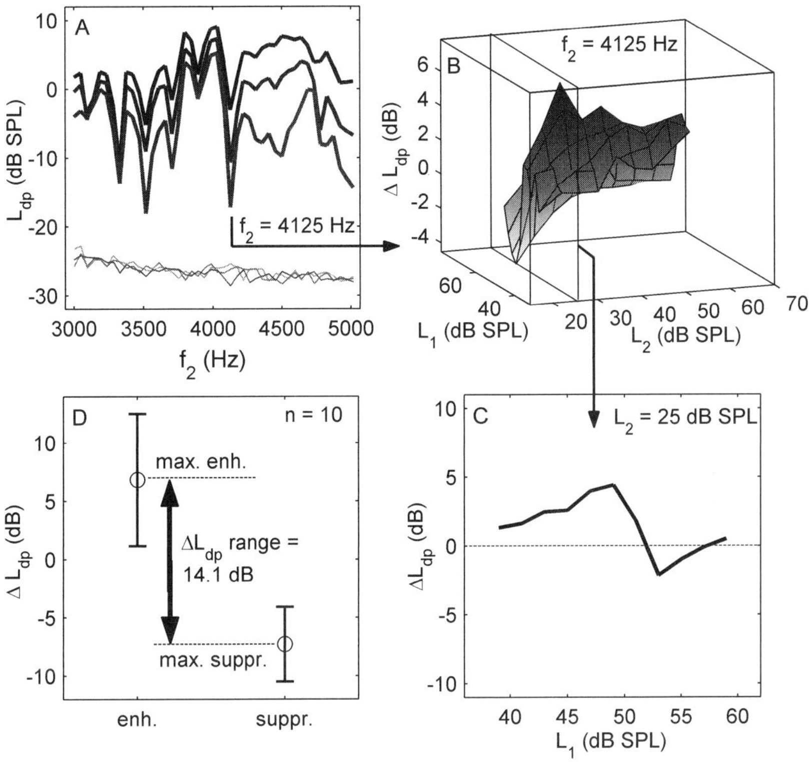

FIGURE 13.12. Procedure for evaluating efferent reflex strength at frequencies corresponding to a dip in the DPOAE-gram. Panels **A**, **B**, and **C** are data for one subject. (**A**) DPOAE-gram data with a frequency resolution of 47 Hz. Pronounced dips and peaks occur in the DPOAE-gram. The deepest dip is located at $f_2 = 4125$ Hz. When measuring contralateral DPOAE suppression across a wide $L_1|L_2$ primary-tone level range and at $f_2 = 4125$ Hz (**B**), the DPOAE level difference between DPOAE measurement in the presence and in the absence of contralateral acoustic stimulation (CAS), ΔL_{dp}, shows distinct regions of suppression (L_{dp} decreases due to CAS) and enhancement (L_{dp} increases due to CAS). Cross sections with fixed L_2 show a transition from enhancement to suppression within a small L_2 level range. Panel C shows the cross section with fixed $L_2 = 25$ dB SPL along with a ΔL_{dp} *range* (difference between maximum enhancement and maximum suppression) amounting to 6.5 dB. Panel **D** shows mean and standard deviation of maximum suppression (suppr.) and enhancement (enh.) averaged across normally hearing subjects (10 ears) for measurements conducted at their individual dip frequencies. The resulting ΔL_{dp} *range* amounts to 14.1 dB! (Adapted from Müller and Janssen 2005, Copyright 2005, with permission from Acoustical Society of America.)

However, a close resemblance was achieved only if a DPOAE normalization strategy was implemented to compensate for the highly varying absolute DPOAE level across subjects, which is presumed to be due to interindividual variations in ear canal length and/or middle-ear impedance.

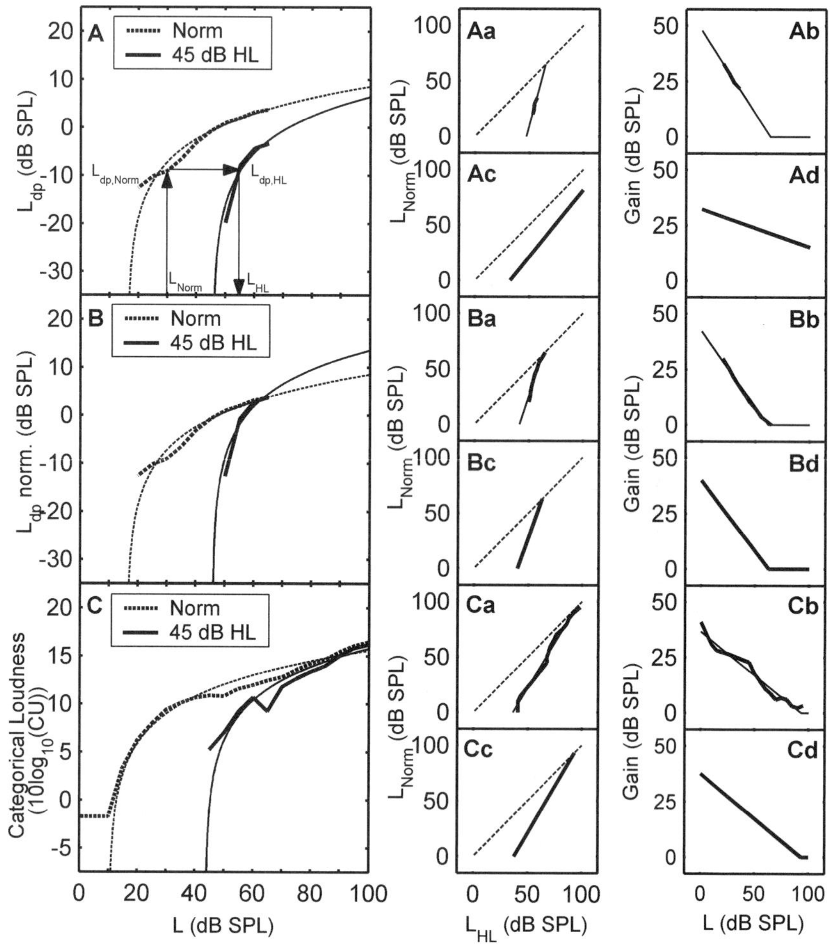

FIGURE 13.13. Data of a case example with a hearing loss of 45 dB HL at 3 kHz is shown in comparison to normative data. (**A**) DPOAEs, (**B**) normalized DPOAEs, and (**C**) categorical loudness data on a logarithmic scale. *Bold lines* represent discrete data, while *thin lines* indicate extrapolated data. Panels **Aa**, **Ba**, and **Ca** show discrete values of L_{Norm} plotted above L_{HL} (relative growth function) for the respective discrete measurement data (*bold solid lines*) and extrapolations of the discrete relative growth functions (*thin solid lines*). Panels **Ac**, **Bc**, and **Cc** show relative growth functions for the respective extrapolated measurement data (*bold solid lines*). The thin dashed lines represent the respective normative function $L_{\text{Norm}} = L_{\text{HL}}$. Panels **Ab**, **Bb**, and **Cb** show gain functions for the respective discrete measurement data (*bold solid lines*) and extrapolations of the discrete gain functions (*thin solid lines*). Panels **Ad**, **Bd**, and **Cd** show gain functions for the respective extrapolated measurement data. (Reprinted Müller and Janssen 2004, p. 3086, Copyright 2004, with permission from Acoustical Society of America.)

A patient case example, which illustrates the effects described above, is shown in Figure 13.13. Data of a subject exhibiting a hearing loss of 45 dB HL at 3 kHz is plotted in comparison to normative data. Panel A shows DPOAE I/O functions, panel B normalized DPOAE I/O functions, and panel C categorical loudness data on a logarithmic scale. Bold lines represent discrete, and thin lines extrapolated data. Both DPOAE and categorical loudness I/O functions were likewise less compressive, consequently steeper, and thus yielded higher response thresholds. In order to estimate level-dependent gain, which might serve as a parameter in hearing-aid fitting, the normally hearing and the hearing-impaired subject data were compared relative to each other for both DPOAE growth and categorical loudness. The procedure applied was derived from a strategy used by Steinberg and Gardner (1937) and is shown schematically in Figure 13.13A. The approach for categorical loudness data is analagous to this. For each stimulus level, designated as L_{Norm}, the respective corresponding DPOAE level of the normally hearing group $L_{\mathrm{dp,Norm}}$ was compared to the DPOAE level data of the hearing-impaired subject in order to determine the stimulus level, designated as L_{HL}, that was required to elicit the same DPOAE level $L_{\mathrm{dp,HL}}(=L_{\mathrm{dp,Norm}})$ in the hearing-impaired subject. The required gain at the level $L = L_{\mathrm{Norm}}$ is then the difference between L_{HL} and L_{Norm}. Functions with L_{Norm} plotted above L_{HL} for discrete and extrapolated data are shown in panels Aa, Ba, Ca, and Ac, Bc, Cc, respectively. Gain functions for discrete and extrapolated data are shown in panels Ab, Bb, Cb, and Ad, Bd, and Cd, respectively. Gain functions for normalized DPOAE and categorical loudness show a similar behavior, suggesting that gain functions derived from DPOAE data have the ability to serve as a future fitting parameter in dynamic-compression hearing aids for compensating loss of cochlear sensitivity and compression.

7. Conclusions

The CA is an essential sound-processing element in that the auditory end organ considerably enhances sensitivity and frequency selectivity. Being a by-product of cochlear amplification, OAEs directly reflect CA function. Whereas TEOAEs more qualitatively assess cochlear function, DPOAEs provide quantitative information about the CA range and operational characteristics. DPOAEs, when elicited by stimulus features that account for the different compression of the two primary tones at the DPOAE generation site, mirror the nonlinear basilar membrane motion found in animal studies. The different DPOAE measures, i.e., DPOAE I/O functions, DPOAE isosuppression tuning curves, ipsilateral DPOAE adaptation, and contralateral DPOAE suppression, provide information about CA sensitivity and compression and the control mechanisms of cochlear efferents. OHCs—the motor elements of the CA—are most sensitive to extrinsic and intrinsic contaminants and can be irreversibly impaired. OAEs are a noninvasive tool for assessing OHC dysfunction. The main OAE applications in clinical diagnostics are newborn hearing screening, topological diagnostics, quantitative evaluation of hearing loss and recruitment, detecting the beginning stages of cochlear impairment during noise exposure or ototoxic drug administration, and monitoring cochlear function during recovery from cochlear dysfunction. Because OAEs describe detailed characteristics of the impaired ear, hearing-aid adjustment can be improved by OAE measurements, especially in infants in which subjective audiometric tests fail to quantitatively determine cochlear dysfunction. Possibly, in the future too, cochlear vulnerability can be evaluated in order to protect workers from noise-induced hearing impairment.

References

Abdala C (1998) A developmental study of distortion product otoacoustic emission (2f1−f2) suppression in humans. Hear Res 121:125–138.

Abdala C (2000) Distortion product otoacoustic emission $2f1-f2$ amplitude growth in human adults and neonates. J Acoust Soc Am 107:446–456.

Abdala C (2001) Maturation of the human cochlear amplifier: distortion product otoacoustic emission suppression tuning curves recorded at low and high primary-tone levels. J Acoust Soc Am 110:1465–1476.

Abdala C (2003) A longitudinal study of distortion product otoacoustic emission ipsilateral suppression and input/output characteristics in human neonates. J Acoust Soc Am 114:3239–3250.

Abdala C, Chatterjee M (2003) Maturation of cochlear nonlinearity as measured by distortion product otoacoustic emission suppression growth in humans. J Acoust Soc Am 114:932–943.

Abdala C, Sininger Y, Ekelid M, Zeng FG (1996) Distortion product otoacoustic emission suppression tuning curves in human adults and neonates. Hear Res 98:38–53.

Agrama MT, Waxman GM, Stagner BB, Martin GK, Lonsbury-Martin BL (1998) Effects of efferent activation on distortion-product otoacoustic emissions in normal humans using ipsilateral acoustic stimulation. Abst Assoc Res Otolaryngol 21:152.

Barker SE, Lesperance MM, Kileny PR (2000) Outcome of newborn hearing screening by ABR compared with four different DPOAE pass criteria. Am J Audiol 9:1–7.

Bassim MK, Miller RL, Buss E, Smith DW (2003) Rapid adaptation of the $2f_1-f_2$ DPOAE in humans: binaural and contralateral stimulation effects. Hear Res 182: 140–152.

Berg AL, Spitzer JB, Garvin JH (1999) Ototoxic impact of cisplatin in pediatric oncology patients. Laryngoscope 109:1806–1814.

Bilger R, Matthies ML, Hammel DR, Demorest ME (1990) Genetic implications of gender differences in the prevalence of spontaneous otoacoustic emissions. J Speech Hear Res 33:418–432.

Boege P, Janssen T (2002) Pure-tone threshold estimation from extrapolated distortion product otoacoustic emission I/O functions in normal and cochlear hearing-loss ears. J Acoust Soc Am 111:1810–1818.

Boettcher FA, Salvi RJ (1991) Salicylate ototoxicity—review and synthesis. Am J Otolaryngol 12:33–47.

Bonfils P, Uziel A (1989) Clinical applications of evoked acoustic emissions—results in normally hearing and hearing-impaired subjects. Ann Otol Rhinol Laryngol 98: 326–331.

Bray PJ (1989) Click evoked otoacoustic emissions and the development of a clinical otoacoustic hearing test instrument. Dissertation at the University College and Middlesex School of Medicine, London.

Brown AM, Williams DM, Gaskill SA (1993) The effects of aspirin on cochlear mechanical tuning. J Acoust Soc Am 93:3298–3307.

Brown AM, Harris FP, Beveridge HA (1996) Two sources of acoustic distortion products from the human cochlea. J Acoust Soc Am 100:3260–3267.

Brownell WE, Bader CR, Bertrand D, de Ribaupierre Y (1985) Evoked mechanical responses in isolated cochlear outer hair cells. Science 227:194–196.

Burns EM, Keefe DH, Ling R (1998) Energy reflectance in the ear canal can exceed unity near spontaneous otoacoustic emission frequencies. J Acoust Soc Am 103:462–474.

Chéry-Croze S, Moulin A, Collet L (1993) Effect of contralateral sound stimulation on the distortion product $2f_1 - f_2$ in humans: evidence of a frequency specificity. Hear Res 68:53–58.

Collet L, Kemp DT, Veuillet E, Duclaux R, Moulin A, Morgon A (1990) Effect of contralateral auditory stimuli on active cochlear micro-mechanical properties in human subjects. Hear Res 43:251–261.

Dallos P (1992) The active cochlea. J Neurosci 12:4575–4585.

Davis H (1983) An active process in cochlear mechanics. Hear Res 9:79–90.

Dorn PA, Konrad-Martin D, Neely ST, Keefe DH, Cyr E, Gorga MP (2001) Distortion product otoacoustic emission input/output functions in normal-hearing and hearing-impaired human ears. J Acoust Soc Am 110:3119–3131.

Doyle KJ, Sininger Y, Starr A (1998) Auditory neuropathy in childhood. Laryngoscope 108:1374–1377.

Eggermont JJ, Brown DK, Ponton CW, Kimberley BP (1996) Comparison of distortion product otoacoustic emission (DPOAE) and auditory brain stem response (ABR) traveling wave delay measurements suggests frequency-specific synapse maturation. Ear Hear 17:386–394.

EPA (1981) In: Noise in America: The Extent of the Noise Problem. United States Environmental Protection Agency.

Fausti S, Larson VD, Noffsinger D, Wilson RH, Phillips DS, Fowler CG (1994) High-frequency audiometric monitoring strategies for early detection of ototoxicity. Ear Hear 15:232–239.

Frank AM, Alexiou C, Hulin P, Janssen T, Arnold W, Trappe AE (2000) Non-invasive measurement of intracranial pressure changes by otoacoustic emissions (OAEs)—a report of preliminary data. Zentralbl Neurochir 61:177–180.

Gates GA, Mills D, Nam BH, D'Agostino R, Rubel EW (2002) Effects of age on the distortion product otoacoustic emission growth functions. Hear Res 163:53–60.

Gehr DD, Janssen T, Michaelis CE, Deingruber K, Lamm K (2004) Middle-ear and cochlear disorders result in different DPOAE growth behaviour: implications for the differentiation of sound-conductive and cochlear hearing-loss. Hear Res 193:9–19.

Giebel A (2001) Applying signal statistical analysis to TEOAE measurements. Scand Audiol 30:130–132.

Glattke TJ, Robinette MS (2002) Transient evoked otoacoustic emissions. In: Robinette MS, Glattke TJ (eds) Otoacoustic Emissions—Clinical Applications, 2nd edn. New York: Thieme, pp. 95–115.

Gorga MP, Stover L, Neely ST, Montoya D (1996) The use of cumulative distributions to determine critical values and levels of confidence for clinical distortion product otoacoustic emission measurements. J Acoust Soc Am 100:968–977.

Gorga MP, Nelson K, Davis T, Dorn PA, Neely ST (2000a) Distortion product otoacoustic emission test performance when both $2f_1 - f_2$ and $2f_2 - f_1$ are used to predict auditory status. J Acoust Soc Am 107:2128–2135.

Gorga MP, Norton SJ, Sininger Y, Cone-Wesson B, Folsom RC, Vohr BR, Widen JE, Neely ST (2000b) Identification of neonatal hearing impairment: distortion product otoacoustic emissions during the perinatal period. Ear Hear 21:400–424.

Gorga MP, Neely ST, Dorn PA, Hoover BM (2003a) Further efforts to predict pure-tone thresholds from distortion product otoacoustic emission input/output functions. J Acoust Soc Am 113:3275–3284.

Gorga MP, Neely ST, Dierking DM, Dorn PA, Hoover BM, Fitzpatrick DF (2003b) Distortion product otoacoustic emission suppression tuning curves in normal-hearing and hearing-impaired human ears. J Acoust Soc Am 114:263–278.

Grandori F, Lutman ME (1999) Screening for neonatal hearing defects—European Consensus Statement. Eur J Pediatr 158:95–96.

Grandori F, Ravazzani P (1993) Non-linearities of click-evoked otoacoustic emissions and the derived non-linear technique. Br J Audiol 27:97–102.

Harris FP, Probst R, Wenger R (1991) Repeatability of transiently evoked otoacoustic emissions in normally hearing humans. Audiology 30:135–141.

Hatzopoulos S, Petrucelli J, Morlet T, Martini A (2003) TEOAE recording protocols revised: data from adult subjects. Int J Audiol 42:339–347.

He NJ, Schmiedt RA (1993) Fine-structure of the $2f1 - f2$ acoustic distortion product—changes with primary level. J Acoust Soc Am 94:2659–2669.

He NJ, Schmiedt RA (1996) Effects of aging on the fine structure of the $2f1 - f2$ acoustic distortion product. J Acoust Soc Am 99:1002–1015.

He NJ, Schmiedt RA (1997) Fine structure of the $2f1 - f2$ acoustic distortion product: Effects of primary level and frequency ratios. J Acoust Soc Am 101:3554–3565.

Heitmann J, Waldmann B, Schnitzler HU, Plinkert PK, and Zenner HP (1998) Suppression of distortion product otoacoustic emissions (DPOAE) near $2f_1 - f_2$ removes DP-gram fine structure—Evidence for a secondary generator. J Acoust Soc Am 103:1527–1531.

Hoth S (2005) On a possible prognostic value of otoacoustic emissions. A study on patients with sudden hearing-loss. Eur Arch Oto-Rhino-Laryn 262:217–224.

Janssen T (2001) Otoakustische Emissionen. In: Lehnhardt E, Laszig R (eds) Praxis der Audiometrie. New York: Thieme, pp. 79–107.

Janssen T, Kummer P, Arnold W (1998) Growth behavior of the $2f_1 - f_2$ distortion product otoacoustic emission in tinnitus. J Acoust Soc Am 103:3418–3430.

Janssen T, Boege P, Oestreicher E, Arnold W (2000) Tinnitus and 2f1-f2 distortion product otoacoustic emissions following salicylate overdose. J Acoust Soc Am 107:1790–1792.

Janssen T, Gehr DD, Kevanishvili Z (2003) Contralateral DPOAE suppression in humans at very low sound intensities. In: Gummer AW (ed) Biophysics of the Cochlea: From Molecule to Models. Hackensack, NJ: World Scientific, pp. 498–505.

Janssen T, Boege P, von Mikusch-Buchberg J, Raczek J (2005a) Investigation of potential effects of cellular phones on human auditory function by means of distortion product otoacoustic emissions. J Acoust Soc Am 117:1241–1247.

Janssen T, Gehr DD, Klein A, Müller J (2005b) Distortion product otoacoustic emissions for hearing threshold estimation and differentiation between middle-ear and cochlear disorders in neonates. J Acoust Soc Am 117:2969–2979.

Johnsen NJ, Elberling C (1982a) Evoked acoustic emissions from the human ear. 1. Equipment and response parameters. Scand Audiol 11:3–12.

Johnsen NJ, Elberling C (1982b) Evoked acoustic emissions from the human ear. 2. Normative data in young-adults and influence of posture. Scand Audiol 11:69–77.

Joint Committee on Infant Hearing (JCIH) (2000) Position statement: Principles and Guidelines for Early Detection and Intervention Programs. Am J Audiol 9:9–29.

Jülicher F, Camalet S, Prost J, Duke TAJ (2003) Active amplification by critical oscillations. In: Gummer AW (ed) Biophysics of the Cochlea: From Molecule to Models. Hackensack, NJ: World Scientific, pp. 16–27.

Keefe DH, Bulen JC, Arehart KH, Burns EM (1993) Ear-canal impedance and reflection coefficient in human infants and adults. J Acoust Soc Am 94:2617–2637.

Kemp DT (1978) Stimulated acoustic emissions from within the human auditory system. J Acoust Soc Am 64:1386–1391.

Kemp DT (1979) Evidence of mechanical nonlinearity and frequency selective wave amplification in the cochlea. Eur Arch Oto-Rhino-Laryn 224:37–45.

Kemp DT, Chum R (1980) Properties of the generator of stimulated acoustic emissions. Hear Res 2:213–232.

Kemp DT, Ryan S (1991) Otoacoustic emission tests in neonatal screening programs. Acta Otolaryngol Suppl 482:73–84.

Kemp DT, Bray P, Alexander L, Brown AM (1986) Acoustic emission cochleography—practical aspects. Scand Audiol Suppl 25:71–94.

Kemp DT, Ryan S, Bray P (1990a) A guide to the effective use of otoacoustic emissions. Ear Hear 11:93–105.

Kemp DT, Ryan S, Bray P (1990b) Otoacoustic emission analysis and interpretation for clinical purposes. In: Grandori F, Ciafrone G, Kemp DT (eds) Cochlear Mechanisms and Otoacoustic Emissions. Karger, pp. 77–98.

Kim DO, Dorn PA, Neely ST, Gorga MP (2001) Adaptation of distortion product otoacoustic emissions in humans. J Assoc Res Otolaryngol 2:31–40.

Kopelman J, Budnick AS, Kramer MB, Sessions RB, Wong GY (1988) Ototoxicity of high-dose cisplatin by bolus administration in patients with advanced cancers and normal hearing. Laryngoscope 98:858–864.

Kummer P, Janssen T, Arnold W (1995) Suppression tuning characteristics of the $2f_1 - f_2$ distortion-product otoacoustic emission in humans. J Acoust Soc Am 98:197–210.

Kummer P, Janssen T, Arnold W (1998) The level and growth behavior of the $2f_1 - f_2$ distortion product otoacoustic emission and its relationship to auditory sensitivity in normal-hearing and cochlear hearing-loss. J Acoust Soc Am 103:3431–3444.

Kummer P, Janssen T, Hulin P, Arnold W (2000) Optimal $L_1 - L_2$ primary tone level separation remains independent of test frequency in humans. Hear Res 146:47–56.

Lasky RE (1998a) Distortion product otoacoustic emissions in human newborns and adults. I. Frequency effects. J Acoust Soc Am 103:981–991.

Lasky RE (1998b) Distortion product otoacoustic emissions in human newborns and adults. II. Level effects. J Acoust Soc Am 103:992–1000.

Liberman MC, Dodds LW (1984) Single-neuron labeling and chronic cochlear pathology. III. Stereocilia damage and alterations of threshold tuning curves. Hear Res 16:55–74.

Liberman MC, Puria S, Guinan JJ Jr. (1996) The ipsilaterally evoked olivocochlear reflex causes rapid adaptation of the $2f_1 - f_2$ distortion product otoacoustic emission. J Acoust Soc Am 99:3572–3584.

Long GR, Tubis A (1988) Modification of spontaneous and evoked otoacoustic emissions and associated psychoacoustic micro-structure by aspirin consumption. J Acoust Soc Am 84:1343–1353.

Lonsbury-Martin BL, Martin GK (2001) Evoked otoacoustic emissions as objective screeners for ototoxicity. Semin Hear 22:377–392.

Luebke AE, Foster PK (2002) Variation in inter-animal susceptibility to noise damage is associated with $\alpha 9$ acetylcholine receptor subunit expression level. J Neurosci 22: 4241–4247.

Maison SF, Liberman MC (2000) Predicting vulnerability to acoustic injury with a noninvasive assay of olivocochlear reflex strength. J Neurosci 20:4701–4707.

Maison S, Micheyl C, Andeol G, Gallego S, Collet L (2000) Activation of medial olivocochlear efferent system in human: influence of stimulus bandwidth. Hear Res 140:111–125.

Margolis RH (2002) Influence of middle-ear disease on otoacoustic emissions. In: Robinette MS, Glattke TJ (eds), Otoacoustic Emissions—Clinical Applications, 2nd ed. New York: Thieme, pp. 190–212.

Mauermann M, Kollmeier B (2004) Distortion product otoacoustic emission (DPOAE) input/output functions and the influence of the second DPOAE source. J Acoust Soc Am 116:2199–2212.

McFadden D, Pasanen EG (1994) Otoacoustic emissions and quinine sulfate. J Acoust Soc Am 95:3460–3474.

McFadden D, Plattsmier HS (1984) Aspirin abolishes spontaneous otoacoustic emissions. J Acoust Soc Am 76:443–448.

Meinke DK, Stagner BB, Martin GK, Lonsbury-Martin BL (2005) Human efferent adaptation of DPOAEs in the L_1, L_2 space. Hear Res 208:89–100.

Mills DM, Rubel EW (1994) Variation of distortion product otoacoustic emissions with furosemide injection. Hear Res 77:183–199.

Mills DM, Rubel EW (1998) Development of the base of the cochlea: place code shift in the gerbil. Hear Res 122:82–96.

Moulin A, Collet L, Duclaux R (1993) Contralateral auditory stimulation alters acoustic distortion products in humans. Hear Res 65:193–210.

Müller J, Janssen T (2004) Similarity in loudness and distortion product otoacoustic emission input/output functions: implications for an objective hearing aid adjustment. J Acoust Soc Am 115:3081–3091.

Müller J, Janssen T, Heppelmann G, Wagner W (2005) Evidence for a bipolar change in distortion product otoacoustic emissions during contralateral acoustic stimulation in humans. J Acoust Soc Am 118:3747–3756.

Myers EN, Bernstein JM (1965) Salicylate ototoxicity—a clinical and experimental study. Arch Otolaryngol 82:483–493.

National Institutes of Health (NIH) (1993) Early identification of hearing impairment in infants and young children. NIH Consensus Statement 11:1–24.

Neely ST, Gorgy MP, Dorn PA (2003) Cochlear compression estimates from measurements of distortion-product otoacoustic emissions. J Acoust Soc Am 114:1499–1507.

Norton SJ (1992) The effects of being a newborn on otoacoustic emissions. J Acoust Soc Am 91:2409.

Norton SJ, Gorga MP, Widen JE, Folsom RC, Sininger Y, Cone-Wesson B, Vohr BR, Mascher K, Fletcher K (2000a) Identification of neonatal hearing impairment: Evaluation of transient evoked otoacoustic emission, distortion product otoacoustic emission, and auditory brain stem response test performance. Ear Hear 21:508–528.

Norton SJ, Gorga MP, Widen JE, Folsom RC, Sininger Y, Cone-Wesson B, Vohr BR, Mascher K, Fletcher K (2000b) Identification of neonatal hearing impairment: Summary and recommendations. Ear Hear 21:529–535.

O'Brien A (1994) Temperature dependency of the frequency and level of a spontaneous otoacoustic emission during fever. Brit J Audiol 28:281–290.

Penner MJ, Zhang T (1997) Prevalence of spontaneous otoacoustic emissions in adults revisited. Hear Res 103:28–34.

Penner MJ, Glotzbach L, Huang T (1993) Spontaneous otoacoustic emissions: measurement and data. Hear Res 68:229–237.

Probst R, Lonsbury-Martin BL, Martin GK, Coats AC (1987) Otoacoustic emissions in ears with hearing-loss. Am J Otolaryngol 8:73–81.

Puel JL, Rebillard G (1990) Effects of contralateral sound stimulation on the distortion product $2f_1 - f_2$: evidence that the medial efferent system is involved. J Acoust Soc Am 87:1630–1635.

Puria S, Guinan JJ Jr, Liberman MC (1996) Olivocochlear reflex assays: effects of contralateral sound on compound action potentials versus ear-canal distortion products. J Acoust Soc Am 99:500–507.

Rhodes MC, Margolis RH, Hirsch JE, Napp AP (1999) Hearing screening in the newborn intensive care nursery: comparison of methods. Otolaryngol Head Neck Surg 120: 799–808.

Ruggero MA, Rich NC, Recio A, Narayan SS, Robles L (1997) Basilar-membrane responses to tones at the base of the chinchilla cochlea. J Acoust Soc Am 101: 2151–2163.

Shera CA, Guinan JJ Jr (1999) Evoked otoacoustic emissions arise by two fundamentally different mechanisms: a taxonomy for mammalian OAEs. J Acoust Soc Am 105: 782–798.

Siegel JH (1994) Ear-canal standing waves and high-frequency sound calibration using otoacoustic emission probes. J Acoust Soc Am 195:2589–2597.

Sininger Y, Cone-Wesson B (2004) Asymmetric cochlear processing mimics hemispheric specialization. Science 305:1581.

Skellett RA, Crist JR, Fallon M, Bobbin RP (1996) Chronic low-level noise exposure alters distortion product otoacoustic emissions. Hear Res 98:68–76.

Smurzynski J, Kim DO (1992) Distortion-product and click-evoked otoacoustic emissions of normally-hearing adults. Hear Res 58:227–240.

Starr A, Picton TW, Sininger Y, Hood LJ, Berlin CI (1996) Auditory neuropathy. Brain 119:741–753.

Stavroulaki P, Apostolopoulos N, Segas J, Tsakanikos M, Adamopoulos G (2001) Evoked otoacoustic emissions—an approach for monitoring cisplatin induced ototoxicity in children. Int J Pediatr Otorhinolaryngol 59:47–57.

Steinberg JC, Gardner MB (1937) The dependence of hearing impairment on sound intensity. J Acoust Soc Am 9:11–23.

Talmadge CL, Long GR, Tubis A, and Dhar S (1999) Experimental confirmation of the two-source interference model for the fine structure of distortion product otoacoustic emissions. J Acoust Soc Am 105:275–292.

Topolska MM, Hassman E, Baczek M (2000) The effects of chronic otitis media with effusion on the measurement of distortion products of otoacoustic emissions: presurgical and postsurgical examination. Clin Otolaryngol 25:315–320.

Tunstall MJ, Gale JE, Ashmore JF (1995) Action of salicylate on membrane capacitance of outer hair cells from guinea-pig cochlea. J Physiol 485:739–752.

Veuillet E, Collet L, Duclaux R (1991) Effect of contralateral acoustic stimulation on active cochlear micromechanical properties in human subjects—dependence on stimulus variables. J Neurophysiol 65:724–735.

von Specht H, Ganz M, Pethe J, Leuschner S, Pytel J (2001) Linear versus non-linear recordings of transiently-evoked otoacoustic emissions—methodological considerations. Scand Audiol Suppl 52:116–118.

Whitehead ML, Lonsbury-Martin BL, Martin GK (1992) Evidence for two discrete sources of $2f_1 - f_2$ distortion-product otoacoustic emissions in rabbit: I. Differential dependence on stimulus parameters. J Acoust Soc Am 91:1587–1607.

Whitehead ML, McCoy MJ, Lonsbury-Martin BL, Martin GK (1995a) Dependence of distortion-product otoacoustic emissions in primary levels in normal and impaired ears. I. Effects of decreasing L_2 below L_1. J Acoust Soc Am 97:2346–2358.

Whitehead ML, Stagner BB, McCoy MJ, Lonsbury-Martin BL, Martin GK (1995b) Dependence of distortion-product otoacoustic emissions in primary levels in normal and impaired ears. II. Asymmetry in L_1, L_2 space. J Acoust Soc Am 97:2359–2377.

Whitehead ML, Stagner BB, Lonsbury-Martin BL, Martin GK (1995c) Effects of ear-canal standing waves on measurements of distortion-product otoacoustic emissions. J Acoust Soc Am 98:3200–3214.

Wier CC, Pasanen EG, McFadden D (1988) Partial dissociation of spontaneous otoacoustic emissions and distortion products during aspirin use in humans J Acoust Soc Am 84:230–237.

Williams DM, Brown AM (1995) Contralateral and ipsilateral suppression of the $2f_1 - f_2$ distortion product in human subjects. J Acoust Soc Am 97:1130–1140.

Wilson JMG, Jungner G (1968) Principles and Practice of Screening for Disease. World Health Organization (WHO).

Zhang M, Zwislocki JJ (1995) OHC response recruitment and its correlation with loudness recruitment. Hear Res 85:1–10.

Zwicker E, Schloth E (1984) Interrelation of different oto-acoustic emissions. J Acoust Soc Am 75:1148–1154.

14 Future Directions in the Study of Active Processes and Otoacoustic Emissions

GEOFFREY A. MANLEY AND WILLIAM E. BROWNELL

1. General Introduction

As with any area of human endeavor, scientific research has its own inertia that is both beneficial and detrimental. It helps to organize thought, identify salient findings, and to demonstrate how they fit into emerging concepts. The same inertia encumbers research, however, because it can delay the impact of discoveries that do not quite fit in to the "conventional wisdom." Chapters 1 (Kemp) and 2 (Cooper et al.) describe early discoveries that took years to have a significant impact on the hearing sciences. Researchers are inclined to dismiss or, even worse, ignore findings not immediately and obviously relevant to their own work, which in turn reduces the probability that important new data will be recognized. Additionally, it is sometimes technically impossible to test some ideas, and experimental verification must wait for advances in equipment design, sensitivity, and speed.

The goal of this chapter is to provide an assessment of the current situation and of future directions of research on active processes in hair cells and on otoacoustic emissions as based on the data and discussions of the previous chapters of this book. There is no doubt that the last 30 years witnessed an unprecedented revolution in auditory research, during which our perceptions of the function of vertebrate hearing organs were dramatically changed. This book provides the opportunity to take stock, access what has been achieved, and attempt to identify unanswered questions on which to base future research.

2. The Dust of History

At a time (1982) when it had finally been established beyond doubt that the movements of the organ of Corti are nonlinear in such a way that a negative impedance was needed to explain it, Brownell (1983) recognized a serendipitous finding for what it was: a potentially highly important discovery. At that time,

Brownell was studying the electrophysiological properties of hair-cell membranes, and the normal technique of the day was to use the capacity-compensation switch of the amplifier (which when turned too far, causes the electrode to vibrate) to help the electrode penetrate the membrane cleanly. Brownell noticed that outer hair cells, but not other cells (including inner hair cells), often showed a length contraction when the electrode was "buzzed" in this way. Rather than ignoring this as an artifact, Brownell was able to show that this was a specific response of outer hair cells to changes of cell potential and that it was found only in cells that were still living and was graded according to the magnitude of the injected current. His findings were widely recognized by the hearing-science community, for they provided a reasonable cellular basis for the active cochlear process and helped to explain the long-standing question as to the function of the outer hair cells. The discovery motivated a host of experiments that, together, have led to our current understanding about the role of membrane-based mechanisms for outer-hair-cell electromotility (see Hallworth and Jensen-Smith, Chapter 5; Manley and van Dijk, Chapter 7; Martin and Lonsbury-Martin, Chapter 8; Shera and Guinan, Chapter 9; Russell and Lukashkin, Chapter 10).

Weiss (1982) was the first to propose a cellular mechanism that could underlie hair-cell amplification before there was any experimental evidence of such mechanisms at the hair-cell level. He assumed that hair-cell bundles would move under the influence of electrical potential changes of the cell membrane. Because such potential changes could result from sound transduction, the resulting motion of the hair-cell bundle would set up a feedback mechanism that could amplify the stimulus signal. In 1985, experiments on turtle hair cells demonstrated that the hair-cell bundles were in some way active (Crawford and Fettiplace 1985). This phenomenon was further studied in many experiments on the turtle ear, but also intensively in hair cells of the frog saccule, by Hudspeth's group (Martin, Chapter 4). However, the vast majority of auditory research is motivated toward understanding the human ear, mostly studied using other mammalian species. Research based on nonmammalian ears was often simply ignored, or at best acknowledged, under the assumption that it was interesting but irrelevant to human hearing. Thus the active mechanisms of hair-cell bundles, although incorporated into some of the earliest models, did not play a central role until it was demonstrated (by groups that normally work on turtle ears and frog saccules) that mammalian hair-cell bundles also show active motility (Martin, Chapter 4).

In view of these facts, we wish to examine the previous chapters of this book with regard to their main conclusions but also for findings whose significance may go unrecognized but should be examined in more detail.

2.1 Why Are Active Processes Necessary?

Vertebrate hearing organs are fluid filled, which imposes the requirement for an active process or processes in order to overcome viscous damping (Cooper et al., Chapter 2; Duke and Jülicher, Chapter 3). Viscous damping is directly proportional to velocity and therefore to frequency. The force required to maintain differential movements of vertebrate inner-ear structures at 10 kHz is 10 times that at 1 kHz. Acoustic localization of predator or prey results in an evolutionary pressure for increasing the frequency range of hearing. A mechanism that works at one frequency may not be able to handle the increased force required to operate at a higher frequency, and a different mechanism may be required in order to process higher frequencies. Measurements of stereocilial bundle movement in frog sacculus, and in turtle and lizard hair cells reveals tuning that must result from a nonlinear active process (Aranyosi and Freeman 2004; Martin, Chapter 4). The tuning also reveals a continued roll-off at frequencies above 10 kHz, which is not surprising, as the fluid environment would be expected to entrain the movements of all the inner-ear structures at higher frequencies if the negative damping force cannot keep up (Brownell 2006). Because the number of stereocilia generally increases toward higher frequencies, and thus the number of motor elements in the bundle increases, this would tend to compensate (e.g., Manley 1990).

2.2 How Common are Active Processes in Mechanoreceptors?

The discovery of active processes in an insect, as described by Göpfert and Robert (Chapter 6), raises the possibility that active processes are a universal feature of mechanoreceptor cells. Because TRP ("transient receptor potential") channels are the basis for stimulus detection in a number of mechanoreceptor cells (Gillespie and Walker 2001), it is likely that at least some of the active processes observed in vertebrate hair cells may be associated with this family of integral membrane proteins. Depending on the context, including the required speed of operation of the channel, it is conceivable that different components may be differently expressed. It would be very interesting to find some indications of a plesiomorphic active mechanism that arose perhaps with the earliest ancestors of vertebrate hair cells (Manley and Ladher 2007). While somatic motility is understood to be membrane-based, the membrane-based nature of TRP and other mechanoreceptors is not as widely appreciated. Many TRP and other mechanoreceptive ion membrane channels respond to changes in membrane tension and curvature. Like all molecules that control the movement of substrates into and/or out of cells, they do not function outside the membrane. While the mechanoelectrical transduction channels of vertebrate hair bundles may be modulated by slow cytoskeletal-based mechanisms, the role of direct lipid-protein interactions has not yet been fully explored. Scanning and transmission electron-microscopic images of stereocilia reveal the membrane to be folded (Furness and Hackney 2006). The excess membrane has not been considered in conventional stereocilia transduction models. It is possible that active membrane-based motor mechanisms are truly ancient, traceable back to problems of ion transport faced by the earliest life forms (Brownell 2006).

As noted by Göpfert and Robert (Chapter 6), the ears of insects, at least those located in the antennae of flies, are readily accessible for experimentation, and it is likely that in the near future much more will be learned about their function. Of particular interest is the tractability of the fruit fly *Drosophila* for genetic manipulations, which may make it possible in the future to study the effect of subtle changes in channel structure. The curious presence of a prestin in the antennal receptor needs to be followed up in order to ascertain the function it plays, as it may not be involved in any way with hearing.

2.3 How Do the Different Active Processes Interact in the Mammalian Cochlea?

It has become clear in recent years not only that mammalian outer hair cells (OHCs) have somatic motility, but that active processes found in the hair-cell bundles of nonmammals are also found in mammalian inner and outer hair cells (Chan and Hudspeth 2005; Kennedy et al. 2005). In addition, it has been shown (e.g., Martin; Chapter 4) that bundle motility is obviously not a unitary process, but can be divided into three components. There are two myosin-dependent components (known as fast and slow adaptation) and a fast, calcium-dependent, channel-reclosure component. Not only have mammals maintained the ancient bundle motors, but these exist together with, and presumably work simultaneously with, the uniquely mammalian feature of OHC somatic motility.

The proposed active processes in hair cells were incorporated in a general way into models of cochlear functions since soon after their discovery, giving somewhat different functions to them. Kim (1986) proposed that hair-bundle motility of OHCs amplifies motion of the cochlear partition, while somatic motility generated by the cell membrane optimizes the operating point of the OHCs. The function of efferent innervation would then be to modulate cochlear amplifier gain by shifting the operating point of the OHCs via changes in their cell potential (Kim 1986; Kim et al. 2003). This latter function would presumably require only relatively slow movements of OHCs compared to the frequencies of stimulation. Recent studies of somatic motility, in contrast, emphasize that it can operate at very high frequencies (Brownell 2006). A recent model of the OHC reveals that stretch-activated ion channels with kinetics similar to those proposed for TRP mechanoreceptors in the stereocilia can boost the transmembrane potential response in an intermediate frequency range (Spector et al. 2005) below 10 kHz. Whereas it is widely accepted that prestin-based OHC motility is the major contributor of force in the active processes of the mammalian cochlea, there are in fact still significant problems with this hypothesis (Hallworth and Jensen-Smith, Chapter 5). A thorough understanding of the respective roles of the different types of hair-cell motilities thus represents one of the most important goals of research in this field, and their study in a coordinated way represents an enormous technical challenge to researchers.

Some of the main questions are: What is the relative contribution of somatic OHC motility and bundle motility of the two hair-cell groups in mammals? Are the different motors operating over the same frequency range, and if so, when do they operate in phase and when out of phase? Do these motors both amplify and—through out-of-phase activity—dampen motion of the cochlear partition? Is the main effect of efferent fibers via cochlear amplifier processes, and if so, which of these processes are affected most? Relevant to the role of efferent fiber activity on hearing is the observation that the specialized cholinergic receptors found in the mammalian cochlea appear to have coevolved with prestin (Franchini and Elgoyhen 2006).

2.4 What Is the Role of Hair-Cell Adaptation Motors?

In their contributions, both Duke and Jülicher (Chapter 3) and Martin (Chapter 4) emphasize the important role of hair-cell adaptation motors. As noted in early discussions (Hudspeth 1989, 1997), the presence of adaptation motors is an extremely elegant solution to the problem of maintaining the optimal position of the transduction channels for high sensitivity. If the hair-cell bundle is shifted steadily toward the taller stereovilli (e.g., in vestibular detectors of linear acceleration by a change in head angle), the channel links come under tension. This leads to their adaptational plaques slipping down the actin core of the stereovilli until positions are reached at which the normal channel-open probabilities (10 % to 20 %) are restored. If the bundle displacement is steadily in the other direction, myosin motors drag the adaptational plaque up the actin until again, the channel-open probabilities are restored to normal. This adjustment would also bring the channels into the optimal region of negative stiffness (Le Goff 2005) or "critical point" (Duke and Jülicher, Chapter 3) required for the best operation of the channel motor. This mechanism of slow adaptation likely evolved first but also laid the foundation for the evolution of the second, faster, myosin mechanism.

Hair-bundle myosins not only are involved in such relatively slow but critically important adjustments of tip-link tension, but are part of the transduction-channel complex involved in cochlear amplification (Martin, Chapter 4) by the process of "fast adaptation." Fast adaptation produces an active movement that is in the opposite direction to that of the applied force. But is this true at all frequencies of stimulation? Can it be that the effect of fast adaptation changes from opposing motion to aiding it, depending on the frequency? Here again, further studies are urgently needed.

A major consideration in the involvement of myosins in active processes of hair-cell bundles has always been the necessary speeds involved (Manley and Gallo 1997). However, even in insect flight systems, which involve structures that are considerably larger than hair cells, myosin contraction-cycle frequencies beyond 1 kHz are possible. As Martin (Chapter 4) points out, if only a small proportion of the myosins are active during a given cycle and if the myosins only rock back and forth, rather than going through attachment–detachment cycles, considerably higher frequencies are conceivable. Obviously, further research on myosin 1c and its role in fast adaptation in nonmammalian and mammalian hair cells is necessary.

2.5 What Is the Role of Rapid Channel Reclosure?

The transduction channels of hair cells apparently belong to the TRP channel family, although which TRP types are involved is not yet known (see Martin, Chapter 4). These channels admit calcium ions when open, and it has been shown that the state of the transduction channel is acutely sensitive to the concentration of calcium. The calcium concentration in the region of the channel determines their open probabilities and the frequency both of spontaneous bundle oscillations (Martin and Hudspeth 1999) and of otoacoustic emissions of the bundle (Manley et al. 2004). Calcium entering the channel is postulated to close the channel with a submillisecond time scale (Chan and Hudspeth 2005). Importantly, the process has a limited dynamic range and thus needs the myosin-dependent processes to be held poised at the optimal position. To quote Martin (Chapter 4), "A nonlinear and dynamic tradeoff between opposing forces that are produced by movements of the adaptation motor on the one end and those exerted by channel gating on the other end probably accounts for the complexity of active hair-bundle motility."

Chapters 3 (Duke and Jülicher) and 4 (Martin) invoke a Hopf oscillator model for inner-ear mechanisms because it is one of the few mathematical tools that can generate oscillations with nonlinear behavior. The facility with which the Hopf approach may be applied to some aspects of hearing has generated enthusiasm because in some, but not all, cases it provides an adequate numerical description of the ear's nonlinear behavior. The descriptions are often an improvement over and certainly easier to use than physically based approaches with their difficult-to-manipulate partial differential equations. The general excitement in this research area is reminiscent of that seen when Wiener kernel analysis entered biology over 25 years ago. The Hopf model requires an oscillator, and because of historical linkages, most papers identify the locus of the oscillator as the hair-cell stereociliary bundle. The bundle is not, of course, the only oscillator in the mammalian inner ear. OHCs feed energy back into the system and contribute to oscillations of the cochlear partition. Microimpedance spectroscopy has revealed resonance behavior in the electrical admittance of individual OHCs (Rabbitt et al. 2005). There may even be active processes in the synaptic release machinery at the base of the hair cells (Brownell et al. 2003). Any model of mammalian hearing must address the possibility that the hair bundle may or may not contribute to high-frequency hearing. The other resonant mechanisms should be systematically investigated along with the traveling wave as contributing to the overall behavior of the system. It is possible that early Hopf models may be extended to permit exploration of multiple forcing functions and perhaps interacting oscillators.

Although very important advances have been made in the last few years, questions remain as to the interaction of these three hair-cell-bundle active processes. What are the precise roles of hair-cell adaptational motors in slow and fast adaptation? How much does fast adaptation contribute to cochlear amplification? What molecular mechanism causes the conformational change associated with channel gating? How much energy does rapid channel closure contribute to cochlear amplification? How do these mechanisms interact and what is their phase dependency?

2.6 What Determines the Patterns of Spontaneous Otoacoustic Emission (SOAE) Frequencies?

As Duke and Jülicher (Chapter 3) note, "In the cochlea, the noisy motion of many critical oscillators, vibrating at different frequencies, would lead to the generation of many wavelike excitations that travel along the basilar membrane and interact via nonlinearities. In such a situation, certain spectral components could be selected as eigenmodes of the cochlear cavity, and these frequencies would appear in the spectrum of spontaneous emissions (Shera 2003; Shera and Guinan, Chapter 9)." It is, however, still unclear which factors determine the frequencies at which SOAEs can be measured and the frequencies at which there are no measurable emissions in normal ears (Manley 2006b). Two kinds of explanation have been offered for the generation of SOAEs at particular frequencies (see also Shera and Guinan Chapter 9).

In the "point source" or local oscillator model, only local irregularities in the properties of the hair cells and/or the larger morphological irregularities in the cochlea (e.g., Manley 1983) are important in determining where SOAE energy peaks.

In the "global standing wave" model, the frequencies of SOAE peaks are determined only by the wavelengths of emitted sound that is reflected back into the cochlea from the stapes footplate. At particular wavelengths, returning waves would add constructively to existing oscillations and thus "allow" SOAEs; at other frequencies, the returning waves would add destructively.

However, the assumptions of these two mutually exclusive models are too simplistic, since if only the global standing wave model applied to mammals, there should be a very strict rule governing the distances between SOAE peaks. This is, however, only roughly true, and this model cannot explain "odd" frequency intervals. The global standing wave concept also cannot explain why most laboratory mammals have no SOAEs at all (see also Lonsbury-Martin and Martin, Chapter 8). If the hair cells of laboratory mammals have active processes, and they certainly do, their activity must lead to standing waves.

Shera's 2003 proposed test to distinguish between the two models is also too simplistic, since the assumption is that a change in the boundary conditions of the cochlea (produced by a change in air pressure) would affect only standing-wave-generated SOAEs. Point sources would, however, also be sensitive to such changes. In addition, SOAE patterns from species other than mammals are very difficult to explain using the global standing wave concept alone. It is difficult to see how standing waves could explain the complex patterns of the spectra of different lizard basilar papillae, for example spectral patterns that differ characteristically according to cochlear structure and not cochlear dimensions (Manley 1997)—especially those in which the spectrum is not stable over time (Manley 2006a).

On the other hand, some features of lizard SOAEs are clearly compatible with global standing waves playing a role. For example, feedback times for putative standing waves in small lizard cochleae must be extremely short. This is compatible with the broad bandwidths of lizard emissions compared to those of mammals, since rapid feedback leads to frequency instability. At present, the SOAE data can be interpreted to indicate that in all cochleae, local structural features determine whether the conditions for an SOAE peak prevail and wave feedback then contributes to its maintenance. Thus both point sources and global influences could determine SOAE patterning. This is one area of research on SOAEs that obviously requires more attention in the future.

The gender difference in the magnitude of SOAEs is another area that deserves attention. SOAEs are larger in human and rhesus monkey females than in males (Bilger et al. 1990; McFadden 1993a,b; McFadden et al. 2006a). The larger SOAEs may be correlated with the better hearing thresholds for females of the same species (McFadden 1993a,b). Part of the reason for the gender difference arises from different hormonal exposure during development and later, but the mechanism by which hormone exposure modulates SOAE magnitude has not been determined (McFadden 1993a,b; McFadden et al. 1996, 2006a,b). Possibilities include the full range of mechanisms described in this book. There could be a sex difference in OHC electromotility and/or in the mechanism(s) responsible for stereociliary bundle motility. Both could, in principle, be the result of gender differences in membrane lipid profiles that would alter lipid–protein interactions. Another possibility is the shorter length of female cochleae (Sato et al. 1991) or gender differences in the size of the middle ear. It is also possible that the difference arises centrally and the efferent control mechanisms described in Chapter 10 (Russell and Lukashkin) play a role in tuning cochlear function. An efferent influence may also help to explain the systematic difference between the magnitude of left and right ear SOAE in humans.

2.7 What Are the Mechanisms by Which the Efferent Fibers Affect Cochlear Responses?

In Chapter 10, Russell and Lukashkin discuss the role of the efferent system in the context of active processes in hair cells. It is a remarkable fact that although at the level of single cells the effects of efferent-fiber activation are strong and obvious, previous attempts to show an effect on behavior and perception have mostly turned up disappointing results. Our current understanding of the active hair cell provides a new environment conducive to understanding the efferent effects as having their main influence on these active processes. Efferents certainly attenuate cochlear amplification, but how this effect is achieved—and the differences between slow and fast efferent effects—clearly needs to be the subject of further research.

2.8 What Are the Future Applications of OAEs in a Clinical Context?

The uses of OAE measurements as diagnostic tools (see Janssen and Müller, Chapter 13) have such obvious advantages (especially in their objectivity and restriction to the cochlea) that they have been rapidly incorporated as clinical tools. This is especially true of distortion-product (DPOAEs) and click-evoked OAEs, which are independent of whether an ear produces SOAEs. The objective nature of these tests is especially useful in newborn and young children, but also in, for example, cases in which a determination of cochlear impairment is a legal matter (as in noise-induced hearing loss). Many of the diagnostic tools need greater standardization and careful protocols, however, and there is room for improvement in this direction in the future.

As pointed out by Janssen and Müller in Chapter 13, there are still some uncertainties with regard to the function of the cochlear amplifier in humans that hamper both research and clinical studies. One of these is the uniformity of the mechanisms across the frequency-response range of the human cochlea. It has been known for many years that there are substantial differences between apical, low-frequency and basal, high-frequency responses, and these are seen at the level of basilar membrane motion, implying that the cochlear amplifier(s) are not manifest uniformly along the cochlear epithelium.

3. General Conclusions and Speculations on Unexplored Territory

Research in active mechanisms in the cochlea and in otoacoustic emissions has matured, and we now understand a great deal more about the mechanisms that lead to sensitive and frequency-selective hearing over a broad range of intensities. Concepts that only 10 years ago were met with skepticism and disbelief are now fundamental to our global understanding of the function of the ear in vertebrates. The future will clear up the questions as to how the diverse mechanisms are integrated into a functional whole, especially in mammals, with their specialized membrane protein that enhances electromotility. Another striking difference between mammals and other vertebrates is the magnitude of the endocochlear potential. The transmembrane electric field across the stereocilia membrane is larger, often double, that of nonmammals. Does this facilitate electrodiffusion of calcium or does it enhance a direct electromechanical effect in the bundle? Yet another species difference is with regard to the tectorial membrane. Insect sensory cells insert into the cuticle, and at the other extreme, portions of some auditory papillae (in lizards) have no tectorial membrane. The molecular nature of the tectorial membrane has not been explored in detail outside of mammals. It would be of interest to know what species differences exist in the types of collagen and glycoproteins that are present. Many of these are highly charged and could provide the basis for yet another electromechanical interaction that would affect hearing. In this case the interaction would be with electric fields associated with the cochlear microphonic and not the hair-cell receptor potential. A better understanding of the differences between species will lead to clearer concepts of why mammals, birds, lizards, amphibians, fish, and insects differ in their hearing in the ways they do. The concepts should in turn allow unprecedented access to clever diagnostic tools to the benefit of human health.

References

Aranyosi AJ, Freeman DM (2004) Sound-induced motions of individual cochlear hair bundles. Biophys J 87:3536–3546.

Bilger RC, Matthies ML, Hammel DR, Demorest ME (1990) Genetic implications of gender differences in the prevalence of spontaneous otoacoustic emissions. J Speech Hear Res 33:418–432.

Brownell WE (1983) Observations of a motile response in isolated outer hair cells. In: Webster WR, Aitkin LM (eds) Mechanisms of Hearing. Clayton, Vic., Australia: Monash University Press, pp. 5–10.

Brownell WE (2006) The piezoelectric hair cell. In: Eatock RE, Fay RR, Popper AN (eds) Vertebrate Hair Cells, New York: Springer, pp. 313–347.

Brownell WE, Farrell B, Raphael RM (2003) Membrane electromechanics at hair cell synapses. In: Gummer AW, Dalhoff E, Nowotny M, Scherer MP (eds) Biophysics of the Cochlea: From Molecules to Models. Singapore: World Scientific, pp. 169–176.

Chan DK, Hudspeth AJ (2005) Ca^{2+} current-driven nonlinear amplification by the mammalian cochlea in vitro. Nat Neurosci 8:149–155.

Crawford AC, Fettiplace R (1985) The mechanical properties of ciliary bundles of turtle cochlear hair cells. J Physiol 364:359–379.

Franchini LF, Elgoyhen AB (2006) Adaptive evolution in mammalian proteins involved in cochlear outer hair cell electromotility. Mol Phylogenet Evol 41:622–635.

Furness DN, Hackney CM (2006) The structure and composition of the stereociliary bundle of vertebrate hair cells. In: Eatock RE, Fay RR, Popper AN (eds) Vertebrate Hair Cells. New York: Springer, pp. 95–153.

Gillespie PG, Walker, RG (2001) Molecular basis of mechanosensory transduction. Nature 413:194–202.

Hudspeth AJ (1989) How the ear's works work. Nature 341:397–404.

Hudspeth AJ (1997) How hearing happens. Neuron 19:947–950.

Kennedy HJ, Crawford AC, Fettiplace R (2005) Force generation by mammalian hair bundles supports a role in cochlear amplification. Nature 433:880–883.

Kim DO (1986) Active and nonlinear cochlear biomechanics and the role of outer-hair-cell subsystem in the mammalian auditory system. Hear Res 22:105–114.

Kim DO, Yang XM, Neely ST (2003) Effects of the medial olivocochlear reflex on cochlear mechanics: Experimental and modeling studies of DPOAE. In: Gummer AW, Dalhoff E, Nowotny M, Scherer MP (eds), Biophysics of the Cochlea: From Molecules to Models. Singapore: World Scientific, pp. 506–516.

Le Goff L, Bozovic D, Hudspeth AJ (2005) Adaptive shift in the domain of negative stiffness during spontaneous oscillation by hair bundles from the internal ear. Proc Natl Acad Sci USA 102:16996–17001.

Manley GA (1983) Frequency spacing of acoustic emissions: a possible explanation. in Webster WR, Aitken LM (eds) Mechanisms of Hearing. Clayton, vic: Monash University press pp. 36–39.

Manley GA (1990) Peripheral Hearing Mechanisms in Reptiles and Birds. Heidelberg: Springer-Verlag.

Manley GA (1997) Diversity in hearing-organ structure and the characteristics of spontaneous otoacoustic emissions in lizards. In: Lewis E, Long GR, Lyon RF, Narins PM, Steele CR (eds) Diversity in Auditory Mechanics. Singapore: World Scientific, pp. 32–38.

Manley GA (2006a) Spontaneous otoacoustic emissions from free-standing stereovillar bundles of ten species of lizard with small papillae. Hear Res 212:33–47.

Manley GA (2006b) Spontaneous otoacoustic emissions in lizards, air pressure effects on them and the question of point sources and global standing waves. In: Nuttall A, de Boer E, Gillespie P, Ren T, Grosh K (eds) Auditory Mechanisms: Processes and Models. Singapore: World Scientific, pp. 369–376.

Manley GA, Gallo L (1997) Otoacoustic emissions, hair cells, and myosin motors. J Acoust Soc Am 102:1049–1055.

Manley GA, Ladher R (2007) Phylogeny and evolution of ciliated mechano-receptor cells. In: Dallos P, Oertel D (ed.) The Senses: A Comprehensive References Amsterdam: Elsevier. In press: Volume 2, Audition.

Manley GA, Sienknecht U, Köppl C (2004) Calcium modulates the frequency and amplitude of spontaneous otoacoustic emissions in the bobtail skink. J Neurophysiol 92:2685–2693.

Martin P, Hudspeth AJ (1999) Active hair-bundle movements can amplify a hair cell's response to oscillatory mechanical stimuli. Proc Natl Acad Sci USA 96:14306–14311.

McFadden D (1993a) A masculinizing effect on the auditory systems of human females having male co-twins. Proc Natl Acad Sci USA 90:11900–11904.

McFadden D (1993b) A speculation about the parallel ear asymmetries and sex differences in hearing sensitivity and otoacoustic emissions. Hear Res 68:143–151.

McFadden D, Loehlin JC, Pasanen EG (1996) Additional findings on heritability and prenatal masculinization of cochlear mechanisms: click-evoked otoacoustic emissions. Hear Res 97:102–119.

McFadden D, Pasanen EG, Raper J, Lange HS, Wallen K (2006a) Sex differences in otoacoustic emissions measured in rhesus monkeys (*Macaca mulatta*). Horm Behav 50:274–284.

McFadden D, Pasanen EG, Weldele ML, Glickman SE, Place NJ (2006b) Masculinized otoacoustic emissions in female spotted hyenas (*Crocuta crocuta*). Horm Behav 50:285–292.

Rabbitt RD, Ayliffe HE, Christensen D, Pamarthy K, Durney C, Clifford S, Brownell WE (2005) Evidence of piezoelectric resonance in isolated outer hair cells. Biophys J 88:2257–2265.

Sato H, Sando I, Takahashi H (1991) Sexual dimorphism and development of the human cochlea. Computer 3-D measurement. Acta Otolaryngol 111:1037–1040.

Shera C (2003) mammalian spontaneous otoacoustic emissions are amplitude-stabilized cochlear standing waves. J Acoust Soc Am 114:244–262.

Spector AA, Popel AS, Eatock RA, Brownell WE (2005) Mechanosensitive channels in the lateral wall can enhance the cochlear outer hair cell frequency response. Ann Biomed Eng 33:991–1002.

Weiss TF (1982) Bidirectional transduction in vertebrate hair cells: a mechanism for coupling mechanical and electrical vibrations. Hear Res 7:353–360.

Index

Springer Handbook of Auditory Research *(continued from page ii)*

Volume 22: Evolution of the Vertebrate Auditory System
Edited by Geoffrey A. Manley, Arthur N. Popper, and Richard R. Fay

Volume 23: Plasticity of the Auditory System
Edited by Thomas N. Parks, Edwin W. Rubel, Arthur N. Popper, and Richard R. Fay

Volume 24: Pitch: Neural Coding and Perception
Edited by Christopher J. Plack, Andrew J. Oxenham, Richard R. Fay, and Arthur N. Popper

Volume 25: Sound Source Localization
Edited by Arthur N. Popper and Richard R. Fay

Volume 26: Development of the Inner Ear
Edited by Matthew W. Kelley, Doris K. Wu, Arthur N. Popper, and Richard R. Fay

Volume 27: Vertebrate Hair Cells
Edited by Ruth Anne Eatock, Richard R. Fay, and Arthur N. Popper

Volume 28: Hearing and Sound Communication in Amphibians
Edited by Peter M. Narins, Albert S. Feng, Richard R. Fay, and Arthur N. Popper

Volume 29: Auditory Perception of Sound Sources
Edited by William A. Yost, Arthur N. Popper and Richard R. Fay

Volume 30: Active Processes and Otoacoustic Emissions in Hearing
Edited by Geoffrey A. Manely, Arthur N. Popper and Richard R. Fay

For more information about the series, please visit www.springer-ny.com/shar.

Printed in the United States of America